Introduction to Probability Models

Seventh Edition

Introduction to Probability Models

Seventh Edition

Sheldon M. Ross

*Department of Industrial Engineering
and Operations Research
University of California,
Berkeley, California*

A Harcourt Science and Technology Company

San Diego   San Francisco   New York   Boston
London   Toronto   Sydney   Tokyo

ACADEMIC PRESS
A Harcourt Science and Technology Company
525 B Street, Suite 1900, San Diego, CA 92101-4495, USA
http://www.academicpress

Academic Press
Harcourt Place, 32 Jamestown Road, London, NW1 7BY, UK
http://www.hbuk.co.uk/ap/

Harcourt/Academic Press
200 Wheeler Road, Burlington, MA 01803
http://www.harcourt-ap.com

Library of Congress Catalog Card Number: 99-68566
International Standard Book Number: 0-12-598475-8

Printed in the United States of America
99 00 01 02 03 EB 9 8 7 6 5 4 3 2 1

Contents

v

5. The Exponential Distribution and the Poisson Process 241

6. Continuous-Time Markov Chains 313

7. Renewal Theory and Its Applications 363

8. Queueing Theory 427

9. Reliability Theory 499

10. Brownian Motion and Stationary Processes 549

Preface to the Fifth Edition

This text is intended as an introduction to elementary probability theory and stochastic processes. It is particularly well suited for those wanting to see how probability theory can be applied to the study of phenomena in fields such as engineering, computer science, management science, the physical and social sciences, and operations research.

It is generally felt that there are two approaches to the study of probability theory. One approach is heuristic and nonrigorous and attempts to develop in the student an intuitive feel for the subject which enables him or her to "think probabilistically." The other approach attempts a rigorous development of probability by using the tools of measure theory. It is the first approach that is employed in this text. However, because it is extremely important in both understanding and applying probability theory to be able to "think probabilistically," this text should also be useful to students interested primarily in the second approach.

Chapters 1 and 2 deal with basic ideas of probability theory. In Chapter 1 an axiomatic framework is presented, while in Chapter 2 the important concept of a random variable is introduced.

Chapter 3 is concerned with the subject matter of conditional probability and conditional expectation. "Conditioning" is one of the key tools of probability theory, and it is stressed throughout the book. When properly used, conditioning often enables us to easily solve problems that at first glance seem quite difficult. The final section of this chapter presents applications to (1) a computer list problem, (2) a random graph, and (3) the Polya urn model and its relation to the Bose–Einstein distribution.

In Chapter 4 we come into contact with our first random, or stochastic, process, known as a Markov chain, which is widely applicable to the study of many real-world phenomena. Applications to genetics and production processes are presented. The concept of time reversibility is introduced and its usefulness illustrated. In the final section we consider a model for optimally making decisions known as a Markovian decision process.

In Chapter 5 we are concerned with a type of stochastic process known as a counting process. In particular, we study a kind of counting process known as a Poisson process. The intimate relationship between this process and the exponential distribution is discussed. Examples relating to analyzing greedy algorithms, minimizing highway encounters, collecting coupons, and tracking the AIDS virus, as well as material on compound Poisson processes are included in this chapter.

Chapter 6 considers Markov chains in continuous time with an emphasis on birth and death models. Time reversibility is shown to be a useful concept, as it is in the study of discrete-time Markov chains. The final section presents the computationally important technique of uniformization.

Chapter 7, the renewal theory chapter, is concerned with a type of counting process more general than the Poisson. By making use of renewal reward processes, limiting results are obtained and applied to various fields.

Chapter 8 deals with queueing, or waiting line, theory. After some preliminaries dealing with basic cost identities and types of limiting probabilities, we consider exponential queueing models and show how such models can be analyzed. Included in the models we study is the important class known as a network of queues. We then study models in which some of the distributions are allowed to be arbitrary.

Chapter 9 is concerned with reliability theory. This chapter will probably be of greatest interest to the engineer and operations researcher.

Chapter 10 is concerned with Brownian motion and its applications. The theory of options pricing is discussed. Also, the arbitrage theorem is presented and its relationship to the duality theorem of linear program is indicated. We show how the arbitrage theorem leads to the Black–Scholes option pricing formula.

Chapter 11 deals with simulation, a powerful tool for analyzing stochastic models that are analytically intractable. Methods for generating the values of arbitrarily distributed random variables are discussed, as are variance reduction methods for increasing the efficiency of the simulation.

Ideally, this text would be used in a one-year course in probability models. Other possible courses would be a one-semester course in introductory probability theory (involving Chapters 1–3 and parts of others) or a course in elementary stochastic processes. It is felt that the textbook is flexible enough to be used in a variety of possible courses. For example, I have used Chapters 5 and 8, with smatterings from Chapters 4 and 6, as the basis of an introductory course in queueing theory.

Many examples are worked out throughout the text, and there are also a large number of problems to be worked by students.

Preface to the Sixth Edition

The sixth edition includes additional material in all chapters. For instance,

- Section 2.6.1 gives a simple derivation of the joint distribution of the sample mean and sample variance of a normal data sample.
- Section 3.6.4 presents k-record values and the surprising Ignatov's theorem.
- Section 4.5.3 presents an analysis, based on random walk theory, of a probabilistic algorithm for the satisfiability problem.
- Section 4.6 deals with the mean times spent in transient states by a Markov chain.
- Section 4.9 introduces Markov chain Monte Carlo methods.
- Section 5.2.4 gives a simple derivation of the convolution of exponential random variables.
- Section 7.9 presents new results concerning the distribution of time until a certain pattern occurs when a sequence of independent and identically distributed random variables is observed. In Section 7.9.1, we show how renewal theory can be used to derive both the mean and the variance of the length of time until a specified pattern appears, as well as the mean time until one of a finite number of specified patterns appears. In Section 7.9.2, we suppose that the random variables are equally likely to take on any of m possible values, and compute an expression for the mean time until a run of m distinct values occurs. In Section 7.9.3, we suppose the random variables are continuous and derive an expression for the mean time until a run of m consecutive increasing values occurs.
- Section 9.6.1 illustrates a method for determining an upper bound for the expected life of a parallel system of not necessarily independent components.

- Section 11.6.4 introduces the important simulation technique of importance sampling, and indicates the usefulness of tilted distributions when applying this method.

Among the new examples are ones relating to

- Random walks on circles (Example 2.52).
- The matching rounds problem (Example 3.13).
- The best prize problem (Example 3.21).
- A probabilistic characterization of e (Example 3.24).
- Ignatov's theorem (Example 3.25).

We have added a large number of new exercises, so that there are now approximately 570 exercises (most consisting of multiple parts). More than 100 of these exercises have been starred and their solutions provided at the end of the text. These starred problems can be used by students for independent study and test preparation. An Instructor's Manual, containing solutions to all exercises, is available free of charge to instructors who adopt the book for class.

We would like to acknowledge with thanks the helpful suggestions made by the many reviewers of the text, including:

Garth Isaak, Lehigh University

Galen Shorack, University of Washington, Seattle

Amarjot Kaur, Pennsylvania State University

Marlin Thomas, Purdue University

Zhenyuan Wang, State University of New York, Binghampton

The reviewers' comments have been critical in our attempt to continue to improve this textbook in its sixth edition.

Preface to the Seventh Edition

The seventh edition continues this evolving text along its path. New examples and exercises have been added, and additional clarifications of some of the existing arguments have been made. Among the more significant additions are new derivations for the Poisson and nonhomogeneous Poisson processes; a subsection (8.6.3) dealing with an optimization problem concerning a single server, general service time queue; a section (8.8) concerned with a single server, general service time queue in which the arrival source is a finite number of potential users; and a subsection (9.7.1) analyzing a series structure reliability model in which components enter a state of suspended animation when one of their cohorts fails.

We would like to thank all those who have commented on the text, particularly

Jay Devore
California Polytechnic Institute

Galen Shorack
University of Washington

Osnat Stramer
University of Iowa

Ramesh Gupta
University of Maine

George Michailidis
University of Michigan

Introduction to Probability Theory

1

♦♦♦

1.1. Introduction

Any realistic model of a real-world phenomenon must take into account the possibility of randomness. That is, more often than not, the quantities we are interested in will not be predictable in advance but, rather, will exhibit an inherent variation that should be taken into account by the model. This is usually accomplished by allowing the model to be probabilistic in nature. Such a model is, naturally enough, referred to as a probability model.

The majority of the chapters of this book will be concerned with different probability models of natural phenomena. Clearly, in order to master both the "model building" and the subsequent analysis of these models, we must have a certain knowledge of basic probability theory. The remainder of this chapter, as well as the next two chapters, will be concerned with a study of this subject.

1.2. Sample Space and Events

Suppose that we are about to perform an experiment whose outcome is not predictable in advance. However, while the outcome of the experiment will not be known in advance, let us suppose that the set of all possible outcomes is known. This set of all possible outcomes of an experiment is known as the *sample space* of the experiment and is denoted by S.

Some examples are the following.

1. If the experiment consists of the flipping of a coin, then

$$S = \{H, T\}$$

where H means that the outcome of the toss is a head and T that it is a tail.

2. If the experiment consists of rolling a die, then the sample space is

$$S = \{1, 2, 3, 4, 5, 6\}$$

where the outcome i means that i appeared on the die, $i = 1, 2, 3, 4, 5, 6$.

3. If the experiments consists of flipping two coins, then the sample space consists of the following four points:

$$S = \{(H, H), (H, T), (T, H), (T, T)\}$$

The outcome will be (H, H) if both coins come up heads; it will be (H, T) if the first coin comes up heads and the second comes up tails; it will be (T, H) if the first comes up tails and the second heads; and it will be (T, T) if both coins come up tails.

4. If the experiment consists of rolling two dice, then the sample space consists of the following 36 points:

$$S = \begin{Bmatrix} (1, 1) \ (1, 2), (1, 3), (1, 4), (1, 5), (1, 6) \\ (2, 1), (2, 2), (2, 3), (2, 4), (2, 5), (2, 6) \\ (3, 1), (3, 2), (3, 3), (3, 4), (3, 5), (3, 6) \\ (4, 1), (4, 2), (4, 3), (4, 4), (4, 5), (4, 6) \\ (5, 1), (5, 2), (5, 3), (5, 4), (5, 5), (5, 6) \\ (6, 1), (6, 2), (6, 3), (6, 4), (6, 5), (6, 6) \end{Bmatrix}$$

where the outcome (i, j) is said to occur if i appears on the first die and j on the second die.

5. If the experiment consists of measuring the lifetime of a car, then the sample space consists of all nonnegative real numbers. That is,

$$S = [0, \infty)^* \quad \blacklozenge$$

Any subset E of the sample space S is known as an *event*. Some examples of events are the following.

1'. In Example (1), if $E = \{H\}$, then E is the event that a head appears on the flip of the coin. Similarly, if $E = \{T\}$, then E would be the event that a tail appears.

2'. In Example (2), if $E = \{1\}$, then E is the event that one appears on the roll of the die. If $E = \{2, 4, 6\}$, then E would be the event that an even number appears on the roll.

---

*The set (a, b) is defined to consist of all points x such that $a < x < b$. The set $[a, b]$ is defined to consist of all points x such that $a \leqslant x \leqslant b$. The sets $(a, b]$ and $[a, b)$ are defined, respectively, to consist of all points x such that $a < x \leqslant b$ and all points x such that $a \leqslant x < b$.

3'. In Example (3), if $E = \{(H, H), (H, T)\}$, then E is the event that a head appears on the first coin.

4'. In Example (4), if $E = \{(1, 6), (2, 5), (3, 4), (4, 3), (5, 2), (6, 1)\}$, then E is the event that the sum of the dice equals seven.

5'. In Example (5), if $E = (2, 6)$, then E is the event that the car lasts between two and six years.   ✦

For any two events E and F of a sample space S we define the new event $E \cup F$ to consist of all outcomes that are either in E or in F or in both E and F. That is, the event $E \cup F$ will occur if *either* E or F occurs. For example, in (1) if $E = \{H\}$ and $F = \{T\}$, then

$$E \cup F = \{H, T\}$$

That is, $E \cup F$ would be the whole sample space S. In (2) if $E = \{1, 3, 5\}$ and $F = \{1, 2, 3\}$, then

$$E \cup F = \{1, 2, 3, 5\}$$

and thus $E \cup F$ would occur if the outcome of the die is 1 or 2 or 3 or 5. The event $E \cup F$ is often referred to as the *union* of the event E and the event F.

For any two events E and F, we may also define the new event EF, referred to as the *intersection* of E and F, as follows. EF consists of all outcomes which are *both* in E and in F. That is, the event EF will occur only if both E and F occur. For example, in (2) if $E = \{1, 3, 5\}$ and $F = \{1, 2, 3\}$, then

$$EF = \{1, 3\}$$

and thus EF would occur if the outcome of the die is either 1 or 3. In Example (1) if $E = \{H\}$ and $F = \{T\}$, then the event EF would not consist of any outcomes and hence could not occur. To give such an event a name, we shall refer to it as the null event and denote it by $\varnothing$. (That is, $\varnothing$ refers to the event consisting of no outcomes.) If $EF = \varnothing$, then E and F are said to be *mutually exclusive*.

We also define unions and intersections of more than two events in a similar manner. If $E_1, E_2, \ldots$ are events, then the union of these events, denoted by $\bigcup_{n=1}^{\infty} E_n$, is defined to be that event which consists of all outcomes that are in E_n for at least one value of $n = 1, 2, \ldots$. Similarly, the intersection of the events E_n, denoted by $\bigcap_{n=1}^{\infty} E_n$, is defined to be the event consisting of those outcomes that are in all of the events E_n, $n = 1, 2, \ldots$.

Finally, for any event E we define the new event E^c, referred to as the *complement* of E, to consist of all outcomes in the sample space S that are not in E. That is, E^c will occur if and only if E does not occur. In Example

(4) if $E = \{(1, 6), (2, 5), (3, 4), (4, 3), (5, 2), (6, 1)\}$, then E^c will occur if the sum of the dice does not equal seven. Also note that since the experiment must result in some outcome, it follows that $S^c = \varnothing$.

1.3. Probabilities Defined on Events

Consider an experiment whose sample space is S. For each event E of the sample space S, we assume that a number $P(E)$ is defined and satisfies the following three conditions:

  (i) $0 \leqslant P(E) \leqslant 1$.
  (ii) $P(S) = 1$.
  (iii) For any sequence of events $E_1, E_2, \ldots$ that are mutually exclusive, that is, events for which $E_n E_m = \varnothing$ when $n \neq m$, then

$$P\left(\bigcup_{n=1}^{\infty} E_n\right) = \sum_{n=1}^{\infty} P(E_n)$$

We refer to $P(E)$ as the probability of the event E.

Example 1.1   In the coin tossing example, if we assume that a head is equally likely to appear as a tail, then we would have

$$P(\{H\}) = P(\{T\}) = \tfrac{1}{2}$$

On the other hand, if we had a biased coin and felt that a head was twice as likely to appear as a tail, then we would have

$$P(\{H\}) = \tfrac{2}{3}, \qquad P(\{T\}) = \tfrac{1}{3} \quad \blacklozenge$$

Example 1.2   In the die tossing example, if we supposed that all six numbers were equally likely to appear, then we would have

$$P(\{1\}) = P(\{2\}) = P(\{3\}) = P(\{4\}) = P(\{5\}) = P(\{6\}) = \tfrac{1}{6}$$

From (iii) it would follow that the probability of getting an even number would equal

$$P(\{2, 4, 6\}) = P(\{2\}) + P(\{4\}) + P(\{6\})$$
$$= \tfrac{1}{2} \quad \blacklozenge$$

Remark   We have chosen to give a rather formal definition of probabilities as being functions defined on the events of a sample space. However, it turns out that these probabilities have a nice intuitive property. Namely, if our experiment is repeated over and over again then (with probability 1) the proportion of time that event E occurs will just be $P(E)$.

Since the events E and E^c are always mutually exclusive and since $E \cup E^c = S$ we have by (ii) and (iii) that

$$1 = P(S) = P(E \cup E^c) = P(E) + P(E^c)$$

or

$$P(E^c) = 1 - P(E) \tag{1.1}$$

In words, Equation (1.1) states that the probability that an event does not occur is one minus the probability that it does occur.

We shall now derive a formula for $P(E \cup F)$, the probability of all outcomes either in E or in F. To do so, consider $P(E) + P(F)$, which is the probability of all outcomes in E plus the probability of all points in F. Since any outcome that is in both E and F will be counted twice in $P(E) + P(F)$ and only once in $P(E \cup F)$, we must have

$$P(E) + P(F) = P(E \cup F) + P(EF)$$

or equivalently

$$P(E \cup F) = P(E) + P(F) - P(EF) \tag{1.2}$$

Note that when E and F are mutually exclusive (that is, when $EF = \varnothing$), then Equation (1.2) states that

$$P(E \cup F) = P(E) + P(F) - P(\varnothing)$$
$$= P(E) + P(F)$$

a result which also follows from condition (iii). [Why is $P(\varnothing) = 0$?]

Example 1.3 Suppose that we toss two coins, and suppose that we assume that each of the four outcomes in the sample space

$$S = \{(H, H), (H, T), (T, H), (T, T)\}$$

is equally likely and hence has probability $\frac{1}{4}$. Let

$$E = \{(H, H), (H, T)\} \qquad \text{and} \qquad F = \{(H, H), (T, H)\}$$

That is, E is the event that the first coin falls heads, and F is the event that the second coin falls heads.

By Equation (1, 2) we have that $P(E \cup F)$, the probability that either the first or the second coin falls heads, is given by

$$P(E \cup F) = P(E) + P(F) - P(EF)$$
$$= \tfrac{1}{2} + \tfrac{1}{2} - P(\{H, H\})$$
$$= 1 - \tfrac{1}{4} = \tfrac{3}{4}$$

This probability could, of course, have been computed directly since

$$P(E \cup F) = P(\{H, H), (H, T), (T, H)\}) = \tfrac{3}{4} \quad \blacklozenge$$

We may also calculate the probability that any one of the three events E or F or G occurs. This is done as follows:

$$P(E \cup F \cup G) = P((E \cup F) \cup G)$$

which by Equation (1.2) equals

$$P(E \cup F) + P(G) - P((E \cup F)G)$$

Now we leave it for the reader to show that the events $(E \cup F)G$ and $EG \cup FG$ are equivalent, and hence the above equals

$P(E \cup F \cup G)$

$$= P(E) + P(F) - P(EF) + P(G) - P(EG \cup FG)$$
$$= P(E) + P(F) - P(EF) + P(G) - P(EG) - P(FG) + P(EGFG)$$
$$= P(E) + P(F) + P(G) - P(EF) - P(EG) - P(FG) + P(EFG) \qquad (1.3)$$

In fact, it can be shown by induction that, for any n events E_1, E_2, $E_3, \ldots, E_n$,

$P(E_1 \cup E_2 \cup \cdots \cup E_n)$

$$= \sum_i P(E_i) - \sum_{i<j} P(E_i E_j) + \sum_{i<j<k} P(E_i E_j E_k)$$
$$- \sum_{i<j<k<l} P(E_i E_j E_k E_l) + \cdots + (-1)^{n+1} P(E_1 E_2 \cdots E_n) \qquad (1.4)$$

In words, Equation (1.4) states that the probability of the union of n events equals the sum of the probabilities of these events taken one at a time minus the sum of the probabilities of these events taken two at a time plus the sum of the probabilities of these events taken three at a time, and so on.

1.4 Conditional Probabilities

Suppose that we toss two dice and that each of the 36 possible outcomes is equally likely to occur and hence has probability $\tfrac{1}{36}$. Suppose that we observe that the first die is a four. Then, given this information, what is the probability that the sum of the two dice equals six? To calculate this probability we reason as follows: Given that the initial die is a four, it follows that there can be at most six possible outcomes of our experiment, namely, (4, 1), (4, 2), (4, 3), (4, 4), (4, 5), and (4, 6). Since each of these outcomes originally had the same probability of occurring, they should still have equal probabilities. That is, given that the first die is a four, then the (conditional) probability of each of the outcomes (4, 1), (4, 2), (4, 3), (4, 4), (4, 5), (4, 6) is $\frac{1}{6}$ while the (conditional) probability of the other 30 points in the sample space is 0. Hence, the desired probability will be $\frac{1}{6}$.

If we let E and F denote respectively the event that the sum of the dice is six and the event that the first die is a four, then the probability just obtained is called the conditional probability that E occurs given that F has occurred and is denoted by

$$P(E \mid F)$$

A general formula for $P(E \mid F)$ which is valid for all events E and F is derived in the same manner as above. Namely, if the event F occurs, then in order for E to occur it is necessary for the actual occurrence to be a point in both E and in F, that is, it must be in EF. Now, because we know that F has occurred, it follows that F becomes our new sample space and hence the probability that the event EF occurs will equal the probability of EF relative to the probability of F. That is,

$$P(E \mid F) = \frac{P(EF)}{P(F)} \tag{1.5}$$

Note that Equation (1.5) is only well defined when $P(F) > 0$ and hence $P(E \mid F)$ is only defined when $P(F) > 0$.

Example 1.4 Suppose cards numbered one through ten are placed in a hat, mixed up, and then one of the cards is drawn. If we are told that the number on the drawn card is at least five, then what is the conditional probability that it is ten?

Solution: Let E denote the event that the number of the drawn card is ten, and let F be the event that it is at least five. The desired probability is $P(E \mid F)$. Now, from Equation (1.5)

$$P(E \mid F) = \frac{P(EF)}{P(F)}$$

However, $EF = E$ since the number of the card will be both ten and at least five if and only if it is number ten. Hence,

$$P(E \mid F) = \frac{\frac{1}{10}}{\frac{6}{10}} = \frac{1}{6} \quad \blacklozenge$$

Example 1.5 A family has two children. What is the conditional probability that both are boys given that at least one of them is a boy? Assume that the sample space S is given by $S = \{(b, b), (b, g), (g, b), (g, g)\}$, and all outcomes are equally likely. [(b, g) means, for instance, that the older child is a boy and the younger child a girl.]

Solution: Letting E denote the event that both children are boys, and F the event that at least one of them is a boy, then the desired probability is given by

$$P(E \mid F) = \frac{P(EF)}{P(F)}$$

$$= \frac{P(\{(b, b)\})}{P(\{(b, b), (b, g), (g, b)\})} = \frac{\frac{1}{4}}{\frac{3}{4}} = \frac{1}{3} \quad \blacklozenge$$

Example 1.6 Bev can either take a course in computers or in chemistry. If Bev takes the computer course, then she will receive an A grade with probability $\frac{1}{2}$, while if she takes the chemistry course then she will receive an A grade with probability $\frac{1}{3}$. Bev decides to base her decision on the flip of a fair coin. What is the probability that Bev will get an A in chemistry?

Solution: If we let F be the event that Bev takes chemistry and E denote the event that she receives an A in whatever course she takes, then the desired probability is $P(EF)$. This is calculated by using Equation (1.5) as follows:

$$P(EF) = P(F)P(E \mid F)$$

$$= \frac{1}{2}\frac{1}{3} = \frac{1}{6} \quad \blacklozenge$$

Example 1.7 Suppose an urn contains seven black balls and five white balls. We draw two balls from the urn without replacement. Assuming that each ball in the urn is equally likely to be drawn, what is the probability that both drawn balls are black?

Solution: Let F and E denote respectively the events that the first and second balls drawn are black. Now, given that the first ball selected is black, there are six remaining black balls and five white balls, and so $P(E|F) = \frac{6}{11}$. As $P(F)$ is clearly $\frac{7}{12}$, our desired probability is

$$P(EF) = P(F)P(E|F)$$

$$= \frac{7}{12}\frac{6}{11} = \frac{42}{132} \quad \blacklozenge$$

Example 1.8 Suppose that each of three men at a party throws his hat into the center of the room. The hats are first mixed up and then each man randomly selects a hat. What is the probability that none of the three men selects his own hat?

Solution: We shall solve the above by first calculating the complementary probability that at least one man selects his own hat. Let us denote by E_i, $i = 1, 2, 3$, the event that the ith man selects his own hat. To calculate the probability $P(E_1 \cup E_2 \cup E_3)$, we first note that

$$P(E_i) = \tfrac{1}{3}, \qquad i = 1, 2, 3$$

$$P(E_iE_j) = \tfrac{1}{6}, \qquad i \neq j \qquad\qquad (1.6)$$

$$P(E_1E_2E_3) = \tfrac{1}{6}$$

To see why Equation (1.6) is correct, consider first

$$P(E_iE_j) = P(E_i)P(E_j|E_i)$$

Now $P(E_i)$, the probability that the ith man selects his own hat, is clearly $\tfrac{1}{3}$ since he is equally likely to select any of the three hats. On the other hand, given that the ith man has selected his own hat, then there remain two hats that the jth man may select, and as one of these two is his own hat, it follows that with probability $\tfrac{1}{2}$ he will select it. That is, $P(E_j|E_i) = \tfrac{1}{2}$ and so

$$P(E_iE_j) = P(E_i)P(E_j|E_i) = \tfrac{1}{3}\tfrac{1}{2} = \tfrac{1}{6}$$

To calculate $P(E_1E_2E_3)$ we write

$$P(E_1E_2E_3) = P(E_1E_2)P(E_3|E_1E_2)$$

$$= \tfrac{1}{6}P(E_3|E_1E_2)$$

However, given that the first two men get their own hats it follows that the third man must also get his own hat (since there are no other hats left). That is, $P(E_3|E_1E_2) = 1$ and so

$$P(E_1E_2E_3) = \tfrac{1}{6}$$

Now, from Equation (1.4) we have that

$$P(E_1 \cup E_2 \cup E_3) = P(E_1) + P(E_2) + P(E_3) - P(E_1 E_2)$$
$$- P(E_1 E_3) - P(E_2 E_3) + P(E_1 E_2 E_3)$$
$$= 1 - \tfrac{1}{2} + \tfrac{1}{6}$$
$$= \tfrac{2}{3}$$

Hence, the probability that none of the men selects his own hat is $1 - \tfrac{2}{3} = \tfrac{1}{3}$. ✦

1.5. Independent Events

Two events E and F are said to be *independent* if

$$P(EF) = P(E)P(F)$$

By Equation (1.5) this implies that E and F are independent if

$$P(E \mid F) = P(E)$$

[which also implies that $P(F \mid E) = P(F)$]. That, is, E and F are independent if knowledge that F has occurred does not affect the probability that E occurs. That is, the occurrence of E is independent of whether or not F occurs.

Two events E and F that are not independent are said to be *dependent*.

Example 1.9 Suppose we toss two fair dice. Let E_1 denote the event that the sum of the dice is six and F denote the event that the first die equals four. Then

$$P(E_1 F) = P(\{4, 2\}) = \tfrac{1}{36}$$

while

$$P(E_1)P(F) = \tfrac{5}{36}\tfrac{1}{6} = \tfrac{5}{216}$$

and hence E_1 and F are not independent. Intuitively, the reason for this is clear for if we are interested in the possibility of throwing a six (with two dice), then we will be quite happy if the first die lands four (or any of the numbers 1, 2, 3, 4, 5) because then we still have a possibility of getting a total of six. On the other hand, if the first die landed six, then we would be unhappy as we would no longer have a chance of getting a total of six. In other words, our chance of getting a total of six depends on the outcome of the first die and hence E_1 and F cannot be independent.

Let E_2 be the event that the sum of the dice equals seven. Is E_2 independent of F? The answer is yes since

$$P(E_2F) = P(\{(4, 3)\}) = \tfrac{1}{36}$$

while

$$P(E_2)P(F) = \tfrac{1}{6}\tfrac{1}{6} = \tfrac{1}{36}$$

We leave it for the reader to present the intuitive argument why the event that the sum of the dice equals seven is independent of the outcome on the first die. ✦

The definition of independence can be extended to more than two events. The events $E_1, E_2, \ldots, E_n$ are said to be independent if for every subset $E_{1'}, E_{2'}, \ldots, E_{r'}, r \leqslant n$, of these events

$$P(E_{1'}E_{2'} \cdots E_{r'}) = P(E_{1'})P(E_{2'}) \cdots P(E_{r'})$$

Intuitively, the events $E_1, E_2, \ldots, E_n$ are independent if knowledge of the occurrence of any of these events has no effect on the probability of any other event.

Example 1.10 (Pairwise Independent Events That Are Not Independent): Let a ball be drawn from an urn containing four balls, numbered 1, 2, 3, 4. Let $E = \{1, 2\}$, $F = \{1, 3\}$, $G = \{1, 4\}$. If all four outcomes are assumed equally likely, then

$$P(EF) = P(E)P(F) = \tfrac{1}{4},$$
$$P(EG) = P(E)P(G) = \tfrac{1}{4},$$
$$P(FG) = P(F)P(G) = \tfrac{1}{4}$$

However,

$$\tfrac{1}{4} = P(EFG) \neq P(E)P(F)P(G)$$

Hence, even though the events E, F, G are pairwise independent, they are not jointly independent. ✦

Suppose that a sequence of experiments, each of which results in either a "success" or a "failure," is to be performed. Let E_i, $i \geqslant 1$, denote the event that the ith experiment results in a success. If, for all $i_1, i_2, \ldots, i_n$,

$$P(E_{i_1}E_{i_2} \cdots E_{i_n}) = \prod_{j=1}^{n} P(E_{i_j})$$

we say that the sequence of experiments consists of *independent trials.*

Example 1.11 The successive flips of a coin consist of independent trials if we assume (as is usually done) that the outcome on any flip is not influenced by the outcomes on earlier flips. A "success" might consist of the outcome heads and a "failure" tails, or possibly the reverse.   ✦

1.6. Bayes' Formula

Let E and F be events. We may express E as

$$E = EF \cup EF^c$$

because in order for a point to be in E, it must either be in both E and F, or it must be in E and not in F. Since EF and EF^c are obviously mutually exclusive, we have that

$$P(E) = P(EF) + P(EF^c)$$

$$= P(E\,|\,F)P(F) + P(E\,|\,F^c)P(F^c)$$

$$= P(E\,|\,F)P(F) + P(E\,|\,F^c)(1 - P(F)) \qquad (1.7)$$

Equation (1.7) states that the probability of the event E is a weighted average of the conditional probability of E given that F has occurred and the conditional probability of E given that F has not occurred, each conditional probability being given as much weight as the event it is conditioned on has of occurring.

Example 1.12 Consider two urns. The first contains two white and seven black balls, and the second contains five white and six black balls. We flip a fair coin and then draw a ball from the first urn or the second urn depending on whether the outcome was heads or tails. What is the conditional probability that the outcome of the toss was heads given that a white ball was selected?

Solution: Let W be the event that a white ball is drawn, and let H be the event that the coin comes up heads. The desired probability $P(H\,|\,W)$ may be calculated as follows:

$$P(H\,|\,W) = \frac{P(HW)}{P(W)} = \frac{P(W\,|\,H)P(H)}{P(W)}$$

$$= \frac{P(W \mid H)P(H)}{P(W \mid H)P(H) + P(W \mid H^c)P(H^c)}$$

$$= \frac{\frac{2}{9}\frac{1}{2}}{\frac{2}{9}\frac{1}{2} + \frac{5}{11}\frac{1}{2}} = \frac{22}{67} \quad \blacklozenge$$

Example 1.13 In answering a question on a multiple-choice test a student either knows the answer or guesses. Let p be the probability that she knows the answer and $1 - p$ the probability that she guesses. Assume that a student who guesses at the answer will be correct with probability $1/m$, where m is the number of multiple-choice alternatives. What is the conditional probability that a student knew the answer to a question given that she answered it correctly?

Solution: Let C and K denote respectively the event that the student answers the question correctly and the event that she actually knows the answer. Now

$$P(K \mid C) = \frac{P(KC)}{P(C)} = \frac{P(C \mid K)P(K)}{P(C \mid K)P(K) + P(C \mid K^c)P(K^c)}$$

$$= \frac{p}{p + (1/m)(1 - p)}$$

$$= \frac{mp}{1 + (m - 1)p}$$

Thus, for example, if $m = 5$, $p = \frac{1}{2}$, then the probability that a student knew the answer to a question she correctly answered is $\frac{5}{6}$. $\quad \blacklozenge$

Example 1.14 A laboratory blood test is 95 percent effective in detecting a certain disease when it is, in fact, present. However, the test also yields a "false positive" result for 1 percent of the healthy persons tested. (That is, if a healthy person is tested, then, with probability 0.01, the test result will imply he has the disease.) If 0.5 percent of the population actually has the disease, what is the probability a person has the disease given that his test result is positive?

Solution: Let D be the event that the tested person has the disease, and E the event that his test result is positive. The desired probability $P(D \mid E)$ is obtained by

$$P(D \mid E) = \frac{P(DE)}{P(E)} = \frac{P(E \mid D)P(D)}{P(E \mid D)P(D) + P(E \mid D^c)P(D^c)}$$

$$= \frac{(0.95)(0.005)}{(0.95)(0.005) + (0.01)(0.995)}$$

$$= \frac{95}{294} \approx 0.323$$

Thus, only 32 percent of those persons whose test results are positive actually have the disease.   ✦

Equation (1.7) may be generalized in the following manner. Suppose that $F_1, F_2, \ldots, F_n$ are mutually exclusive events such that $\bigcup_{i=1}^{n} F_i = S$. In other words, exactly one of the events $F_1, F_2, \ldots, F_n$ will occur. By writing

$$E = \bigcup_{i=1}^{n} EF_i$$

and using the fact that the events EF_i, $i = 1, \ldots, n$, are mutually exclusive, we obtain that

$$P(E) = \sum_{i=1}^{n} P(EF_i)$$

$$= \sum_{i=1}^{n} P(E \mid F_i)P(F_i) \tag{1.8}$$

Thus, Equation (1.8) shows how, for given events $F_1, F_2, \ldots, F_n$ of which one and only one must occur, we can compute $P(E)$ by first "conditioning" upon which one of the F_i occurs. That is, it states that $P(E)$ is equal to a weighted average of $P(E \mid F_i)$, each term being weighted by the probability of the event on which it is conditioned.

Suppose now that E has occurred and we are interested in determining which one of the F_j also occurred. By Equation (1.8) we have that

$$P(F_j \mid E) = \frac{P(EF_j)}{P(E)}$$

$$= \frac{P(E \mid F_j)P(F_j)}{\sum_{i=1}^{n} P(E \mid F_i)P(F_i)} \tag{1.9}$$

Equation (1.9) is known as *Bayes' formula*.

Example 1.15   You know that a certain letter is equally likely to be in any one of three different folders. Let α_i be the probability that you will find your letter upon making a quick examination of folder i if the letter is, in fact, in folder i, $i = 1, 2, 3$. (We may have $\alpha_i < 1$.) Suppose you look in folder 1 and do not find the letter. What is the probability that the letter is in folder 1?

Solution: Let F_i, $i = 1, 2, 3$, be the event that the letter is in folder i; and let E be the event that a search of folder 1 does not come up with the letter. We desire $P(F_1 | E)$. From Bayes' formula we obtain

$$P(F_1 | E) = \frac{P(E | F_1)P(F_1)}{\Sigma_{i=1}^{3} P(E | F_i)P(F_i)}$$

$$= \frac{(1 - \alpha_1)\frac{1}{3}}{(1 - \alpha_1)\frac{1}{3} + \frac{1}{3} + \frac{1}{3}} = \frac{1 - \alpha_1}{3 - \alpha_1} \quad \blacklozenge$$

Exercises

1.   A box contains three marbles: one red, one green, and one blue. Consider an experiment that consists of taking one marble from the box then replacing it in the box and drawing a second marble from the box. What is the sample space? If, at all times, each marble in the box is equally likely to be selected, what is the probability of each point in the sample space?

*2.   Repeat Exercise 1 when the second marble is drawn without replacing the first marble.

3.   A coin is to be tossed until a head appears twice in a row. What is the sample space for this experiment? If the coin is fair, what is the probability that it will be tossed exactly four times?

4.   Let E, F, G be three events. Find expressions for the events that of E, F, G

    (a) only F occurs,
    (b) both E and F but not G occur,
    (c) at least one event occurs,
    (d) at least two events occur,
    (e) all three events occur,
    (f) none occurs,
    (g) at most one occurs,
    (h) at most two occur.

*5.   An individual uses the following gambling system at Las Vegas. He bets \$1 that the roulette wheel will come up red. If he wins, he quits. If he loses then he makes the same bet a second time only this time he bets \$2; and then regardless of the outcome, quits. Assuming that he has a probability of $\frac{1}{2}$ of winning each bet, what is the probability that he goes home a winner? Why is this system not used by everyone?

6.   Show that $E(F \cup G) = EF \cup EG$.

7.   Show that $(E \cup F)^c = E^c F^c$.

8.   If $P(E) = 0.9$ and $P(F) = 0.8$, show that $P(EF) \geqslant 0.7$. In general, show that

$$P(EF) \geqslant P(E) + P(F) - 1$$

This is known as Bonferroni's inequality.

*9.   We say that $E \subset F$ if every point in E is also in F. Show that if $E \subset F$, then

$$P(F) = P(E) + P(FE^c) \geqslant P(E)$$

10.   Show that

$$P\left(\bigcup_{i=1}^{n} E_i\right) \leqslant \sum_{i=1}^{n} P(E_i)$$

This is known as Boole's inequality.

Hint: Either use Equation (1.2) and mathematical induction, or else show that $\bigcup_{i=1}^{n} E_i = \bigcup_{i=1}^{n} F_i$, where $F_1 = E_1$, $F_i = E_i \prod_{j=1}^{i-1} E_j^c$, and use property (iii) of a probability.

11.   If two fair dice are tossed, what is the probability that the sum is i, $i = 2, 3, \ldots, 12$?

12.   Let E and F be mutually exclusive events in the sample space of an experiment. Suppose that the experiment is repeated until either event E or event F occurs. What does the sample space of this new super experiment look like? Show that the probability that event E occurs before event F is $P(E)/[P(E) + P(F)]$.

Hint: Argue that the probability that the original experiment is performed n times and E appears on the nth time is $P(E) \times (1 - p)^{n-1}$, $n = 1, 2, \ldots$, where $p = P(E) + P(F)$. Add these probabilities to get the desired answer.

13.   The dice game craps is played as follows. The player throws two dice, and if the sum is seven or eleven, then he wins. If the sum is two, three, or twelve, then he loses. If the sum is anything else, then he continues throwing until he either throws that number again (in which case he wins) or he throws a seven (in which case he loses). Calculate the probability that the player wins.

14. The probability of winning on a single toss of the dice is p. A starts, and if he fails, he passes the dice to B, who then attempts to win on her toss. They continue tossing the dice back and forth until one of them wins. What are their respective probabilities of winning?

15. Argue that $E = EF \cup EF^c, E \cup F = E \cup FE^c$.

16. Use Exercise 15 to show that $P(E \cup F) = P(E) + P(F) - P(EF)$.

*17. Suppose each of three persons tosses a coin. If the outcome of one of the tosses differs from the other outcomes, then the game ends. If not, then the persons start over and retoss their coins. Assuming fair coins, what is the probability that the game will end with the first round of tosses? If all three coins are biased and have a probability $\frac{1}{4}$ of landing heads, then what is the probability that the game will end at the first round?

18. Assume that each child who is born is equally likely to be a boy or a girl. If a family has two children, what is the probability that both are girls given that (a) the eldest is a girl, (b) at least one is a girl?

*19. Two dice are rolled. What is the probability that at least one is a six? If the two faces are different, what is the probability that at least one is a six?

20. Three dice are thrown. What is the probability the same number appears on exactly two of the three dice?

21. Suppose that 5 percent of men and 0.25 percent of women are color-blind. A color-blind person is chosen at random. What is the probability of this person being male? Assume that there are an equal number of males and females.

22. A and B play until one has 2 more points than the other. Assuming that each point is independently won by A with probability p, what is the probability they will play a total of $2n$ points? What is the probability that A will win?

23. For events $E_1, E_2, \ldots, E_n$ show that

$$P(E_1 E_2 \cdots E_n) = P(E_1)P(E_2 \mid E_1)P(E_3 \mid E_1 E_2) \cdots P(E_n \mid E_1 \cdots E_{n-1})$$

24. In an election, candidate A receives n votes and candidate B receives m votes, where $n > m$. Assume that in the count of the votes all possible orderings of the $n + m$ votes are equally likely. Let $P_{n,m}$ denote the probability that from the first vote on A is always in the lead. Find

(a) $P_{2,1}$ (b) $P_{3,1}$ (c) $P_{n,1}$ (d) $P_{3,2}$ (e) $P_{4,2}$
(f) $P_{n,2}$ (g) $P_{4,3}$ (h) $P_{5,3}$ (i) $P_{5,4}$
(j) Make a conjecture as to the value of $P_{n,m}$.

*25. Two cards are randomly selected from a deck of 52 playing cards.

(a) What is the probability they constitute a pair (that is, that they are of the same denomination)?
(b) What is the conditional probability they constitute a pair given that they are of different suits?

26. A deck of 52 playing cards, containing all 4 aces, is randomly divided into 4 piles of 13 cards each. Define events E_1, E_2, E_3, and E_4 as follows:

$$E_1 = \{\text{the first pile has exactly 1 ace}\},$$
$$E_2 = \{\text{the second pile has exactly 1 ace}\},$$
$$E_3 = \{\text{the third pile has exactly 1 ace}\},$$
$$E_4 = \{\text{the fourth pile has exactly 1 ace}\}$$

Use Exercise 23 to find $P(E_1 E_2 E_3 E_4)$, the probability that each pile has an ace.

*27. Suppose in Exercise 26 we had defined the events E_i, $i = 1, 2, 3, 4$, by

$E_1 = \{\text{one of the piles contains the ace of spades}\},$
$E_2 = \{\text{the ace of spades and the ace of hearts are in different piles}\},$
$E_3 = \{\text{the ace of spades, the ace of hearts, and the}$
     $\quad\text{ace of diamonds are in different piles}\},$
$E_4 = \{\text{all 4 aces are in different piles}\}$

Now use Exercise 23 to find $P(E_1 E_2 E_3 E_4)$, the probability that each pile has an ace. Compare your answer with the one you obtained in Exercise 26.

28. If the occurrence of B makes A more likely, does the occurrence of A make B more likely?

29. Suppose that $P(E) = 0.6$. What can you say about $P(E|F)$ when

(a) E and F are mutually exclusive?
(b) $E \subset F$?
(c) $F \subset E$?

*30. Bill and George go target shooting together. Both shoot at a target at the same time. Suppose Bill hits the target with probability 0.7, whereas

George, independently, hits the target with probability 0.4.

(a) Given that exactly one shot hit the target, what is the probability that it was George's shot?
(b) Given that the target is hit, what is the probability that George hit it?

31.   What is the conditional probability that the first die is six given that the sum of the dice is seven?

*32.   Suppose all n men at a party throw their hats in the center of the room. Each man then randomly selects a hat. Show that the probability that none of the n men selects his own hat is

$$\frac{1}{2!} - \frac{1}{3!} + \frac{1}{4!} - + \cdots \frac{(-1)^n}{n!}$$

Note that as $n \to \infty$ this converges to e^{-1}. Is this surprising?

33.   In a class there are four freshman boys, six freshman girls, and six sophomore boys. How many sophomore girls must be present if sex and class are to be independent when a student is selected at random?

34.   Mr. Jones has devised a gambling system for winning at roulette. When he bets, he bests on red, and places a bet only when the ten previous spins of the roulette have landed on a black number. He reasons that his chance of winning is quite large since the probability of eleven consecutive spins resulting in black is quite small. What do you think of this system?

35.   A fair coin is continually flipped. What is the probability that the first four flips are

(a) H, H, H, H?
(b) T, H, H, H?
(c) What is the probability that the pattern T, H, H, H occurs before the pattern H, H, H, H?

36.   Consider two boxes, one containing one black and one white marble, the other, two black and one white marble. A box is selected at random and a marble is drawn at random from the selected box. What is the probability that the marble is black?

37.   In Exercise 36, what is the probability that the first box was the one selected given that the marble is white?

38.   Urn 1 contains two white balls and one black ball, while urn 2 contains one white ball and five black balls. One ball is drawn at random from urn 1 and placed in urn 2. A ball is then drawn from urn 2. It happens to be white. What is the probability that the transferred ball was white?

39. Stores *A*, *B*, and *C* have 50, 75, 100 employees, and respectively 50, 60, and 70 percent of these are women. Resignations are equally likely among all employees, regardless of sex. One employee resigns and this is a woman. What is the probability that she works in store *C*?

*40. (a) A gambler has in his pocket a fair coin and a two-headed coin. He selects one of the coins at random, and when he flips it, it shows heads. What is the probability that it is the fair coin? (b) Suppose that he flips the same coin a second time and again it shows heads. Now what is the probability that it is the fair coin? (c) Suppose that he flips the same coin a third time and it shows tails. Now what is the probability that it is the fair coin?

41. In a certain species of rats, black dominates over brown. Suppose that a black rat with two black parents has a brown sibling.

(a) What is the probability that this rat is a pure black rat (as opposed to being a hybrid with one black and one brown gene)?
(b) Suppose that when the black rat is mated with a brown rat, all five of their offspring are black. Now, what is the probability that the rat is a pure black rat?

42. There are three coins in a box. One is a two-headed coin, another is a fair coin, and the third is a biased coin that comes up heads 75 percent of the time. When one of the three coins is selected at random and flipped, it shows heads. What is the probability that it was the two-headed coin?

43. Suppose we have ten coins which are such that if the *i*th one is flipped then heads will appear with probability $i/10$, $i = 1, 2, \ldots, 10$. When one of the coins is randomly selected and flipped, it shows heads. What is the conditional probability that it was the fifth coin?

44. Urn 1 has five white and seven black balls. Urn 2 has three white and twelve black balls. We flip a fair coin. If the outcome is heads, then a ball from urn 1 is selected, while if the outcome is tails, then a ball from urn 2 is selected. Suppose that a white ball is selected. What is the probability that the coin landed tails?

*45. An urn contains *b* black balls and *r* red balls. One of the balls is drawn at random, but when it is put back in the urn *c* additional balls of the same color are put in with it. Now suppose that we draw another ball. Show that the probability that the first ball drawn was black given that the second ball drawn was red is $b/(b + r + c)$.

46. Three prisoners are informed by their jailer that one of them has been chosen at random to be executed, and the other two are to be freed. Prisoner *A* asks the jailer to tell him privately which of his fellow prisoners will be set free, claiming that there would be no harm in divulging this information, since he already knows that at least one will go free. The jailer refuses to answer this question, pointing out that if *A* knew which of his fellows were to be set free, then his own probability of being executed would rise from $\frac{1}{3}$ to $\frac{1}{2}$, since he would then be one of two prisoners. What do you think of the jailer's reasoning?

References

Reference [2] provides a colorful introduction to some of the earliest developments in probability theory. References [3], [4], and [7] are all excellent introductory texts in modern probability theory. Reference [5] is the definitive work which established the axiomatic foundation of modern mathematical probability theory. Reference [6] is a nonmathematical introduction to probability theory and its applications, written by one of the greatest mathematicians of the eighteenth century.

1. L. Breiman, "Probability," Addison-Wesley, Reading, Massachusetts, 1968.
2. F. N. David, "Games, Gods, and Gambling," Hafner, New York, 1962.
3. W. Feller, "An Introduction to Probability Theory and Its Applications," Vol. I, John Wiley, New York, 1957.
4. B. V. Gnedenko, "Theory of Probability," Chelsea, New York, 1962.
5. A. N. Kolmogorov, "Foundations of the Theory of Probability," Chelsea, New York, 1956.
6. Marquis de Laplace, "A Philosophical Essay on Probabilities," 1825 (English Translation), Dover, New York, 1951.
7. S. Ross, "A First Course in Probability," Fifth Edition, Prentice Hall, New Jersey, 1998.

Random Variables

2

$\leftarrow$

2.1. Random Variables

It frequently occurs that in performing an experiment we are mainly interested in some functions of the outcome as opposed to the outcome itself. For instance, in tossing dice we are often interested in the sum of the two dice and are not really concerned about the actual outcome. That is, we may be interested in knowing that the sum is seven and not be concerned over whether the actual outcome was (1, 6) or (2, 5) or (3, 4) or (4, 3) or (5, 2) or (6, 1). These quantities of interest, or more formally, these real-valued functions defined on the sample space, are known as *random variables*.

Since the value of a random variable is determined by the outcome of the experiment, we may assign probabilities to the possible values of the random variable.

Example 2.1 Letting X denote the random variable that is defined as the sum of two fair dice; then

$$
\begin{aligned}
P\{X = \ 2\} &= P\{(1, 1)\} = \tfrac{1}{36}, \\
P\{X = \ 3\} &= P\{(1, 2), (2, 1)\} = \tfrac{2}{36}, \\
P\{X = \ 4\} &= P\{(1, 3), (2, 2), (3, 1)\} = \tfrac{3}{36}, \\
P\{X = \ 5\} &= P\{(1, 4), (2, 3), (3, 2), (4, 1)\} = \tfrac{4}{36}, \\
P\{X = \ 6\} &= P\{(1, 5), (2, 4), (3, 3), (4, 2), (5, 1)\} = \tfrac{5}{36}, \\
P\{X = \ 7\} &= P\{(1, 6), (2, 5), (3, 4), (4, 3), (5, 2), (6, 1)\} = \tfrac{6}{36}, \\
P\{X = \ 8\} &= P\{(2, 6), (3, 5), (4, 4), (5, 3), (6, 2)\} = \tfrac{5}{36}, \\
P\{X = \ 9\} &= P\{(3, 6), (4, 5), (5, 4), (6, 3)\} = \tfrac{4}{36}, \\
P\{X = 10\} &= P\{(4, 6), (5, 5), (6, 4)\} = \tfrac{3}{36}, \\
P\{X = 11\} &= P\{(5, 6), (6, 5)\} = \tfrac{2}{36}, \\
P\{X = 12\} &= P\{(6, 6)\} = \tfrac{1}{36}
\end{aligned}
\tag{2.1}
$$

In other words, the random variable X can take on any integral value between two and twelve, and the probability that it takes on each value is given by Equation (2.1). Since X must take on one of the values two through twelve, we must have that

$$1 = P\left\{ \bigcup_{i=2}^{12} \{X = n\} \right\} = \sum_{n=2}^{12} P\{X = n\}$$

which may be checked from Equation (2.1). ✦

Example 2.2   For a second example, suppose that our experiment consists of tossing two fair coins. Letting Y denote the number of heads appearing, then Y is a random variable taking on one of the values 0, 1, 2 with respective probabilities

$$P\{Y = 0\} = P\{(T, T)\} = \tfrac{1}{4},$$
$$P\{Y = 1\} = P\{(T, H), (H, T)\} = \tfrac{2}{4},$$
$$P\{Y = 2\} = P\{(H, H)\} = \tfrac{1}{4}$$

Of course, $P\{Y = 0\} + P\{Y = 1\} + P\{Y = 2\} = 1$. ✦

Example 2.3   Suppose that we toss a coin having a probability p of coming up heads, until the first head appears. Letting N denote the number of flips required, then assuming that the outcome of successive flips are independent, N is a random variable taking on one of the values 1, 2, 3, …, with respective probabilities

$$P\{N = 1\} = P\{H\} = p,$$
$$P\{N = 2\} = P\{(T, H)\} = (1 - p)p,$$
$$P\{N = 3\} = P\{(T, T, H)\} = (1 - p)^2 p,$$
$$\vdots$$
$$P\{N = n\} = P\{(\underbrace{T, T, \ldots,}_{n-1} T, H)\} = (1 - p)^{n-1}p, \qquad n \geqslant 1$$

As a check, note that

$$P\left\{ \bigcup_{n=1}^{\infty} \{N = n\} \right\} = \sum_{n=1}^{\infty} P\{N = n\}$$

$$= p \sum_{n=1}^{\infty} (1 - p)^{n-1}$$

$$= \frac{p}{1 - (1 - p)}$$

$$= 1 \quad ✦$$

Example 2.4 Suppose that our experiment consists of seeing how long a battery can operate before wearing down. Suppose also that we are not primarily interested in the actual lifetime of the battery but are only concerned about whether or not the battery lasts at least two years. In this case, we may define the random variable I by

$$I = \begin{cases} 1, & \text{if the lifetime of the battery is two or more years} \\ 0, & \text{otherwise} \end{cases}$$

If E denotes the event that the battery lasts two or more years, then the random variable I is known as the *indicator* random variable for event E. (Note that I equals 1 or 0 depending on whether or not E occurs.) ◆

Example 2.5 Suppose that independent trials, each of which results in any of m possible outcomes with respective probabilities $p_1, \ldots, p_m$, $\sum_{i=1}^{m} p_i = 1$, are continually performed. Let X denote the number of trials needed until each outcome has occurred at least once.

Rather than directly considering $P\{X = n\}$ we will first determine $P\{X > n\}$, the probability that at least one of the outcomes has not yet occurred after n trials. Letting A_i denote the event that outcome i has not yet occurred after the first n trials, $i = 1, \ldots, m$, then

$$P\{X > n\} = P\left(\bigcup_{i=1}^{m} A_i\right)$$

$$= \sum_{i=1}^{m} P(A_i) - \sum\sum_{i<j} P(A_i A_j)$$

$$+ \sum\sum\sum_{i<j<k} P(A_i A_j A_k) - \cdots + (-1)^{m+1} P(A_1 \cdots A_m)$$

Now, $P(A_i)$ is the probability that each of the first n trials results in a non-i outcome, and so by independence

$$P(A_i) = (1 - p_i)^n$$

Similarly, $P(A_i A_j)$ is the probability that the first n trials all result in a non-i and non-j outcome, and so

$$P(A_i A_j) = (1 - p_i - p_j)^n$$

As all of the other probabilities are similar, we see that

$$P\{X > n\} = \sum_{i=1}^{m} (1 - p_i)^n - \sum\sum_{i<j} (1 - p_i - p_j)^n$$

$$+ \sum\sum\sum_{i<j<k} (1 - p_i - p_j - p_k)^n - \cdots$$

Since $P\{X = n\} = P\{X > n - 1\} - P\{X > n\}$, we see, upon using the algebraic identity $(1 - a)^{n-1} - (1 - a)^n = a(1 - a)^{n-1}$, that

$$P\{X = n\} = \sum_{i=1}^{m} p_i(1 - p_i)^{n-1} - \sum\sum_{i<j}(p_i + p_j)(1 - p_i - p_j)^{n-1}$$
$$+ \sum\sum\sum_{i<j<k}(p_i + p_j + p_k)(1 - p_i - p_j - p_k)^{n-1} - \cdots \quad ✦$$

In all of the preceding examples, the random variables of interest took on either a finite or a countable number of possible values. Such random variables are called *discrete*. However, there also exist random variables that take on a continuum of possible values. These are known as *continuous* random variables. One example is the random variable denoting the lifetime of a car, when the car's lifetime is assumed to take on any value in some interval (a, b).

The *cumulative distribution function* (cdf) (or more simply the *distribution function*) $F(\cdot)$ of the random variable X is defined for any real number $b, -\infty < b < \infty$, by

$$F(b) = P\{X \leqslant b\}$$

In words, $F(b)$ denotes the probability that the random variable X takes on a value which will be less than or equal to b. Some properties of the cdf F are

(i) $F(b)$ is a nondecreasing function of b,
(ii) $\lim_{b \to \infty} F(b) = F(\infty) = 1$.
(iii) $\lim_{b \to -\infty} F(b) = F(-\infty) = 0$.

Property (i) follows since for $a < b$ the event $\{X \leqslant a\}$ is contained in the event $\{X \leqslant b\}$, and so it must have a smaller probability. Properties (ii) and (iii) follow since X must take on some finite value.

All probability questions about X can be answered in terms of the cdf $F(\cdot)$. For example,

$$P\{a < X \leqslant b\} = F(b) - F(a) \qquad \text{for all } a < b$$

This follows since we may calculate $P\{a < X \leqslant b\}$ by first computing the probability that $X \leqslant b$ [that is, $F(b)$] and then subtracting from this the probability that $X \leqslant a$ [that is, $F(a)$].

If we desire the probability that X is strictly smaller than b, we may calculate this probability by

$$P\{X < b\} = \lim_{h \to 0^+} P\{X \leqslant b - h\}$$
$$= \lim_{h \to 0^+} F(b - h)$$

where $\lim_{h \to 0^+}$ means that we are taking the limit as h decreases to 0. Note that $P\{X < b\}$ does not necessarily equal $F(b)$ since $F(b)$ also includes the probability that X equals b.

2.2. Discrete Random Variables

As was previously mentioned, a random variable that can take on at most a countable number of possible values is said to be *discrete*. For a discrete random variable X, we define the *probability mass function* $p(a)$ of X by

$$p(a) = P\{X = a\}$$

The probability mass function $p(a)$ is positive for at most a countable number of values of a. That is, if X must assume one of the values $x_1, x_2, \ldots$, then

$$p(x_i) > 0, \qquad i = 1, 2, \ldots$$
$$p(x) = 0, \qquad \text{all other values of } x$$

Since X must take on one of the values x_i, we have

$$\sum_{i=1}^{\infty} p(x_i) = 1$$

The cumulative distribution function F can be expressed in terms of $p(a)$ by

$$F(a) = \sum_{\text{all } x_i \leqslant a} p(x_i)$$

For instance, suppose X has a probability mass function given by

$$p(1) = \tfrac{1}{2}, \qquad p(2) = \tfrac{1}{3}, \qquad p(3) = \tfrac{1}{6}$$

then, the cumulative distribution function F of X is given by

$$F(a) = \begin{cases} 0, & a < 1 \\ \tfrac{1}{2}, & 1 \leqslant a < 2 \\ \tfrac{5}{6}, & 2 \leqslant a < 3 \\ 1, & 3 \leqslant a \end{cases}$$

This is graphically presented in Figure 2.1.

Discrete random variables are often classified according to their probability mass function. We now consider some of these random variables.

2.2.1. The Bernoulli Random Variable

Suppose that a trial, or an experiment, whose outcome can be classified as either a "success" or as a "failure" is performed. If we let X equal 1 if the

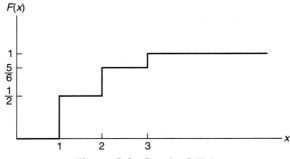

Figure 2.1. Graph of $F(x)$.

outcome is a success and 0 if it is a failure, then the probability mass function of X is given by

$$p(0) = P\{X = 0\} = 1 - p,$$

$$p(1) = P\{X = 1\} = p \tag{2.2}$$

where p, $0 \leqslant p \leqslant 1$, is the probability that the trial is a "success."

A random variable X is said to be a *Bernoulli* random variable if its probability mass function is given by Equation (2.2) for some $p \in (0, 1)$.

2.2.2.   The Binomial Random Variable

Suppose that n independent trials, each of which results in a "success" with probability p and in a "failure" with probability $1 - p$, are to be performed. If X represents the number of successes that occur in the n trials, then X is said to be a *binomial* random variable with parameters (n, p).

The probability mass function of a binomial random variable having parameters (n, p) is given by

$$p(i) = \binom{n}{i} p^i (1 - p)^{n-i}, \qquad i = 0, 1, \ldots, n \tag{2.3}$$

where

$$\binom{n}{i} = \frac{n!}{(n - i)! i!}$$

equals the number of different groups of i objects that can be chosen from a set of n objects. The validity of Equation (2.3) may be verified by first noting that the probability of any particular sequence of the n outcomes containing i successes and $n - i$ failures is, by the assumed independence of trials, $p^i(1 - p)^{n-i}$. Equation (2.3) then follows since there are $\binom{n}{i}$ different sequences of the n outcomes leading to i successes and $n - i$ failures. For instance, if $n = 3$, $i = 2$, then there are $\binom{3}{2} = 3$ ways in which the three trials can result in two successes. Namely, any one of the three outcomes (s, s, f), (s, f, s), (f, s, s), where the outcome (s, s, f) means that the first two trials are successes and the third a failure. Since each of the three outcomes (s, s, f), (s, f, s), (f, s, s) has a probability $p^2(1 - p)$ of occurring the desired probability is thus $\binom{3}{2} p^2(1 - p)$.

Note that, by the binomial theorem, the probabilities sum to one, that is,

$$\sum_{i=0}^{\infty} p(i) = \sum_{i=0}^{n} \binom{n}{i} p^i(1 - p)^{n-i} = (p + (1 - p))^n = 1$$

Example 2.6 Four fair coins are flipped. If the outcomes are assumed independent, what is the probability that two heads and two tails are obtained?

Solution: Letting X equal the number of heads ("successes") that appear, then X is a binomial random variable with parameters $(n = 4, p = \frac{1}{2})$. Hence, by Equation (2.3),

$$P\{X = 2\} = \binom{4}{2}\left(\frac{1}{2}\right)^2\left(\frac{1}{2}\right)^2 = \frac{3}{8} \quad \blacklozenge$$

Example 2.7 It is known that all items produced by a certain machine will be defective with probability 0.1, independently of each other. What is the probability that in a sample of three items, at most one will be defective?

Solution: If X is the number of defective items in the sample, then X is a binomial random variable with parameters $(3, 0.1)$. Hence, the desired probability is given by

$$P\{X = 0\} + P\{X = 1\} = \binom{3}{0}(0.1)^0(0.9)^3 + \binom{3}{1}(0.1)^1(0.9)^2 = 0.972 \quad \blacklozenge$$

Example 2.8 Suppose that an airplane engine will fail, when in flight, with probability $1 - p$ independently from engine to engine; suppose that the airplane will make a successful flight if at least 50 percent of its engines remain operative. For what values of p is a four-engine plane preferable to a two-engine plane?

Solution: Because each engine is assumed to fail or function independently of what happens with the other engines, it follows that the number of engines remaining operative is a binomial random variable. Hence, the probability that a four-engine plane makes a successful flight is

$$\binom{4}{2}p^2(1-p)^2 + \binom{4}{3}p^3(1-p) + \binom{4}{4}p^4(1-p)^0$$

$$= 6p^2(1-p)^2 + 4p^3(1-p) + p^4$$

whereas the corresponding probability for a two-engine plane is

$$\binom{2}{1}p(1-p) + \binom{2}{2}p^2 = 2p(1-p) + p^2$$

Hence the four-engine plane is safer if

$$6p^2(1-p)^2 + 4p^3(1-p) + p^4 \geqslant 2p(1-p) + p^2$$

or equivalently if

$$6p(1-p)^2 + 4p^2(1-p) + p^3 \geqslant 2 - p$$

which simplifies to

$$3p^3 - 8p^2 + 7p - 2 \geqslant 0 \qquad \text{or} \qquad (p-1)^2(3p-2) \geqslant 0$$

which is equivalent to

$$3p - 2 \geqslant 0 \qquad \text{or} \qquad p \geqslant \tfrac{2}{3}$$

Hence, the four-engine plane is safer when the engine success probability is at least as large as $\tfrac{2}{3}$, whereas the two-engine plane is safer when this probability falls below $\tfrac{2}{3}$. ✦

Example 2.9   Suppose that a particular trait of a person (such as eye color or left handedness) is classified on the basis of one pair of genes and suppose that d represents a dominant gene and r a recessive gene. Thus a person with dd genes is pure dominance, one with rr is pure recessive, and one with rd is hybrid. The pure dominance and the hybrid are alike in appearance. Children receive one gene from each parent. If, with respect to a particular trait, two hybrid parents have a total of four children, what is the probability that exactly three of the four children have the outward appearance of the dominant gene?

Solution: If we assume that each child is equally likely to inherit either of two genes from each parent, the probabilities that the child of two hybrid parents will have dd, rr, or rd pairs of genes are, respectively, $\tfrac{1}{4}$, $\tfrac{1}{4}$,

$\frac{1}{2}$. Hence, because an offspring will have the outward appearance of the dominant gene if its gene pair is either *dd* or *rd*, it follows that the number of such children is binomially distributed with parameters $(4, \frac{3}{4})$. Thus the desired probability is

$$\binom{4}{3}\left(\frac{3}{4}\right)^3\left(\frac{1}{4}\right)^1 = \frac{27}{64} \quad \blacklozenge$$

Remark on Terminology If X is a binomial random variable with parameters (n, p), then we say that X has a binomial distribution with parameters (n, p).

2.2.3. The Geometric Random Variable

Suppose that independent trials, each having a probability p of being a success, are performed until a success occurs. If we let X be the number of trials required until the first success, then X is said to be a *geometric* random variable with parameter p. Its probability mass function is given by

$$p(n) = P\{X = n\} = (1 - p)^{n-1}p, \qquad n = 1, 2, \ldots \qquad (2.4)$$

Equation (2.4) follows since in order for X to equal n it is necessary and sufficient that the first $n - 1$ trials be failures and the nth trial a success. Equation (2.4) follows since the outcomes of the successive trials are assumed to be independent.

To check that $p(n)$ is a probability mass function, we note that

$$\sum_{n=1}^{\infty} p(n) = p \sum_{1}^{\infty} (1 - p)^{n-1} = 1$$

2.2.4. The Poisson Random Variable

A random variable X, taking on one of the values $0, 1, 2, \ldots$, is said to be a *Poisson* random variable with parameter λ, if for some $\lambda > 0$,

$$p(i) = P\{X = i\} = e^{-\lambda}\frac{\lambda^i}{i!}, \qquad i = 0, 1, \ldots \qquad (2.5)$$

Equation (2.5) defines a probability mass function since

$$\sum_{i=0}^{\infty} p(i) = e^{-\lambda} \sum_{i=0}^{\infty} \frac{\lambda^i}{i!} = e^{-\lambda}e^{\lambda} = 1$$

The Poisson random variable has a wide range of applications in a diverse number of areas, as will be seen in Chapter 5.

An important property of the Poisson random variable is that it may be used to approximate a binomial random variable when the binomial parameter n is large and p is small. To see this, suppose that X is a binomial random variable with parameters (n, p), and let $\lambda = np$. Then

$$P\{X = i\} = \frac{n!}{(n - i)!i!} p^i(1 - p)^{n-i}$$

$$= \frac{n!}{(n - i)!i!} \left(\frac{\lambda}{n}\right)^i \left(1 - \frac{\lambda}{n}\right)^{n-i}$$

$$= \frac{n(n - 1) \cdots (n - i + 1)}{n^i} \frac{\lambda^i}{i!} \frac{(1 - \lambda/n)^n}{(1 - \lambda/n)^i}$$

Now, for n large and p small

$$\left(1 - \frac{\lambda}{n}\right)^n \approx e^{-\lambda}, \qquad \frac{n(n - 1) \cdots (n - i + 1)}{n^i} \approx 1, \qquad \left(1 - \frac{\lambda}{n}\right)^i \approx 1$$

Hence, for n large and p small,

$$P\{X = i\} \approx e^{-\lambda}\frac{\lambda^i}{i!}$$

Example 2.10 Suppose that the number of typographical errors on a single page of this book has a Poisson distribution with parameter $\lambda = 1$. Calculate the probability that there is at least one error on this page.

Solution:

$$P\{X \geqslant 1\} = 1 - P\{X = 0\} = 1 - e^{-1} \approx 0.633 \quad \blacklozenge$$

Example 2.11 If the number of accidents occurring on a highway each day is a Poisson random variable with parameter $\lambda = 3$, what is the probability that no accidents occur today?

Solution:

$$P\{X = 0\} = e^{-3} \approx 0.05 \quad \blacklozenge$$

Example 2.12 Consider an experiment that consists of counting the number of α-particles given off in a one-second interval by one gram of radioactive material. If we know from past experience that, on the average, 3.2 such α-particles are given off, what is a good approximation to the probability that no more than 2 α-particles will appear?

Solution: If we think of the gram of radioactive material as consisting of a large number n of atoms each of which has probability $3.2/n$ of disintegrating and sending off an α-particle during the second considered, then we see that, to a very close approximation, the number of α-particles given off will be a Poison random variable with parameter $\lambda = 3.2$. Hence the desired probability is

$$P\{X \leqslant 2\} = e^{-3.2} + 3.2e^{-3.2} + \frac{(3.2)^2}{2} e^{-3.2} \approx 0.382 \quad \blacklozenge$$

2.3. Continuous Random Variables

In this section, we shall concern ourselves with random variables whose set of possible values is uncountable. Let X be such a random variable. We say that X is a *continuous* random variable if there exists a nonnegative function $f(x)$, defined for all real $x \in (-\infty, \infty)$, having the property that for any set B of real numbers

$$P\{X \in B\} = \int_B f(x)\, dx \qquad (2.6)$$

The function $f(x)$ is called the *probability density function* of the random variable X.

In words, Equation (2.6) states that the probability that X will be in B may be obtained by integrating the probability density function over the set B. Since X must assume some value, $f(x)$ must satisfy

$$1 = P\{X \in (-\infty, \infty)\} = \int_{-\infty}^{\infty} f(x)\, dx$$

All probability statements about X can be answered in terms of $f(x)$. For instance, letting $B = [a, b]$, we obtain from Equation (2.6) that

$$P\{a \leqslant X \leqslant b\} = \int_a^b f(x)\, dx \qquad (2.7)$$

If we let $a = b$ in the preceding, then

$$P\{X = a\} = \int_a^a f(x)\, dx = 0$$

In words, this equation states that the probability that a continuous random variable will assume any *particular* value is zero.

The relationship between the cumulative distribution $F(\cdot)$ and the probability density $f(\cdot)$ is expressed by

$$F(a) = P\{X \in (-\infty, a)\} = \int_{-\infty}^{a} f(x)\, dx$$

Differentiating both sides of the preceding yields

$$\frac{d}{da} F(a) = f(a)$$

That is, the density is the derivative of the cumulative distribution function. A somewhat more intuitive interpretation of the density function may be obtained from Equation (2.7) as follows:

$$P\left\{a - \frac{\varepsilon}{2} \leqslant X \leqslant a + \frac{\varepsilon}{2}\right\} = \int_{a-\varepsilon/2}^{a+\varepsilon/2} f(x)\, dx \approx \varepsilon f(a)$$

when ε is small. In other words, the probability that X will be contained in an interval of length ε around the point a is approximately $\varepsilon f(a)$. From this, we see that $f(a)$ is a measure of how likely it is that the random variable will be near a.

There are several important continuous random variables that appear frequently in probability theory. The remainder of this section is devoted to a study of certain of these random variables.

2.3.1. The Uniform Random Variable

A random variable is said to be uniformly distributed over the interval $(0, 1)$ if its probability density function is given by

$$f(x) = \begin{cases} 1, & 0 < x < 1 \\ 0, & \text{otherwise} \end{cases}$$

Note that the preceding is a density function since $f(x) \geqslant 0$ and

$$\int_{-\infty}^{\infty} f(x)\, dx = \int_{0}^{1} dx = 1$$

Since $f(x) > 0$ only when $x \in (0, 1)$, it follows that X must assume a value in $(0, 1)$. Also, since $f(x)$ is constant for $x \in (0, 1)$, X is just as likely to be "near" any value in $(0, 1)$ as any other value. To check this, note that, for any $0 < a < b < 1$,

$$P\{a \leqslant X \leqslant b\} = \int_{a}^{b} f(x)\, dx = b - a$$

In other words, the probability that X is in any particular subinterval of $(0, 1)$ equals the length of that subinterval.

In general, we say that X is a uniform random variable on the interval (α, β) if its probability density function is given by

$$f(x) = \begin{cases} \dfrac{1}{\beta - \alpha}, & \text{if } \alpha < x < \beta \\ 0, & \text{otherwise} \end{cases} \tag{2.8}$$

Example 2.13 Calculate the cumulative distribution function of a random variable uniformly distributed over (α, β).

Solution: Since $F(a) = \int_{-\infty}^{a} f(x)\,dx$, we obtain from Equation (2.8) that

$$F(a) = \begin{cases} 0, & a \leqslant \alpha \\ \dfrac{a-\alpha}{\beta-\alpha}, & \alpha < a < \beta \\ 1, & a \geqslant \beta \end{cases} \blacklozenge$$

Example 2.14 If X is uniformly distributed over $(0, 10)$, calculate the probability that (a) $X < 3$, (b) $X > 7$, (c) $1 < X < 6$.

Solution:

$$P\{X < 3\} = \frac{\int_0^3 dx}{10} = \frac{3}{10},$$

$$P\{X > 7\} = \frac{\int_7^{10} dx}{10} = \frac{3}{10},$$

$$P\{1 < X < 6\} = \frac{\int_1^6 dx}{10} = \frac{1}{2} \quad \blacklozenge$$

2.3.2. Exponential Random Variables

A continuous random variable whose probability density function is given, for some $\lambda > 0$, by

$$f(x) = \begin{cases} \lambda e^{-\lambda x} & \text{if } x \geqslant 0 \\ 0, & \text{if } x < 0 \end{cases}$$

is said to be an exponential random variable with parameter λ. These random variables will be extensively studied in Chapter 5, so we will content ourselves here with just calculating the cumulative distribution function F:

$$F(a) = \int_0^a \lambda e^{-\lambda x} = 1 - e^{-\lambda a}, \qquad a \geqslant 0$$

Note that $F(\infty) = \int_0^\infty \lambda e^{-\lambda x}\,dx = 1$, as, of course, it must.

2.3.3. Gamma Random Variables

A continuous random variable whose density is given by

$$f(x) = \begin{cases} \dfrac{\lambda e^{-\lambda x}(\lambda x)^{\alpha-1}}{\Gamma(\alpha)}, & \text{if } x \geqslant 0 \\ 0, & \text{if } x < 0 \end{cases}$$

for some $\lambda > 0$, $\alpha > 0$ is said to be a gamma random variable with parameters α, λ. The quantity $\Gamma(\alpha)$ is called the gamma function and is defined by

$$\Gamma(\alpha) = \int_0^\infty e^{-x} x^{\alpha-1}\, dx$$

It is easy to show by induction that for integral α, say, $\alpha = n$,

$$\Gamma(n) = (n-1)!$$

2.3.4. Normal Random Variables

We say that X is a normal random variable (or simply that X is normally distributed) with parameters μ and σ^2 if the density of X is given by

$$f(x) = \frac{1}{\sqrt{2\pi}\,\sigma} e^{-(x-\mu)^2/2\sigma^2}, \qquad -\infty < x < \infty$$

This density function is a bell-shaped curve that is symmetric around μ (see Figure 2.2).

An important fact about normal random variables is that if X is normally distributed with parameters μ and σ^2 then $Y = \alpha X + \beta$ is normally distributed with parameters $\alpha\mu + \beta$ and $\alpha^2\sigma^2$. To prove this, suppose first that $\alpha > 0$ and note that $F_Y(\cdot)^*$ the cumulative distribution function of the random variable Y is given by

$$
\begin{aligned}
F_Y(a) &= P\{Y \leqslant a\} \\
&= P\{\alpha X + \beta \leqslant a\} \\
&= P\left\{ X \leqslant \frac{a-\beta}{\alpha} \right\} \\
&= F_X\left(\frac{a-\beta}{\alpha} \right) \\
&= \int_{-\infty}^{(a-\beta)/\alpha} \frac{1}{\sqrt{2\pi}\,\sigma} e^{-(x-\mu)^2/2\sigma^2}\, dx \\
&= \int_{-\infty}^{a} \frac{1}{\sqrt{2\pi}\,\alpha\sigma} \exp\left\{ \frac{-(v-(\alpha\mu+\beta))^2}{2\alpha^2\sigma^2} \right\} dv
\end{aligned}
\tag{2.9}
$$

---

* When there is more than one random variable under consideration, we shall denote the cumulative distribution function of a random variable Z by $F_Z(\cdot)$. Similarly, we shall denote the density of Z by $f_Z(\cdot)$.

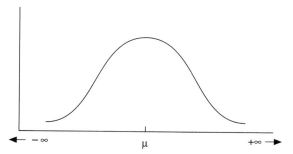

Figure 2.2. Normal density function.

where the last equality is obtained by the change in variables $v = \alpha x + \beta$. However, since $F_Y(a) = \int_{-\infty}^{a} f_Y(v)\, dv$, it follows from Equation (2.9) that the probability density function $f_Y(\cdot)$ is given by

$$f_Y(v) = \frac{1}{\sqrt{2\pi}\,\alpha\sigma} \exp\left\{ \frac{-(v - (\alpha\mu + \beta))^2}{2(\alpha\sigma)^2} \right\}, \qquad -\infty < v < \infty$$

Hence, Y is normally distributed with parameters $\alpha\mu + \beta$ and $(\alpha\sigma)^2$. A similar result is also true when $\alpha < 0$.

One implication of the preceding result is that if X is normally distributed with parameters μ and σ^2 then $Y = (X - \mu)/\sigma$ is normally distributed with parameters 0 and 1. Such a random variable Y is said to have the *standard* or *unit* normal distribution.

2.4.   Expectation of a Random Variable

2.4.1.   The Discrete Case

If X is a discrete random variable having a probability mass function $p(x)$, then the *expected value* of X is defined by

$$E[X] = \sum_{x:p(x)>0} xp(x)$$

In other words, the expected value of X is a weighted average of the possible values that X can take on, each value being weighted by the probability that X assumes that value. For example, if the probability mass function of X is given by

$$p(1) = \tfrac{1}{2} = p(2)$$

then

$$E[X] = 1(\tfrac{1}{2}) + 2(\tfrac{1}{2}) = \tfrac{3}{2}$$

is just an ordinary average of the two possible values 1 and 2 that X can assume. On the other hand, if

$$p(1) = \tfrac{1}{3}, \qquad p(2) = \tfrac{2}{3}$$

then

$$E[X] = 1(\tfrac{1}{3}) + 2(\tfrac{2}{3}) = \tfrac{5}{3}$$

is a weighted average of the two possible values 1 and 2 where the value 2 is given twice as much weight as the value 1 since $p(2) = 2p(1)$.

Example 2.15   Find $E[X]$ where X is the outcome when we roll a fair die.

Solution: Since $p(1) = p(2) = p(3) = p(4) = p(5) = p(6) = \tfrac{1}{6}$, we obtain

$$E[X] = 1(\tfrac{1}{6}) + 2(\tfrac{1}{6}) + 3(\tfrac{1}{6}) + 4(\tfrac{1}{6}) + 5(\tfrac{1}{6}) + 6(\tfrac{1}{6}) = \tfrac{7}{2} \quad \blacklozenge$$

Example 2.16   (Expectation of a Bernoulli Random Variable): Calculate $E[X]$ when X is a Bernoulli random variable with parameter p.

Solution: Since $p(0) = 1 - p$, $p(1) = p$, we have

$$E[X] = 0(1 - p) + 1(p) = p$$

Thus, the expected number of successes in a single trial is just the probability that the trial will be a success. ◆

Example 2.17   (Expectation of a Binomial Random Variable): Calculate $E[X]$ when X is binomially distributed with parameters n and p.

Solution:

$$E[X] = \sum_{i=0}^{n} ip(i)$$

$$= \sum_{i=0}^{n} i \binom{n}{i} p^i (1 - p)^{n-i}$$

$$= \sum_{i=1}^{n} \frac{in!}{(n-i)!i!} p^i (1 - p)^{n-i}$$

$$= \sum_{i=1}^{n} \frac{n!}{(n-i)!(i-1)!} p^i (1 - p)^{n-i}$$

$$= np \sum_{i=1}^{n} \frac{(n-1)!}{(n-i)!(i-1)!} p^{i-1}(1-p)^{n-i}$$

$$= np \sum_{k=0}^{n-1} \binom{n-1}{k} p^{k}(1-p)^{n-1-k}$$

$$= np[p + (1-p)]^{n-1}$$

$$= np$$

where the second from the last equality follows by letting $k = i - 1$. Thus, the expected number of successes in n independent trials is n multiplied by the probability that a trial results in a success. ✦

Example 2.18 (Expectation of a Geometric Random Variable): Calculate the expectation of a geometric random variable having parameter p.

Solution: By Equation (2.4), we have

$$E[X] = \sum_{n=1}^{\infty} np(1-p)^{n-1}$$

$$= p \sum_{n=1}^{\infty} nq^{n-1}$$

where $q = 1 - p$,

$$E[X] = p \sum_{n=1}^{\infty} \frac{d}{dq}(q^n)$$

$$= p \frac{d}{dq}\left(\sum_{n=1}^{\infty} q^n \right)$$

$$= p \frac{d}{dq}\left(\frac{q}{1-q} \right)$$

$$= \frac{p}{(1-q)^2}$$

$$= \frac{1}{p}$$

In words, the expected number of independent trials we need to perform until we attain our first success equals the reciprocal of the probability that any one trial results in a success. ✦

Example 2.19   (Expectation of a Poisson Random Variable): Calculate $E[X]$ if X is a Poisson random variable with parameter λ.

Solution: From Equation (2.5), we have

$$E[X] = \sum_{i=0}^{\infty} \frac{ie^{-\lambda}\lambda^{i}}{i!}$$

$$= \sum_{i=1}^{\infty} \frac{e^{-\lambda}\lambda^{i}}{(i-1)!}$$

$$= \lambda e^{-\lambda} \sum_{i=1}^{\infty} \frac{\lambda^{i-1}}{(i-1)!}$$

$$= \lambda e^{-\lambda} \sum_{k=0}^{\infty} \frac{\lambda^{k}}{k!}$$

$$= \lambda e^{-\lambda}e^{\lambda}$$

$$= \lambda$$

where we have used the identity $\sum_{k=0}^{\infty} \lambda^{k}/k! = e^{\lambda}$.   ✦

2.4.2. The Continuous Case

We may also define the expected value of a continuous random variable. This is done as follows. If X is a continuous random variable having a probability density function $f(x)$, then the expected value of X is defined by

$$E[X] = \int_{-\infty}^{\infty} xf(x)\,dx$$

Example 2.20   (Expectation of a Uniform Random Variable): Calculate the expectation of a random variable uniformly distributed over (α, β).

Solution: From Equation (2.8) we have

$$E[X] = \int_{\alpha}^{\beta} \frac{x}{\beta - \alpha}\,dx$$

$$= \frac{\beta^2 - \alpha^2}{2(\beta - \alpha)}$$

$$= \frac{\beta + \alpha}{2}$$

In other words, the expected value of a random variable uniformly distributed over the interval (α, β) is just the midpoint of the interval. ✦

Example 2.21   (Expectation of an Exponential Random Variable): Let X be exponentially distributed with parameter λ. Calculate $E[X]$.

Solution:

$$E[X] = \int_0^\infty x\lambda e^{-\lambda x}\, dx$$

Integrating by parts yields

$$E[X] = -xe^{-\lambda x}\Big|_0^\infty + \int_0^\infty e^{-\lambda x}\, dx$$

$$= 0 - \frac{e^{-\lambda x}}{\lambda}\Big|_0^\infty$$

$$= \frac{1}{\lambda} \quad ✦$$

Example 2.22   (Expectation of a Normal Random Variable): Calculate $E[X]$ when X is normally distributed with parameters μ and σ^2.

Solution:

$$E[X] = \frac{1}{\sqrt{2\pi}\,\sigma} \int_{-\infty}^\infty x e^{-(x-\mu)^2/2\sigma^2}\, dx$$

Writing x as $(x - \mu) + \mu$ yields

$$E[X] = \frac{1}{\sqrt{2\pi}\,\sigma} \int_{-\infty}^\infty (x - \mu) e^{-(x-\mu)^2/2\sigma^2}\, dx$$

$$+ \mu \frac{1}{\sqrt{2\pi}\,\sigma} \int_{-\infty}^\infty e^{-(x-\mu)^2/2\sigma^2}\, dx$$

Letting $y = x - \mu$ leads to

$$E[X] = \frac{1}{\sqrt{2\pi}\,\sigma} \int_{-\infty}^\infty y e^{-y^2/2\sigma^2}\, dy + \mu \int_{-\infty}^\infty f(x)\, dx$$

where $f(x)$ is the normal density. By symmetry, the first integral must be 0, and so

$$E[X] = \mu \int_{-\infty}^\infty f(x)\, dx = \mu \quad ✦$$

2.4.3.   Expectation of a Function of a Random Variable

Suppose now that we are given a random variable X and its probability distribution (that is, its probability mass function in the discrete case or its probability density function in the continuous case). Suppose also that we are interested in calculating, not the expected value of X, but the expected value of some function of X, say, $g(X)$. How do we go about doing this? One way is as follows. Since $g(X)$ is itself a random variable, it must have a probability distribution, which should be computable from a knowledge of the distribution of X. Once we have obtained the distribution of $g(X)$, we can then compute $E[g(X)]$ by the definition of the expectation.

Example 2.23   Suppose X has the following probability mass function:

$$p(0) = 0.2, \qquad p(1) = 0.5, \qquad p(2) = 0.3$$

Calculate $E[X^2]$.

Solution: Letting $Y = X^2$, we have that Y is a random variable that can take on one of the values 0^2, 1^2, 2^2 with respective probabilities

$$p_Y(0) = P\{Y = 0^2\} = 0.2,$$
$$p_Y(1) = P\{Y = 1^2\} = 0.5,$$
$$p_Y(4) = P\{Y = 2^2\} = 0.3$$

Hence,

$$E[X^2] = E[Y] = 0(0.2) + 1(0.5) + 4(0.3) = 1.7$$

Note that

$$1.7 = E[X^2] \neq (E[X])^2 = 1.21 \quad \blacklozenge$$

Example 2.24   Let X be uniformly distributed over $(0, 1)$. Calculate $E[X^3]$.

Solution: Letting $Y = X^3$, we calculate the distribution of Y as follows. For $0 \leqslant a \leqslant 1$,

$$F_Y(a) = P\{Y \leqslant a\}$$
$$= P\{X^3 \leqslant a\}$$
$$= P\{X \leqslant a^{1/3}\}$$
$$= a^{1/3}$$

where the last equality follows since X is uniformity distributed over $(0, 1)$. By differentiating $F_Y(a)$, we obtain the density of Y, namely,

$$f_Y(a) = \tfrac{1}{3}a^{-2/3}, \qquad 0 \leqslant a \leqslant 1$$

Hence,

$$
\begin{aligned}
E[X^3] = E[Y] &= \int_{-\infty}^{\infty} a f_Y(a)\, da \\
&= \int_0^1 a \tfrac{1}{3} a^{-2/3}\, da \\
&= \tfrac{1}{3} \int_0^1 a^{1/3}\, da \\
&= \tfrac{1}{3} \tfrac{3}{4} a^{4/3} \big|_0^1 \\
&= \tfrac{1}{4} \quad \blacklozenge
\end{aligned}
$$

While the foregoing procedure will, in theory, always enable us to compute the expectation of any function of X from a knowledge of the distribution of X, there is, fortunately, an easier way to do this. The following proposition shows how we can calculate the expectation of $g(X)$ without first determining its distribution.

Proposition 2.1 (a) If X is a discrete random variable with probability mass function $p(x)$, then for any real-valued function g,

$$E[g(X)] = \sum_{x:p(x) > 0} g(x)p(x)$$

(b) If X is a continuous random variable with probability density function $f(x)$, then for any real-valued function g,

$$E[g(X)] = \int_{-\infty}^{\infty} g(x)f(x)\, dx \quad \blacklozenge$$

Example 2.25 Applying the proposition to Example 2.23 yields

$$E[X^2] = 0^2(0.2) + (1^2)(0.5) + (2^2)(0.3) = 1.7$$

which, of course, checks with the result derived in Example 2.23. $\quad \blacklozenge$

Example 2.26   Applying the proposition to Example 2.24 yields

$$E[X^3] = \int_0^1 x^3 \, dx \qquad \text{(since } f(x) = 1, 0 < x < 1)$$

$$= \tfrac{1}{4} \quad \blacklozenge$$

A simple corollary of Proposition 2.1 is the following.

Corollary 2.2   If a and b are constants, then

$$E[aX + b] = aE[X] + b$$

Proof   In the discrete case,

$$E[aX + b] = \sum_{x:p(x)>0} (ax + b)p(x)$$

$$= a \sum_{x:p(x)>0} xp(x) + b \sum_{x:p(x)>0} p(x)$$

$$= aE[X] + b$$

In the continuous case,

$$E[aX + b] = \int_{-\infty}^{\infty} (ax + b)f(x) \, dx$$

$$= a \int_{-\infty}^{\infty} xf(x) \, dx + b \int_{-\infty}^{\infty} f(x) \, dx$$

$$= aE[X] + b \quad \blacklozenge$$

The expected value of a random variable X, $E[X]$, is also referred to as the *mean* or the first *moment* of X. The quantity $E[X^n]$, $n \geqslant 1$, is called the nth moment of X. By Proposition 2.1, we note that

$$E[X^n] = \begin{cases} \displaystyle\sum_{x:p(x)>0} x^n p(x), & \text{if } X \text{ is discrete} \\[2ex] \displaystyle\int_{-\infty}^{\infty} x^n f(x) \, dx, & \text{if } X \text{ is continuous} \end{cases}$$

Another quantity of interest is the variance of a random variable X, denoted by $\text{Var}(X)$, which is defined by

$$\text{Var}(X) = E[(X - E[X])^2]$$

Thus, the variance of X measures the expected square of the deviation of X from its expected value.

Example 2.27 (Variance of the Normal Random Variable): Let X be normally distributed with parameters μ and σ^2. Find $\text{Var}(X)$.

Solution: Recalling (see Example 2.22) that $E[X] = \mu$, we have that

$$\text{Var}(X) = E[(X - \mu)^2]$$

$$= \frac{1}{\sqrt{2\pi}\,\sigma} \int_{-\infty}^{\infty} (x - \mu)^2 e^{-(x-\mu)^2/2\sigma^2}\, dx$$

Substituting $y = (x - \mu)/\sigma$ yields

$$\text{Var}(X) = \frac{\sigma^2}{\sqrt{2\pi}} \int_{-\infty}^{\infty} y^2 e^{-y^2/2}\, dy$$

We now employ an identity that can be found in many integration tables, namely, $\int_{-\infty}^{\infty} y^2 e^{-y^2/2}\, dy = \sqrt{2\pi}$. Hence,

$$\text{Var}(X) = \sigma^2$$

Another derivation of $\text{Var}(X)$ will be given in Example 2.42. ✦

Suppose that X is continuous with density f, and let $E[X] = \mu$. Then,

$$\text{Var}(X) = E[(X - \mu)^2]$$

$$= E[X^2 - 2\mu X + \mu^2]$$

$$= \int_{-\infty}^{\infty} (x^2 - 2\mu x + \mu^2) f(x)\, dx$$

$$= \int_{-\infty}^{\infty} x^2 f(x)\, dx - 2\mu \int_{-\infty}^{\infty} x f(x)\, dx + \mu^2 \int_{-\infty}^{\infty} f(x)\, dx$$

$$= E[X^2] - 2\mu\mu + \mu^2$$

$$= E[X^2] - \mu^2$$

A similar proof holds in the discrete case, and so we obtain the useful identity

$$\text{Var}(X) = E[X^2] - (E[X])^2$$

Example 2.28 Calculate $\text{Var}(X)$ when X represents the outcome when a fair die is rolled.

Solution: As previously noted in Example 2.15, $E[X] = \frac{7}{2}$. Also,

$$E[X^2] = 1(\tfrac{1}{6}) + 2^2(\tfrac{1}{6}) + 3^2(\tfrac{1}{6}) + 4^2(\tfrac{1}{6}) + 5^2(\tfrac{1}{6}) + 6^2(\tfrac{1}{6}) = (\tfrac{1}{6})(91)$$

Hence,

$$\text{Var}(X) = \tfrac{91}{6} - (\tfrac{7}{2})^2 = \tfrac{35}{12} \quad \blacklozenge$$

2.5. Jointly Distributed Random Variables

2.5.1.   Joint Distribution Functions

Thus far, we have concerned ourselves with the probability distribution of a single random variable. However, we are often interested in probability statements concerning two or more random variables. To deal with such probabilities, we define, for any two random variables X and Y, the *joint cumulative probability distribution function* of X and Y by

$$F(a, b) = P\{X \leqslant a, Y \leqslant b\}, \qquad -\infty < a, b < \infty$$

The distribution of X can be obtained from the joint distribution of X and Y as follows:

$$F_X(a) = P\{X \leqslant a\}$$
$$= P\{X \leqslant a, Y \leqslant \infty\}$$
$$= F(a, \infty)$$

Similarly, the cumulative distribution function of Y is given by

$$F_Y(b) = P\{Y \leqslant b\} = F(\infty, b)$$

In the case where X and Y are both discrete random variables, it is convenient to define the *joint probability mass function* of X and Y by

$$p(x, y) = P\{X = x, Y = y\}$$

The probability mass function of X may be obtained from $p(x, y)$ by

$$p_X(x) = \sum_{y : p(x,y) > 0} p(x, y)$$

Similarly,

$$p_Y(y) = \sum_{x : p(x,y) > 0} p(x, y)$$

We say that X and Y are *jointly continuous* if there exists a function $f(x, y)$, defined for all real x and y, having the property that for all sets A and B of real numbers

$$P\{X \in A, Y \in B\} = \int_B \int_A f(x, y) \, dx \, dy$$

The function $f(x, y)$ is called the *joint probability density function* of X and Y. The probability density of X can be obtained from a knowledge of $f(x, y)$ by the following reasoning:

$$P\{X \in A\} = P\{X \in A, Y \in (-\infty, \infty)\}$$

$$= \int_{-\infty}^{\infty} \int_{A} f(x, y) \, dx \, dy$$

$$= \int_{A} f_X(x) \, dx$$

where

$$f_X(x) = \int_{-\infty}^{\infty} f(x, y) \, dy$$

is thus the probability density function of X. Similarly, the probability density function of Y is given by

$$f_Y(y) = \int_{-\infty}^{\infty} f(x, y) \, dx$$

A variation of Proposition 2.1 states that if X and Y are random variables and g is a function of two variables, then

$$E[g(X, Y)] = \sum_y \sum_x g(x, y)p(x, y) \qquad \text{in the discrete case}$$

$$= \int_{-\infty}^{\infty} \int_{-\infty}^{\infty} g(x, y)f(x, y) \, dx \, dy \qquad \text{in the continuous case}$$

For example, if $g(X, Y) = X + Y$, then, in the continuous case,

$$E[X + Y] = \int_{-\infty}^{\infty} \int_{-\infty}^{\infty} (x + y)f(x, y) \, dx \, dy$$

$$= \int_{-\infty}^{\infty} \int_{-\infty}^{\infty} xf(x, y) \, dx \, dy + \int_{-\infty}^{\infty} \int_{-\infty}^{\infty} yf(x, y) \, dx \, dy$$

$$= E[X] + E[Y]$$

where the first integral is evaluated by using the variation of Proposition 2.1 with $g(x, y) = x$, and the second with $g(x, y) = y$.

The same result holds in the discrete case and, combined with the corollary in Section 2.4.3, yields that for any constants a, b

$$E[aX + bY] = aE[X] + bE[Y] \tag{2.10}$$

Joint probability distributions may also be defined for n random variables. The details are exactly the same as when $n = 2$ and are left as an exercise. The corresponding result to Equation (2.10) states that if $X_1, X_2, \ldots, X_n$ are n random variables, then for any n constants $a_1, a_2, \ldots, a_n$,

$$E[a_1 X_1 + a_2 X_2 + \cdots + a_n X_n]$$
$$= a_1 E[X_1] + a_2 E[X_2] + \cdots + a_n E[X_n] \qquad (2.11)$$

Example 2.29   Calculate the expected sum obtained when three fair dice are rolled.

Solution: Let X denote the sum obtained. Then $X = X_1 + X_2 + X_3$ where X_i represents the value of the ith die. Thus,

$$E[X] = E[X_1] + E[X_2] + E[X_3] = 3(\tfrac{7}{2}) = \tfrac{21}{2} \quad \blacklozenge$$

Example 2.30   As another example of the usefulness of Equation (2.11), let us use it to obtain the expectation of a binomial random variable having parameters n and p. Recalling that such a random variable X represents the number of successes in n trials when each trial has probability p of being a success, we have that

$$X = X_1 + X_2 + \cdots + X_n$$

where

$$X_i = \begin{cases} 1, & \text{if the } i\text{th trial is a success} \\ 0, & \text{if the } i\text{th trial is a failure} \end{cases}$$

Hence, X_i is a Bernoulli random variable having expectation $E[X_i] = 1(p) + 0(1 - p) = p$. Thus,

$$E[X] = E[X_1] + E[X_2] + \cdots + E[X_n] = np$$

This derivation should be compared with the one presented in Example 2.17. $\blacklozenge$

Example 2.31   At a party N men throw their hats into the center of a room. The hats are mixed up and each man randomly selects one. Find the expected number of men who select their own hats.

Solution: Letting X denote the number of men that select their own hats, we can best compute $E[X]$ by noting that

$$X = X_1 + X_2 + \cdots + X_N$$

where

$$X_i = \begin{cases} 1, & \text{if the } i\text{th man selects his own hat} \\ 0, & \text{otherwise} \end{cases}$$

Now, because the ith man is equally likely to select any of the N hats, it follows that

$$P\{X_i = 1\} = P\{i\text{th man selects his own hat}\} = \frac{1}{N}$$

and so

$$E[X_i] = 1P\{X_i = 1\} + 0P\{X_i = 0\} = \frac{1}{N}$$

Hence, from Equation (2.11) we obtain that

$$E[X] = E[X_1] + \cdots + E[X_N] = \left(\frac{1}{N}\right)N = 1$$

Hence, no matter how many people are at the party, on the average exactly one of the men will select his own hat. ✦

Example 2.32 Suppose there are 25 different types of coupons and suppose that each time one obtains a coupon, it is equally likely to be any one of the 25 types. Compute the expected number of different types that are contained in a set of 10 coupons.

Solution: Let X denote the number of different types in the set of 10 coupons. We compute $E[X]$ by using the representation

$$X = X_1 + \cdots + X_{25}$$

where

$$X_i = \begin{cases} 1, & \text{if at least one type } i \text{ coupon is in the set of 10} \\ 0, & \text{otherwise} \end{cases}$$

Now,

$$E[X_i] = P\{X_i = 1\}$$

$$= P\{\text{at least one type } i \text{ coupon is in the set of 10}\}$$

$$= 1 - P\{\text{no type } i \text{ coupons are in the set of 10}\}$$

$$= 1 - (\tfrac{24}{25})^{10}$$

when the last equality follows since each of the 10 coupons will (independently) not be a type i with probability $\frac{24}{25}$. Hence,

$$E[X] = E[X_1] + \cdots + E[X_{25}] = 25[1 - (\tfrac{24}{25})^{10}] \quad \blacklozenge$$

2.5.2. Independent Random Variables

The random variables X and Y are said to be *independent* if, for all a, b,

$$P\{X \leqslant a, Y \leqslant b\} = P\{X \leqslant a\}P\{Y \leqslant b\} \tag{2.12}$$

In other words, X and Y are independent if, for all a and b, the events $E_a = \{X \leqslant a\}$ and $F_b = \{Y \leqslant b\}$ are independent.

In terms of the joint distribution function F of X and Y, we have that X and Y are independent if

$$F(a, b) = F_X(a)F_Y(b) \qquad \text{for all } a, b$$

When X and Y are discrete, the condition of independence reduces to

$$p(x, y) = p_X(x)p_Y(y) \tag{2.13}$$

while if X and Y are jointly continuous, independence reduces to

$$f(x, y) = f_X(x)f_Y(y) \tag{2.14}$$

To prove this statement, consider first the discrete version, and suppose that the joint probability mass function $p(x, y)$ satisfies Equation (2.13). Then

$$
\begin{aligned}
P\{X \leqslant a, Y \leqslant b\} &= \sum_{y \leqslant b} \sum_{x \leqslant a} p(x, y) \\
&= \sum_{y \leqslant b} \sum_{x \leqslant a} p_X(x)p_Y(y) \\
&= \sum_{y \leqslant b} p_Y(y) \sum_{x \leqslant a} p_X(x) \\
&= P\{Y \leqslant b\}P\{X \leqslant a\}
\end{aligned}
$$

and so X and Y are independent. That Equation (2.14) implies independence in the continuous case is proven in the same manner and is left as an exercise.

An important result concerning independence is the following.

Proposition 2.3 If X and Y are independent, then for any functions h and g

$$E[g(X)h(Y)] = E[g(X)]E[h(Y)]$$

Proof Suppose that X and Y are jointly continuous. Then

$$E[g(X)h(Y)] = \int_{-\infty}^{\infty} \int_{-\infty}^{\infty} g(x)h(y)f(x, y)\,dx\,dy$$

$$= \int_{-\infty}^{\infty} \int_{-\infty}^{\infty} g(x)\text{h}(y)f_X(x)f_Y(y)\,dx\,dy$$

$$= \int_{-\infty}^{\infty} h(y)f_Y(y)\,dy \int_{-\infty}^{\infty} g(x)f_X(x)\,dx$$

$$= E[h(Y)]E[g(X)]$$

The proof in the discrete case is similar. ✦

2.5.3. Covariance and Variance of Sums of Random Variables

The covariance of any two random variables X and Y, denoted by $\text{Cov}(X, Y)$, is defined by

$$\text{Cov}(X, Y) = E[(X - E[X])(Y - E[Y])]$$

$$= E[XY - YE[X] - XE[Y] + E[X]E[Y]]$$

$$= E[XY] - E[Y]E[X] - E[X]E[Y] + E[X]E[Y]$$

$$= E[XY] - E[X]E[Y]$$

Note that if X and Y are independent, then by Proposition 2.3 it follows that $\text{Cov}(X, Y) = 0$.

Let us consider now the special case where X and Y are indicator variables for whether or not the events A and B occur. That is, for events A and B, define

$$X = \begin{cases} 1, & \text{if } A \text{ occurs} \\ 0, & \text{otherwise} \end{cases}, \qquad Y = \begin{cases} 1, & \text{if } B \text{ occurs} \\ 0, & \text{otherwise} \end{cases}$$

Then,

$$\text{Cov}(X, Y) = E[XY] - E[X]E[Y]$$

and, because XY will equal 1 or 0 depending on whether or not both X and Y equal 1, we see that

$$\text{Cov}(X, Y) = P\{X = 1, Y = 1\} - P\{X = 1\}P\{Y = 1\}$$

From this we see that

$$\text{Cov}(X, Y) > 0 \Leftrightarrow P\{X = 1, Y = 1\} > P\{X = 1\}P\{Y = 1\}$$

$$\Leftrightarrow \frac{P\{X = 1, Y = 1\}}{P\{X = 1\}} > P\{Y = 1\}$$

$$\Leftrightarrow P\{Y = 1 \mid X = 1\} > P\{Y = 1\}$$

That is, the covariance of X and Y is positive if the outcome $X = 1$ makes it more likely that $Y = 1$ (which, as is easily seen by symmetry, also implies the reverse).

In general it can be shown that a positive value of $\text{Cov}(X, Y)$ is an indication that Y tends to increase as X does, whereas a negative value indicates that Y tends to decrease as X increases.

The following are important properties of covariance.

Properties of Covariance

For any random variables X, Y, Z and constant c,

1. $\text{Cov}(X, X) = \text{Var}(X)$,
2. $\text{Cov}(X, Y) = \text{Cov}(Y, X)$,
3. $\text{Cov}(cX, Y) = c\,\text{Cov}(X, Y)$,
4. $\text{Cov}(X, Y + Z) = \text{Cov}(X, Y) + \text{Cov}(X, Z)$.

Whereas the first three properties are immediate, the final one is easily proven as follows:

$$\text{Cov}(X, Y + Z) = E[X(Y + Z)] - E[X]E[Y + Z]$$

$$= E[XY] - E[X]E[Y] + E[XZ] - E[X]E[Z]$$

$$= \text{Cov}(X, Y) + \text{Cov}(X, Z)$$

The fourth property listed easily generalizes to give the following result:

$$\text{Cov}\left(\sum_{i=1}^{n} X_i, \sum_{j=1}^{m} Y_j\right) = \sum_{i=1}^{n} \sum_{j=1}^{m} \text{Cov}(X_i, Y_j) \tag{2.15}$$

A useful expression for the variance of the sum of random variables can be obtained from Equation (2.15) as follows:

$$\text{Var}\left(\sum_{i=1}^{n} X_i\right) = \text{Cov}\left(\sum_{i=1}^{n} X_i, \sum_{j=1}^{n} X_j\right)$$

$$= \sum_{i=1}^{n} \sum_{j=1}^{n} \text{Cov}(X_i, X_j)$$

$$= \sum_{i=1}^{n} \text{Cov}(X_i, X_i) + \sum_{i=1}^{n} \sum_{j \neq i} \text{Cov}(X_i, X_j)$$

$$= \sum_{i=1}^{n} \text{Var}(X_i) + 2 \sum_{i=1}^{n} \sum_{j < i} \text{Cov}(X_i, X_j) \tag{2.16}$$

If X_i, $i = 1, \ldots, n$ are independent random variables, then Equation (2.16) reduces to

$$\text{Var}\left(\sum_{i=1}^{n} X_i\right) = \sum_{i=1}^{n} \text{Var}(X_i)$$

Definition 2.1 If $X_1, \ldots, X_n$ are independent and identically distributed, then the random variable $\bar{X} = \sum_{i=1}^{n} X_i/n$ is called the *sample mean*.

The following proposition shows that the covariance between the sample mean and a deviation from that sample mean is zero. It will be needed in Section 2.6.1.

Proposition 2.4 Suppose that $X_1, \ldots, X_n$ are independent and identically distributed with expected value μ and variance σ^2. Then,

(a) $E[\bar{X}] = \mu$.
(b) $\text{Var}(\bar{X}) = \sigma^2/n$.
(c) $\text{Cov}(\bar{X}, X_i - \bar{X}) = 0$, $i = 1, \ldots, n$.

Proof Parts (a) and (b) are easily established as follows:

$$E[\bar{X}] = \frac{1}{n} \sum_{i=1}^{m} E[X_i] = \mu$$

$$\text{Var}(\bar{X}) = \left(\frac{1}{n}\right)^2 \text{Var}\left(\sum_{i=1}^{n} X_i\right) = \left(\frac{1}{n}\right)^2 \sum_{i=1}^{n} \text{Var}(X_i) = \frac{\sigma^2}{n}$$

To establish part (c) we reason as follows:

$$\text{Cov}(\bar{X}, X_i - \bar{X}) = \text{Cov}(\bar{X}, X_i) - \text{Cov}(\bar{X}, \bar{X})$$

$$= \frac{1}{n} \text{Cov}\left(X_i + \sum_{j \neq i} X_j, X_i\right) - \text{Var}(\bar{X})$$

$$= \frac{1}{n} \text{Cov}(X_i, X_i) - \frac{1}{n} \text{Cov}\left(\sum_{j \neq i} X_j, X_i\right) - \frac{\sigma^2}{n}$$

$$= \frac{\sigma^2}{n} - \frac{\sigma^2}{n} = 0$$

where the final equality used the fact that X_i and $\Sigma_{j \neq i} X_j$ are independent and thus have covariance 0.   ✦

Equation (2.16) is often useful when computing variances.

Example 2.33   (Variance of a Binomial Random Variable): Compute the variance of a binomial random variable X with parameters n and p.

Solution: Since such a random variable represents the number of successes in n independent trials when each trial has a common probability p of being a success, we may write

$$X = X_1 + \cdots + X_n$$

where the X_i are independent Bernoulli random variables such that

$$X_i = \begin{cases} 1, & \text{if the } i\text{th trial is a success} \\ 0, & \text{otherwise} \end{cases}$$

Hence, from Equation (2.16) we obtain

$$\text{Var}(X) = \text{Var}(X_1) + \cdots + \text{Var}(X_n)$$

But

$$\begin{aligned} \text{Var}(X_i) &= E[X_i^2] - (E[X_i])^2 \\ &= E[X_i] - (E[X_i])^2 \qquad \text{since } X_i^2 = X_i \\ &= p - p^2 \end{aligned}$$

and thus

$$\text{Var}(X) = np(1-p) \quad ✦$$

Example 2.34   (Sampling from a Finite Population: The Hypergeometric): Consider a population of N individuals, some of whom are in favor of a certain proposition. In particular suppose that Np of them are in favor and $N - Np$ are opposed, where p is assumed to be unknown. We are interested in estimating p, the fraction of the population that is for the proposition, by randomly choosing and then determining the positions of n members of the population.

In such situations as described in the preceding, it is common to use the fraction of the sampled population that is in favor of the proposition as an estimator of p. Hence, if we let

$$X_i = \begin{cases} 1, & \text{if the } i\text{th person chosen is in favor} \\ 0, & \text{otherwise} \end{cases}$$

then the usual estimator of p is $\Sigma_{i=1}^n X_i/n$. Let us now compute its mean and variance. Now

$$E\left[\sum_{i=1}^n X_i\right] = \sum_1^n E[X_i]$$

$$= np$$

where the final equality follows since the ith person chosen is equally likely to be any of the N individuals in the population and so has probability Np/N of being in favor.

$$\text{Var}\left(\sum_1^n X_i\right) = \sum_1^n \text{Var}(X_i) + 2\sum\sum_{i<j} \text{Cov}(X_i, X_j)$$

Now, since X_i is a Bernoulli random variable with mean p, it follows that

$$\text{Var}(X_i) = p(1-p)$$

Also, for $i \neq j$,

$$\text{Cov}(X_i, X_j) = E[X_i X_j] - E[X_i]E[X_j]$$

$$= P\{X_i = 1, X_j = 1\} - p^2$$

$$= P\{X_i = 1\}P\{X_j = 1 \mid X_i = 1\} - p^2$$

$$= \frac{Np}{N}\frac{(Np-1)}{N-1} - p^2$$

where the last equality follows since if the ith person to be chosen is in favor, then the jth person chosen is equally likely to be any of the other $N-1$ of which $Np-1$ are in favor. Thus, we see that

$$\text{Var}\left(\sum_1^n X_i\right) = np(1-p) + 2\binom{n}{2}\left[\frac{p(Np-1)}{N-1} - p^2\right]$$

$$= np(1-p) - \frac{n(n-1)p(1-p)}{N-1}$$

and so the mean and variance of our estimator are given by

$$E\left[\sum_1^n \frac{X_i}{n}\right] = p,$$

$$\text{Var}\left[\sum_1^n \frac{X_i}{n}\right] = \frac{p(1-p)}{n} - \frac{(n-1)p(1-p)}{n(N-1)}$$

Some remarks are in order: As the mean of the estimator is the unknown value p, we would like its variance to be as small as possible (why is this?), and we see by the preceding that, as a function of the population size N, the variance increases as N increases. The limiting value, as $N \to \infty$, of the variance is $p(1 - p)/n$, which is not surprising since for N large each of the X_i will be (approximately) independent random variables, and thus $\sum_1^n X_i$ will have an (approximately) binomial distribution with parameters n and p.

The random variable $\sum_1^n X_i$ can be thought of as representing the number of white balls obtained when n balls are randomly selected from a population consisting of Np white and $N - Np$ black balls. (Identify a person who favors the proposition with a white ball and one against with a black ball.) Such a random variable is called *hypergeometric* and has a probability mass function given by

$$P\left\{ \sum_1^n X_i = k \right\} = \frac{\binom{Np}{k}\binom{N - Np}{n - k}}{\binom{N}{n}} \quad \blacklozenge$$

It is often important to be able to calculate the distribution of $X + Y$ from the distributions of X and Y when X and Y are independent. Suppose first that X and Y are continuous, X having probability density f and Y having probability density g. Then, letting $F_{X+Y}(a)$ be the cumulative distribution function of $X + Y$, we have

$$F_{X+Y}(a) = P\{X + Y \leqslant a\}$$

$$= \iint_{x+y \leqslant a} f(x)g(y)\, dx\, dy$$

$$= \int_{-\infty}^{\infty} \int_{-\infty}^{a-y} f(x)g(y)\, dx\, dy$$

$$= \int_{-\infty}^{\infty} \left(\int_{-\infty}^{a-y} f(x)\, dx \right) g(y)\, dy$$

$$= \int_{-\infty}^{\infty} F_X(a - y)g(y)\, dy \qquad (2.17)$$

The cumulative distribution function F_{X+Y} is called the *convolution* of the distributions F_X and F_Y (the cumulative distribution functions of X and Y, respectively).

By differentiating Equation (2.17), we obtain that the probability density function $f_{X+Y}(a)$ of $X + Y$ is given by

$$f_{X+Y}(a) = \frac{d}{da} \int_{-\infty}^{\infty} F_X(a - y)g(y)\,dy$$

$$= \int_{-\infty}^{\infty} \frac{d}{da}(F_X(a - y))g(y)\,dy$$

$$= \int_{-\infty}^{\infty} f(a - y)g(y)\,dy \qquad (2.18)$$

Example 2.35 (Sum of Two Independent Uniform Random Variables): If X and Y are independent random variables both uniformly distributed on $(0, 1)$, then calculate the probability density of $X + Y$.

Solution: From Equation (2.18), since

$$f(a) = g(a) = \begin{cases} 1, & 0 < a < 1 \\ 0, & \text{otherwise} \end{cases}$$

we obtain

$$f_{X+Y}(a) = \int_0^1 f(a - y)\,dy$$

For $0 \leqslant a \leqslant 1$, this yields

$$f_{X+Y}(a) = \int_0^a dy = a$$

For $1 < a < 2$, we get

$$f_{X+Y}(a) = \int_{a-1}^1 dy = 2 - a$$

Hence,

$$f_{X+Y}(a) = \begin{cases} a, & 0 \leqslant a \leqslant 1 \\ 2 - a, & 1 < a < 2 \\ 0, & \text{otherwise} \end{cases} \quad \blacklozenge$$

Rather than deriving a general expression for the distribution of $X + Y$ in the discrete case, we shall consider an example.

Example 2.36   (Sums of Independent Poisson Random Variables): Let X and Y be independent Poisson random variables with respective means λ_1 and λ_2. Calculate the distribution of $X + Y$.

Solution: Since the event $\{X + Y = n\}$ may be written as the union of the disjoint events $\{X = k, Y = n - k\}$, $0 \leqslant k \leqslant n$, we have

$$P\{X + Y = n\} = \sum_{k=0}^{n} P\{X = k, Y = n - k\}$$

$$= \sum_{k=0}^{n} P\{X = k\}P\{Y = n - k\}$$

$$= \sum_{k=0}^{n} e^{-\lambda_1} \frac{\lambda_1^k}{k!} e^{-\lambda_2} \frac{\lambda_2^{n-k}}{(n-k)!}$$

$$= e^{-(\lambda_1 + \lambda_2)} \sum_{k=0}^{n} \frac{\lambda_1^k \lambda_2^{n-k}}{k!(n-k)!}$$

$$= \frac{e^{-(\lambda_1 + \lambda_2)}}{n!} \sum_{k=0}^{n} \frac{n!}{k!(n-k)!} \lambda_1^k \lambda_2^{n-k}$$

$$= \frac{e^{-(\lambda_1 + \lambda_2)}}{n!} (\lambda_1 + \lambda_2)^n$$

In words, $X_1 + X_2$ has a Poisson distribution with mean $\lambda_1 + \lambda_2$.   ✦

The concept of independence may, of course, be defined for more than two random variables. In general, the n random variables $X_1, X_2, \ldots, X_n$ are said to be independent if, for all values $a_1, a_2, \ldots, a_n$,

$$P\{X_1 \leqslant a_1, X_2 \leqslant a_2, \ldots, X_n \leqslant a_n\} = P\{X_1 \leqslant a_1\}P\{X_2 \leqslant a_2\} \cdots P\{X_n \leqslant a_n\}$$

Example 2.37   Let $X_1, \ldots, X_n$ be independent and identically distributed continuous random variables with probability distribution F and density function $F' = f$. If we let $X_{(i)}$ denote the ith smallest of these random variables, then $X_{(1)}, \ldots, X_{(n)}$ are called the *order statistics*. To obtain the distribution of $X_{(i)}$, note that $X_{(i)}$ will be less than or equal to x if and only if at least i of the n random variables $X_1, \ldots, X_n$ are less than or equal to x. Hence,

$$P\{X_{(i)} \leqslant x\} = \sum_{k=i}^{n} \binom{n}{k} (F(x))^k (1 - F(x))^{n-k}$$

Differentiation yields that the density function of $X_{(i)}$ is as follows:

$$f_{X_{(i)}}(x) = f(x) \sum_{k=i}^{n} \binom{n}{k} k(F(x))^{k-1}(1 - F(x))^{n-k}$$

$$- f(x) \sum_{k=i}^{n} \binom{n}{k}(n-k)(F(x))^{k}(1 - F(x))^{n-k-1}$$

$$= f(x) \sum_{k=i}^{n} \frac{n!}{(n-k)!(k-1)!} (F(x))^{k-1}(1 - F(x))^{n-k}$$

$$- f(x) \sum_{k=i}^{n-1} \frac{n!}{(n-k-1)!k!} (F(x))^{k}(1 - F(x))^{n-k-1}$$

$$= f(x) \sum_{k=i}^{n} \frac{n!}{(n-k)!(k-1)!} (F(x))^{k-1}(1 - F(x))^{n-k}$$

$$- f(x) \sum_{j=i+1}^{n} \frac{n!}{(n-j)!(j-1)!} (F(x))^{j-1}(1 - F(x))^{n-j}$$

$$= \frac{n!}{(n-i)!(i-1)!} f(x)(F(x))^{i-1}(1 - F(x))^{n-i}$$

The preceding density is quite intuitive, since in order for $X_{(i)}$ to equal x, $i - 1$ of the n values $X_1, \ldots, X_n$ must be less than x; $n - i$ of them must be greater than x; and one must be equal to x. Now, the probability density that every member of a specified set of $i - 1$ of the X_j is less than x, every member of another specified set of $n - i$ is greater than x, and the remaining value is equal to x is $(F(x))^{i-1}(1 - F(x))^{n-i}f(x)$. Therefore, since there are $n!/[(i - 1)!(n - i)!]$ different partitions of the n random variables into the three groups, we obtain the preceding density function. ✦

2.5.4. Joint Probability Distribution of Functions of Random Variables

Let X_1 and X_2 be jointly continuous random variables with joint probability density function $f(x_1, x_2)$. It is sometimes necessary to obtain the joint distribution of the random variables Y_1 and Y_2 which arise as functions of X_1 and X_2. Specifically, suppose that $Y_1 = g_1(X_1, X_2)$ and $Y_2 = g_2(X_1, X_2)$ for some functions g_1 and g_2.

Assume that the functions g_1 and g_2 satisfy the following conditions:

1. The equations $y_1 = g_1(x_1, x_2)$ and $y_2 = g_2(x_1, x_2)$ can be uniquely solved for x_1 and x_2 in terms of y_1 and y_2 with solutions given by, say, $x_1 = h_1(y_1, y_2)$, $x_2 = h_2(y_1, y_2)$.

2. The functions g_1 and g_2 have continuous partial derivatives at all points (x_1, x_2) and are such that the following 2×2 determinant

$$J(x_1, x_2) = \begin{vmatrix} \dfrac{\partial g_1}{\partial x_1} & \dfrac{\partial g_1}{\partial x_2} \\[2mm] \dfrac{\partial g_2}{\partial x_1} & \dfrac{\partial g_2}{\partial x_2} \end{vmatrix} \equiv \frac{\partial g_1}{\partial x_1}\frac{\partial g_2}{\partial x_2} - \frac{\partial g_1}{\partial x_2}\frac{\partial g_2}{\partial x_1} \neq 0$$

at all points (x_1, x_2).

Under these two conditions it can be shown that the random variables Y_1 and Y_2 are jointly continuous with joint density function given by

$$f_{Y_1, Y_2}(y_1, y_2) = f_{X_1, X_2}(x_1, x_2)|J(x_1, x_2)|^{-1} \qquad (2.19)$$

where $x_1 = h_1(y_1, y_2)$, $x_2 = h_2(y_1, y_2)$.

A proof of Equation (2.19) would proceed along the following lines:

$$P\{Y_1 \leqslant y_1, Y_2 \leqslant y_2\} = \iint_{\substack{(x_1, x_2): \\ g_1(x_1, x_2) \leqslant y_1 \\ g_2(x_1, x_2) \leqslant y_2}} f_{X_1, X_2}(x_1, x_2)\, dx_1\, dx_2 \qquad (2.20)$$

The joint density function can now be obtained by differentiating Equation (2.20) with respect to y_1 and y_2. That the result of this differentiation will be equal to the right-hand side of Equation (2.19) is an exercise in advanced calculus whose proof will not be presented in the present text.

Example 2.38 If X and Y are independent gamma random variables with parameters (α, λ) and (β, λ), respectively, compute the joint density of $U = X + Y$ and $V = X/(X + Y)$.

Solution: The joint density of X and Y is given by

$$f_{X,Y}(x, y) = \frac{\lambda e^{-\lambda x}(\lambda x)^{\alpha - 1}}{\Gamma(\alpha)} \frac{\lambda e^{-\lambda y}(\lambda y)^{\beta - 1}}{\Gamma(\beta)}$$

$$= \frac{\lambda^{\alpha + \beta}}{\Gamma(\alpha)\Gamma(\beta)} e^{-\lambda(x + y)} x^{\alpha - 1} y^{\beta - 1}$$

Now, if $g_1(x, y) = x + y$, $g_2(x, y) = x/(x + y)$, then

$$\frac{\partial g_1}{\partial x} = \frac{\partial g_1}{\partial y} = 1, \qquad \frac{\partial g_2}{\partial x} = \frac{y}{(x + y)^2}, \qquad \frac{\partial g_2}{\partial y} = -\frac{x}{(x + y)^2}$$

and so

$$J(x, y) = \begin{vmatrix} 1 & 1 \\ \dfrac{y}{(x+y)^2} & \dfrac{-x}{(x+y)^2} \end{vmatrix} = -\frac{1}{x+y}$$

Finally, because the equations $u = x + y$, $v = x/(x + y)$ have as their solutions $x = uv$, $y = u(1 - v)$, we see that

$$f_{U,V}(u, v) = f_{X,Y}[uv, u(1 - v)]u$$

$$= \frac{\lambda e^{-\lambda u}(\lambda u)^{\alpha + \beta - 1}}{\Gamma(\alpha + \beta)} \frac{v^{\alpha - 1}(1 - v)^{\beta - 1}\Gamma(\alpha + \beta)}{\Gamma(\alpha)\Gamma(\beta)}$$

Hence $X + Y$ and $X/(X + Y)$ are independent, with $X + Y$ having a gamma distribution with parameters $(\alpha + \beta, \lambda)$ and $X/(X + Y)$ having density function

$$f_V(v) = \frac{\Gamma(\alpha + \beta)}{\Gamma(\alpha)\Gamma(\beta)} v^{\alpha - 1}(1 - v)^{\beta - 1}, \qquad 0 < v < 1$$

This is called the beta density with parameters (α, β).

The above result is quite interesting. For suppose there are $n + m$ jobs to be performed, with each (independently) taking an exponential amount of time with rate λ for performance, and suppose that we have two workers to perform these jobs. Worker I will do jobs $1, 2, \ldots, n$, and worker II will do the remaining m jobs. If we let X and Y denote the total working times of workers I and II, respectively, then upon using the above result it follows that X and Y will be independent gamma random variables having parameters (n, λ) and (m, λ), respectively. Then the above result yields that independently of the working time needed to complete all $n + m$ jobs (that is, of $X + Y$), the proportion of this work that will be performed by worker I has a beta distribution with parameters (n, m). ◆

When the joint density function of the n random variables $X_1, X_2, \ldots, X_n$ is given and we want to compute the joint density function of $Y_1, Y_2, \ldots, Y_n$, where

$$Y_1 = g_1(X_1, \ldots, X_n), \qquad Y_2 = g_2(X_1, \ldots, X_n), \ldots,$$

$$Y_n = g_n(X_1, \ldots, X_n)$$

the approach is the same. Namely, we assume that the functions g_i have continuous partial derivatives and that the Jacobian determinant

$J(x_1, \ldots, x_n) \neq 0$ at all points $(x_1, \ldots, x_n)$, where

$$
J(x_1, \ldots, x_n) = \begin{vmatrix}
\dfrac{\partial g_1}{\partial x_1} & \dfrac{\partial g_1}{\partial x_2} & \cdots & \dfrac{\partial g_1}{\partial x_n} \\[2mm]
\dfrac{\partial g_2}{\partial x_1} & \dfrac{\partial g_2}{\partial x_2} & \cdots & \dfrac{\partial g_2}{\partial x_n} \\[2mm]
\dfrac{\partial g_n}{\partial x_1} & \dfrac{\partial g_n}{\partial x_2} & \cdots & \dfrac{\partial g_n}{\partial x_n}
\end{vmatrix}
$$

Furthermore, we suppose that the equations $y_1 = g_1(x_1, \ldots, x_n)$, $y_2 = g_2(x_1, \ldots, x_n), \ldots, y_n = g_n(x_1, \ldots, x_n)$ have a unique solution, say, $x_1 = h_1(y_1, \ldots, y_n), \ldots, x_n = h_n(y_1, \ldots, y_n)$. Under these assumptions the joint density function of the random variables Y_i is given by

$$
f_{Y_1, \ldots, Y_n}(y_1, \ldots, y_n) = f_{X_1, \ldots, X_n}(x_1, \ldots, x_n) \, |J(x_1, \ldots, x_n)|^{-1}
$$

where $x_i = h_i(y_1, \ldots, y_n)$, $i = 1, 2, \ldots, n$.

2.6. Moment Generating Functions

The moment generating function $\phi(t)$ of the random variable X is defined for all values t by

$$
\phi(t) = E[e^{tX}]
$$

$$
= \begin{cases}
\displaystyle\sum_x e^{tx} p(x), & \text{if } X \text{ is discrete} \\[4mm]
\displaystyle\int_{-\infty}^{\infty} e^{tx} f(x)\, dx, & \text{if } X \text{ is continuous}
\end{cases}
$$

We call $\phi(t)$ the moment generating function because all of the moments of X can be obtained by successively differentiating $\phi(t)$. For example,

$$
\phi'(t) = \frac{d}{dt} E[e^{tX}]
$$

$$
= E\left[\frac{d}{dt} (e^{tX}) \right]
$$

$$
= E[X e^{tX}]
$$

Hence,

$$
\phi'(0) = E[X]
$$

Similarly,

$$\phi''(t) = \frac{d}{dt} \phi'(t)$$

$$= \frac{d}{dt} E[Xe^{tX}]$$

$$= E\left[\frac{d}{dt}(Xe^{tX})\right]$$

$$= E[X^2 e^{tX}]$$

and so

$$\phi''(0) = E[X^2]$$

In general, the nth derivative of $\phi(t)$ evaluated at $t = 0$ equals $E[X^n]$, that is,

$$\phi^n(0) = E[X^n], \qquad n \geqslant 1$$

We now compute $\phi(t)$ for some common distributions.

Example 2.39   (The Binomial Distribution with Parameters n and p):

$$\phi(t) = E[e^{tX}]$$

$$= \sum_{k=0}^{n} e^{tk} \binom{n}{k} p^k (1-p)^{n-k}$$

$$= \sum_{k=0}^{n} \binom{n}{k} (pe^t)^k (1-p)^{n-k}$$

$$= (pe^t + 1 - p)^n$$

Hence,

$$\phi'(t) = n(pe^t + 1 - p)^{n-1} pe^t$$

and so

$$E[X] = \phi'(0) = np$$

which checks with the result obtained in Example 2.17. Differentiating a second time yields

$$\phi''(t) = n(n-1)(pe^t + 1 - p)^{n-2}(pe^t)^2 + n(pe^t + 1 - p)^{n-1} pe^t$$

and so

$$E[X^2] = \phi''(0) = n(n-1)p^2 + np$$

Thus, the variance of X is given

$$\text{Var}(X) = E[X^2] - (E[X])^2$$
$$= n(n-1)p^2 + np - n^2p^2$$
$$= np(1-p) \quad \blacklozenge$$

Example 2.40   (The Poisson Distribution with Mean λ):

$$\phi(t) = E[e^{tX}]$$
$$= \sum_{n=0}^{\infty} \frac{e^{tn}e^{-\lambda}\lambda^n}{n!}$$
$$= e^{-\lambda} \sum_{n=0}^{\infty} \frac{(\lambda e^t)^n}{n!}$$
$$= e^{-\lambda}e^{\lambda e^t}$$
$$= \exp\{\lambda(e^t - 1)\}$$

Differentiation yields

$$\phi'(t) = \lambda e^t \exp\{\lambda(e^t - 1)\},$$
$$\phi''(t) = (\lambda e^t)^2 \exp\{\lambda(e^t - 1)\} + \lambda e^t \exp\{\lambda(e^t - 1)\}$$

and so

$$E[X] = \phi'(0) = \lambda,$$
$$E[X^2] = \phi''(0) = \lambda^2 + \lambda,$$
$$\text{Var}(X) = E[X^2] - (E[X])^2$$
$$= \lambda$$

Thus, both the mean and the variance of the Poisson equal λ. $\blacklozenge$

Example 2.41   (The Exponential Distribution with Parameter λ):

$$\phi(t) = E[e^{tX}]$$
$$= \int_0^{\infty} e^{tx}\lambda e^{-\lambda x}\,dx$$
$$= \lambda \int_0^{\infty} e^{-(\lambda - t)x}\,dx$$
$$= \frac{\lambda}{\lambda - t} \qquad \text{for } t < \lambda$$

We note by the preceding derivation that, for the exponential distribution, $\phi(t)$ is only defined for values of t less than λ. Differentiation of $\phi(t)$ yields

$$\phi'(t) = \frac{\lambda}{(\lambda - t)^2}, \qquad \phi''(t) = \frac{2\lambda}{(\lambda - t)^3}$$

Hence,

$$E[X] = \phi'(0) = \frac{1}{\lambda}, \qquad E[X^2] = \phi''(0) = \frac{2}{\lambda^2}$$

The variance of X is thus given by

$$\mathrm{Var}(X) = E[X^2] - (E[X])^2 = \frac{1}{\lambda^2} \quad \blacklozenge$$

Example 2.42 (The Normal Distribution with Parameters μ and σ^2):

$$\phi(t) = E[e^{tX}]$$

$$= \frac{1}{\sqrt{2\pi}\,\sigma} \int_{-\infty}^{\infty} e^{tx} e^{-(x-\mu)^2/2\sigma^2}\,dx$$

$$= \frac{1}{\sqrt{2\pi}\,\sigma} \int_{-\infty}^{\infty} \exp\left\{\frac{-(x^2 - 2\mu x + \mu^2 - 2\sigma^2 tx)}{2\sigma^2}\right\} dx$$

Now writing

$$x^2 - 2\mu x + \mu^2 - 2\sigma^2 tx = x^2 - 2(\mu + \sigma^2 t)x + \mu^2$$

$$= (x - (\mu + \sigma^2 t))^2 - (\mu + \sigma^2 t)^2 + \mu^2$$

$$= (x - (\mu + \sigma^2 t))^2 - \sigma^4 t^2 - 2\mu\sigma^2 t$$

we have

$$\phi(t) = \frac{1}{\sqrt{2\pi}\,\sigma} \exp\left\{\frac{\sigma^4 t^2 + 2\mu\sigma^2 t}{2\sigma^2}\right\} \int_{-\infty}^{\infty} \exp\left\{\frac{-(x - (\mu + \sigma^2 t))^2}{2\sigma^2}\right\} dx$$

$$= \exp\left\{\frac{\sigma^2 t^2}{2} + \mu t\right\} \frac{1}{\sqrt{2\pi}\,\sigma} \int_{-\infty}^{\infty} \exp\left\{\frac{-(x - (\mu + \sigma^2 t))^2}{2\sigma^2}\right\} dx$$

However,

$$\frac{1}{\sqrt{2\pi}\,\sigma} \int_{-\infty}^{\infty} \exp\left\{\frac{-(x - (\mu + \sigma^2 t))^2}{2\sigma^2}\right\} dx = P\{-\infty < \bar{X} < \infty\} = 1$$

where $\bar{X}$ is a normally distributed random variable having parameters $\bar{\mu} = \mu + \sigma^2 t$ and $\bar{\sigma}^2 = \sigma^2$. Thus, we have shown that

$$\phi(t) = \exp\left\{\frac{\sigma^2 t^2}{2} + \mu t\right\}$$

By differentiating we obtain

$$\phi'(t) = (\mu + t\sigma^2)\exp\left\{\frac{\sigma^2 t^2}{2} + \mu t\right\}$$

$$\phi''(t) = (\mu + t\sigma^2)^2 \exp\left\{\frac{\sigma^2 t^2}{2} + \mu t\right\} + \sigma^2 \exp\left\{\frac{\sigma^2 t^2}{2} + \mu t\right\}$$

and so

$$E[X] = \phi'(0) = \mu,$$
$$E[X^2] = \phi''(0) = \mu^2 + \sigma^2$$

implying that

$$\text{Var}(X) = E[X^2] - E([X])^2$$
$$= \sigma^2 \quad \blacklozenge$$

Tables 2.1 and 2.2 give the moment generating function for some common distributions.

An important property of moment generating functions is that the *moment generating function of the sum of independent random variables is just the product of the individual moment generating functions*. To see this, suppose that X and Y are independent and have moment generating functions $\phi_X(t)$

Table 2.1

Discrete probability distribution	Probability mass function, $p(x)$	Moment generating function, $\phi(t)$	Mean	Variance
Binomial with parameters n, p $0 \leqslant p \leqslant 1$	$\binom{n}{p} p^x (1-p)^{n-x},$ $x = 0, 1, \ldots, n$	$(pe^t + (1-p))^n$	np	$np(1-p)$
Poisson with parameter $\lambda > 0$	$e^{-\lambda}\dfrac{\lambda^x}{x!},$ $x = 0, 1, 2, \ldots$	$\exp\{\lambda(e^t - 1)\}$	λ	λ
Geometric with parameter $0 \leqslant p \leqslant 1$	$p(1-p)^{x-1},$ $x = 1, 2, \ldots$	$\dfrac{pe^t}{1 - (1-p)e^t}$	$\dfrac{1}{p}$	$\dfrac{1-p}{p^2}$

Table 2.2

Continuous probability distribution	Probability density function, $f(x)$	Moment generating function, $\phi(t)$	Mean	Variance
Uniform over (a, b)	$f(x) = \begin{cases} \dfrac{1}{b-a}, & a < x < b \\ 0, & \text{otherwise} \end{cases}$	$\dfrac{e^{tb} - e^{ta}}{t(b-a)}$	$\dfrac{a+b}{2}$	$\dfrac{(b-a)^2}{12}$
Exponential with parameter $\lambda > 0$	$f(x) = \begin{cases} \lambda e^{-\lambda x}, & x > 0 \\ 0, & x < 0 \end{cases}$	$\dfrac{\lambda}{\lambda - t}$	$\dfrac{1}{\lambda}$	$\dfrac{1}{\lambda^2}$
Gamma with parameters (n, λ) $\lambda > 0$	$f(x) = \begin{cases} \dfrac{\lambda e^{-\lambda x}(\lambda x)^{n-1}}{(n-1)!}, & x \geqslant 0 \\ 0, & x < 0 \end{cases}$	$\left(\dfrac{\lambda}{\lambda - t}\right)^n$	$\dfrac{n}{\lambda}$	$\dfrac{n}{\lambda^2}$
Normal with parameters (μ, σ^2)	$f(x) = \dfrac{1}{\sqrt{2\pi}\,\sigma} e^{-(x-\mu)^2/2\sigma^2},$ $-\infty < x < \infty$	$\exp\left\{\mu t + \dfrac{\sigma^2 t^2}{2}\right\}$	μ	σ^2

and $\phi_Y(t)$, respectively. Then $\phi_{X+Y}(t)$, the moment generating function of $X + Y$, is given by

$$\phi_{X+Y}(t) = E[e^{t(X+Y)}]$$
$$= E[e^{tX}e^{tY}]$$
$$= E[e^{tX}]E[e^{tY}]$$
$$= \phi_X(t)\phi_Y(t)$$

where the next to the last equality follows from Proposition 2.3 since X and Y are independent.

Another important result is that the *moment generating function uniquely determines the distribution.* That is, there exists a one-to-one correspondence between the moment generating function and the distribution function of a random variable.

Example 2.43 Suppose the moment generating function of a random variable X is given by $\phi(t) = e^{3(e^t - 1)}$. What is $P\{X = 0\}$?

Solution: We see from Table 2.1 that $\phi(t) = e^{3(e^t - 1)}$ is the moment generating function of a Poisson random variable with mean 3. Hence, by the one-to-one correspondence between moment generating functions and distribution functions, it follows that X must be a Poisson random variable with mean 3. Thus, $P\{X = 0\} = e^{-3}$. ✦

Example 2.44   (Sums of Independent Binomial Random Variables): If X and Y are independent binomial random variables with parameters (n, p) and (m, p), respectively, then what is the distribution of $X + Y$?

Solution: The moment generating function of $X + Y$ is given by

$$\phi_{X+Y}(t) = \phi_X(t)\phi_Y(t) = (pe^t + 1 - p)^n (pe^t + 1 - p)^m$$
$$= (pe^t + 1 - p)^{m+n}$$

But $(pe^t + (1 - p))^{m+n}$ is just the moment generating function of a binomial random variable having parameters $m + n$ and p. Thus, this must be the distribution of $X + Y$. ✦

Example 2.45   (Sums of Independent Poisson Random Variables): Calculate the distribution of $X + Y$ when X and Y are independent Poisson random variables with means λ_1 and λ_2, respectively.

Solution:

$$\phi_{X+Y}(t) = \phi_X(t)\phi_Y(t)$$
$$= e^{\lambda_1(e^t - 1)}e^{\lambda_2(e^t - 1)}$$
$$= e^{(\lambda_1 + \lambda_2)(e^t - 1)}$$

Hence, $X + Y$ is Poisson distributed with mean $\lambda_1 + \lambda_2$, verifying the result given in Example 2.36. ✦

Example 2.46   (Sums of Independent Normal Random Variables): Show that if X and Y are independent normal random variables with parameters (μ_1, σ_1^2) and (μ_2, σ_2^2), respectively, then $X + Y$ is normal with mean $\mu_1 + \mu_2$ and variance $\sigma_1^2 + \sigma_2^2$.

Solution:

$$\phi_{X+Y}(t) = \phi_X(t)\phi_Y(t)$$
$$= \exp\left\{\frac{\sigma_1^2 t^2}{2} + \mu_1 t\right\}\exp\left\{\frac{\sigma_2^2 t^2}{2} + \mu_2 t\right\}$$
$$= \exp\left\{\frac{(\sigma_1^2 + \sigma_2^2)t^2}{2} + (\mu_1 + \mu_2)t\right\}$$

which is the moment generating function of a normal random variable with mean $\mu_1 + \mu_2$ and variance $\sigma_1^2 + \sigma_2^2$. Hence, the result follows since the moment generating function uniquely determines the distribution. ◆

Remark For a nonnegative random variable X, it is often convenient to define its *Laplace transform* $g(t)$, $t \geqslant 0$, by

$$g(t) = \phi(-t) = E[e^{-tX}]$$

That is, the Laplace transform evaluated at t is just the moment generating function evaluated at $-t$. The advantage of dealing with the Laplace transform, rather than the moment generating function, when the random variable is nonnegative is that if $X \geqslant 0$ and $t \geqslant 0$, then

$$0 \leqslant e^{-tX} \leqslant 1$$

That is, the Laplace transform is always between 0 and 1. As in the case of moment generating functions, it remains true that nonnegative random variables that have the same Laplace transform must also have the same distribution. ◆

It is also possible to define the joint moment generating function of two or more random variables. This is done as follows. For any n random variables $X_1, \ldots, X_n$, the joint moment generating function, $\phi(t_1, \ldots, t_n)$, is defined for all real values of $t_1, \ldots, t_n$ by

$$\phi(t_1, \ldots, t_n) = E[e^{(t_1 X_1 + \cdots + t_n X_n)}]$$

It can be shown that $\phi(t_1, \ldots, t_n)$ uniquely determines the joint distribution of $X_1, \ldots, X_n$.

Example 2.47 (The Multivariate Normal Distribution): Let $Z_1, \ldots, Z_n$ be a set of n independent unit normal random variables. If, for some constants a_{ij}, $1 \leqslant i \leqslant m$, $1 \leqslant j \leqslant n$, and μ_i, $1 \leqslant i \leqslant m$,

$$X_1 = a_{11}Z_1 + \cdots + a_{1n}Z_n + \mu_1,$$

$$X_2 = a_{21}Z_1 + \cdots + a_{2n}Z_n + \mu_2,$$

$$\vdots$$

$$X_i = a_{i1}Z_1 + \cdots + a_{in}Z_n + \mu_i,$$

$$\vdots$$

$$X_m = a_{m1}Z_1 + \cdots + a_{mn}Z_n + \mu_m$$

then the random variables $X_1, \ldots, X_m$ are said to have a multivariate normal distribution.

It follows from the fact that the sum of independent normal random variables is itself a normal random variable that each X_i is a normal random variable with mean and variance given by

$$E[X_i] = \mu_i,$$

$$\text{Var}(X_i) = \sum_{j=1}^{n} a_{ij}^2$$

Let us now determine

$$\phi(t_1, \ldots, t_m) = E[\exp\{t_1 X_1 + \cdots + t_m X_m\}]$$

the joint moment generating function of $X_1, \ldots, X_m$. The first thing to note is that since $\sum_{i=1}^{m} t_i X_i$ is itself a linear combination of the independent normal random variables $Z_1, \ldots, Z_n$, it is also normally distributed. Its mean and variance are respectively

$$E\left[\sum_{i=1}^{m} t_i X_i\right] = \sum_{i=1}^{m} t_i \mu_i$$

and

$$\text{Var}\left(\sum_{i=1}^{m} t_i X_i\right) = \text{Cov}\left(\sum_{i=1}^{m} t_i X_i, \sum_{j=1}^{m} t_j X_j\right)$$

$$= \sum_{i=1}^{m} \sum_{j=1}^{m} t_i t_j \text{Cov}(X_i, X_j)$$

Now, if Y is a normal random variable with mean μ and variance σ^2, then

$$E[e^Y] = \phi_Y(t)|_{t=1} = e^{\mu + \sigma^2/2}$$

Thus, we see that

$$\phi(t_1, \ldots, t_m) = \exp\left\{\sum_{i=1}^{m} t_i \mu_i + \tfrac{1}{2} \sum_{i=1}^{m} \sum_{j=1}^{m} t_i t_j \text{Cov}(X_i, X_j)\right\}$$

which shows that the joint distribution of $X_1, \ldots, X_m$ is completely determined from a knowledge of the values of $E[X_i]$ and $\text{Cov}(X_i, X_j)$, $i, j = 1, \ldots, m$. ◆

2.6.1. The Joint Distribution of the Sample Mean and Sample Variance from a Normal Population

Let $X_1, \ldots, X_n$ be independent and identically distributed random variables, each with mean μ and variance σ^2. The random variable S^2 defined by

$$S^2 = \sum_{i=1}^{n} \frac{(X_i - \bar{X})^2}{n - 1}$$

is called the *sample variance* of these data. To compute $E[S^2]$ we use the identity

$$\sum_{i=1}^{n} (X_i - \bar{X})^2 = \sum_{i=1}^{n} (X_i - \mu)^2 - n(\bar{X} - \mu)^2 \tag{2.21}$$

which is proven as follows:

$$\sum_{i=1}^{n} (X_i - \bar{X}) = \sum_{i=1}^{n} (X_i - \mu + \mu - \bar{X})^2$$

$$= \sum_{i=1}^{n} (X_i - \mu)^2 + n(\mu - \bar{X})^2 + 2(\mu - \bar{X}) \sum_{i=1}^{n} (X_i - \mu)$$

$$= \sum_{i=1}^{n} (X_i - \mu)^2 + n(\mu - \bar{X})^2 + 2(\mu - \bar{X})(n\bar{X} - n\mu)$$

$$= \sum_{i=1}^{n} (X_i - \mu)^2 + n(\mu - \bar{X})^2 - 2n(\mu - \bar{X})^2$$

and identity (2.21) follows.
   Using identity (2.21) gives

$$E[(n - 1)S^2] = \sum_{i=1}^{n} E[(X_i - \mu)^2] - nE[(\bar{X} - \mu)^2]$$

$$= n\sigma^2 - n \operatorname{Var}(\bar{X})$$

$$= (n - 1)\sigma^2 \qquad \text{from Proposition 2.4(b)}$$

Thus, we obtain from the preceding that

$$E[S^2] = \sigma^2$$

   We will now determine the joint distribution of the sample mean $\bar{X} = \Sigma_{i=1}^{n} X_i/n$ and the sample variance S^2 when the X_i have a normal distribution. To begin we need the concept of a chi-squared random variable.

Definition 2.2   If $Z_1, \ldots, Z_n$ are independent standard normal random variables, then the random variable $\Sigma_{i=1}^{n} Z_i^2$ is said to be a *chi-squared random variable with* n *degrees of freedom*.

We shall now compute the moment generating function of $\Sigma_{i=1}^{n} Z_i^2$. To begin, note that

$$E[\exp\{tZ_i^2\}] = \frac{1}{\sqrt{2\pi}} \int_{-\infty}^{\infty} e^{tx^2} e^{-x^2/2} \, dx$$

$$= \frac{1}{\sqrt{2\pi}} \int_{-\infty}^{\infty} e^{-x^2/2\sigma^2} \, dx \qquad \text{where } \sigma^2 = (1-2t)^{-1}$$

$$= \sigma$$

$$= (1-2t)^{-1/2}$$

Hence,

$$E\left[\exp\left\{t \sum_{i=1}^{n} Z_i^2\right\}\right] = \prod_{i=1}^{n} E[\exp\{tZ_i^2\}] = (1-2t)^{-n/2}$$

Now, let $X_1, \ldots, X_n$ be independent normal random variables, each with mean μ and variance σ^2, and let $\bar{X} = \Sigma_{i=1}^{n} X_i/n$ and S^2 denote their sample mean and sample variance. Since the sum of independent normal random variables is also a normal random variable, it follows that $\bar{X}$ is a normal random variable with expected value μ and variance σ^2/n. In addition, from Proposition 2.4,

$$\text{Cov}(\bar{X}, X_i - \bar{X}) = 0, \qquad i = 1, \ldots, n \qquad (2.22)$$

Also, since $\bar{X}, X_1 - \bar{X}, X_2 - \bar{X}, \ldots, X_n - \bar{X}$ are all linear combinations of the independent standard normal random variables $(X_i - \mu)/\sigma$, $i = 1, \ldots, n$, it follows that the random variables $\bar{X}, X_1 - \bar{X}, X_2 - \bar{X}, \ldots, X_n - \bar{X}$ have a joint distribution that is multivariate normal. However, if we let Y be a normal random variable with mean μ and variance σ^2/n that is independent of $X_1, \ldots, X_n$, then the random variables $Y, X_1 - \bar{X}, X_2 - \bar{X}, \ldots, X_n - \bar{X}$ also have a multivariate normal distribution, and by Equation (2.22), they have the same expected values and covariances as the random variables $\bar{X}, X_i - \bar{X}$, $i = 1, \ldots, n$. Thus, since a multivariate normal distribution is completely determined by its expected values and covariances, we can conclude that the random vectors $Y, X_1 - \bar{X}, X_2 - \bar{X}, \ldots, X_n - \bar{X}$ and $\bar{X}, X_1 - \bar{X}, X_2 - \bar{X}, \ldots, X_n - \bar{X}$ have the same joint distribution; thus showing that $\bar{X}$ is independent of the sequence of deviations $X_i - \bar{X}$, $i = 1, \ldots, n$.

Since $\bar{X}$ is independent of the sequence of deviations $X_i - \bar{X}$, $i = 1, \ldots, n$, it follows that it is also independent of the sample variance

$$S^2 \equiv \sum_{i=1}^{n} \frac{(X_i - \bar{X})^2}{n-1}$$

To determine the distribution of S^2, use the identity (2.21) to obtain

$$(n-1)S^2 = \sum_{i=1}^{n} (X_i - \mu)^2 - n(\bar{X} - \mu)^2$$

Dividing both sides of this equation by σ^2 yields

$$\frac{(n-1)S^2}{\sigma^2} + \left(\frac{\bar{X}-\mu}{\sigma/\sqrt{n}}\right)^2 = \sum_{i=1}^{n} \frac{(X_i - \mu)^2}{\sigma^2} \tag{2.23}$$

Now, $\sum_{i=1}^{n}(X_i - \mu)^2/\sigma^2$ is the sum of the squares of n independent standard normal random variables, and so is a chi-squared random variable with n degrees of freedom; it thus has moment generating function $(1 - 2t)^{-n/2}$. Also $[(\bar{X} - \mu)/(\sigma/\sqrt{n})]^2$ is the square of a standard normal random variable and so is a chi-squared random variable with one degree of freedom; and thus has moment generating function $(1 - 2t)^{-1/2}$. In addition, we have previously seen that the two random variables on the left side of Equation (2.23) are independent. Therefore, because the moment generating function of the sum of independent random variables is equal to the product of their individual moment generating functions, we obtain that

$$E[e^{t(n-1)S^2/\sigma^2}](1 - 2t)^{-1/2} = (1 - 2t)^{-n/2}$$

or

$$E[e^{t(n-1)S^2/\sigma^2}] = (1 - 2t)^{-(n-1)/2}$$

But because $(1 - 2t)^{-(n-1)/2}$ is the moment generating function of a chi-squared random variable with $n - 1$ degrees of freedom, we can conclude, since the moment generating function uniquely determines the distribution of the random variable, that this is the distribution of $(n - 1)S^2/\sigma^2$.

Summing up, we have shown the following.

Proposition 2.5 If $X_1, \ldots, X_n$ are independent and identically distributed normal random variables with mean μ and variance σ^2, then the sample mean $\bar{X}$ and the sample variance S^2 are independent. $\bar{X}$ is a normal random variable with mean μ and variance σ^2/n; $(n - 1)S^2/\sigma^2$ is a chi-squared random variable with $n - 1$ degrees of freedom.

2.7. Limit Theorems

We start this section by proving a result known as Markov's inequality.

Proposition 2.6 (Markov's Inequality). If X is a random variable that takes only nonnegative values, then for any value $a > 0$

$$P\{X \geq a\} \leq \frac{E[X]}{a}$$

Proof   We give a proof for the case where X is continuous with density f:

$$E[X] = \int_0^\infty xf(x)\,dx$$

$$= \int_0^a xf(x)\,dx + \int_a^\infty xf(x)\,dx$$

$$\geqslant \int_a^\infty xf(x)\,dx$$

$$\geqslant \int_a^\infty af(x)\,dx$$

$$= a\int_a^\infty f(x)\,dx$$

$$= aP\{X \geqslant a\}$$

and the result is proven.   ✦

As a corollary, we obtain the following.

Proposition 2.7   (Chebyshev's Inequality). If X is a random variable with mean μ and variance σ^2, then, for any value $k > 0$,

$$P\{|X - \mu| \geqslant k\} \leqslant \frac{\sigma^2}{k^2}$$

Proof   Since $(X - \mu)^2$ is a nonnegative random variable, we can apply Markov's inequality (with $a = k^2$) to obtain

$$P\{(X - \mu)^2 \geqslant k^2\} \leqslant \frac{E[(X - \mu)^2]}{k^2}$$

But since $(X - \mu)^2 \geqslant k^2$ if and only if $|X - \mu| \geqslant k$, the preceding is equivalent to

$$P\{|X - \mu| \geqslant k\} \geqslant \frac{E[(X - \mu)^2]}{k^2} = \frac{\sigma^2}{k^2}$$

and the proof is complete.   ✦

The importance of Markov's and Chebyshev's inequalities is that they enable us to derive bounds on probabilities when only the mean, or both the mean and the variance, of the probability distribution are known. Of course, if the actual distribution were known, then the desired probabilities could be exactly computed, and we would not need to resort to bounds.

Example 2.48 Suppose we know that the number of items produced in a factory during a week is a random variable with mean 500.

(a) What can be said about the probability that this week's production will be at least 1000?

(b) If the variance of a week's production is known to equal 100, then what can be said about the probability that this week's production will be between 400 and 600?

Solution: Let X be the number of items that will be produced in a week.

(a) By Markov's inequality,

$$P\{X \geqslant 1000\} \leqslant \frac{E[X]}{1000} = \frac{500}{1000} = \frac{1}{2}$$

(b) By Chebyshev's inequality,

$$P\{|X - 500| \geqslant 100\} \leqslant \frac{\sigma^2}{(100)^2} = \frac{1}{100}$$

Hence,

$$P\{|X - 500| < 100\} \geqslant 1 - \frac{1}{100} = \frac{99}{100}$$

and so the probability that this week's production will be between 400 and 600 is at least 0.99. ✦

The following theorem, known as the *strong law of large numbers*, is probably the most well-known result in probability theory. It states that the average of a sequence of independent random variables having the same distribution will, with probability 1, converge to the mean of that distribution.

Theorem 2.1 (Strong Law of Large Numbers). Let $X_1, X_2, \ldots$ be a sequence of independent random variables having a common distribution, and let $E[X_i] = \mu$. Then, with probability 1,

$$\frac{X_1 + X_2 + \cdots + X_n}{n} \to \mu \qquad \text{as } n \to \infty$$

As an example of the preceding, suppose that a sequence of independent trials is performed. Let E be a fixed event and denote by $P(E)$ the probability that E occurs on any particular trial. Letting

$$X_i = \begin{cases} 1, & \text{if } E \text{ occurs on the } i\text{th trial} \\ 0, & \text{if } E \text{ does not occur on the } i\text{th trial} \end{cases}$$

we have by the strong law of large numbers that, with probability 1,

$$\frac{X_1 + \cdots + X_n}{n} \to E[X] = P(E) \tag{2.24}$$

Since $X_1 + \cdots + X_n$ represents the number of times that the event E occurs in the first n trials, we may interpret Equation (2.24) as stating that, with probability 1, the limiting proportion of time that the event E occurs is just $P(E)$.

Running neck and neck with the strong law of large numbers for the honor of being probability theory's number one result is the *central limit theorem*. Besides its theoretical interest and importance, this theorem provides a simple method for computing approximate probabilities for sums of independent random variables. It also explains the remarkable fact that the empirical frequencies of so many natural "populations" exhibit a bell-shaped (that is, normal) curve.

Theorem 2.2   (Central Limit Theorem). Let $X_1, X_2, \ldots$ be a sequence of independent, identically distributed random variables each with mean μ and variance σ^2. Then the distribution of

$$\frac{X_1 + X_2 + \cdots + X_n - n\mu}{\sigma\sqrt{n}}$$

tends to the standard normal as $n \to \infty$. That is,

$$P\left\{\frac{X_1 + X_2 + \cdots + X_n - n\mu}{\sigma\sqrt{n}} \leqslant a\right\} \to \frac{1}{\sqrt{2\pi}} \int_{-\infty}^{a} e^{-x^2/2}\, dx$$

as $n \to \infty$.

Note that like the other results of this section, this theorem holds for *any* distribution of the X_is; herein lies its power.

If X is binomially distributed with parameters n and p, then X has the same distribution as the sum of n independent Bernoulli random variables each with parameter p. (Recall that the Bernoulli random variable is just a binomial random variable whose parameter n equals 1.) Hence, the distribution of

$$\frac{X - E[X]}{\sqrt{\text{Var}(X)}} = \frac{X - np}{\sqrt{np(1 - p)}}$$

approaches the standard normal distribution as n approaches ∞. The normal approximation will, in general, be quite good for values of n satisfying $np(1 - p) \geq 10$.

Example 2.49 (Normal Approximation to the Binomial): Let X be the number of times that a fair coin, flipped 40 times, lands heads. Find the probability that $X = 20$. Use the normal approximation and then compare it to the exact solution.

Solution: Since the binomial is a discrete random variable, and the normal a continuous random variable, it leads to a better approximation to write the desired probability as

$$P\{X = 20\} = P\{19.5 < X < 20.5\}$$

$$= P\left\{\frac{19.5 - 20}{\sqrt{10}} < \frac{X - 20}{\sqrt{10}} < \frac{20.5 - 20}{\sqrt{10}}\right\}$$

$$= P\left\{-0.16 < \frac{X - 20}{\sqrt{10}} < 0.16\right\}$$

$$\approx \Phi(0.16) - \Phi(-0.16)$$

where $\Phi(x)$, the probability that the standard normal is less than x is given by

$$\Phi(x) = \frac{1}{\sqrt{2\pi}} \int_{-\infty}^{x} e^{-y^2/2}\, dy$$

By the symmetry of the standard normal distribution

$$\Phi(-0.16) = P\{N(0, 1) > 0.16\} = 1 - \Phi(0.16)$$

where $N(0, 1)$ is a standard normal random variable. Hence, the desired probability is approximated by

$$P\{X = 20\} \approx 2\Phi(0.16) - 1$$

Using Table 2.3, we obtain that

$$P\{X = 20\} \approx 0.1272$$

Table 2.3   Area $\Phi(x)$ under the Standard Normal Curve to the Left of x

x	0.00	0.01	0.02	0.03	0.04	0.05	0.06	0.07	0.08	0.09
0.0	0.5000	0.5040	0.5080	0.5120	0.5160	0.5199	0.5239	0.5279	0.5319	0.5359
0.1	0.5398	0.5438	0.5478	0.5517	0.5557	0.5597	0.5636	0.5675	0.5714	0.5753
0.2	0.5793	0.5832	0.5871	0.5910	0.5948	0.5987	0.6026	0.6064	0.6103	0.6141
0.3	0.6179	0.6217	0.6255	0.6293	0.6331	0.6368	0.6406	0.6443	0.6480	0.6517
0.4	0.6554	0.6591	0.6628	0.6664	0.6700	0.6736	0.6772	0.6808	0.6844	0.6879
0.5	0.6915	0.6950	0.6985	0.7019	0.7054	0.7088	0.7123	0.7157	0.7190	0.7224
0.6	0.7257	0.7291	0.7324	0.7357	0.7389	0.7422	0.7454	0.7486	0.7517	0.7549
0.7	0.7580	0.7611	0.7642	0.7673	0.7704	0.7734	0.7764	0.7794	0.7823	0.7852
0.8	0.7881	0.7910	0.7939	0.7967	0.7995	0.8023	0.8051	0.8078	0.8106	0.8133
0.9	0.8159	0.8186	0.8212	0.8238	0.8264	0.8289	0.8315	0.8340	0.8365	0.8389
1.0	0.8413	0.8438	0.8461	0.8485	0.8508	0.8531	0.8554	0.8557	0.8599	0.8621
1.1	0.8643	0.8665	0.8686	0.8708	0.8729	0.8749	0.8770	0.8790	0.8810	0.8830
1.2	0.8849	0.8869	0.8888	0.8907	0.8925	0.8944	0.8962	0.8980	0.8997	0.9015
1.3	0.9032	0.9049	0.9066	0.9082	0.9099	0.9115	0.9131	0.9147	0.9162	0.9177
1.4	0.9192	0.9207	0.9222	0.9236	0.9251	0.9265	0.9279	0.9292	0.9306	0.9319
1.5	0.9332	0.9345	0.9357	0.9370	0.9382	0.9394	0.9406	0.9418	0.9429	0.9441
1.6	0.9452	0.9463	0.9474	0.9484	0.9495	0.9505	0.9515	0.9525	0.9535	0.9545
1.7	0.9554	0.9564	0.9573	0.9582	0.9591	0.9599	0.9608	0.9616	0.9625	0.9633
1.8	0.9641	0.9649	0.9656	0.9664	0.9671	0.9678	0.9686	0.9693	0.9699	0.9706
1.9	0.9713	0.9719	0.9726	0.9732	0.9738	0.9744	0.9750	0.9756	0.9761	0.9767
2.0	0.9772	0.9778	0.9783	0.9788	0.9793	0.9798	0.9803	0.9808	0.9812	0.9817
2.1	0.9821	0.9826	0.9830	0.9834	0.9838	0.9842	0.9846	0.9850	0.9854	0.9857
2.2	0.9861	0.9864	0.9868	0.9871	0.9875	0.9878	0.9881	0.9884	0.9887	0.9890
2.3	0.9893	0.9896	0.9898	0.9901	0.9904	0.9906	0.9909	0.9911	0.9913	0.9916
2.4	0.9918	0.9920	0.9922	0.9925	0.9927	0.9929	0.9931	0.9932	0.9934	0.9936
2.5	0.9938	0.9940	0.9941	0.9943	0.9945	0.9946	0.9948	0.9949	0.9951	0.9952
2.6	0.9953	0.9955	0.9956	0.9957	0.9959	0.9960	0.9961	0.9962	0.9963	0.9964
2.7	0.9965	0.9966	0.9967	0.9968	0.9969	0.9970	0.9971	0.9972	0.9973	0.9974
2.8	0.9974	0.9975	0.9976	0.9977	0.9977	0.9978	0.9979	0.9979	0.9980	0.9981
2.9	0.9981	0.9982	0.9982	0.9983	0.9984	0.9984	0.9985	0.9985	0.9986	0.9986
3.0	0.9987	0.9987	0.9987	0.9988	0.9988	0.9989	0.9989	0.9989	0.9990	0.9990
3.1	0.9990	0.9991	0.9991	0.9991	0.9992	0.9992	0.9992	0.9992	0.9993	0.9993
3.2	0.9993	0.9993	0.9994	0.9994	0.9994	0.9994	0.9994	0.9995	0.9995	0.9995
3.3	0.9995	0.9995	0.9995	0.9996	0.9996	0.9996	0.9996	0.9996	0.9996	0.9997
3.4	0.9997	0.9997	0.9997	0.9997	0.9997	0.9997	0.9997	0.9997	0.9997	0.9998

The exact result is

$$P\{X = 20\} = \binom{40}{20}\left(\frac{1}{2}\right)^{40}$$

which, after some calculation, can be shown to equal 0.1268.   ✦

Example 2.50   Let $X_i, i = 1, 2, \ldots, 10$ be independent random variables, each being uniformly distributed over $(0, 1)$. Calculate $P\{\Sigma_1^{10} X_i > 7\}$.

Solution: Since $E[X_i] = \frac{1}{2}$, $\text{Var}(X_i) = \frac{1}{12}$ we have by the central limit theorem that

$$P\left\{\sum_1^{10} X_i > 7\right\} = P\left\{\frac{\Sigma_1^{10} X_i - 5}{\sqrt{10(\frac{1}{12})}} > \frac{7 - 5}{\sqrt{10(\frac{1}{12})}}\right\}$$

$$\approx 1 - \Phi(2.2)$$

$$= 0.0139 \quad \blacklozenge$$

Example 2.51   The lifetime of a special type of battery is a random variable with mean 40 hours and standard deviation 20 hours. A battery is used until it fails, at which point it is replaced by a new one. Assuming a stockpile of 25 such batteries the lifetimes of which are independent, approximate the probability that over 1100 hours of use can be obtained.

Solution: If we let X_i denote the lifetime of the ith battery to be put in use, then we desire $p = P\{X_1 + \cdots + X_{25} > 1100\}$, which is approximated as follows:

$$p = P\left\{\frac{X_1 + \cdots + X_{25} - 1000}{20\sqrt{25}} > \frac{1100 - 1000}{20\sqrt{25}}\right\}$$

$$\approx P\{N(0, 1) > 1\}$$

$$= 1 - \Phi(1)$$

$$\approx 0.1587 \quad \blacklozenge$$

2.8.   Stochastic Processes

A *stochastic process* $\{X(t), t \in T\}$ is a collection of random variables. That is, for each $t \in T$, $X(t)$ is a random variable. The index t is often interpreted as time and, as a result, we refer to $X(t)$ as the *state* of the process at time t. For example, $X(t)$ might equal the total number of customers that have entered a supermarket by time t; or the number of customers in the supermarket at time t; or the total amount of sales that have been recorded in the market by time t; etc.

The set T is called the *index* set of the process. When T is a countable set the stochastic process is said to be *discrete-time* process. If T is an interval of the real line, the stochastic process is said to be a *continuous-time* process. For instance, $\{X_n, n = 0, 1, \ldots\}$ is a discrete-time stochastic process indexed by the nonnegative integers; while $\{X(t), t \geqslant 0\}$ is a continuous-time stochastic process indexed by the nonnegative real numbers.

The *state space* of a stochastic process is defined as the set of all possible values that the random variables $X(t)$ can assume.

Thus, a stochastic process is a family of random variables that describes the evolution through time of some (physical) process. We shall see much of stochastic processes in the following chapters of this text.

Example 2.52 Consider a particle that moves along a set of $m + 1$ nodes, labeled $0, 1, \ldots, m$, that are arranged around a circle (see Figure 2.3). At each step the particle is equally likely to move one position in either the clockwise or counterclockwise direction. That is, if X_n is the position of the particle after its nth step then

$$P\{X_{n+1} = i + 1 \mid X_n = i\} = P\{X_{n+1} = i - 1 \mid X_n = i\} = \tfrac{1}{2}$$

where $i + 1 \equiv 0$ when $i = m$, and $i - 1 \equiv m$ when $i = 0$. Suppose now that the particle starts at 0 and continues to move around according to the preceding rules until all the nodes $1, 2, \ldots, m$ have been visited. What is the probability that node $i, i = 1, \ldots, m$, is the last one visited?

Solution: Surprisingly enough, the probability that node i is the last node visited can be determined without any computations. To do so, consider the first time that the particle is at one of the two neighbors of node i, that is, the first time that the particle is at one of the nodes $i - 1$ or $i + 1$ (with $m + 1 \equiv 0$). Suppose it is at node $i - 1$ (the argument in

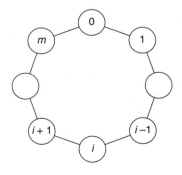

Figure 2.3. Particle moving around a circle.

the alternative situation is identical). Since neither node i nor $i + 1$ has yet been visited, it follows that i will be the last node visited if and only if $i + 1$ is visited before i. This is so because in order to visit $i + 1$ before i the particle will have to visit all the nodes on the counterclockwise path from $i - 1$ to $i + 1$ before it visits i. But the probability that a particle at node $i - 1$ will visit $i + 1$ before i is just the probability that a particle will progress $m - 1$ steps in a specified direction before progressing one step in the other direction. That is, it is equal to the probability that a gambler who starts with one unit, and wins one when a fair coin turns up heads and loses one when it turns up tails, will have his fortune go up by $m - 1$ before he goes broke. Hence, because the preceding implies that the probability that node i is the last node visited is the same for all i, and because these probabilities must sum to 1, we obtain

$$P\{i \text{ is the last node visited}\} = 1/m, \qquad i = 1, \ldots, m \quad \blacklozenge$$

Remark   The argument used in Example 2.52 also shows that a gambler who is equally likely to either win or lose one unit on each gamble will lose n before winning 1 with probability $1/(n + 1)$; or equivalently

$$P\{\text{gambler is up 1 before being down } n\} = \frac{n}{n + 1}$$

Suppose now we want the probability that the gambler is up 2 before being down n. Upon conditioning on whether he reaches up 1 before down n, we obtain that

$$P\{\text{gambler is up 2 before being down } n\}$$

$$= P\{\text{up 2 before down } n \mid \text{up 1 before down } n\} \frac{n}{n + 1}$$

$$= P\{\text{up 1 before down } n + 1\} \frac{n}{n + 1}$$

$$= \frac{n + 1}{n + 2} \frac{n}{n + 1} = \frac{n}{n + 2}$$

Repeating this argument yields that

$$P\{\text{gambler is up } k \text{ before being down } n\} = \frac{n}{n + k}$$

Exercises

1. An urn contains five red, three orange, and two blue balls. Two balls are randomly selected. What is the sample space of this experiment? Let X represent the number of orange balls selected. What are the possible values of X? Calculate $P\{X = 0\}$.

2. Let X represent the difference between the number of heads and the number of tails obtained when a coin is tossed n times. What are the possible values of X?

3. In Exercise 2, if the coin is assumed fair, then, for $n = 2$, what are the probabilities associated with the values that X can take on?

*4. Suppose a die is rolled twice. What are the possible values that the following random variables can take on?

  (i) The maximum value to appear in the two rolls.
  (ii) The minimum value to appear in the two rolls.
  (iii) The sum of the two rolls.
  (iv) The value of the first roll minus the value of the second roll.

5. If the die in Exercise 4 is assumed fair, calculate the probabilities associated with the random variables in (i)–(iv).

6. Suppose five fair coins are tossed. Let E be the event that all coins land heads. Define the random variable I_E

$$I_E = \begin{cases} 1, & \text{if } E \text{ occurs} \\ 0, & \text{if } E^c \text{ occurs} \end{cases}$$

For what outcomes in the original sample space does I_E equal 1? What is $P\{I_E = 1\}$?

7. Suppose a coin having probability 0.7 of coming up heads is tossed three times. Let X denote the number of heads that appear in the three tosses. Determine the probability mass function of X.

8. Suppose the distribution function of X is given by

$$F(b) = \begin{cases} 0, & b < 0 \\ \frac{1}{2}, & 0 \leqslant b < 1 \\ 1, & 1 \leqslant b < \infty \end{cases}$$

What is the probability mass function of X?

9. If the distribution function of F is given by

$$F(b) = \begin{cases} 0, & b < 0 \\ \frac{1}{2}, & 0 \leqslant b < 1 \\ \frac{3}{5}, & 1 \leqslant b < 2 \\ \frac{4}{5}, & 2 \leqslant b < 3 \\ \frac{9}{10}, & 3 \leqslant b < 3.5 \\ 1, & b \geqslant 3.5 \end{cases}$$

calculate the probability mass function of X.

10. Suppose three fair dice are rolled. What is the probability that at most one six appears?

*11. A ball is drawn from an urn containing three white and three black balls. After the ball is drawn, it is then replaced and another ball is drawn. This goes on indefinitely. What is the probability that of the first four balls drawn, exactly two are white?

12. On a multiple-choice exam with three possible answers for each of the five questions, what is the probability that a student would get four or more correct answers just by guessing?

13. An individual claims to have extrasensory perception (ESP). As a test, a fair coin is flipped ten times, and he is asked to predict in advance the outcome. Our individual gets seven out of ten correct. What is the probability he would have done at least this well if he had no ESP? (Explain why the relevant probability is $P\{X \geqslant 7\}$ and not $P\{X = 7\}$.)

14. Suppose X has a binomial distribution with parameters 6 and $\frac{1}{2}$. Show that $X = 3$ is the most likely outcome.

15. Let X be binomially distributed with parameters n and p. Show that as k goes from 0 to n, $P(X = k)$ increases monotonically, then decreases monotonically reaching its largest value

(a) in the case that $(n + 1)p$ is an integer, when k equals either $(n + 1)p - 1$ or $(n + 1)p$,
(b) in the case that $(n + 1)p$ is not an integer, when k satisfies $(n + 1)p - 1 < k < (n + 1)p$.

Hint: Consider $P\{X = k\}/P\{X = k - 1\}$ and see for what values of k it is greater or less than 1.

*16. An airline knows that 5 percent of the people making reservations on a certain flight will not show up. Consequently, their policy is to sell 52 tickets for a flight that can only hold 50 passengers. What is the probability that there will be a seat available for every passenger who shows up?

17. Suppose that an experiment can result in one of r possible outcomes, the ith outcome having probability p_i, $i = 1, \ldots, r$, $\Sigma_{i=1}^r p_i = 1$. If n of these experiments are performed, and if the outcome of any one of the n does not affect the outcome of the other $n - 1$ experiments, then show that the probability that the first outcome appears x_1 times, the second x_2 times, and the rth x_r times is

$$\frac{n!}{x_1! x_2! \cdots x_r!} p_1^{x_1} p_2^{x_2} \cdots p_r^{x_r} \qquad \text{when } x_1 + x_2 + \cdots + x_r = n$$

This is known as the *multinomial* distribution.

18. Show that when $r = 2$ the multinomial reduces to the binomial.

19. In Exercise 17, let X_i denote the number of times the ith outcome appears, $i = 1, \ldots, r$. What is the probability mass function of $X_1 + X_2 + \cdots + X_k$?

20. A television store owner figures that 50 percent of the customers entering his store will purchase an ordinary television set, 20 percent will purchase a color television set, and 30 percent will just be browsing. If five customers enter his store on a certain day, what is the probability that two customers purchase color sets, one customer purchases an ordinary set, and two customers purchase nothing?

21. In Exercise 20, what is the probability that our store owner sells three or more televisions on that day?

22. If a fair coin is successively flipped, find the probability that a head first appears on the fifth trial.

*23. A coin having probability p of coming up heads is successively flipped until the rth head appears. Argue that X, the number of flips required, will be n, $n \geq r$, with probability

$$P\{X = n\} = \binom{n-1}{r-1} p^r (1 - p)^{n-r}, \qquad n \geq r$$

This is known as the negative binomial distribution.

Hint: How many successes must there be in the first $n - 1$ trials?

24. The probability mass function of X is given by

$$p(k) = \binom{r + k - 1}{r - 1} p^r (1 - p)^k, \qquad k = 0, 1, \ldots$$

Give a possible interpretation of the random variable X.

Hint: See Exercise 23.

In Exercises 25 and 26, suppose that two teams are playing a series of games, each of which is independently won by team A with probability p and by team B with probability $1 - p$. The winner of the series is the first team to win i games.

25. If $i = 4$, find the probability that a total of 7 games are played. Also show that this probability is maximized when $p = 1/2$.

26. Find the expected number of games that are played when

(a) $i = 2$.
(b) $i = 3$.

In both cases, show that this number is maximized when $p = 1/2$.

*27. A fair coin is independently flipped n times, k times by A and $n - k$ times by B. Show that the probability that A and B flip the same number of heads is equal to the probability that there are a total of k heads.

28. Suppose that we want to generate a random variable X that is equally likely to be either 0 or 1, and that all we have at our disposal is a biased coin that, when flipped, lands on heads with some (unknown) probability p. Consider the following procedure:

1. Flip the coin, and let 0_1, either heads or tails, be the result.
2. Flip the coin again, and let 0_2 be the result.
3. If 0_1 and 0_2 are the same, return to step 1.
4. If 0_2 is heads, set $X = 0$, otherwise set $X = 1$.

(a) Show that the random variable X generated by this procedure is equally likely to be either 0 or 1.
(b) Could we use a simpler procedure that continues to flip the coin until the last two flips are different, and then sets $X = 0$ if the final flip is a head, and sets $X = 1$ if it is a tail?

29. Consider n independent flips of a coin having probability p of landing heads. Say a changeover occurs whenever an outcome differs from the one preceding it. For instance, if the results of the flips are $H H T H T H H T$, then there are a total of 5 changeovers. If $p = 1/2$, what is the probability there are k changeovers?

30. Let X be a Poisson random variable with parameter λ. Show that $P\{X = i\}$ increases monotonically and then decreases monotonically as i increases, reaching its maximum when i is the largest integer not exceeding λ.

Hint: Consider $P\{X = i\}/P\{X = i - 1\}$.

31.   Compare the Poisson approximation with the correct binomial probability for the following cases:

(i) $P\{X = 2\}$ when $n = 8$, $p = 0.1$.
(ii) $P\{X = 9\}$ when $n = 10$, $p = 0.95$.
(iii) $P\{X = 0\}$ when $n = 10$, $p = 0.1$.
(iv) $P\{X = 4\}$ when $n = 9$, $p = 0.2$.

32.   If you buy a lottery ticket in 50 lotteries, in each of which your chance of winning a prize is $\frac{1}{100}$, what is the (approximate) probability that you will win a prize (a) at least once, (b) exactly once, (c) at least twice?

33.   Let X be a random variable with probability density

$$f(x) = \begin{cases} c(1 - x^2), & -1 < x < 1 \\ 0, & \text{otherwise} \end{cases}$$

(a) What is the value of c?
(b) What is the cumulative distribution function of X?

34.   Let the probability density of X be given by

$$f(x) = \begin{cases} c(4x - 2x^2), & 0 < x < 2 \\ 0, & \text{otherwise} \end{cases}$$

(a) What is the value of c?
(b) $P\{\frac{1}{2} < X < \frac{3}{2}\} = ?$

35.   The density of X is given by

$$f(x) = \begin{cases} 10/x^2, & \text{for } x > 10 \\ 0, & \text{for } x \leqslant 10 \end{cases}$$

What is the distribution of X? Find $P\{X > 20\}$.

36.   A point is uniformly distributed within the disk of radius 1. That is, its density is

$$f(x, y) = C, \qquad 0 \leqslant x^2 + y^2 \leqslant 1$$

Find the probability that its distance from the origin is less than x, $0 \leqslant x \leqslant 1$.

37.   Let $X_1, X_2, \ldots, X_n$ be independent random variables, each having a uniform distribution over $(0, 1)$. Let $M = $ maximum $(X_1, X_2, \ldots, X_n)$. Show that the distribution function of $M, F_M(\cdot)$, is given by

$$F_M(x) = x^n, \qquad 0 \leqslant x \leqslant 1$$

What is the probability density function of M?

*38 If the density function of X equals

$$f(x) = \begin{cases} ce^{-2x}, & 0 < x < \infty \\ 0, & x < 0 \end{cases}$$

find c. What is $P\{X > 2\}$?

39. The random variable X has the following probability mass function

$$p(1) = \tfrac{1}{2}, \qquad p(2) = \tfrac{1}{3}, \qquad p(24) = \tfrac{1}{6}$$

Calculate $E[X]$.

40. Suppose that two teams are playing a series of games, each of which is independently won by team A with probability p and by team B with probability $1 - p$. The winner of the series is the first team to win 4 games. Find the expected number of games that are played, and evaluate this quantity when $p = 1/2$.

41. Consider the case of arbitrary p in Exercise 29. Compute the expected number of changeovers.

42. Suppose that each coupon obtained is, independent of what has been previously obtained, equally likely to be any of m different types. Find the expected number of coupons one needs to obtain in order to have at least one of each type.

Hint: Let X be the number needed. It is useful to represent X by

$$X = \sum_{i=1}^{m} X_i$$

where each X_i is a geometric random variable.

43. An urn contains $n + m$ balls, of which n are red and m are black. They are withdrawn from the urn, one at a time and without replacement. Let X be the number of red balls removed before the first black ball is chosen. We are interested in determining $E[X]$. To obtain this quantity, number the red balls from 1 to n. Now define the random variables $X_i, i = 1, \ldots, n$, by

$$X_i = \begin{cases} 1, & \text{if red ball } i \text{ is taken before any black ball is chosen} \\ 0, & \text{otherwise} \end{cases}$$

(a) Express X in terms of the X_i.
(b) Find $E[X]$.

44.   In Exercise 43, let Y denote the number of red balls chosen after the first but before the second black ball has been chosen.

(a) Express Y as the sum of n random variables, each of which is equal to either 0 or 1.
(b) Find $E[Y]$.
(c) Compare $E[Y]$ to $E[X]$ obtained in Exercise 43.
(d) Can you explain the result obtained in part (c)?

45.   A total of r keys are to be put, one at a time, in k boxes, with each key independently being put in box i with probability p_i, $\Sigma_{i=1}^k p_i = 1$. Each time a key is put in a nonempty box, we say that a collision occurs. Find the expected number of collisions.

*46.   Consider three trials, each of which is either a success or not. Let X denote the number of successes. Suppose that $E[X] = 1.8$.

(a) What is the largest possible value of $P\{X = 3\}$?
(b) What is the smallest possible value of $P\{X = 3\}$?

In both cases, construct a probability scenario that results in $P\{X = 3\}$ having the desired value.

47.   If X is uniformly distributed over $(0, 1)$, calculate $E[X^2]$.

*48.   Prove that $E[X^2] \geqslant (E[X])^2$. When do we have equality?

49.   Let c be a constant. Show that

(i) $\text{Var}(cX) = c^2 \, \text{Var}(X)$.
(ii) $\text{Var}(c + X) = \text{Var}(X)$.

50.   A coin, having probability p of landing heads, is flipped until the head appears for the rth time. Let N denote the number of flips required. Calculate $E[N]$.

Hint: There is an easy way of doing this. It involves writing N as the sum of r geometric random variables.

51.   Calculate the variance of the Bernoulli random variable.

52.   (a) Calculate $E[X]$ for the maximum random variable of Exercise 37.
(b) Calculate $E[X]$ for X as in Exercise 33.
(c) Calculate $E[X]$ for X as in Exercise 34.

53.   If X is uniform over $(0, 1)$, calculate $E[X^n]$ and $\text{Var}(X^n)$.

54. Let X and Y each take on either the value 1 or -1. Let

$$p(1, 1) = P\{X = 1, Y = 1\},$$
$$p(1, -1) = P\{X = 1, Y = -1\},$$
$$p(-1, 1) = P\{X = -1, Y = 1\},$$
$$p(-1, -1) = P\{X = -1, Y = -1\}$$

Suppose that $E[X] = E[Y] = 0$. Show that

(a) $p(1, 1) = p(-1, -1)$
(b) $p(1, -1) = p(-1, 1)$

Let $p = 2p(1, 1)$. Find

(c) $\text{Var}(X)$
(d) $\text{Var}(Y)$
(e) $\text{Cov}(X, Y)$

55. Let X be a positive random variable having density function $f(x)$. If $f(x) \leqslant c$ for all x, show that, for $a > 0$,

$$P\{X > a\} \geqslant 1 - ac$$

*56. Calculate, without using moment generating functions, the variance of a binomial random variable with parameters n and p.

57. Suppose that X and Y are independent binomial random variables with parameters (n, p) and (m, p). Argue probabilistically (no computations necessary) that $X + Y$ is binomial with parameters $(n + m, p)$.

58. Suppose that X and Y are independent continuous random variables. Show that

$$P\{X \leqslant Y\} = \int_{-\infty}^{\infty} F_X(y) f_Y(y)\, dy$$

59. Let X_1, X_2, X_3, and X_4 be independent continuous random variables with a common distribution function F and let

$$p = P\{X_1 < X_2 > X_3 < X_4\}$$

(a) Argue that the value of p is the same for all continuous distribution functions F.

(b) Find p by integrating the joint density function over the appropriate region.

(c) Find p by using the fact that all 4! possible orderings of $X_1, \ldots, X_4$ are equally likely.

60. Calculate the moment generating function of the uniform distribution on $(0, 1)$. Obtain $E[X]$ and $\text{Var}[X]$ by differentiating.

61. Suppose that X takes on each of the values 1, 2, 3 with probability $\frac{1}{3}$. What is the moment generating function? Derive $E[X]$, $E[X^2]$, and $E[X^3]$ by differentiating the moment generating function and then compare the obtained result with a direct derivation of these moments.

62. Suppose the density of X is given by

$$f(x) = \begin{cases} \frac{1}{4}xe^{-x/2}, & x > 0 \\ 0, & \text{otherwise} \end{cases}$$

Calculate the moment generating function, $E[X]$, and $\text{Var}(X)$.

63. Calculate the moment generating function of a geometric random variable.

***64.** Show that the sum of independent identically distributed exponential random variables has a gamma distribution.

65. Consider Example 2.47. Find $\text{Cov}(X_i, X_j)$ in terms of the a_{rs}.

66. Use Chebyshev's inequality to prove the *weak law of large numbers*. Namely, if $X_1, X_2, \ldots$ are independent and identically distributed with mean μ and variance σ^2 then, for any $\varepsilon > 0$,

$$P\left\{\left|\frac{X_1 + X_2 + \cdots + X_n}{n} - \mu\right| > \varepsilon\right\} \to 0 \qquad \text{as } n \to \infty$$

67. Suppose that X is a random variable with mean 10 and variance 15. What can we say about $P\{5 < X < 15\}$?

68. Let $X_1, X_2, \ldots, X_{10}$ be independent Poisson random variables with mean 1.

   (i) Use the Markov inequality to get a bound on $P\{X_1 + \cdots + X_{10} \geqslant 15\}$.
   (ii) Use the central limit theorem to approximate $P\{X_1 + \cdots + X_{10} \geqslant 15\}$.

69. If X is normally distributed with mean 1 and variance 4, use the tables to find $P\{2 < X < 3\}$.

***70.** Show that

$$\lim_{n \to \infty} e^{-n} \sum_{k=0}^{n} \frac{n^k}{k!} = \frac{1}{2}$$

Hint: Let X_n be Poisson with mean n. Use the central limit theorem to show that $P\{X_n \leqslant n\} \to \frac{1}{2}$.

71.   Let X denote the number of white balls selected when k balls are chosen at random from an urn containing n white and m black balls.

(i) Compute $P\{X = i\}$.
(ii) Let, for $i = 1, 2, \ldots, k; j = 1, 2, \ldots, n$,

$$X_i = \begin{cases} 1, & \text{if the } i\text{th ball selected is white} \\ 0, & \text{otherwise} \end{cases}$$

$$Y_j = \begin{cases} 1, & \text{if the } j\text{th white ball is selected} \\ 0, & \text{otherwise} \end{cases}$$

Compute $E[X]$ in two ways by expressing X first as a function of the X_is and then of the Y_j's.

*72.   Show that $\text{Var}(X) = 1$ when X is the number of men that select their own hats in Example 2.31.

73.   For the multinomial distribution (Exercise 17), let N_i denote the number of times outcome i occurs. Find

(i) $E[N_i]$
(ii) $\text{Var}(N_i)$
(iii) $\text{Cov}(N_i, N_j)$
(iv) Compute the expected number of outcomes which do not occur.

74.   Let $X_1, X_2, \ldots$ be a sequence of independent identically distributed continuous random variables. We say that a record occurs at time n if $X_n > \max(X_1, \ldots, X_{n-1})$. That is, X_n is a record if it is larger than each of $X_1, \ldots, X_{n-1}$. Show

(i) $P\{\text{a record occurs at time } n\} = 1/n$
(ii) $E[\text{number of records by time } n] = \sum_{i=1}^{n} 1/i$
(iii) $\text{Var}(\text{number of records by time } n) = \sum_{i=1}^{n} (i-1)/i^2$
(iv) Let $N = \min\{n : n > 1 \text{ and a record occurs at time } n\}$. Show $E[N] = \infty$.

Hint: For (ii) and (iii) represent the number of records as the sum of indicator (that is, Bernoulli) random variables.

75.   Let $a_1 < a_2 < \cdots < a_n$ denote a set of n numbers, and consider any permutation of these numbers. We say that there is an inversion of a_i and a_j in the permutation if $i < j$ and a_j precedes a_i. For instance the permutation 4, 2, 1, 5, 3 has 5 inversions—(4, 2), (4, 1), (4, 3), (2, 1), (5, 3). Consider now a random permutation of $a_1, a_2, \ldots, a_n$—in the sense that each of the $n!$ permutations is equally likely to be chosen—and let N denote the number of inversions in this permutation. Also, let

$$N_i = \text{number of } k: k < i, a_i \text{ precedes } a_k \text{ in the permutation}$$

and note that $N = \Sigma_{i=1}^n N_i$.

  (i) Show that $N_1, \ldots, N_n$ are independent random variables.
  (ii) What is the distribution of N_i?
  (iii) Compute $E[N]$ and $\text{Var}(N)$.

76. Let X and Y be independent random variables with means μ_x and μ_y and variances σ_x^2 and σ_y^2. Show that

$$\text{Var}(XY) = \sigma_x^2 \sigma_y^2 + \mu_y^2 \sigma_x^2 + \mu_x^2 \sigma_y^2$$

77. Let X and Y be independent normal random variables each having parameters μ and σ^2. Show that $X + Y$ is independent of $X - Y$.

78. Let $\phi(t_1, \ldots, t_n)$ denote the joint moment generating function of $X_1, \ldots, X_n$.

  (a) Explain how the moment generating function of X_i, $\phi_{X_i}(t_i)$, can be obtained from $\phi(t_1, \ldots, t_n)$.
  (b) Show that $X_1, \ldots, X_n$ are independent if and only if

$$\phi(t_1, \ldots, t_n) = \phi_{X_1}(t_1) \cdots \phi_{X_n}(t_n)$$

References

1. W. Feller, "An Introduction to Probability Theory and Its Applications," Vol. I, John Wiley, New York, 1957.
2. M. Fisz, "Probability Theory and Mathematical Statistics," John Wiley, New York, 1963.
3. E. Parzen, "Modern Probability Theory and Its Applications," John Wiley, New York, 1960.
4. S. Ross, "A First Course in Probability," Fifth Edition, Prentice Hall, New Jersey, 1998.

Conditional Probability and Conditional Expectation

3

✦✦✦

3.1. Introduction

One of the most useful concepts in probability theory is that of conditional probability and conditional expectation. The reason is twofold. First, in practice, we are often interested in calculating probabilities and expectations when some partial information is available; hence, the desired probabilities and expectations are conditional ones. Secondly, in calculating a desired probability or expectation it is often extremely useful to first "condition" on some appropriate random variable.

3.2. The Discrete Case

Recall that for any two events E and F, the conditional probability of E given F is defined, as long as $P(F) > 0$, by

$$P(E \mid F) = \frac{P(EF)}{P(F)}$$

Hence, if X and Y are discrete random variables, then it is natural to define the *conditional probability mass function* of X given that $Y = y$, by

$$p_{X\mid Y}(x \mid y) = P\{X = x \mid Y = y\}$$
$$= \frac{P\{X = x, Y = y\}}{P\{Y = y\}}$$
$$= \frac{p(x, y)}{p_Y(y)}$$

for all values of y such that $P\{Y=y\} > 0$. Similarly, the conditional probability distribution function of X given that $Y=y$ is defined, for all y such that $P\{Y=y\} > 0$, by

$$F_{X|Y}(x|y) = P\{X \leqslant x \,|\, Y = y\}$$
$$= \sum_{a \leqslant x} p_{X|Y}(a|y)$$

Finally, the conditional expectation of X given that $Y=y$ is defined by

$$E[X \,|\, Y = y] = \sum_{x} xP\{X = x \,|\, Y = y\}$$
$$= \sum_{x} xp_{X|Y}(x|y)$$

In other words, the definitions are exactly as before with the exception that everything is now conditional on the event that $Y=y$. If X is independent of Y, then the conditional mass function, distribution, and expectation are the same as the unconditional ones. This follows, since if X is independent of Y, then

$$p_{X|Y}(x|y) = P\{X = x \,|\, Y = y\}$$
$$= \frac{P\{X = x, Y = y\}}{P\{Y = y\}}$$
$$= \frac{P\{X = x\}P\{Y = y\}}{P\{Y = y\}}$$
$$= P\{X = x\}$$

Example 3.1   Suppose that $p(x, y)$, the joint probability mass function of X and Y, is given by

$$p(1, 1) = 0.5, \qquad p(1, 2) = 0.1, \qquad p(2, 1) = 0.1, \qquad p(2, 2) = 0.3$$

Calculate the probability mass function of X given that $Y = 1$.

Solution: We first note that

$$p_Y(1) = \sum_{x} p(x, 1) = p(1, 1) + p(2, 1) = 0.6$$

Hence,

$$p_{X|Y}(1|1) = P\{X = 1 \,|\, Y = 1\}$$
$$= \frac{P\{X = 1, Y = 1\}}{P\{Y = 1\}}$$
$$= \frac{p(1, 1)}{p_Y(1)}$$
$$= \tfrac{5}{6}$$

Similarly,

$$p_{X|Y}(2|1) = \frac{p(2, 1)}{p_Y(1)} = \frac{1}{6} \quad \blacklozenge$$

Example 3.2 If X and Y are independent Poisson random variables with respective means λ_1 and λ_2, then calculate the conditional expected value of X given that $X + Y = n$.

Solution: Let us first calculate the conditional probability mass function of X given that $X + Y = n$. We obtain

$$P\{X = k \mid X + Y = n\} = \frac{P\{X = k, X + Y = n\}}{P\{X + Y = n\}}$$

$$= \frac{P\{X = k, Y = n - k\}}{P\{X + Y = n\}}$$

$$= \frac{P\{X = k\}P\{Y = n - k\}}{P\{X + Y = n\}}$$

where the last equality follows from the assumed independence of X and Y. Recalling (see Example 2.36) that $X + Y$ has a Poisson distribution with mean $\lambda_1 + \lambda_2$, the preceding equation equals

$$P\{X = k \mid X + Y = n\} = \frac{e^{-\lambda_1}\lambda_1^k}{k!} \frac{e^{-\lambda_2}\lambda_2^{n-k}}{(n-k)!} \left[\frac{e^{-(\lambda_1 + \lambda_2)}(\lambda_1 + \lambda_2)^n}{n!} \right]^{-1}$$

$$= \frac{n!}{(n-k)!k!} \frac{\lambda_1^k \lambda_2^{n-k}}{(\lambda_1 + \lambda_2)^n}$$

$$= \binom{n}{k} \left(\frac{\lambda_1}{\lambda_1 + \lambda_2} \right)^k \left(\frac{\lambda_2}{\lambda_1 + \lambda_2} \right)^{n-k}$$

In other words, the conditional distribution of X given that $X + Y = n$, is the binomial distribution with parameters n and $\lambda_1/(\lambda_1 + \lambda_2)$. Hence,

$$E\{X \mid X + Y = n\} = n\frac{\lambda_1}{\lambda_1 + \lambda_2} \quad \blacklozenge$$

Example 3.3 If X and Y are independent binomial random variables with identical parameters n and p, calculate the conditional probability mass function of X given that $X + Y = m$.

Solution: For $k \leqslant \min(n, m)$,

$$P\{X = k \mid X + Y = m\} = \frac{P\{X = k, X + Y = m\}}{P\{X + Y = m\}}$$

$$= \frac{P\{X = k, Y = m - k\}}{P\{X + Y = m\}}$$

$$= \frac{P\{X = k\}P\{Y = m - k\}}{P\{X + Y = m\}}$$

$$= \frac{\binom{n}{k} p^k (1 - p)^{n-k} \binom{n}{m - k} p^{m-k} (1 - p)^{n-m+k}}{\binom{2n}{m} p^m (1 - p)^{2n-m}}$$

where we have used the fact (see Example 2.44) that $X + Y$ is binomially distributed with parameters $(2n, p)$. Hence, the conditional probability mass function of X given that $X + Y = m$ is given by

$$P\{X = k \mid X + Y = m\} = \frac{\binom{n}{k}\binom{n}{m - k}}{\binom{2n}{m}}, \qquad k = 0, 1, \ldots, \min(m, n) \qquad (3.1)$$

The distribution given in Equation (3.1) is known as the *hypergeometric* distribution. It arises as the distribution of the number of black balls that are chosen when a sample of m balls is randomly selected from an urn containing n white and n black balls.   ✦

Example 3.4   Consider an experiment which results in one of three possible outcomes with outcome i occurring with probability p_i, $i = 1, 2, 3$, $\Sigma_{i=1}^{3} p_i = 1$. Suppose that n independent replications of this experiment are performed and let X_i, $i = 1, 2, 3$, denote the number of times outcome i appears. Determine the conditional distribution of X_1 given that $X_2 = m$.

Solution: For $k \leqslant n - m$,

$$P\{X_1 = k \mid X_2 = m\} = \frac{P\{X_1 = k, X_2 = m\}}{P\{X_2 = m\}}$$

Now if $X_1 = k$ and $X_2 = m$, then it follows that $X_3 = n - k - m$.

However,

$$P\{X_1 = k, X_2 = m, X_3 = n - k - m\} = \frac{n!}{k!m!(n-k-m)!} p_1^k p_2^m p_3^{(n-k-m)}$$

(3.2)

This follows since any particular sequence of the n experiments having outcome 1 appearing k times, outcome 2 m times, and outcome 3 $(n - k - m)$ times has probability $p_1^k p_2^m p_3^{(n-k-m)}$ of occurring. Since there are $n!/[k!m!(n - k - m)!]$ such sequences, Equation (3.2) follows.

Therefore, we have

$$P\{X_1 = k \,|\, X_2 = m\} = \frac{\dfrac{n!}{k!m!(n - k - m)!} p_1^k p_2^m p_3^{(n-k-m)}}{\dfrac{n!}{m!(n - m)!} p_2^m (1 - p_2)^{n-m}}$$

where we have used the fact that X_2 has a binomial distribution with parameters n and p_2. Hence,

$$P\{X_1 = k \,|\, X_2 = m\} = \frac{(n - m)!}{k!(n - m - k)!} \left(\frac{p_1}{1 - p_2}\right)^k \left(\frac{p_3}{1 - p_2}\right)^{n-m-k}$$

or equivalently, writing $p_3 = 1 - p_1 - p_2$,

$$P\{X_1 = k \,|\, X_2 = m\} = \binom{n - m}{k} \left(\frac{p_1}{1 - p_2}\right)^k \left(1 - \frac{p_1}{1 - p_2}\right)^{n-m-k}$$

In other words, the conditional distribution of X_1, given that $X_2 = m$, is binomial with parameters $n - m$ and $p_1/(1 - p_2)$. ◆

Remarks (i) The desired conditional probability in Example 3.4 could also have been computed in the following manner. Consider the $n - m$ experiments that did not result in outcome 2. For each of these experiments, the probability that outcome 1 was obtained is given by

$$P\{\text{outcome 1} \,|\, \text{not outcome 2}\} = \frac{P\{\text{outcome 1, not outcome 2}\}}{P\{\text{not outcome 2}\}}$$

$$= \frac{p_1}{1 - p_2}$$

It therefore follows that, given $X_2 = m$, the number of times outcome 1 occurs is binomially distributed with parameters $n - m$ and $p_1/(1 - p_2)$.

(ii) Conditional expectations possess all of the properties of ordinary expectations. For instance, such identities as

$$E\left[\sum_{i=1}^{n} X_i \mid Y = y\right] = \sum_{i=1}^{n} E[X_i \mid Y = y]$$

remain valid.

Example 3.5   Consider $n + m$ independent trials, each of which results in a success with probability p. Compute the expected number of successes in the first n trials given that there are k successes in all.

Solution: Letting Y denote the total number of successes, and

$$X_i = \begin{cases} 1, & \text{if the } i\text{th trial is a success} \\ 0, & \text{otherwise} \end{cases}$$

the desired expectation is $E[\sum_{i=1}^{n} X_i \mid Y = k]$ which is obtained as

$$E\left[\sum_{1}^{n} X_i \mid Y = k\right] = \sum_{1}^{n} E[X_i \mid Y = k]$$

$$= n \frac{k}{n + m}$$

where the last equality follows since if there are a total of k successes, then any individual trial will be a success with probability $k/(n + m)$. That is,

$$E[X_i \mid Y = k] = P\{X_i = 1 \mid Y = k\}$$

$$= \frac{k}{n + m} \quad \blacklozenge$$

3.3.   The Continuous Case

If X and Y have a joint probability density function $f(x, y)$, then the *conditional probability density function* of X, given that $Y = y$, is defined for all values of y such that $f_Y(y) > 0$, by

$$f_{X \mid Y}(x \mid y) = \frac{f(x, y)}{f_Y(y)}$$

To motivate this definition, multiply the left side by dx and the right side by $(dx\,dy)/dy$ to get

$$f_{X|Y}(x\,|\,y)\,dx = \frac{f(x, y)\,dx\,dy}{f_Y(y)\,dy}$$

$$\approx \frac{P\{x \leqslant X \leqslant x + dx, y \leqslant Y \leqslant y + dy\}}{P\{y \leqslant Y \leqslant y + dy\}}$$

$$= P\{x \leqslant X \leqslant x + dx\,|\,y \leqslant Y \leqslant y + dy\}$$

In other words, for small values dx and dy, $f_{X|Y}(x\,|\,y)\,dx$ represents the conditional probability that X is between x and $x + dx$ given that Y is between y and $y + dy$.

The *conditional expectation* of X, given that $Y = y$, is defined for all values of y such that $f_Y(y) > 0$, by

$$E[X\,|\,Y = y] = \int_{-\infty}^{\infty} x f_{X|Y}(x\,|\,y)\,dx$$

Example 3.6   Suppose the joint density of X and Y is given by

$$f(x, y) = \begin{cases} 6xy(2 - x - y), & 0 < x < 1, 0 < y < 1 \\ 0, & \text{otherwise} \end{cases}$$

Compute the conditional expectation of X given that $Y = y$, where $0 < y < 1$.

Solution: We first compute the conditional density

$$f_{X|Y}(x\,|\,y) = \frac{f(x, y)}{f_Y(y)}$$

$$= \frac{6xy(2 - x - y)}{\int_0^1 6xy(2 - x - y)\,dx}$$

$$= \frac{6xy(2 - x - y)}{y(4 - 3y)}$$

$$= \frac{6x(2 - x - y)}{4 - 3y}$$

Hence,

$$E[X \mid Y = y] = \int_0^1 \frac{6x^2(2 - x - y)\,dx}{4 - 3y}$$

$$= \frac{(2 - y)2 - \frac{6}{4}}{4 - 3y}$$

$$= \frac{5 - 4y}{8 - 6y} \quad \blacklozenge$$

Example 3.7   Suppose the joint density of X and Y is given by

$$f(x, y) = \begin{cases} 4y(x - y)e^{-(x+y)}, & 0 < x < \infty, 0 \leqslant y \leqslant x \\ 0, & \text{otherwise} \end{cases}$$

Compute $E[X \mid Y = y]$.

Solution: The conditional density of X, given that $Y = y$, is given by

$$f_{X|Y}(x \mid y) = \frac{f(x, y)}{f_Y(y)}$$

$$= \frac{4y(x - y)e^{-(x+y)}}{\int_y^\infty 4y(x - y)e^{-(x+y)}\,dx}, \qquad x > y$$

$$= \frac{(x - y)e^{-(x+y)}}{\int_y^\infty (x - y)e^{-(x+y)}\,dx}$$

$$= \frac{(x - y)e^{-x}}{\int_y^\infty (x - y)e^{-x}\,dx}$$

Integrating by parts shows that the above gives

$$f_{X|Y}(x \mid y) = \frac{(x - y)e^{-x}}{e^{-y}}$$

$$= (x - y)e^{-(x-y)}, \qquad x > y$$

Therefore,

$$E[X \mid Y = y] = \int_{-\infty}^\infty x f_{X|Y}(x \mid y)\,dx$$

$$= \int_y^\infty x(x - y)e^{-(x-y)}\,dx$$

Integration by parts yields

$$E[X \mid Y = y] = -x(x - y)e^{-(x-y)}\Big|_y^\infty + \int_y^\infty (2x - y)e^{-(x-y)}\,dx$$

$$= \int_y^\infty (2x - y)e^{-(x-y)}\,dx$$

$$= -(2x - y)e^{-(x-y)}\Big|_y^\infty + 2\int_y^\infty e^{-(x-y)}\,dx$$

$$= y + 2 \quad \blacklozenge$$

Example 3.8 The joint density of X and Y is given by

$$f(x, y) = \begin{cases} \frac{1}{2}ye^{-xy}, & 0 < x < \infty, 0 < y < 2 \\ 0, & \text{otherwise} \end{cases}$$

What is $E[e^{X/2} \mid Y = 1]$?

Solution: The conditional density of X, given that $Y = 1$, is given by

$$f_{X \mid Y}(x \mid 1) = \frac{f(x, 1)}{f_Y(1)}$$

$$= \frac{\frac{1}{2}e^{-x}}{\int_0^\infty \frac{1}{2}e^{-x}\,dx} = e^{-x}$$

Hence, by Proposition 2.1,

$$E[e^{X/2} \mid Y = 1] = \int_0^\infty e^{x/2}f_{X \mid Y}(x \mid 1)\,dx$$

$$= \int_0^\infty e^{x/2}e^{-x}\,dx$$

$$= 2 \quad \blacklozenge$$

3.4. Computing Expectations by Conditioning

Let us denote by $E[X \mid Y]$ that function of the random variable Y whose value at $Y = y$ is $E[X \mid Y = y]$. Note that $E[X \mid Y]$ is itself a random variable. An extremely important property of conditional expectation is that for all random variables X and Y

$$E[X] = E[E[X \mid Y]] \tag{3.3}$$

If Y is a discrete random variable, then Equation (3.3) states that

$$E[X] = \sum_y E[X \mid Y = y]P\{Y = y\} \tag{3.3a}$$

while if Y is continuous with density $f_Y(y)$, then Equation (3.3) says that

$$E[X] = \int_{-\infty}^{\infty} E[X \mid Y = y]f_Y(y)\,dy \tag{3.3b}$$

We now give a proof of Equation (3.3) in the case where X and Y are both discrete random variables.

Proof of Equation (3.3) When *X* and *Y* Are Discrete   We must show that

$$E[X] = \sum_y E[X \mid Y = y]P\{Y = y\} \tag{3.4}$$

Now, the right side of the preceding can be written

$$\sum_y E[X \mid Y = y]P\{Y = y\} = \sum_y \sum_x xP\{X = x \mid Y = y\}P\{Y = y\}$$

$$= \sum_y \sum_x x\frac{P\{X = x, Y = y\}}{P\{Y = y\}}P\{Y = y\}$$

$$= \sum_y \sum_x xP\{X = x, Y = y\}$$

$$= \sum_x x\sum_y P\{X = x, Y = y\}$$

$$= \sum_x xP\{X = x\}$$

$$= E[X]$$

and the result is obtained.   ✦

One way to understand Equation (3.4) is to interpret it as follows. It states that to calculate $E[X]$ we may take a weighted average of the conditional expected value of X given that $Y = y$, each of the terms $E[X \mid Y = y]$ being weighted by the probability of the event on which it is conditioned.

The following examples will indicate the usefulness of Equation (3.3).

Example 3.9   Sam will read either one chapter of his probability book or one chapter of his history book. If the number of misprints in a chapter of his probability book is Poisson distributed with mean 2 and if the number of misprints in his history chapter is Poisson distributed with mean 5, then assuming Sam is equally likely to choose either book, what is the expected number of misprints that Sam will come across?

Solution: Letting X denote the number of misprints and letting

$$Y = \begin{cases} 1, & \text{if Sam chooses his history book} \\ 2, & \text{if Sam chooses his probability book} \end{cases}$$

then

$$E[X] = E[X \mid Y = 1]P\{Y = 1\} + E[X \mid Y = 2]P\{Y = 2\}$$
$$= 5(\tfrac{1}{2}) + 2(\tfrac{1}{2})$$
$$= \tfrac{7}{2} \quad \blacklozenge$$

Example 3.10   (The Expectation of the Sum of a Random Number of Random Variables): Suppose that the expected number of accidents per week at an industrial plant is four. Suppose also that the numbers of workers injured in each accident are independent random variables with a common mean of 2. Assume also that the number of workers injured in each accident is independent of the number of accidents that occur. What is the expected number of injuries during a week?

Solution: Letting N denote the number of accidents and X_i the number injured in the ith accident, $i = 1, 2, \ldots$, then the total number of injuries can be expressed as $\Sigma_{i=1}^{N} X_i$. Now

$$E\left[\sum_{1}^{N} X_i\right] = E\left[E\left[\sum_{1}^{N} X_i \mid N\right]\right]$$

But

$$E\left[\sum_{1}^{N} X_i \mid N = n\right] = E\left[\sum_{1}^{n} X_i \mid N = n\right]$$
$$= E\left[\sum_{1}^{n} X_i\right] \quad \text{by the independence of } X_i \text{ and } N$$
$$= nE[X]$$

which yields that

$$E\left[\sum_{i=1}^{N} X_i \mid N\right] = NE[X]$$

and thus

$$E\left[\sum_{i=1}^{N} X_i\right] = E[NE[X]] = E[N]E[X]$$

Therefore, in our example, the expected number of injuries during a week equals $4 \times 2 = 8$. ✦

Example 3.11 (The Mean of a Geometric Distribution): A coin, having probability p of coming up heads, is to be successively flipped until the first head appears. What is the expected number of flips required?

Solution: Let N be the number of flips required, and let

$$Y = \begin{cases} 1, & \text{if the first flip results in a head} \\ 0, & \text{if the first flip results in a tail} \end{cases}$$

Now

$$\begin{aligned} E[N] &= E[N \mid Y = 1]P\{Y = 1\} + E[N \mid Y = 0]P\{Y = 0\} \\ &= pE[N \mid Y = 1] + (1 - p)E[N \mid Y = 0] \end{aligned} \tag{3.5}$$

However,

$$E[N \mid Y = 1] = 1, \qquad E[N \mid Y = 0] = 1 + E[N] \tag{3.6}$$

To see why Equation (3.6) is true, consider $E[N \mid Y = 1]$. Since $Y = 1$, we know that the first flip resulted in heads and so, clearly, the expected number of flips required is 1. On the other hand if $Y = 0$, then the first flip resulted in tails. However, since the successive flips are assumed independent, it follows that, after the first tail, the expected additional number of flips until the first head is just $E[N]$. Hence $E[N \mid Y = 0] = 1 + E[N]$. Substituting Equation (3.6) into Equation (3.5) yields

$$E[N] = p + (1 - p)(1 + E[N])$$

or

$$E[N] = 1/p \quad ✦$$

Because the random variable N is a geometric random variable with probability mass function $p(n) = p(1 - p)^{n-1}$, its expectation could easily have been computed from $E[N] = \sum_{1}^{\infty} np(n)$ without recourse to conditional expectation. However, if the reader attempts to obtain the solution to our next example without using conditional expectation, he will quickly learn what a useful technique "conditioning" can be.

Example 3.12   A miner is trapped in a mine containing three doors. The first door leads to a tunnel that takes him to safety after two hours of travel. The second door leads to a tunnel that returns him to the mine after three hours of travel. The third door leads to a tunnel that returns him to his mine after five hours. Assuming that the miner is at all times equally likely to choose any one of the doors, what is the expected length of time until the miner reaches safety?

Solution: Let X denote the time until the miner reaches safety, and let Y denote the door he initially chooses. Now

$$E[X] = E[X \mid Y = 1]P\{Y = 1\} + E[X \mid Y = 2]P\{Y = 2\}$$
$$+ E[X \mid Y = 3]P\{Y = 3\}$$
$$= \tfrac{1}{3}(E[X \mid Y = 1] + E[X \mid Y = 2] + E[X \mid Y = 3])$$

However,

$$E[X \mid Y = 1] = 2$$
$$E[X \mid Y = 2] = 3 + E[X]$$
$$E[X \mid Y = 3] = 5 + E[X] \qquad\qquad (3.7)$$

To understand why the above is correct consider, for instance, $E[X \mid Y = 2]$, and reason as follows. If the miner chooses the second door, then he spends three hours in the tunnel and then returns to the mine. But once he returns to the mine the problem is as before, and hence his expected additional time until safety is just $E[X]$. Hence $E[X \mid Y = 2] = 3 + E[Y]$. The argument behind the other equalities in Equation (3.7) is similar. Hence

$$E[X] = \tfrac{1}{3}(2 + 3 + E[X] + 5 + E[X]) \qquad \text{or} \qquad E[X] = 10 \quad \blacklozenge$$

Example 3.13   (The Matching Rounds Problem): Suppose in Example 2.31 that those choosing their own hats depart, while the others (those without a match) put their selected hats in the center of the room, mix them up, and then reselect. Also, suppose that this process continues until each individual has his or her own hat.

(a) Find $E[R_n]$ where R_n is the number of rounds that are necessary when n individuals are initially present.
(b) Find $E[S_n]$ where S_n is the total number of selections made by the n individuals, $n \geqslant 2$.
(c) Find the expected number of false selections made by one of the n people, $n \geqslant 2$.

Solution: (a) It follows from the results of Example 2.31 that no matter how many people remain there will, on average, be one match per round. Hence, one might suggest that $E[R_n] = n$. This turns out to be true, and an induction proof will now be given. Because it is obvious that $E[R_1] = 1$, assume that $E[R_k] = k$ for $k = 1, \ldots, n - 1$. To compute $E[R_n]$, start by conditioning on X_n, the number of matches that occur in the first round. This gives

$$E[R_n] = \sum_{i=0}^{n} E[R_n \mid X_n = i] P\{X_n = i\}$$

Now, given a total of i matches in the initial round, the number of rounds needed will equal 1 plus the number of rounds that are required when $n - i$ persons are to be matched with their hats. Therefore,

$$E[R_n] = \sum_{i=0}^{n} (1 + E[R_{n-i}]) P\{X_n = i\}$$

$$= 1 + E[R_n] P\{X_n = 0\} + \sum_{i=1}^{n} E[R_{n-i}] P\{X_n = i\}$$

$$= 1 + E[R_n] P\{X_n = 0\} + \sum_{i=1}^{n} (n - i) P\{X_n = i\}$$

by the induction hypothesis

$$= 1 + E[R_n] P\{X_n = 0\} + n(1 - P\{X_n = 0\}) - E[X_n]$$

$$= E[R_n] P\{X_n = 0\} + n(1 - P\{X_n = 0\})$$

where the final equality used the result, established in Example 2.31, that $E[X_n] = 1$. Since the preceding equation implies that $E[R_n] = n$, the result is proven.

(b) For $n \geqslant 2$, conditioning on X_n, the number of matches in round 1, gives

$$E[S_n] = \sum_{i=0}^{n} E[S_n \mid X_n = i] P\{X_n = i\}$$

$$= \sum_{i=0}^{n} (n + E[S_{n-i}]) P\{X_n = i\}$$

$$= n + \sum_{i=0}^{n} E[S_{n-i}] P\{X_n = i\}$$

where $E[S_0] = 0$. To solve the preceding equation, rewrite it as

$$E[S_n] = n + E[S_{n-X_n}]$$

Now, if there were *exactly* one match in each round, then it would take a total of $1 + 2 + \cdots + n = n(n+1)/2$ selections. Thus, let us try a solution of the form $E[S_n] = an + bn^2$. For the preceding equation to be satisfied by a solution of this type, for $n \geq 2$, we need

$$an + bn^2 = n + E[a(n - X_n) + b(n - X_n)^2]$$

or, equivalently,

$$an + bn^2 = n + a(n - E[X_n]) + b(n^2 - 2nE[X_n] + E[X_n^2])$$

Now, using the results of Example 2.31 and Exercise 72 of Chapter 2 that $E[X_n] = \text{Var}(X_n) = 1$, the preceding will be satisfied if

$$an + bn^2 = n + an - a + bn^2 - 2nb + 2b$$

and this will be valid provided that $b = 1/2$, $a = 1$. That is,

$$E[S_n] = n + n^2/2$$

satisfies the recursive equation for $E[S_n]$.

The formal proof that $E[S_n] = n + n^2/2$, $n \geq 2$, is obtained by induction on n. It is true when $n = 2$ (since, in this case, the number of selections is twice the number of rounds and the number of rounds is a geometric random variable with parameter $p = 1/2$). Now, the recursion gives that

$$E[S_n] = n + E[S_n]P\{X_n = 0\} + \sum_{i=1}^{n} E[S_{n-i}]P\{X_n = i\}$$

Hence, upon assuming that $E[S_0] = E[S_1] = 0$, $E[S_k] = k + k^2/2$, for $k = 2, \ldots, n - 1$ and using that $P\{X_n = n - 1\} = 0$, we see that

$$E[S_n] = n + E[S_n]P\{X_n = 0\} + \sum_{i=1}^{n} [n - i + (n - i)^2/2]P\{X_n = i\}$$

$$= n + E[S_n]P\{X_n = 0\} + (n + n^2/2)(1 - P\{X_n = 0\})$$

$$- (n + 1)E[X_n] + E[X_n^2]/2$$

Substituting the identities $E[X_n] = 1$, $E[X_n^2] = 2$ in the preceding shows that

$$E[S_n] = n + n^2/2$$

and the induction proof is complete.

(c) If we let C_j denote the number of hats chosen by person $j, j = 1, \ldots, n$ then

$$\sum_{j=1}^{n} C_j = S_n$$

Taking expectations, and using the fact that each C_j has the same mean, yields the result

$$E[C_j] = E[S_n]/n = 1 + n/2$$

Hence, the expected number of false selections by person j is

$$E[C_j - 1] = n/2. \quad \blacklozenge$$

Example 3.14 Independent trials, each of which is a success with probability p, are performed until there are k consecutive successes. What is the mean number of necessary trials?

Solution: Let N_k denote the number of necessary trials to obtain k consecutive successes, and let M_k denote its mean. We will obtain a recursive equation for M_k by conditioning on N_{k-1}, the number of trials needed for $k - 1$ consecutive successes. This yields

$$M_k = E[N_k] = E[E[N_k \mid N_{k-1}]]$$

Now,

$$E[N_k \mid N_{k-1}] = N_{k-1} + 1 + (1 - p)E[N_k]$$

where the above follows since if it takes N_{k-1} trials to obtain $k - 1$ consecutive successes, then either the next trial is a success and we have our k in a row or it is a failure and we must begin anew. Taking expectations of both sides of the above yields

$$M_k = M_{k-1} + 1 + (1 - p)M_k$$

or

$$M_k = \frac{1}{p} + \frac{M_{k-1}}{p}$$

Since N_1, the time of the first success, is geometric with parameter p, we see that

$$M_1 = \frac{1}{p}$$

and, recursively

$$M_2 = \frac{1}{p} + \frac{1}{p^2},$$

$$M_3 = \frac{1}{p} + \frac{1}{p^2} + \frac{1}{p^3}$$

and, in general,

$$M_k = \frac{1}{p} + \frac{1}{p^2} + \cdots + \frac{1}{p^k} \quad \blacklozenge$$

Example 3.15 (Analyzing the Quick-Sort Algorithm): Suppose we are given a set of n distinct values — $x_1, \ldots, x_n$ — and we desire to put these values in increasing order or, as it is commonly called, to *sort* them. An efficient procedure for accomplishing this is the quick-sort algorithm which is defined recursively as follows: When $n = 2$ the algorithm compares the 2 values and puts them in the appropriate order. When $n > 2$ it starts by choosing at random one of the n values — say, x_i — and then compares each of the other $n - 1$ values with x_i, noting which are smaller and which are larger than x_i. Letting S_i denote the set of elements smaller than x_i, and $\bar{S}_i$ the set of elements greater than x_i, the algorithm now sorts the set S_i and the set $\bar{S}_i$. The final ordering, therefore, consists of the ordered set of the elements in S_i, then x_i, and then the ordered set of the elements in $\bar{S}_i$. For instance, suppose that the set of elements is 10, 5, 8, 2, 1, 4, 7. We start by choosing one of these values at random (that is, each of the 7 values has probability of $\frac{1}{7}$ of being chosen). Suppose, for instance, that the value 4 is chosen. We then compare 4 with each of the other 6 values to obtain

$$\{2, 1\}, 4, \{10, 5, 8, 7\}$$

We now sort the set $\{2, 1\}$ to obtain

$$1, 2, 4, \{10, 5, 8, 7\}$$

Next we choose a value at random from $\{10, 5, 8, 7\}$ — say, 7 is chosen — and compare each of the other 3 values with 7 to obtain

$$1, 2, 4, 5, 7, \{10, 8\}$$

Finally, we sort $\{10, 8\}$ to end up with

$$1, 2, 4, 5, 7, 8, 10$$

One measure of the effectiveness of this algorithm is the expected number of comparisons that it makes. Let us denote by M_n the expected number of comparisons needed by the quick-sort algorithm to sort a set of n distinct values. To obtain a recursion for M_n we condition on the rank of the initial value selected to obtain:

$$M_n = \sum_{j=1}^{n} E[\text{number of comparisons} \mid \text{value selected is } j\text{th smallest}] \frac{1}{n}$$

Now if the initial value selected is the jth smallest, then the set of values smaller than it is of size $j - 1$, and the set of values greater than it is of size $n - j$. Hence, as $n - 1$ comparisons with the initial value chosen must be made, we see that

$$M_n = \sum_{j=1}^{n} (n - 1 + M_{j-1} + M_{n-j}) \frac{1}{n}$$

$$= n - 1 + \frac{2}{n} \sum_{k=1}^{n-1} M_k \qquad (\text{since } M_0 = 0)$$

or, equivalently,

$$nM_n = n(n - 1) + 2 \sum_{k=1}^{n-1} M_k$$

To solve the preceding, note that upon replacing n by $n + 1$ we obtain

$$(n + 1)M_{n+1} = (n + 1)n + 2 \sum_{k=1}^{n} M_k$$

Hence, upon subtraction,

$$(n + 1)M_{n+1} - nM_n = 2n + 2M_n$$

or

$$(n + 1)M_{n+1} = (n + 2)M_n + 2n$$

Therefore,

$$\frac{M_{n+1}}{n + 2} = \frac{2n}{(n + 1)(n + 2)} + \frac{M_n}{n + 1}$$

Iterating this gives

$$\frac{M_{n+1}}{n + 2} = \frac{2n}{(n + 1)(n + 2)} + \frac{2(n - 1)}{n(n + 1)} + \frac{M_{n-1}}{n}$$

$$\vdots$$

$$= 2 \sum_{k=0}^{n-1} \frac{n - k}{(n + 1 - k)(n + 2 - k)} \qquad \text{since } M_1 = 0$$

Hence,

$$M_{n+1} = 2(n+2) \sum_{k=0}^{n-1} \frac{n-k}{(n+1-k)(n+2-k)}$$

$$= 2(n+2) \sum_{i=1}^{n} \frac{i}{(i+1)(i+2)}, \qquad n \geq 1$$

Using the identity $i/(i+1)(i+2) = 2/(i+2) - 1/(i+1)$, we can approximate M_{n+1} for large n as follows:

$$M_{n+1} = 2(n+2) \left[\sum_{i=1}^{n} \frac{2}{i+2} - \sum_{i=1}^{n} \frac{1}{i+1} \right]$$

$$\sim 2(n+2) \left[\int_{3}^{n+2} \frac{2}{x} dx - \int_{2}^{n+1} \frac{1}{x} dx \right]$$

$$= 2(n+2)[2 \log(n+2) - \log(n+1) + \log 2 - 2 \log 3]$$

$$= 2(n+2) \left[\log(n+2) + \log \frac{n+2}{n+1} + \log 2 - 2 \log 3 \right]$$

$$\sim 2(n+2) \log(n+2) \quad \blacklozenge$$

The conditional expectation is often useful in computing the variance of a random variable. In particular, we have that

$$\text{Var}(X) = E[X^2] - (E[X])^2$$

$$= E[E[X^2 \mid Y]] - (E[E[X \mid Y]])^2$$

Example 3.16 (The Variance of a Random Number of Random Variables): In Example 3.10 we showed that if $X_1, X_2, \ldots$ are independent and identically distributed, and if N is a nonnegative integer valued random variable independent of the Xs, then

$$E\left[\sum_{i=1}^{N} X_i \right] = E[N]E[X]$$

What can we say about $\text{Var}(\sum_{i=1}^{N} X_i)$?

Solution:

$$\text{Var}\left(\sum_{i=1}^{N} X_i \right) = E\left[\left(\sum_{i=1}^{N} X_i \right)^2 \right] - \left(E\left[\sum_{i=1}^{N} X_i \right] \right)^2 \tag{3.8}$$

To compute each of the individual terms, we condition on N:

$$E\left[\left(\sum_{i=1}^{N} X_i\right)^2\right] = E\left[E\left[\left(\sum_{i=1}^{N} X_i\right)^2 \middle| N\right]\right]$$

Now, given that $N = n$, $(\sum_{i=1}^{N} X_i)^2$ is distributed as the square of the sum of n independent and identically distributed random variables. Hence, using the identity $E[Z^2] = \text{Var}(Z) + (E[Z])^2$, we have that

$$E\left[\left(\sum_{i=1}^{N} X_i\right)^2 \middle| N = n\right] = \text{Var}\left(\sum_{i=1}^{n} X_i\right) + \left(E\left[\sum_{i=1}^{n} X_i\right]\right)^2$$

$$= n\,\text{Var}(X) + (nE[X])^2$$

Therefore,

$$E\left[\left(\sum_{i=1}^{N} X_i\right)^2 \middle| N\right] = N\,\text{Var}(X) + N^2(E[X])^2$$

Taking expectations of both sides of the above equation yields that

$$E\left[\left(\sum_{i=1}^{N} X_i\right)^2\right] = E[N]\,\text{Var}(X) + E[N^2](E[X])^2$$

Hence, from Equation (3.8) we obtain

$$\text{Var}\left(\sum_{i=1}^{N} X_i\right) = E[N]\,\text{Var}(X) + E[N^2](E[X])^2 - \left(E\left[\sum_{i=1}^{N} X_i\right]\right)^2$$

$$= E[N]\,\text{Var}(X) + E[N^2](E[X])^2 - (E[N]E[X])^2$$

$$= E[N]\,\text{Var}(X) + (E[X])^2(E[N^2] - (E[N])^2)$$

$$= E[N]\,\text{Var}(X) + (E[X])^2\,\text{Var}(N)$$

If N is a Poisson random variable, then $\sum_{i=1}^{N} X_i$ is called a *compound Poisson* random variable. Since the variance of a Poisson random variable is equal to its mean, it follows that for a compound Poisson random variable with $E[N] = \lambda$,

$$\text{Var}\left(\sum_{i=1}^{N} X_i\right) = \lambda\,\text{Var}(X) + \lambda(E[X])^2 = \lambda E[X^2] \quad \blacklozenge$$

Example 3.17 (Variance of the Geometric Distribution): Independent trials, each resulting in a success with probability p, are successively performed. Let N be the time of the first success. Find $\text{Var}(N)$.

Solution: Let $Y = 1$ if the first trial results in a success, and $Y = 0$ otherwise.

$$\text{Var}(N) = E(N^2) - (E[N])^2$$

To calculate $E[N^2]$ and $E[N]$ we condition on Y. For instance,

$$E[N^2] = E[E[N^2 \mid Y]]$$

However,

$$E[N^2 \mid Y = 1] = 1$$

$$E[N^2 \mid Y = 0] = E[(1 + N)^2]$$

These two equations are true since if the first trial results in a success, then clearly $N = 1$ and so $N^2 = 1$. On the other hand, if the first trial results in a failure, then the total number of trials necessary for the first success will equal one (the first trial that results in failure) plus the necessary number of additional trials. Since this latter quantity has the same distribution as N, we get that $E[N^2 \mid Y = 0] = E[(1 + N)^2]$. Hence, we see that

$$E[N^2] = E[N^2 \mid Y = 1]P\{Y = 1\} + E[N^2 \mid Y = 0]P\{Y = 0\}$$

$$= p + E[(1 + N)^2](1 - p)$$

$$= 1 + (1 - p)E[2N + N^2]$$

Since, as was shown in Example 3.11, $E[N] = 1/p$, this yields

$$E[N^2] = 1 + \frac{2(1 - p)}{p} + (1 - p)E[N^2]$$

or

$$E[N^2] = \frac{2 - p}{p^2}$$

Therefore,

$$\text{Var}(N) = E[N^2] - (E[N])^2$$

$$= \frac{2 - p}{p^2} - \left(\frac{1}{p}\right)^2$$

$$= \frac{1 - p}{p^2} \quad \blacklozenge$$

3.5. Computing Probabilities by Conditioning

Not only can we obtain expectations by first conditioning on an appropriate random variable, but we may also use this approach to compute probabilities. To see this, let E denote an arbitrary event and define the indicator random variable X by

$$X = \begin{cases} 1, & \text{if } E \text{ occurs} \\ 0, & \text{if } E \text{ does not occur} \end{cases}$$

It follows from the definition of X that

$$E[X] = P(E),$$

$$E[X \mid Y = y] = P(E \mid Y = y), \qquad \text{for any random variable } Y$$

Therefore, from Equations (3.3a) and (3.3b) we obtain

$$P(E) = \sum_y P(E \mid Y = y)P(Y = y), \qquad \text{if } Y \text{ is discrete}$$

$$= \int_{-\infty}^{\infty} P(E \mid Y = y) f_Y(y)\, dy, \qquad \text{if } Y \text{ is continuous}$$

Example 3.18 Suppose that X and Y are independent continuous random variables having densities f_X and f_Y, respectively. Compute $P\{X < Y\}$.

Solution: Conditioning on the value of Y yields

$$P\{X < Y\} = \int_{-\infty}^{\infty} P\{X < Y \mid Y = y\} f_Y(y)\, dy$$

$$= \int_{-\infty}^{\infty} P\{X < y \mid Y = y\} f_Y(y)\, dy$$

$$= \int_{-\infty}^{\infty} P\{X < y\} f_Y(y)\, dy$$

$$= \int_{-\infty}^{\infty} F_X(y) f_Y(y)\, dy$$

where

$$F_X(y) = \int_{-\infty}^{y} f_X(x)\, dx \quad \blacklozenge$$

Example 3.19 Suppose that X and Y are independent continuous random variables. Find the distribution of $X + Y$.

Solution: By conditioning on the value of Y we obtain

$$P\{X + Y < a\} = \int_{-\infty}^{\infty} P\{X + Y < a \mid Y = y\} f_Y(y) \, dy$$

$$= \int_{-\infty}^{\infty} P\{X + y < a \mid Y = y\} f_Y(y) \, dy$$

$$= \int_{-\infty}^{\infty} P\{X < a - y\} f_Y(y) \, dy$$

$$= \int_{-\infty}^{\infty} F_X(a - y) f_Y(y) \, dy \quad \blacklozenge$$

Example 3.20 Each customer who enters Rebecca's clothing store will purchase a suit with probability p. If the number of customers entering the store is Poisson distributed with mean λ, what is the probability that Rebecca does not sell any suits?

Solution: Let X be the number of suits that Rebecca sells, and let N denote the number of customers who enter the store. By conditioning on N we see that

$$P\{X = 0\} = \sum_{n=0}^{\infty} P\{X = 0 \mid N = n\} P\{N = n\}$$

$$= \sum_{n=0}^{\infty} P\{X = 0 \mid N = n\} e^{-\lambda} \lambda^n / n!$$

Now, given that n customers enter the store, the probability that Rebecca does not sell any suits is just $(1-p)^n$. That is, $P\{X = 0 \mid N = n\} = (1 - p)^n$. Therefore,

$$P\{X = 0\} = \sum_{n=0}^{\infty} \frac{(1 - p)^n e^{-\lambda} \lambda^n}{n!}$$

$$= e^{-\lambda} \sum_{n=0}^{\infty} \frac{(\lambda(1 - p))^n}{n!}$$

$$= e^{-\lambda} e^{\lambda(1 - p)}$$

$$= e^{-\lambda p} \quad \blacklozenge$$

Example 3.20 (continued)   What is the probability that Rebecca sells k suits?

Solution:

$$P\{X = k\} = \sum_{n=0}^{\infty} P\{X = k \mid N = n\} e^{-\lambda} \lambda^n / n!$$

Now, given that $N = n$, X has a binomial distribution with parameters n and p. Hence,

$$P\{X = k \mid N = n\} = \begin{cases} \binom{n}{k} p^k (1-p)^{n-k}, & n \geq k \\ 0, & n < k \end{cases}$$

so that

$$\begin{aligned} P\{X = k\} &= \sum_{n=k}^{\infty} \binom{n}{k} \frac{p^k (1-p)^{n-k} e^{-\lambda} \lambda^n}{n!} \\ &= \sum_{n=k}^{\infty} \frac{n!}{(n-k)! k!} \frac{(\lambda p)^k (\lambda(1-p))^{n-k} e^{-\lambda}}{n!} \\ &= \frac{e^{-\lambda}(\lambda p)^k}{k!} \sum_{n=k}^{\infty} \frac{(\lambda(1-p))^{n-k}}{(n-k)!} \\ &= e^{-\lambda} \frac{(\lambda p)^k}{k!} \sum_{i=0}^{\infty} \frac{(\lambda(1-p))^i}{i!} \\ &= e^{-\lambda} \frac{(\lambda p)^k}{k!} e^{\lambda(1-p)} \\ &= e^{-\lambda p} \frac{(\lambda p)^k}{k!} \end{aligned}$$

In other words, X has a Poisson distribution with mean λp.   ✦

Example 3.21   (The Best Prize Problem): Suppose that we are to be presented with n distinct prizes in sequence. After being presented with a prize we must immediately decide whether to accept it or reject it and consider the next prize. The only information we are given when deciding whether to accept a prize is the relative rank of that prize compared to ones already seen. That is, for instance, when the fifth prize is presented we learn how it compares with the first four prizes already seen. Suppose that once a prize is rejected it is lost, and that our objective is to maximize the probability of obtaining the best prize. Assuming that all $n!$ orderings of the prizes are equally likely, how well can we do?

Solution: Rather surprisingly, we can do quite well. To see this, fix a value $k, 0 \leqslant k < n$, and consider the strategy that rejects the first k prizes and then accepts the first one that is better than all of those first k. Let $P_k(\text{best})$ denote the probability that the best prize is selected when this strategy is employed. To compute this probability, condition on X, the position of the best prize. This gives

$$P_k(\text{best}) = \sum_{i=1}^{n} P_k(\text{best} \mid X = i)P(X = i)$$

$$= \frac{1}{n} \sum_{i=1}^{n} P_k(\text{best} \mid X = i)$$

Now, if the overall best prize is among the first k, then no prize is ever selected under the strategy considered. On the other hand, if the best prize is in position i, where $i > k$, then the best prize will be selected if the best of the first k prizes is also the best of the first $i - 1$ prizes (for then none of the prizes in positions $k + 1, k + 2, \ldots, i - 1$ would be selected). Hence, we see that

$$P_k(\text{best} \mid X = i) = 0, \qquad \text{if } i \leqslant k$$

$$P_k(\text{best} \mid X = i) = P\{\text{best of first } i - 1 \text{ is among the first } k\}$$

$$= k/(i - 1), \qquad \text{if } i > k$$

From the preceding, we obtain that

$$P_k(\text{best}) = \frac{k}{n} \sum_{i=k+1}^{n} \frac{1}{i - 1}$$

$$\approx \frac{k}{n} \int_{k}^{n-1} \frac{1}{x} \, dx$$

$$= \frac{k}{n} \log\left(\frac{n - 1}{k}\right)$$

$$\approx \frac{k}{n} \log\left(\frac{n}{k}\right)$$

Now, if we consider the function

$$g(x) = \frac{x}{n} \log\left(\frac{n}{x}\right)$$

then

$$g'(x) = \frac{1}{n} \log\left(\frac{n}{x}\right) - \frac{1}{n}$$

and so

$$g'(x) = 0 \Rightarrow \log(n/x) = 1 \Rightarrow x = n/e$$

Thus, since $P_k(\text{best}) \approx g(k)$, we see that the best strategy of the type considered is to let the first n/e prizes go by and then accept the first one to appear that is better than all of those. In addition, since $g(n/e) = 1/e$, the probability that this strategy selects the best prize is approximately $1/e \approx 0.36788$.

Remark   Most students are quite surprised by the size of the probability of obtaining the best prize, thinking that this probability would be close to 0 when n is large. However, even without going through the calculations, a little thought reveals that the probability of obtaining the best prize can be made to be reasonably large. Consider the strategy of letting half of the prizes go by, and then selecting the first one to appear that is better than all of those. The probability that a prize is actually selected is the probability that the overall best is among the second half and this is 1/2. In addition, given that a prize is selected, at the time of selection that prize would have been the best of more than $n/2$ prizes to have appeared, and would thus have probability of at least 1/2 of being the overall best. Hence, the strategy of letting the first half of all prizes go by and then accepting the first one that is better than all of those prizes results in a probability greater than 1/4 of obtaining the best prize.   ✦

Example 3.22   At a party n men take off their hats. The hats are then mixed up and each man randomly selects one. We say that a match occurs if a man selects his own hat. What is the probability of no matches? What is the probability of exactly k matches?

Solution: Let E denote the event that no matches occur, and to make explicit the dependence on n, write $P_n = P(E)$. We start by conditioning on whether or not the first man selects his own hat—call these events M and M^c. Then

$$P_n = P(E) = P(E \mid M)P(M) + P(E \mid M^c)P(M^c)$$

Clearly, $P(E \mid M) = 0$, and so

$$P_n = P(E \mid M^c) \frac{n-1}{n} \qquad (3.9)$$

Now, $P(E \mid M^c)$ is the probability of no matches when $n-1$ men select from a set of $n-1$ hats that does not contain the hat of one of these men. This can happen in either of two mutually exclusive ways. Either there are no matches and the extra man does not select the extra hat (this being the hat of the man that chose first), or there are no matches and the extra man does select the extra hat. The probability of the first of these events is just P_{n-1}, which is seen by regarding the extra hat as "belonging" to the extra man. Because the second event has probability $[1/(n-1)]P_{n-2}$, we have

$$P(E \mid M^c) = P_{n-1} + \frac{1}{n-1} P_{n-2}$$

and thus, from Equation (3.9),

$$P_n = \frac{n-1}{n} P_{n-1} + \frac{1}{n} P_{n-2}$$

or, equivalently,

$$P_n - P_{n-1} = -\frac{1}{n}(P_{n-1} - P_{n-2}) \qquad (3.10)$$

However, because P_n is the probability of no matches when n men select among their own hats, we have

$$P_1 = 0, \qquad P_2 = \tfrac{1}{2}$$

and so, from Equation (3.10),

$$P_3 - P_2 = -\frac{(P_2 - P_1)}{3} = -\frac{1}{3!} \quad \text{or} \quad P_3 = \frac{1}{2!} - \frac{1}{3!},$$

$$P_4 - P_3 = -\frac{(P_3 - P_2)}{4} = \frac{1}{4!} \quad \text{or} \quad P_4 = \frac{1}{2!} - \frac{1}{3!} + \frac{1}{4!}$$

and, in general, we see that

$$P_n = \frac{1}{2!} - \frac{1}{3!} + \frac{1}{4!} - \cdots + \frac{(-1)^n}{n!}$$

To obtain the probability of exactly k matches, we consider any fixed group of k men. The probability that they, and only they, select their own hats is

$$\frac{1}{n} \frac{1}{n-1} \cdots \frac{1}{n-(k-1)} P_{n-k} = \frac{(n-k)!}{n!} P_{n-k}$$

where P_{n-k} is the conditional probability that the other $n-k$ men, selecting among their own hats, have no matches. Because there are $\binom{n}{k}$ choices of a set of k men, the desired probability of exactly k matches is

$$\frac{P_{n-k}}{k!} = \frac{\dfrac{1}{2!} - \dfrac{1}{3!} + \cdots + \dfrac{(-1)^{n-k}}{(n-k)!}}{k!}$$

which, for n large, is approximately equal to $e^{-1}/k!$.

Remark   The recursive equation, Equation (3.10), could also have been obtained by using the concept of a cycle, where we say that the sequence of distinct individuals $i_1, i_2, \ldots, i_k$ constitutes a *cycle* if i_1 chooses i_2's hat, i_2 chooses i_3's hat, ..., i_{k-1} chooses i_k's hat, and i_k chooses i_1's hat. Note that every individual is part of a cycle, and that a cycle of size $k = 1$ occurs when someone chooses his or her own hat. With E being, as before, the event that no matches occur, it follows upon conditioning on the size of the cycle containing a specified person, say, person 1, that

$$P_n = P(E) = \sum_{i=1}^{n} P(E \mid C = k)P(C = k) \tag{3.11}$$

where C is the size of the cycle that contains person 1. Now call person 1 the first person, and note that $C = k$ if the first person does not choose 1's hat; the person whose hat was chosen by the first person — call this person the second person — does not choose 1's hat; the person whose hat was chosen by the second person — call this person the third person — does not choose 1's hat; ..., the person whose hat was chosen by the $(k-1)$st person does choose 1's hat. Consequently,

$$P(C = k) = \frac{n-1}{n} \frac{n-2}{n-1} \cdots \frac{n-k+1}{n-k+2} \frac{1}{n-k+1} = \frac{1}{n} \tag{3.12}$$

That is, the size of the cycle that contains a specified person is equally likely to be any of the values $1, 2, \ldots, n$. Moreover, since $C = 1$ means that 1 chooses his or her own hat, it follows that

$$P(E \mid C = 1) = 0 \tag{3.13}$$

On the other hand, if $C = k$, then the set of hats chosen by the k individuals in this cycle is exactly the set of hats of these individuals. Hence, conditional on $C = k$, the problem reduces to determining the probability of no matches when $n - k$ people randomly choose among their own $n - k$ hats. Therefore, for $k > 1$

$$P(E \mid C = k) = P_{n-k} \tag{3.14}$$

Substituting (3.12), (3.13), and (3.14) back into Equation (3.11) gives

$$P_n = \frac{1}{n} \sum_{k=2}^{n} P_{n-k}$$

which is easily shown to be equivalent to Equation (3.10). ◆

Example 3.23 (The Ballot Problem): In an election, candidate A receives n votes, and candidate B receives m votes where $n > m$. Assuming that all orderings are equally likely, show that the probability that A is always ahead in the count of votes is $(n - m)/(n + m)$.

Solution: Let $P_{n,m}$ denote the desired probability. By conditioning on which candidate receives the last vote counted we have

$$P_{n,m} = P\{A \text{ always ahead} \mid A \text{ receives last vote}\} \frac{n}{n + m}$$

$$+ P\{A \text{ always ahead} \mid B \text{ receives last vote}\} \frac{m}{n + m}$$

Now given that A receives the last vote, one can see that the probability that A is always ahead is the same as if A had received a total of $n - 1$ and B a total of m votes. Because a similar result is true when we are given that B receives the last vote, we see from the above that

$$P_{n,m} = \frac{n}{n + m} P_{n-1,m} + \frac{m}{m + n} P_{n,m-1} \tag{3.15}$$

We can now prove that $P_{n,m} = (n - m)/(n + m)$ by induction on $n + m$. As it is true when $n + m = 1$, i.e., $P_{1,0} = 1$, assume it whenever $n + m = k$. Then when $n + m = k + 1$, we have by Equation (3.15) and the induction hypothesis that

$$P_{n,m} = \frac{n}{n + m} \frac{n - 1 - m}{n - 1 + m} + \frac{m}{m + n} \frac{n - m + 1}{n + m - 1}$$

$$= \frac{n - m}{n + m}$$

and the result is proven. ◆

The ballot problem has some interesting applications. For example, consider successive flips of a coin that always land on "heads" with probability p, and let us determine the probability distribution of the first time, after beginning, that the total number of heads is equal to the total number of tails. The probability that the first time this occurs is at time $2n$ can be obtained by first conditioning on the total number of heads in the first $2n$ trials. This yields

$P\{\text{first time equal} = 2n\}$

$$= P\{\text{first time equal} = 2n \mid n \text{ heads in first } 2n\} \binom{2n}{n} p^n (1 - p)^n$$

Now given a total of n heads in the first $2n$ flips, one can see that all possible orderings of the n heads and n tails are equally likely, and thus the preceding conditional probability is equivalent to the probability that in an election, in which each candidate receives n votes, one of the candidates is always ahead in the counting until the last vote (which ties them). But by conditioning on whomever receives the last vote, we see that this is just the probability in the ballot problem when $m = n - 1$. Hence

$$P\{\text{first time equal} = 2n\} = P_{n,n-1} \binom{2n}{n} p^n (1 - p)^n$$

$$= \frac{\binom{2n}{n} p^n (1 - p)^n}{2n - 1}$$

Suppose now that we wanted to determine the probability that the first time there are i more heads than tails occurs after the $(2n + i)$th flip. Now, in order for this to be the case, the following two events must occur:

(a) The first $2n + i$ tosses result in $n + i$ heads and n tails; and
(b) the order in which the $n + i$ heads and n tails occur is such that the number of heads is never i more than the number of tails until after the final flip.

Now, it is easy to see that event (b) will occur if and only if the order of appearance of the $n + i$ heads and n tails is such that starting from the final flip and working backwards heads is always in the lead. For instance, if there are 4 heads and 2 tails ($n = 2, i = 2$), then the outcome _ _ _ _ TH would not suffice because there would have been 2 more heads than tails sometime before the sixth flip (since the first 4 flips resulted in 2 more heads than tails).

Now, the probability of the event specified in (a) is just the binomial probability of getting $n + i$ heads and n tails in $2n + i$ flips of the coin.

We must now determine the conditional probability of the event specified in (b) given that there are $n + i$ heads and n tails in the first $2n + i$ flips. To do so, note first that given that there are a total of $n + i$ heads and n tails in the first $2n + i$ flips, all possible orderings of these flips are equally likely. As a result, the conditional probability of (b) given (a) is just the probability that a random ordering of $n + i$ heads and n tails will, when counted in reverse order, always have more heads than tails. Since all reverse orderings are also equally likely, it follows from the ballot problem that this conditional probability is $i/(2n + i)$.

That is, we have shown that

$$P\{a\} = \binom{2n + i}{n} p^{n + i}(1 - p)^n,$$

$$P\{b \mid a\} = \frac{i}{2n + i}$$

and so

$$P\{\text{first time heads leads by } i \text{ is after flip } 2n+i\} = \binom{2n+i}{n} p^{n+i}(1-p)^n \frac{i}{2n+i}$$

Example 3.24   Let $U_1, U_2, \ldots$ be a sequence of independent uniform $(0, 1)$ random variables, and let

$$N = \min\{n \geqslant 2 : U_n > U_{n-1}\}$$

and

$$M = \min\{n \geqslant 1 : U_1 + \cdots + U_n > 1\}$$

That is, N is the index of the first uniform random variable that is larger than its immediate predecessor, and M is the number of uniform random variables whose sum we need to exceed 1. Surprisingly, N and M have the same probability distribution, and their common mean is e!

Solution: It is easy to find the distribution of N. Since all $n!$ possible orderings of $U_1, \ldots, U_n$ are equally likely, we have

$$P\{N > n\} = P\{U_1 > U_2 > \cdots > U_n\} = 1/n!$$

To show that $P\{M > n\} = 1/n!$, we will use mathematical induction. However, to give ourselves a stronger result to use as the induction hypothesis, we will prove the stronger result that for $0 < x \leqslant 1$, $P\{M(x) > n\} = x^n/n!$, $n \geqslant 1$, where

$$M(x) = \min\{n \geqslant 1 : U_1 + \cdots + U_n > x\}$$

is the minimum number of uniforms that need be summed to exceed x. To prove that $P\{M(x) > n\} = x^n/n!$, note first that it is true for $n = 1$ since

$$P\{M(x) > 1\} = P\{U_1 \leqslant x\} = x$$

So assume that for all $0 < x \leqslant 1$, $P\{M(x) > n\} = x^n/n!$. To determine $P\{M(x) > n + 1\}$, condition on U_1 to obtain:

$$
\begin{aligned}
P\{M(x) > n + 1\} &= \int_0^1 P\{M(x) > n + 1 \mid U_1 = y\}\, dy \\
&= \int_0^x P\{M(x) > n + 1 \mid U_1 = y\}\, dy \\
&= \int_0^x P\{M(x - y) > n\}\, dy \\
&= \int_0^x \frac{(x - y)^n}{n!}\, dy \qquad \text{by the induction hypothesis} \\
&= \int_0^x \frac{u^n}{n!}\, du \\
&= \frac{x^{n+1}}{(n + 1)!}
\end{aligned}
$$

where the third equality of the preceding follows from the fact that given $U_1 = y$, $M(x)$ is distributed as 1 plus the number of uniforms that need be summed to exceed $x - y$. Thus, the induction is complete and we have shown that for $0 < x \leqslant 1$, $n \geqslant 1$,

$$P\{M(x) > n\} = x^n/n!$$

Letting $x = 1$ shows that N and M have the same distribution. Finally, we have that

$$E[M] = E[N] = \sum_{n=0}^{\infty} P\{N > n\} = \sum_{n=0}^{\infty} 1/n! = e \quad \blacklozenge$$

Example 3.25 Let $X_1, X_2, \ldots$ be independent continuous random variables with a common distribution function F and density $f = F'$, and suppose that they are to be observed one at a time in sequence. Let

$$N = \min\{n \geqslant 2 : X_n = \text{second largest of } X_1, \ldots, X_n\}$$

and let

$$M = \min\{n \geqslant 2 : X_n = \text{second smallest of } X_1, \ldots, X_n\}$$

Which random variable—X_N, the first random variable which when observed is the second largest of those that have been seen, or X_M, the first one that on observation is the second smallest to have been seen—tends to be larger?

Solution: To calculate the probability density function of X_N, it is natural to condition on the value of N; so let us start by determining its probability mass function. Now, if we let

$$A_i = \{X_i \neq \text{second largest of } X_1, \ldots, X_i\}, \qquad i \geqslant 2$$

then, for $n \geqslant 2$,

$$P\{N = n\} = P(A_2 A_3 \cdots A_{n-1} A_n^c)$$

Since the X_i are independent and identically distributed it follows that, for any $m \geqslant 1$, knowing the ordering of the variables $X_1, \ldots, X_m$ yields no information about the set of m values $\{X_1, \ldots, X_m\}$. That is, for instance, knowing that $X_1 < X_2$ gives us no information about the values of $\min(X_1, X_2)$ or $\max(X_1, X_2)$. It follows from this that the events $A_i, i \geqslant 2$ are independent. Also, since X_i is equally likely to be the largest, or the second largest, ..., or the ith largest of $X_1 \ldots, X_i$ it follows that $P\{A_i\} = (i-1)/i, i \geqslant 2$. Therefore, we see that

$$P\{N = n\} = \frac{1}{2}\frac{2}{3}\frac{3}{4} \cdots \frac{n-2}{n-1}\frac{1}{n} = \frac{1}{n(n-1)}$$

Hence, conditioning on N yields that the probability density function of X_N is as follows:

$$f_{X_N}(x) = \sum_{n=2}^{\infty} \frac{1}{n(n-1)} f_{X_N|N}(x \mid n)$$

Now, since the ordering of the variables $X_1, \ldots, X_n$ is independent of the set of values $\{X_1, \ldots, X_n\}$, it follows that the event $\{N = n\}$ is independent of $\{X_1, \ldots, X_n\}$. From this, it follows that the conditional distribution of X_N given that $N = n$ is equal to the distribution of the second largest from a set of n random variables having distribution F. Thus, using the results of Example 2.37 concerning the density function of such a random variable, we obtain that

$$f_{X_N}(x) = \sum_{n=2}^{\infty} \frac{1}{n(n-1)} \frac{n!}{(n-2)!1!} (F(x))^{n-2} f(x)(1 - F(x))$$

$$= f(x)(1 - F(x)) \sum_{i=0}^{\infty} (F(x))^i$$

$$= f(x)$$

Thus, rather surprisingly, X_N has the same distribution as X_1, namely, F. Also, if we now let $W_i = -X_i$, $i \geqslant 1$, then W_M will be the value of the first W_i, which on observation is the second largest of all those that have been seen. Hence, by the preceding, it follows that W_M has the same distribution as W_1. That is, $-X_M$ has the same distribution as $-X_1$, and so X_M also has distribution F! In other words, whether we stop at the first random variable that is the second largest of all those presently observed, or we stop at the first one that is the second smallest of all those presently observed, we will end up with a random variable having distribution F.

Whereas the preceding result is quite surprising, it is a special case of a general result known as *Ignatov's theorem*, which yields even more surprises. For instance, for $k \geqslant 1$, let

$$N_k = \min\{n \geqslant k : X_n = k\text{th largest of } X_1, \ldots, X_n\}$$

Therefore, N_2 is what we previously called N, and X_{N_k} is the first random variable that upon observation is the kth largest of all those observed up to this point. It can then be shown by a similar argument as used in the preceding that X_{N_k} has distribution function F for all k (see Exercise 79 at the end of this chapter). In addition, it can be shown that the random variables X_{N_k}, $k \geqslant 1$ are independent. (A statement and proof of Ignatov's theorem in the case of discrete random variables is given in Section 3.6.4.) ✦

The use of conditioning can also result in a more computationally efficient solution than a direct calculation. This is illustrated by our next example.

Example 3.26   Consider n independent trials in which each trial results in one of the outcomes $1, \ldots, k$ with respective probabilities $p_1, \ldots, p_k$, $\sum_{i=1}^{k} p_i = 1$. Suppose further that $n > k$, and that we are interested in determining the probability that each outcome occurs at least once. If we let A_i denote the event that outcome i does not occur in any of the n trials, then the desired probability is $1 - P(\bigcup_{i=1}^{k} A_i)$, and it can be obtained by using the inclusion–exclusion theorem as follows:

$$P\left(\bigcup_{i=1}^{k} A_i\right) = \sum_{i=1}^{k} P(A_i) - \sum_{i}\sum_{j>i} P(A_i A_j)$$

$$+ \sum_{i}\sum_{j>i}\sum_{k>j} P(A_i A_j A_k) - \cdots + (-1)^{k+1} P(A_1 \cdots A_k)$$

where

$$P(A_i) = (1 - p_i)^n$$

$$P(A_i A_j) = (1 - p_i - p_j)^n, \qquad i < j$$

$$P(A_i A_j A_k) = (1 - p_i - p_j - p_k)^n, \qquad i < j < k$$

The difficulty with the preceding solution is that its computation requires the calculation of $2^k - 1$ terms, each of which is a quantity raised to the power n. The preceding solution is thus computationally inefficient when k is large. Let us now see how to make use of conditioning to obtain an efficient solution.

To begin, note that if we start by conditioning on N_k (the number of times that outcome k occurs) then when $N_k > 0$ the resulting conditional probability will equal the probability that all of the outcomes $1, \ldots, k - 1$ occur at least once when $n - N_k$ trials are performed, and each results in outcome i with probability $p_i / \sum_{j=1}^{k-1} p_j$, $i = 1, \ldots, k - 1$. We could then use a similar conditioning step on these terms.

To follow through on the preceding idea, let $A_{m,r}$, for $m \leqslant n, r \leqslant k$, denote the event that each of the outcomes $1, \ldots, r$ occurs at least once when m independent trials are performed, where each trial results in one of the outcomes $1, \ldots, r$ with respective probabilities $p_1/P_r, \ldots, p_r/P_r$, where $P_r = \sum_{j=1}^{r} p_j$. Let $P(m, r) = P(A_{m,r})$ and note that $P(n, k)$ is the desired probability. To obtain an expression for $P(m, r)$, condition on the number of times that outcome r occurs. This gives

$$P(m, r) = \sum_{j=0}^{m} P\{A_{m,r} \mid r \text{ occurs } j \text{ times}\} \binom{m}{j} \left(\frac{p_r}{P_r}\right)^j \left(1 - \frac{p_r}{P_r}\right)^{m-j}$$

$$= \sum_{j=1}^{m-r+1} P(m - j, r - 1) \binom{m}{j} \left(\frac{p_r}{P_r}\right)^j \left(1 - \frac{p_r}{P_r}\right)^{m-j}$$

Starting with

$$P(m, 1) = 1, \qquad \text{if } m \geqslant 1$$

$$P(m, 1) = 0, \qquad \text{if } m = 0$$

we can use the preceding recursion to obtain the quantities $P(m, 2)$, $m = 2, \ldots, n - (k - 2)$, and then the quantities $P(m, 3)$, $m = 3, \ldots, n - (k - 3)$, and so on, up to $P(m, k - 1)$, $m = k - 1, \ldots, n - 1$. At this point we can then use the recursion to compute $P(n, k)$. It is not difficult to check that the amount of computation needed is a polynomial function of k, which will be much smaller than 2^k when k is large. ◆

3.6. Some Applications

3.6.1. A List Model

Consider n elements—$e_1, e_2, \ldots, e_n$—that are initially arranged in some ordered list. At each unit of time a request is made for one of these elements—e_i being requested, independently of the past, with probability P_i. After being requested the element is then moved to the front of the list. That is, for instance, if the present ordering is e_1, e_2, e_3, e_4 and if e_3 is requested, then the next ordering is e_3, e_1, e_2, e_4.

We are interested in determining the expected position of the element requested after this process has been in operation for a long time. However, before computing this expectation, let us note two possible applications of this model. In the first we have a stack of reference books. At each unit of time a book is randomly selected and is then returned to the top of the stack. In the second application we have a computer receiving requests for elements stored in its memory. The request probabilities for the elements may not be known, so to reduce the average time it takes the computer to locate the element requested (which is proportional to the position of the requested element if the computer locates the element by starting at the beginning and then going down the list), the computer is programmed to replace the requested element at the beginning of the list.

To compute the expected position of the element requested, we start by conditioning on which element is selected. This yields

$$E[\text{position of element requested}]$$

$$= \sum_{i=1}^{n} E[\text{position} \mid e_i \text{ is selected}] P_i$$

$$= \sum_{i=1}^{n} E[\text{position of } e_i \mid e_i \text{ is selected}] P_i \qquad (3.16)$$

$$= \sum_{i=1}^{n} E[\text{position of } e_i] P_i$$

Now

$$\text{position of } e_i = 1 + \sum_{j \neq i} I_j$$

where

$$I_j = \begin{cases} 1, & \text{if } e_j \text{ precedes } e_i \\ 0, & \text{otherwise} \end{cases}$$

and so,

$$E[\text{position of } e_i] = 1 + \sum_{j \neq i} E[I_j]$$

$$= 1 + \sum_{j \neq i} P\{e_j \text{ precedes } e_i\} \tag{3.17}$$

To compute $P\{e_j \text{ precedes } e_i\}$, note that e_j will precede e_i if the most recent request for either of them was for e_j. But given that a request is for either e_i or e_j, the probability that it is for e_j is

$$P\{e_j | e_i \text{ or } e_j\} = \frac{P_j}{P_i + P_j}$$

and, thus,

$$P\{e_j \text{ precedes } e_i\} = \frac{P_j}{P_i + P_j}$$

Hence from Equations (3.16) and (3.17) we see that

$$E\{\text{position of element requested}\} = 1 + \sum_{i=1}^{n} P_i \sum_{j \neq i} \frac{P_j}{P_i + P_j}$$

This list model will be further analyzed in Section 4.8, where we will assume a different reordering rule—namely, that the element requested is moved one closer to the front of the list as opposed to being moved to the front of the list as assumed here. We will show there that the average position of the requested element is less under the one-closer rule than it is under the front-of-the-line rule.

3.6.2. A Random Graph

A graph consists of a set V of elements called nodes and a set A of pairs of elements of V called arcs. A graph can be represented graphically by drawing circles for nodes and drawing lines between nodes i and j whenever (i, j) is an arc. For instance if $V = \{1, 2, 3, 4\}$ and $A = \{(1, 2), (1, 4), (2, 3), (1, 2), (3, 3)\}$, then we can represent this graph as shown in Figure 3.1. Note that the arcs have no direction (a graph in which the arcs are ordered pairs of nodes is called a directed graph); and that in the figure there are multiple arcs connecting nodes 1 and 2, and a self-arc (called a self-loop) from node 3 to itself.

We say that there exists a path from node i to node j, $i \neq j$, if there exists a sequence of nodes $i, i_1, \ldots, i_k, j$ such that $(i, i_1), (i_1, i_2), \ldots, (i_k, j)$ are all

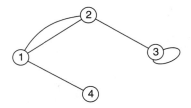

Figure 3.1. A graph.

arcs. If there is a path between each of the $\binom{n}{2}$ distinct pair of nodes we say that the graph is *connected*. The graph in Figure 3.1 is connected but the graph in Figure 3.2 is not. Consider now the following graph where $V = \{1, 2, \ldots, n\}$ and $A = \{(i, X(i)), \ i = 1, \ldots, n\}$ where the $X(i)$ are independent random variables such that

$$P\{X(i) = j\} = \frac{1}{n}, \qquad j = 1, 2, \ldots, n$$

In other words from each node i we select at random one of the n nodes (including possibly the node i itself) and then join node i and the selected node with an arc. Such a graph is commonly referred to as a *random graph*.

We are interested in determining the probability that the random graph so obtained is connected. As a prelude, starting at some node — say, node 1 — let us follow the sequence of nodes, 1, $X(1)$, $X^2(1), \ldots$, where $X^n(1) = X(X^{n-1}(1))$; and define N to equal the first k such that $X^k(1)$ is not a new node. In other words,

$$N = \text{1st } k \text{ such that } X^k(1) \in \{1, X(1), \ldots, X^{k-1}(1)\}$$

We can represent this as shown in Figure 3.3 where the arc from $X^{N-1}(1)$ goes back to a node previously visited.

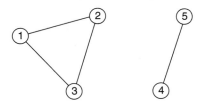

Figure 3.2. A disconnected graph.

Figure 3.3.

To obtain the probability that the graph is connected we first condition on N to obtain

$$P\{\text{graph is connected}\} = \sum_{k=1}^{n} P\{\text{connected} \mid N = k\} P\{N = k\} \quad (3.18)$$

Now given that $N = k$, the k nodes $1, X(1), \ldots, X^{k-1}(1)$ are connected to each other, and there are no other arcs emanating out of these nodes. In other words, if we regard these k nodes as being one supernode, the situation is the same as if we had one supernode and $n - k$ ordinary nodes with arcs emanating from the ordinary nodes—each arc going into the supernode with probability k/n. The solution in this situation is obtained from Lemma 3.1 by taking $r = n - k$.

Lemma 3.1 Given a random graph consisting of nodes $0, 1, \ldots, r$ and r arcs—namely, (i, Y_i), $i = 1, \ldots, r$, where

$$Y_i = \begin{cases} j & \text{with probability } \dfrac{1}{r+k}, \quad j = 1, \ldots, r \\ 0 & \text{with probability } \dfrac{k}{r+k} \end{cases}$$

then

$$P\{\text{graph is connected}\} = \frac{k}{r+k}$$

[In other words, for the preceding graph there are $r + 1$ nodes—r ordinary nodes and one supernode. Out of each ordinary node an arc is chosen. The arc goes to the supernode with probability $k/(r + k)$ and to each of the ordinary ones with probability $1/(r + k)$. There is no arc emanating out of the supernode.]

Proof The proof is by induction on r. As it is true when $r = 1$ for any k, assume it true for all values less than r. Now in the case under consideration, let us first condition on the number of arcs (j, Y_j) for which $Y_j = 0$. This yields

$P\{\text{connected}\}$

$$= \sum_{i=0}^{r} P\{\text{connected} \,|\, i \text{ of the } Y_j = 0\} \binom{r}{i} \left(\frac{k}{r+k}\right)^i \left(\frac{r}{r+k}\right)^{r-i} \quad (3.19)$$

Now given that exactly i of the arcs are into the supernode (see Figure 3.4), the situation for the remaining $r-i$ arcs which do not go into the supernode is the same as if we had $r-i$ ordinary nodes and one supernode with an arc going out of each of the ordinary nodes—into the supernode with probability i/r and into each ordinary node with probability $1/r$. But by the induction hypothesis the probability that this would lead to a connected graph is i/r.

Hence,

$$P\{\text{connected} \,|\, i \text{ of the } Y_j = 0\} = \frac{i}{r}$$

and from Equation (3.19)

$$P\{\text{connected}\} = \sum_{i=0}^{r} \frac{i}{r} \binom{r}{i} \left(\frac{k}{r+k}\right)^i \left(\frac{r}{r+k}\right)^{r-i}$$

$$= \frac{1}{r} E\left[\text{binomial}\left(r, \frac{k}{r+k}\right)\right]$$

$$= \frac{k}{r+k}$$

which completes the proof of the lemma. ✦

Hence as the situation given $N = k$ is exactly as described by Lemma 3.1

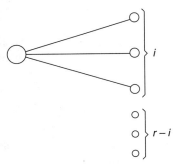

Figure 3.4. The situation given that i of the r arcs are into the supernode.

when $r = n - k$, we see that, for the original graph,

$$P\{\text{graph is connected} \mid N = k\} = \frac{k}{n}$$

and, from Equation (3.18),

$$P\{\text{graph is connected}\} = \frac{E(N)}{n} \qquad (3.20)$$

To compute $E(N)$ we use the identity

$$E(N) = \sum_{i=1}^{\infty} P\{N \geq i\}$$

which can be proved by defining indicator variables I_i, $i \geq 1$, by

$$I_i = \begin{cases} 1, & \text{if } i \leq N \\ 0, & \text{if } i > N \end{cases}$$

Hence,

$$N = \sum_{i=1}^{\infty} I_i$$

and so

$$E(N) = E\left[\sum_{i=1}^{\infty} I_i\right]$$

$$= \sum_{i=1}^{\infty} E[I_i]$$

$$= \sum_{i=1}^{\infty} P\{N \geq i\} \qquad (3.21)$$

Now the event $\{N \geq i\}$ occurs if the nodes $1, X(1), \ldots, X^{i-1}(1)$ are all distinct. Hence,

$$P\{N \geq i\} = \frac{(n-1)}{n} \frac{(n-2)}{n} \cdots \frac{(n-i+1)}{n}$$

$$= \frac{(n-1)!}{(n-i)! \, n^{i-1}}$$

and so, from Equations (3.20) and (3.21),

$$P\{\text{graph is connected}\} = (n-1)! \sum_{i=1}^{n} \frac{1}{(n-i)! \, n^i}$$

$$= \frac{(n-1)!}{n^n} \sum_{j=0}^{n-1} \frac{n^j}{j!} \qquad (\text{by } j = n - i) \quad (3.22)$$

We can also use Equation (3.22) to obtain a simple approximate expression for the probability that the graph is connected when n is large. To do so, we first note that if X is a Poisson random variable with mean n, then

$$P\{X < n\} = e^{-n} \sum_{j=0}^{n-1} \frac{n^j}{j!}$$

Since a Poisson random variable with mean n can be regarded as being the sum of n independent Poisson random variables each with mean 1, it follows from the central limit theorem that for n large such a random variable has approximately a normal distribution and as such has probability $\frac{1}{2}$ of being less than its mean. That is, for n large

$$P\{X < n\} \approx \tfrac{1}{2}$$

and so for n large,

$$\sum_{j=0}^{n-1} \frac{n^j}{j!} \approx \frac{e^n}{2}$$

Hence from Equation (3.22), for n large,

$$P\{\text{graph is connected}\} \approx \frac{e^n(n-1)!}{2n^n}$$

By employing an approximation due to Stirling which states that for n large

$$n! \approx n^{n+1/2} e^{-n} \sqrt{2\pi}$$

we see that, for n large,

$$P\{\text{graph is connected}\} \approx \sqrt{\frac{\pi}{2(n-1)}} \, e \left(\frac{n-1}{n}\right)^n$$

and as

$$\lim_{n \to \infty} \left(\frac{n-1}{n}\right)^n = \lim_{n \to \infty} \left(1 - \frac{1}{n}\right)^n = e^{-1}$$

We see that, for n large,

$$P\{\text{graph is connected}\} \approx \sqrt{\frac{\pi}{2(n-1)}}$$

Now a graph is said to consist of r connected components if its nodes can be partitioned into r subsets so that each of the subsets is connected and there are no arcs between nodes in different subsets. For instance, the graph in Figure 3.5 consists of three connected components—namely, $\{1, 2, 3\}$, $\{4, 5\}$, and $\{6\}$. Let C denote the number of connected components of our

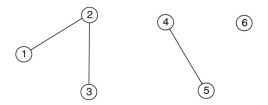

Figure 3.5. A graph having three connected components.

random graph and let

$$P_n(i) = P\{C = i\}$$

where we use the notation $P_n(i)$ to make explicit the dependence on n, the number of nodes. Since a connected graph is by definition a graph consisting of exactly one component, from Equation (3.22) we have

$$P_n(1) = P\{C = 1\}$$

$$= \frac{(n-1)!}{n^n} \sum_{j=0}^{n-1} \frac{n^j}{j!} \qquad (3.23)$$

To obtain $P_n(2)$, the probability of exactly two components, let us first fix attention on some particular node—say, node 1. In order that a given set of $k - 1$ other nodes—say, nodes $2, \ldots, k$—will along with node 1 constitute one connected component, and the remaining $n - k$ a second connected component, we must have

(i) $X(i) \in \{1, 2, \ldots, k\}$, for all $i = 1, \ldots, k$.
(ii) $X(i) \in \{k + 1, \ldots, n\}$, for all $i = k + 1, \ldots, n$.
(iii) The nodes $1, 2, \ldots, k$ form a connected subgraph.
(iv) The nodes $k + 1, \ldots, n$ form a connected subgraph.

The probability of the preceding occurring is clearly

$$\left(\frac{k}{n}\right)^k \left(\frac{n-k}{n}\right)^{n-k} P_k(1) P_{n-k}(1)$$

and because there are $\binom{n-1}{k-1}$ ways of choosing a set of $k - 1$ nodes from the nodes 2 through n, we have

$$P_n(2) = \sum_{k=1}^{n-1} \binom{n-1}{k-1} \left(\frac{k}{n}\right)^k \left(\frac{n-k}{n}\right)^{n-k} P_k(1) P_{n-k}(1)$$

and so $P_n(2)$ can be computed from Equation (3.23). In general, the recursive formula for $P_n(i)$ is given by

$$P_n(i) = \sum_{k=1}^{n-i+1} \binom{n-1}{k-1}\left(\frac{k}{n}\right)^k\left(\frac{n-k}{n}\right)^{n-k} P_k(1)P_{n-k}(i-1)$$

To compute $E[C]$, the expected number of connected components, first note that every connected component of our random graph must contain exactly one cycle [a cycle is a set of arcs of the form (i, i_1), $(i_1, i_2), \ldots,$ (i_{k-1}, i_k), (i_k, i) for distinct nodes $i, i_1, \ldots, i_k$]. For example, Figure 3.6 depicts a cycle.

The fact that every connected component of our random graph must contain exactly one cycle is most easily proved by noting that if the connected component consists of r nodes, then it must also have r arcs and, hence, must contain exactly one cycle (why?). Thus, we see that

$$E[C] = E[\text{number of cycles}]$$

$$= E\left[\sum_S I(S)\right]$$

$$= \sum_S E[I(S)]$$

where the sum is over all subsets $S \subset \{1, 2, \ldots, n\}$ and

$$I(S) = \begin{cases} 1, & \text{if the nodes in } S \text{ are all the nodes of a cycle} \\ 0, & \text{otherwise} \end{cases}$$

Now, if S consists of k nodes, say $1, \ldots, k$, then

$$E[I(S)] = P\{1, X(1), \ldots, X^{k-1}(1) \text{ are all distinct and contained in } 1, \ldots, k \text{ and } X^k(1) = 1\}$$

$$= \frac{k-1}{n}\frac{k-2}{n}\cdots\frac{1}{n}\frac{1}{n} = \frac{(k-1)!}{n^k}$$

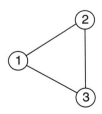

Figure 3.6. A cycle.

Hence, because there are $\binom{n}{k}$ subsets of size k we see that

$$E[C] = \sum_{k=1}^{n} \binom{n}{k} \frac{(k-1)!}{n^k}$$

3.6.3. Uniform Priors, Polya's Urn Model, and Bose–Einstein Statistics

Suppose that n independent trials, each of which is a success with probability p, are performed. If we let X denote the total number of successes, then X is a binomial random variable such that

$$P\{X = k \,|\, p\} = \binom{n}{k} p^k (1-p)^{n-k}, \qquad k = 0, 1, \ldots, n$$

However, let us now suppose that whereas the trials all have the same success probability p, its value is not predetermined but is chosen according to a uniform distribution on $(0, 1)$. (For instance, a coin may be chosen at random from a huge bin of coins representing a uniform spread over all possible values of p, the coin's probability of coming up heads. The chosen coin is then flipped n times.) In this case, by conditioning on the actual value of p, we have that

$$P\{X = k\} = \int_0^1 P\{X = k \,|\, p\} f(p) \, dp$$

$$= \int_0^1 \binom{n}{k} p^k (1-p)^{n-k} \, dp$$

Now, it can be shown that

$$\int_0^1 p^k (1-p)^{n-k} \, dp = \frac{k!(n-k)!}{(n+1)!} \tag{3.24}$$

and thus

$$P\{X = k\} = \binom{n}{k} \frac{k!(n-k)!}{(n+1)!}$$

$$= \frac{1}{n+1}, \qquad k = 0, 1, \ldots, n \tag{3.25}$$

In other words, each of the $n + 1$ possible values of X is equally likely.

As an alternate way of describing the preceding experiment, let us compute the conditional probability that the $(r + 1)$st trial will result in a success given a total of k successes (and $r - k$ failures) in the first r trials.

$P\{(r + 1)\text{st trial is a success} \mid k \text{ successes in first } r\}$

$$= \frac{P\{(r + 1)\text{st is a success}, k \text{ successes in first } r \text{ trials}\}}{P\{k \text{ successes in first } r \text{ trials}\}}$$

$$= \frac{\int_0^1 P\{(r + 1)\text{st is a success}, k \text{ in first } r \mid p\}\, dp}{1/(r + 1)}$$

$$= (r + 1) \int_0^1 \binom{r}{k} p^{k+1}(1 - p)^{r-k}\, dp$$

$$= (r + 1) \binom{r}{k} \frac{(k + 1)!(r - k)!}{(r + 2)!} \qquad \text{by Equation (3.24)}$$

$$= \frac{k + 1}{r + 2} \qquad\qquad\qquad\qquad\qquad\qquad (3.26)$$

That is, if the first r trials result in k successes, then the next trial will be a success with probability $(k + 1)/(r + 2)$.

It follows from Equation (3.26) that an alternative description of the stochastic process of the successive outcomes of the trials can be described as follows: There is an urn which initially contains one white and one black ball. At each stage a ball is randomly drawn and is then replaced along with another ball of the same color. Thus, for instance, if of the first r balls drawn k were white, then the urn at the time of the $(r + 1)$th draw would consist of $k + 1$ white and $r - k + 1$ black, and thus the next ball would be white with probability $(k + 1)/(r + 2)$. If we identify the drawing of a white ball with a successful trial, then we see that this yields an alternate description of the original model. This later urn model is called *Polya's urn model*.

Remarks   (i) In the special case when $k = r$, Equation (3.26) is sometimes called Laplace's rule of succession, after the French mathematician Pierre de Laplace. In Laplace's era, this "rule" provoked much controversy, for people attempted to employ it in diverse situations where its validity was questionable. For instance, it was used to justify such proposition as "If you have dined twice at a restaurant and both meals were good, then the next meal also will be good with probability $\frac{3}{4}$," and "Since the sun has risen the past 1,826,213 days, so will it rise tomorrow with probability 1,826,214/ 1,826,215." The trouble with such claims resides in the fact that it is not at all clear the situation they are describing can be modeled as consisting of independent trials having a common probability of success which is itself uniformly chosen.

(ii) In the original description of the experiment, we referred to the successive trials as being independent, and in fact they are independent when the success probability is known. However, when p is regarded as a random variable, the successive outcomes are no longer independent since knowing whether an outcome is a success or not gives us some information about p, which in turn yields information about the other outcomes.

The preceding can be generalized to situations in which each trial has more than two possible outcomes. Suppose that n independent trials, each resulting in one of m possible outcomes $1, \ldots, m$ with respective probabilities $p_1, \ldots, p_m$ are performed. If we let X_i denote the number of type i outcomes that result in the n trials, $i = 1, \ldots, m$, then the vector $X_1, \ldots, X_m$ will have the multinomial distribution given by

$$P\{X_1 = x_1, X_2 = x_2, \ldots, X_m = x_m \mid \mathbf{p}\} = \frac{n!}{x_1! \cdots x_m!} p_1^{x_1} p_2^{x_2} \cdots p_m^{x_m}$$

where $x_1, \ldots, x_m$ is any vector of nonnegative integers which sum to n. Now let us suppose that the vectors $\mathbf{p} = (p_1, \ldots, p_m)$ is not specified, but instead is chosen by a "uniform" distribution. Such a distribution would be of the form

$$f(p_1, \ldots, p_m) = \begin{cases} c, & 0 \leqslant p_i \leqslant 1, i = 1, \ldots, m, \sum_{1}^{m} p_i = 1 \\ 0, & \text{otherwise} \end{cases}$$

The preceding multivariate distribution is a special case of what is known as the *Dirichlet distribution*, and it is not difficult to show, using the fact that the distribution must integrate to 1, that $c = (m - 1)!$.

The unconditional distribution of the vector $\mathbf{X}$ is given by

$$P\{X_1 = x_1, \ldots, X_m = x_m\} = \int\!\!\int \cdots \int P\{X_1 = x_1, \ldots, X_m = x_m \mid p_1, \ldots, p_m\}$$

$$\times f(p_1, \ldots, p_m) \, dp_1 \cdots dp_m = \frac{(m-1)! n!}{x_1! \cdots x_m!} \int\!\!\int_{\substack{0 \leqslant p_i \leqslant 1 \\ \sum_1^m p_i = 1}} \cdots \int p_1^{x_1} \cdots p_m^{x_m} \, dp_1 \cdots dp_m$$

Now it can be shown that

$$\int\!\!\int_{\substack{0 \leqslant p_i \leqslant 1 \\ \sum_1^m p_i = 1}} \cdots \int p_1^{x_1} \cdots p_m^{x_m} \, dp_1 \cdots dp_m = \frac{x_1! \cdots x_m!}{(\sum_1^m x_i + m - 1)!} \tag{3.27}$$

and thus, using the fact that $\sum_1^m x_i = n$, we see that

$$P\{X_1 = x_1, \ldots, X_m = x_m\} = \frac{n!(m-1)!}{(n+m-1)!}$$

$$= \binom{n+m-1}{m-1}^{-1} \quad (3.28)$$

Hence, all of the $\binom{n+m-1}{m-1}$ possible outcomes [there are $\binom{n+m-1}{m-1}$ possible nonnegative integer valued solutions of $x_1 + \cdots + x_m = n$] of the vector $(X_1, \ldots, X_m)$ are equally likely. The probability distribution given by Equation (3.28) is sometimes called the *Bose–Einstein distribution*.

To obtain an alternative description of the foregoing, let us compute the conditional probability that the $(n+1)$st outcome is of type j if the first n trials have resulted in x_i type i outcomes, $i = 1, \ldots, m$, $\sum_1^m x_i = n$. This is given by

$$P\{(n+1)\text{st is } j \mid x_i \text{ type } i \text{ in first } n, i = 1, \ldots, m\}$$

$$= \frac{P\{(n+1)\text{st is } j, x_i \text{ type } i \text{ in first } n, i = 1, \ldots, m\}}{P\{x_i \text{ type } i \text{ in first } n, i = 1, \ldots, m\}}$$

$$= \frac{\dfrac{n!(m-1)!}{x_1! \cdots x_m!} \displaystyle\iint \cdots \int p_1^{x_1} \cdots p_j^{x_j+1} \cdots p_m^{x_m} \, dp_1 \cdots dp_m}{\dbinom{n+m-1}{m-1}^{-1}}$$

where the numerator is obtained by conditioning on the **p** vector and the denominator is obtained by using Equation (3.28). By Equation (3.27), we have that

$$P\{(n+1)\text{st is } j \mid x_i \text{ type } i \text{ in first } n, i = 1, \ldots, m\}$$

$$= \frac{\dfrac{(x_j+1)n!(m-1)!}{(n+m)!}}{\dfrac{(m-1)!n!}{(n+m-1)!}}$$

$$= \frac{x_j+1}{n+m} \quad (3.29)$$

Using Equation (3.29), we can now present an urn model description of the stochastic process of successive outcomes. Namely, consider an urn which initially contains one of each of m types of balls. Balls are then randomly drawn and are replaced along with another of the same type. Hence, if in the first n drawings there have been a total of x_j type j balls drawn, then the urn immediately before the $(n + 1)$st draw will contain $x_j + 1$ type j balls out of a total of $m + n$, and so the probability of a type j on the $(n + 1)$st draw will be given by Equation (3.29).

Remarks   Consider a situation where n particles are to be distributed at random among m possible regions; and suppose that the regions appear, at least before the experiment, to have the same physical characteristics. It would thus seem that the most likely distribution for the numbers of particles that fall into each of the regions is the multinomial distribution with $p_i \equiv 1/m$. (This, of course, would correspond to each particle, independent of the others, being equally likely to fall in any of the m regions.) Physicists studying how particles distribute themselves observed the behavior of such particles as photons and atoms containing an even number of elementary particles. However, when they studied the resulting data, they were amazed to discover that the observed frequencies did not follow the multinomial distribution but rather seemed to follow the Bose–Einstein distribution. They were amazed because they could not imagine a physical model for the distribution of particles which would result in all possible outcomes being equally likely. (For instance, if 10 particles are to distribute themselves between two regions, it hardly seems reasonable that it is just as likely that both regions will contain 5 particles as it is that all 10 will fall in region 1 or that all 10 will fall in region 2.)

However, from the results of this section we now have a better understanding of the cause of the physicists' dilemma. In fact, two possible hypotheses present themselves. First, it may be that the data gathered by the physicists were actually obtained under a variety of different situations, each having its own characteristic **p** vector which gave rise to a uniform spread over all possible **p** vectors. A second possibility (suggested by the urn model interpretation) is that the particles select their regions sequentially and a given particle's probability of falling in a region is roughly proportional to the fraction of the landed particles that are in that region. (In other words, the particles presently in a region provide an "attractive" force on elements which have not yet landed.)

3.6.4. The k-Record Values of Discrete Random Variables

Let $X_1, X_2, \ldots$ be independent and identically distributed random variables whose set of possible values is the positive integers, and let $P\{X = j\}, j \geq 1$, denote their common probability mass function. Suppose that these random variables are observed in sequence, and say that X_n is a *k-record value* if

$$X_i \geqslant X_n \qquad \text{for exactly } k \text{ of the values } i, i = 1, \ldots, n$$

That is, the nth value in the sequence is a k-record value if exactly k of the first n values (including X_n) are at least as large as it. Let $\mathbf{R}_k$ denote the ordered set of k-record values.

It is a rather surprising result that not only do the sequences of k-record values have *the same probability distributions* for all k, but these sequences are also *independent* of each other. This result is known as Ignatov's theorem.

Ignatov's Theorem $\mathbf{R}_k, k \geqslant 1$, are independent and identically distributed random vectors.

Proof   Define a series of subsequences of the data sequence $X_1, X_2, \ldots$ by letting the ith subsequence consist of all data values that are at least as large as $i, i \geqslant 1$. For instance, if the data sequence is

$$2, 5, 1, 6, 9, 8, 3, 4, 1, 5, 7, 8, 2, 1, 3, 4, 2, 5, 6, 1, \ldots$$

then the subsequences are as follows:

$\geqslant 1$: $2, 5, 1, 6, 9, 8, 3, 4, 1, 5, 7, 8, 2, 1, 3, 4, 2, 5, 6, 1, \ldots$

$\geqslant 2$: $2, 5, 6, 9, 8, 3, 4, 5, 7, 8, 2, 3, 4, 2, 5, 6, \ldots$

$\geqslant 3$: $5, 6, 9, 8, 3, 4, 5, 7, 8, 3, 4, 5, 6, \ldots$

and so on.

Let X_j^i be the jth element of subsequence i. That is, X_j^i is the jth data value that is at least as large as i. An important observation is that i is a k-record value if and only if $X_k^i = i$. That is, i will be a k-record value if and only if the kth value to be at least as large as i is equal to i. (For instance, for the preceding data since the fifth value to be at least as large as 3 is equal to 3 it follows that 3 is a 5-record value.) Now, it is not difficult to see that, independent of which values in the first subsequence are equal to 1, the values in the second subsequence are independent and identically distributed according to the mass function

$$P\{\text{value in second subsequence} = j\} = P\{X = j \,|\, X \geqslant 2\}, \qquad j \geqslant 2$$

Similarly, independent of which values in the first subsequence are equal to 1 and which values in the second subsequence are equal to 2, the values in the third subsequence are independent and identically distributed according to the mass function

$$P\{\text{value in third subsequence} = j\} = P\{X = j \,|\, X \geqslant 3\}, \qquad j \geqslant 3$$

and so on. It therefore follows that the events $\{X_j^i = i\}$, $i \geqslant 1$, $j \geqslant 1$, are independent and

$$P\{i \text{ is a } k\text{-record value}\} = P\{X_k^i = i\} = P\{X = i \mid X \geqslant i\}$$

It now follows from the independence of the events $\{X_k^i = i\}$, $i \geqslant 1$, and the fact that $P\{i \text{ is a } k\text{-record value}\}$ does not depend on k, that $\mathbf{R}_k$ has the same distribution for all $k \geqslant 1$. In addition, it follows from the independence of the events $\{X_k^i = i\}$, that the random vectors $\mathbf{R}_k$, $k \geqslant 1$, are also independent. ✦

Suppose now that the X_i, $i \geqslant 1$ are independent finite-valued random variables with probability mass function

$$p_i = P\{X = i\}, \qquad i = 1, \ldots, m$$

and let

$$T = \min\{n : X_i \geqslant X_n \text{ for exactly } k \text{ of the values } i, i = 1, \ldots, n\}$$

denote the first *k-record index*. We will now determine its mean.

Proposition 3.1 Let $\lambda_i = p_i / \Sigma_{j=i}^m p_j$, $i = 1, \ldots, m$. Then

$$E[T] = k + (k - 1) \sum_{i=1}^{m-1} \lambda_i$$

Proof To begin, suppose that the observed random variables $X_1 X_2, \ldots$ take on one of the values $i, i + 1, \ldots, m$ with respective probabilities

$$P\{X = j\} = \frac{p_j}{p_i + \cdots + p_m}, \qquad j = i, \ldots, m$$

Let T_i denote the first k-record index when the observed data have the preceding mass function, and note that since the each data value is at least i it follows that the k-record value will equal i, and T_i will equal k, if $X_k = i$. As a result,

$$E[T_i \mid X_k = i] = k$$

On the other hand, if $X_k > i$ then the k-record value will exceed i, and so all data values equal to i can be disregarded when searching for the k-record value. In addition, since each data value greater than i will have probability mass function

$$P\{X = j \mid X > i\} = \frac{p_j}{p_{i+1} + \cdots + p_m}, \qquad j = i + 1, \ldots, m$$

it follows that the total number of data values greater than i that need be observed until a k-record value appears has the same distribution as T_{i+1}. Hence,

$$E[T_i | X_k > i] = E[T_{i+1} + N_i | X_k > i]$$

where T_{i+1} is the total number of variables greater than i that we need observe to obtain a k-record, and N_i is the number of values equal to i that are observed in that time. Now, given that $X_k > i$ and that $T_{i+1} = n \ (n \geqslant k)$ it follows that the time to observe T_{i+1} values greater than i has the same distribution as the number of trials to obtain n successes given that trial k is a success and that each trial is independently a success with probability $1 - p_i / \sum_{j \geqslant i} p_j = 1 - \lambda_i$. Thus, since the number of trials needed to obtain a success is a geometric random variable with mean $1/(1 - \lambda_i)$, we see that

$$E[T_i | T_{i+1}, X_k > i] = 1 + \frac{T_{i+1} - 1}{1 - \lambda_i} = \frac{T_{i+1} - \lambda_i}{1 - \lambda_i}$$

Taking expectations gives that

$$E[T_i | X_k > i] = E\left[\frac{T_{i+1} - \lambda_i}{1 - \lambda_i} \,\middle|\, X_k > i \right] = \frac{E[T_{i+1}] - \lambda_i}{1 - \lambda_i}$$

Thus, upon conditioning on whether $X_k = i$, we obtain

$$E[T_i] = E[T_i | X_k = i]\lambda_i + E[T_i | X_k > i](1 - \lambda_i)$$

$$= (k - 1)\lambda_i + E[T_{i+1}]$$

Starting with $E[T_m] = k$, the preceding gives that

$$E[T_{m-1}] = (k - 1)\lambda_{m-1} + k$$

$$E[T_{m-2}] = (k - 1)\lambda_{m-2} + (k - 1)\lambda_{m-1} + k$$

$$= (k - 1) \sum_{j=m-2}^{m-1} \lambda_j + k$$

$$E[T_{m-3}] = (k - 1)\lambda_{m-3} + (k - 1) \sum_{j=m-2}^{m-1} \lambda_j + k$$

$$= (k - 1) \sum_{j=m-3}^{m-1} \lambda_j + k$$

In general,

$$E[T_i] = (k - 1) \sum_{j=i}^{m-1} \lambda_j + k$$

and the result follows since $T = T_1$. ✦

Exercises

1. If X and Y are both discrete, show that $\sum_x p_{X|Y}(x \mid y) = 1$ for all y such that $p_Y(y) > 0$.

*2. Let X_1 and X_2 be independent geometric random variables having the same parameter p. Guess the value of

$$P\{X_1 = i \mid X_1 + X_2 = n\}$$

Hint: Suppose a coin having probability p of coming up heads is continually flipped. If the second head occurs on flip number n, what is the conditional probability that the first head was on flip number i, $i = 1, \ldots, n - 1$?

Verify your guess analytically.

3. The joint probability mass function of X and Y, $p(x, y)$, is given by

$$p(1, 1) = \tfrac{1}{9}, \qquad p(2, 1) = \tfrac{1}{3}, \qquad p(3, 1) = \tfrac{1}{9}$$

$$p(1, 2) = \tfrac{1}{9}, \qquad p(2, 2) = 0, \qquad p(3, 2) = \tfrac{1}{18}$$

$$p(1, 3) = 0, \qquad p(2, 3) = \tfrac{1}{6}, \qquad p(3, 3) = \tfrac{1}{9}$$

Compute $E[X \mid Y = i]$ for $i = 1, 2, 3$.

4. In Exercise 3, are the random variables X and Y independent?

5. An urn contains three white, six red, and five black balls. Six of these balls are randomly selected from the urn. Let X and Y denote respectively the number of white and black balls selected. Compute the conditional probability mass function of X given that $Y = 3$. Also compute $E[X \mid Y = 1]$.

*6. Repeat Exercise 5 but under the assumption that when a ball is selected its color is noted, and it is then replaced in the urn before the next selection is made.

7. Suppose $p(x, y, z)$, the joint probability mass function of the random variables X, Y, and Z, is given by

$$p(1, 1, 1) = \tfrac{1}{8}, \qquad p(2, 1, 1) = \tfrac{1}{4},$$
$$p(1, 1, 2) = \tfrac{1}{8}, \qquad p(2, 1, 2) = \tfrac{3}{16},$$
$$p(1, 2, 1) = \tfrac{1}{16}, \qquad p(2, 2, 1) = 0,$$
$$p(1, 2, 2) = 0, \qquad p(2, 2, 2) = \tfrac{1}{4}$$

What is $E[X \mid Y = 2]$? What is $E[X \mid Y = 2, Z = 1]$?

8. An unbiased die is successively rolled. Let X and Y denote respectively the number of rolls necessary to obtain a six and a five. Find (a) $E[X]$, (b) $E[X \mid Y = 1]$, (c) $E[X \mid Y = 5]$.

9. Show in the discrete case that if X and Y are independent, then

$$E[X \mid Y = y] = E[X] \qquad \text{for all } y$$

10. Suppose X and Y are independent continuous random variables. Show that

$$E[X \mid Y = y] = E[X] \qquad \text{for all } y$$

11. The joint density of X and Y is

$$f(x, y) = \frac{(y^2 - x^2)}{8} e^{-y}, \qquad 0 < y < \infty, \quad -y \leqslant x \leqslant y$$

Show that $E[X \mid Y = y] = 0$.

12. The joint density of X and Y is given by

$$f(x, y) = \frac{e^{-x/y} e^{-y}}{y}, \qquad 0 < x < \infty, \quad 0 < y < \infty$$

Show $E[X \mid Y = y] = y$.

*13. Let X be exponential with mean $1/\lambda$; that is,

$$f_X(x) = \lambda e^{-\lambda x}, \qquad 0 < x < \infty$$

Find $E[X \mid X > 1]$.

14. Let X be uniform over $(0, 1)$. Find $E[X \mid X < \tfrac{1}{2}]$.

15. The joint density of X and Y is given by

$$f(x, y) = \frac{e^{-y}}{y}, \qquad 0 < x < y, \quad 0 < y < \infty$$

Compute $E[X^2 \mid Y = y]$.

16. The random variables X and Y are said to have a bivariate normal distribution if their joint density function is given by

$$f(x, y) = \frac{1}{2\pi\sigma_x\sigma_y\sqrt{1 - \rho^2}} \exp\left\{-\frac{1}{2(1 - \rho^2)}\right.$$

$$\left. \times \left[\left(\frac{x - \mu_x}{\sigma_x}\right)^2 - \frac{2\rho(x - \mu_x)(y - \mu_y)}{\sigma_x\sigma_y} + \left(\frac{y - \mu_y}{\sigma_y}\right)^2\right]\right\}$$

for $-\infty < x < \infty$, $-\infty < y < \infty$, where σ_x, σ_y, μ_x, μ_y, and ρ are constants such that $-1 < \rho < 1$, $\sigma_x > 0$, $\sigma_y > 0$, $-\infty < \mu_x < \infty$, $-\infty < \mu_y < \infty$.

(a) Show that X is normally distributed with mean μ_x and variance σ_x^2, and Y is normally distributed with mean μ_y and variance σ_y^2.
(b) Show that the conditional density of X given that $Y = y$ is normal with mean $\mu_x + (\rho\sigma_x/\sigma_y)(y - \mu_y)$ and variance $\sigma_x^2(1 - \rho^2)$.

The quantity ρ is called the correlation between X and Y. It can be shown that

$$\rho = \frac{E[(X - \mu_x)(Y - \mu_y)]}{\sigma_x\sigma_y}$$

$$= \frac{\text{Cov}(X, Y)}{\sigma_x\sigma_y}$$

17. Let Y be a gamma random variable with parameters (s, α). That is, its density is

$$f_Y(y) = Ce^{-\alpha y}y^{s-1}, \qquad y > 0$$

where C is a constant that does not depend on y. Suppose also that the conditional distribution of X given that $Y = y$ is Poisson with mean y. That is,

$$P\{X = i \mid Y = y\} = e^{-y}y^i/i!, \qquad i \geqslant 0$$

Show that the conditional distribution of Y given that $X = i$ is the gamma distribution with parameters $(s + i, \alpha + 1)$.

18. Let $X_1, \ldots, X_n$ be independent random variables having a common distribution function that is specified up to an unknown parameter θ. Let $T = T(\mathbf{X})$ be a function of the data $\mathbf{X} = (X_1, \ldots, X_n)$. If the conditional distribution of $X_1, \ldots, X_n$ given $T(\mathbf{X})$ does not depend on θ then $T(\mathbf{X})$ is said to be a *sufficient statistic* for θ. In the following cases, show that $T(\mathbf{X}) = \Sigma_{i=1}^{n} X_i$ is a sufficient statistic for θ.

(a) The X_i are normal with mean θ and variance 1.

(b) The density of X_i is $f(x) = \theta e^{-\theta x}$, $x > 0$.
(c) The mass function of X_i is $p(x) = \theta^x(1 - \theta)^{1-x}$, $x = 0, 1, 0 < \theta < 1$.
(d) The X_i are Poisson random variables with mean θ.

*19. Prove that if X and Y are jointly continuous, then

$$E[X] = \int_{-\infty}^{\infty} E[X \mid Y = y] f_Y(y) \, dy$$

20. An individual whose level of exposure to a certain pathogen is x will contract the disease caused by this pathogen with probability $P(x)$. If the exposure level of a randomly chosen member of the population has probability density function f, determine the conditional probability density of the exposure level of that member given that he or she

(a) has the disease,
(b) does not have the disease.
(c) Show that when $P(x)$ increases in x, then the ratio of the density of part (a) to that of part (b) also increases in x.

21. Consider Example 3.12 which refers to a miner trapped in a mine. Let N denote the total number of doors selected before the miner reaches safety. Also, let T_i denote the travel time corresponding to the ith choice, $i \geqslant 1$. Again let X denote the time when the miner reaches safety.

(a) Give an identity that relates X to N and the T_i.
(b) What is $E[N]$?
(c) What is $E[T_N]$?
(d) What is $E[\sum_{i=1}^{N} T_i \mid N = n]$?
(e) Using the preceding, what is $E[X]$?

22. Suppose that independent trials, each of which is equally likely to have any of m possible outcomes, are performed until the same outcome occurs k consecutive times. If N denotes the number of trials, show that

$$E[N] = \frac{m^k - 1}{m - 1}$$

Some people believe that the successive digits in the expansion of $\pi = 3.14159\ldots$ are "uniformly" distributed. That is, they believe that these digits have all the appearance of being independent choices from a distribution that is equally likely to be any of the digits from 0 through 9. Possible evidence against this hypothesis is the fact that starting with the 24,658,601st digit there is a run of nine successive 7s. Is this information consistent with the hypothesis of a uniform distribution?
To answer this, we note from the preceding that if the uniform hypothesis

were correct, then the expected number of digits until a run of nine of the same value occurs is

$$(10^9 - 1)/9 = 111,111,111$$

Thus, the actual value of approximately 25 million is roughly 22 percent of the theoretical mean. However, it can be shown that under the uniformity assumption the standard deviation of N will be approximately equal to the mean. As a result, the observed value is approximately 0.78 standard deviations less than its theoretical mean and is thus quite consistent with the uniformity assumption.

*23. A coin having probability p of coming up heads is successively flipped until 2 of the most recent 3 flips are heads. Let N denote the number of flips. (Note that if the first 2 flips are heads, then $N = 2$.) Find $E[N]$.

24. A coin, having probability p of landing heads, is continually flipped until at least one head and one tail have been flipped.

(a) Find the expected number of flips needed.
(b) Find the expected number of flips that land on heads.
(c) Find the expected number of flips that land on tails.
(d) Repeat part (a) in the case where flipping is continued until a total of at least two heads and one tail have been flipped.

25. A gambler wins each game with probability p. In each of the cases below, determine the expected total number of wins.

(a) The gambler will play n games; if he wins X of these games, then he will play an additional X games before stopping.
(b) The gambler will play until he wins; if it takes him Y games to get this win, then he will play an additional Y games.

26. You have two opponents with whom you alternate play. Whenever you play A, you win with probability p_A; whenever you play B, you win with probability p_B, where $p_B > p_A$. If your objective is to minimize the number of games you need play to win two in a row, should you start with A or with B?

Hint: Let $E[N_i]$ denote the mean number of games needed if you initially play i. Derive an expression for $E[N_A]$ that involves $E[N_B]$; write down the equivalent expression for $E[N_B]$ and then subtract.

27. A coin that comes up heads with probability p is continually flipped until the pattern T, T, H appears. (That is, you stop flipping when the most recent flip landed heads, and the two immediately preceding it landed tails.) Let X denote the number of flips made, and find $E[X]$.

28.   The random variables X and Y have the following joint probability mass function:

$$P\{X = i, Y = j\} = e^{-(a+bi)}\frac{(bi)^j}{j!}\frac{a^i}{i!}, \qquad i \geqslant 0, \ j \geqslant 0$$

(a) What is the conditional distribution of Y given that $X = i$?
(b) Find $\text{Cov}(X, Y)$.

29.   Two players take turns shooting at a target, with each shot by player i hitting the target with probability p_i, $i = 1, 2$. Shooting ends when two consecutive shots hit the target. Let μ_i denote the mean number of shots taken when player i shoots first, $i = 1, 2$.

(a) Find μ_1 and μ_2.
(b) Let h_i denote the mean number of times that the target is hit when player i shoots first, $i = 1, 2$. Find h_1 and h_2.

30.   Let X_i, $i \geqslant 0$ be independent and identically distributed random variables with probability mass function

$$p(j) = P\{X_i = j\}, \qquad j = 1, \ldots, m \qquad \sum_{j=1}^{m} P(j) = 1$$

Find $E[N]$, where $N = \min\{n > 0 : X_n = X_0\}$.

31.   A set of n dice is thrown. All those that land on six are put aside, and the others are again thrown. This is repeated until all the dice have landed on six. Let N denote the number of throws needed. (For instance, suppose that $n = 3$ and that on the initial throw exactly 2 of the dice land on six. Then the other die will be thrown, and if it lands on six, then $N = 2$.) Let $m_n = E[N]$.

(a) Derive a recursive formula for m_n and use it to calculate m_i, $i = 2, 3, 4$, and to show that $m_5 \approx 13.024$.
(b) Let X_i denote the number of dice rolled on the ith throw. Find $E[\sum_{i=1}^{N} X_i]$.

32.   If $E[X \mid Y = y] = c$ for all y, show that $\text{Cov}(X, Y) = 0$.

33.   Suppose that conditional on $Y = y$, the random variables X_1 and X_2 are independent with mean y. Show that

$$\text{Cov}(X_1, X_2) = \text{Var}(Y)$$

34.   An interval of length 1 is broken at a point uniformly distributed over $(0, 1)$. Find the expected length of the subinterval that contains the point x, $0 < x < 1$, and show that it is maximized when $x = \frac{1}{2}$.

35. A manuscript is sent to a typing firm consisting of typists A, B, and C. If it is typed by A, then the number of errors made is a Poisson random variable with mean 2.6; if typed by B, then the number of errors is a Poisson random variable with mean 3; and if typed by C, then it is a Poisson random variable with mean 3.4. Let X denote the number of errors in the typed manuscript. Assume that each typist is equally likely to do the work.

(a) Find $E[X]$
(b) Find $\text{Var}(X)$.

36. Let U be a uniform $(0, 1)$ random variable. Suppose that n trials are to be performed and that conditional on $U = u$ these trials will be independent with a common success probability u. Compute the mean and variance of the number of successes that occur in these trials.

37. A deck of n cards, numbered 1 through n, is randomly shuffled so that all $n!$ possible permutations are equally likely. The cards are then turned over one at a time until card number 1 appears. These upturned cards constitute the first cycle. We now determine (by looking at the upturned cards) the lowest numbered card that has not yet appeared, and we continue to turn the cards face up until that card appears. This new set of cards represents the second cycle. We again determine the lowest numbered of the remaining cards and turn the cards until it appears, and so on until all cards have been turned over. Let m_n denote the mean number of cycles.

(a) Derive a recursive formula for m_n in terms of m_k, $k = 1, \ldots, n - 1$.
(b) Starting with $m_0 = 0$, use the recursion to find m_1, m_2, m_3, and m_4.
(c) Conjecture a general formula for m_n.
(d) Prove your formula by induction on n. That is, show it is valid for $n = 1$, then assume it is true whenever k is any of the values $1, \ldots, n - 1$ and show that this implies it is true when $k = n$.
(e) Let X_i equal 1 if one of the cycles ends with card i, and let it equal 0 otherwise, $i = 1, \ldots, n$. Express the number of cycles in terms of these X_i.
(f) Use the representation in part (e) to determine m_n.
(g) Are the random variables $X_1, \ldots, X_n$ independent? Explain.
(h) Find the variance of the number of cycles.

38. A prisoner is trapped in a cell containing three doors. The first door leads to a tunnel that returns him to his cell after two days of travel. The second leads to a tunnel that returns him to his cell after three days of travel. The third door leads immediately to freedom.

(a) Assuming that the prisoner will always select doors 1, 2, and 3 with probabilities 0.5, 0.3, 0.2, what is the expected number of days until he reaches freedom?

(b) Assuming that the prisoner is always equally likely to choose among those doors that he has not used, what is the expected number of days until he reaches freedom? (In this version, for instance, if the prisoner initially tries door 1, then when he returns to the cell, he will now select only from doors 2 and 3.)

(c) For parts (a) and (b) find the variance of the number of days until the prisoner reaches freedom.

39. A rat is trapped in a maze. Initially he has to choose one of two directions. If he goes to the right, then he will wander around in the maze for three minutes and will then return to his initial position. If he goes to the left, then with probability $\frac{1}{3}$ he will depart the maze after two minutes of traveling, and with probability $\frac{2}{3}$ he will return to his initial position after five minutes of traveling. Assuming that the rat is at all times equally likely to go to the left or the right, what is the expected number of minutes that he will be trapped in the maze?

40. Find the variance of the amount of time the rat spends in the maze in Exercise 39.

41. The number of claims received at an insurance company during a week is a random variable with mean μ_1 and variance σ_1^2. The amount paid in each claim is a random variable with mean μ_2 and variance σ_2^2. Find the mean and variance of the amount of money paid by the insurance company each week. What independence assumptions are you making? Are these assumptions reasonable?

42. The number of customers entering a store on a given day is Poisson distributed with mean $\lambda = 10$. The amount of money spent by a customer is uniformly distributed over $(0, 100)$. Find the mean and variance of the amount of money that the store takes in on a given day.

43. The conditional variance of X, given the random variable Y, is defined by

$$\text{Var}(X \mid Y) = E[[X - E(X \mid Y)]^2 \mid Y]$$

Show that

$$\text{Var}(X) = E[\text{Var}(X \mid Y)] + \text{Var}(E[X \mid Y])$$

*44. Use Exercise 43 to give another proof of the fact that

$$\text{Var}\left(\sum_{i=1}^{N} X_i \right) = E[N] \, \text{Var}(X) + (E[X])^2 \, \text{Var}(N)$$

45. An individual traveling on the real line is trying to reach the origin.

However, the larger the desired step, the greater is the variance in the result of that step. Specifically, whenever the person is at location x, he next moves to a location having mean 0 and variance βx^2. Let X_n denote the position of the individual after having taken n steps. Supposing that $X_0 = x_0$, find

(a) $E[X_n]$
(b) $\text{Var}(X_n)$

46. (a) Show that

$$\text{Cov}(X, Y) = \text{Cov}(X, E[Y|X])$$

(b) Suppose, that, for constants a and b,

$$E[Y|X] = a + bX$$

Show that

$$b = \text{Cov}(X, Y)/\text{Var}(X)$$

*47. If $E[Y|X] = 1$, show that

$$\text{Var}(XY) \geqslant \text{Var}(X)$$

48. Give another proof of Exercise 44 by computing the moment generating function of $\Sigma_{i=1}^{N} X_i$ and then differentiating to obtain its moments.

Hint: Let

$$\phi(t) = E\left[\exp\left(t \sum_{i=1}^{N} X_i\right)\right]$$

$$= E\left[E\left[\exp\left(t \sum_{i=1}^{N} X_i\right)\middle| N\right]\right]$$

Now,

$$E\left[\exp\left(t \sum_{i=1}^{N} X_i\right)\middle| N = n\right] = E\left[\exp\left(t \sum_{i=1}^{n} X_i\right)\right] = (\phi_X(t))^n$$

since N is independent of the Xs where $\phi_X(t) = E[e^{tX}]$ is the moment generating function for the Xs. Therefore,

$$\phi(t) = E[(\phi_X(t))^N]$$

Differentiation yields

$$\phi'(t) = E[N(\phi_X(t))^{N-1}\phi'_X(t)],$$

$$\phi''(t) = E[N(N-1)(\phi_X(t))^{N-2}(\phi'_X(t))^2 + N(\phi_X(t))^{N-1}\phi''_X(t)]$$

Evaluate at $t = 0$ to get the desired result.

49. The number of fish that Elise catches in a day is a Poisson random variable with mean 30. However, on the average, Elise tosses back two out of every three fish she catches. What is the probability that, on a given day, Elise takes home n fish. What is the mean and variance of (a) the number of fish she catches, (b) the number of fish she takes home? (What independence assumptions have you made?)

50. There are three coins in a barrel. These coins, when flipped, will come up heads with respective probabilities 0.3, 0.5, 0.7. A coin is randomly selected from among these three and is then flipped ten times. Let N be the number of heads obtained on the ten flips. Find

(a) $P\{N = 0\}$.
(b) $P\{N = n\}$, $n = 0, 1, \ldots, 10$.
(c) Does N have a binomial distribution?
(d) If you win \$1 each time a head appears and you lose \$1 each time a tail appears, is this a fair game? Explain.

51. Do Exercise 50 under the assumption that each time a coin is flipped, it is then put back in the barrel and another coin is randomly selected. Does N have a binomial distribution now?

52. Explain the relationship between the general formula

$$P(E) = \sum_y P(E \mid Y = y)P(Y = y)$$

and Bayes' formula.

*53. Suppose X is a Poisson random variable with mean λ. The parameter λ is itself a random variable whose distribution is exponential with mean 1. Show that $P\{X = n\} = (\frac{1}{2})^{n+1}$.

54. A coin is randomly selected from a group of ten coins, the nth coin having a probability $n/10$ of coming up heads. The coin is then repeatedly flipped until a head appears. Let N denote the number of flips necessary. What is the probability distribution of N? Is N a geometric random variable? When would N be a geometric random variable; that is, what would have to be done differently?

55. A collection of n coins is flipped. The outcomes are independent, and the ith coin comes up heads with probability α_i, $i = 1, \ldots, n$. Suppose that for some value of j, $1 \leqslant j \leqslant n$, $\alpha_j = \frac{1}{2}$. Find the probability that the total number of heads to appear on the n coins is an even number.

56. Let A and B be mutually exclusive events of an experiment. If independent replications of the experiment are continually performed, what is the probability that A occurs before B?

*57. Two players alternate flipping a coin that comes up heads with probability p. The first one to obtain a head is declared the winner. We are interested in the probability that the first player to flip is the winner. Before determining this probability, which we will call $f(p)$, answer the following questions.

(a) Do you think that $f(p)$ is a monotone function of p? If so, is it increasing or decreasing?
(b) What do you think is the value of $\lim_{p \to 1} f(p)$?
(c) What do you think is the value of $\lim_{p \to 0} f(p)$?
(d) Find $f(p)$.

58. Suppose in Exercise 29 that the shooting ends when the target has been hit twice. Let m_i denote the mean number of shots needed for the first hit when player i shoots first, $i = 1, 2$. Also, let P_i, $i = 1, 2$, denote the probability that the first hit is by player 1, when player i shoots first.

(a) Find m_1 and m_2.
(b) Find P_1 and P_2.

For the remainder of the problem, assume that player 1 shoots first.

(c) Find the probability that the final hit was by 1.
(d) Find the probability that both hits were by 1.
(e) Find the probability that both hits were by 2.
(f) Find the mean number of shots taken.

59. A, B, and C are evenly matched tennis players. Initially A and B play a set, and the winner then plays C. This continues, with the winner always playing the waiting player, until one of the players has won two sets in a row. That player is then declared the overall winner. Find the probability that A is the overall winner.

60. Let X_1 and X_2 be independent geometric random variables with respective parameters p_1 and p_2. Find $P\{|X_1 - X_2| \leq 1\}$.

61. A and B roll a pair of dice in turn, with A rolling first. A's objective is to obtain a sum of 6, and B's is to obtain a sum of 7. The game ends when either player reaches his or her objective, and that player is declared the winner.

(a) Find the probability that A is the winner.
(b) Find the expected number of rolls of the dice.
(c) Find the variance of the number of rolls of the dice.

62. The number of red balls in an urn that contains n balls is a random variable that is equally likely to be any of the values $0, 1, \ldots, n$. That is,

$$P\{i \text{ red, } n - i \text{ non-red}\} = \frac{1}{n + 1}, \qquad i = 0, \ldots, n$$

The n balls are then randomly removed one at a time. Let Y_k denote the number of red balls in the first k selections, $k = 1, \ldots, n$.

(a) Find $P\{Y_n = j\}, j = 0, \ldots, n$.
(b) Find $P\{Y_{n-1} = j\}, j = 0, \ldots, n$.
(c) What do you think is the value of $P\{Y_k = j\}, j = 0, \ldots, n$?
(d) Verify your answer to part (c) by a backwards induction argument. That is, check that your answer is correct when $k = n$, and then show that whenever it is true for k it is also true for $k - 1, k = 1, \ldots, n$.

63. The opponents of soccer team A are of two types: either they are a class 1 or a class 2 team. The number of goals team A scores against a class i opponent is a Poisson random variable with mean λ_i, where $\lambda_1 = 2, \lambda_2 = 3$. This weekend the team has two games against teams they are not very familiar with. Assuming that the first team they play is a class 1 team with probability 0.6 and the second is, independently of the class of the first team, a class 1 team with probability 0.3, determine

(a) the expected number of goals team A will score this weekend.
(b) the probability that team A will score a total of 5 goals.

*64. A coin having probability p of coming up heads is continually flipped. Let $P_j(n)$ denote the probability that a run of j successive heads occurs within the first n flips.

(a) Argue that

$$P_j(n) = P_j(n - 1) + p^j(1 - p)[1 - P_j(n - j - 1)]$$

(b) By conditioning on the first non-head to appear, derive another equation relating $P_j(n)$ to the quantities $P_j(n - k), k = 1, \ldots, j$.

65. In a knockout tennis tournament of 2^n contestants, the players are paired and play a match. The losers depart, the remaining 2^{n-1} players are paired, and they play a match. This continues for n rounds, after which a single player remains unbeaten and is declared the winner. Suppose that the contestants are numbered 1 through 2^n, and that whenever two players contest a match, the lower numbered one wins with probability p. Also suppose that the pairings of the remaining players are always done at random so that all possible pairings for that round are equally likely.

(a) What is the probability that player 1 wins the tournament?
(b) What is the probability that player 2 wins the tournament?

Hint: Imagine that the random pairings are done in advance of the tournament. That is, the first-round pairings are randomly determined; the 2^{n-1} first-round pairs are then themselves randomly paired, with the winners of each pair to play in round 2; these 2^{n-2} groupings (of 4 players each) are then randomly paired, with the winners of each grouping to play in round 3, and so on. Say that players i and j are scheduled to meet in round k if, provided they both win their first $k-1$ matches, they will meet in round k. Now condition on the round in which players 1 and 2 are scheduled to meet.

66. In the match problem, say that (i, j), $i < j$, is a pair if i chooses j's hat and j chooses i's hat.

(a) Find the expected number of pairs.
(b) Let Q_n denote the probability that there are no pairs, and derive a recursive formula for Q_n in terms of Q_j, $j < n$.

Hint: Use the cycle concept.

(c) Use the recursion of part (b) to find Q_8.

67. Let N denote the number of cycles that result in the match problem.

(a) Let $M_n = E[N]$, and derive an equation for M_n in terms of $M_1, \ldots, M_{n-1}$.
(b) Let C_j denote the size of the cycle that contains person j. Argue that

$$N = \sum_{j=1}^{n} 1/C_j$$

and use the preceding to determine $E[N]$.
(c) Find the probability that persons $1, 2, \ldots, k$ are all in the same cycle.
(d) Find the probability that $1, 2, \ldots, k$ is a cycle.

68. Use the equation following (3.14) to obtain Equation (3.10).

Hint: First multiply both sides of that equation (3.14) by n, then write a new equation by replacing n by $n-1$, and then subtract the former from the latter.

69. In Example 3.24 show that the conditional distribution of N given that $U_1 = y$ is the same as the conditional distribution of M given that $U_1 = 1 - y$. Also, show that

$$E[N \mid U_1 = y] = E[M \mid U_1 = 1 - y] = 1 + e^y$$

*70. Suppose that we continually roll a die until the sum of all throws exceeds 100. What is the most likely value of this total when you stop?

71. There are five components. The components act independently, with component i working with probability p_i, $i = 1, 2, 3, 4, 5$. These components form a system as shown in Figure 3.7.

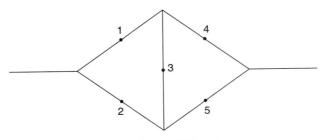

Figure 3.7.

The system is said to work if a signal originating at the left end of the diagram can reach the right end, where it can only pass through a component if that component is working. (For instance, if components 1 and 4 both work, then the system also works.) What is the probability that the system works?

72. This problem will present another proof of the ballot problem of Example 3.23.

(a) Argue that

$$P_{n,m} = 1 - P\{A \text{ and } B \text{ are tied at some point}\}$$

(b) Explain why

$$P\{A \text{ receives first vote and they are eventually tied}\}$$

$$= P\{B \text{ receives first vote and they are eventually tied}\}$$

Hint: Any outcome in which they are eventually tied with A receiving the first vote corresponds to an outcome in which they are eventually tied with B receiving the first vote. Explain this correspondence.

(c) Argue that $P\{\text{eventually tied}\} = 2m/(n + m)$, and conclude that $P_{n,m} = (n - m)/(n + m)$.

73. Consider a gambler who on each bet either wins 1 with probability 18/38 or loses 1 with probability 20/38. (These are the probabilities if the bet is that a roulette wheel will land on a specified color.) The gambler will quit either when he is winning a total of 5 or after 100 plays. What is the probability he or she plays exactly 15 times?

74. Show that

(a) $E[XY|Y = y] = yE[X|Y = y]$
(b) $E[g(X, Y)|Y = y] = E[g(X, y)|Y = y]$
(c) $E[XY] = E[YE[X|Y]]$

75. In the ballot problem (Example 3.23), compute $P\{A \text{ is never behind}\}$.

76. An urn contains n white and m black balls which are removed one at a time. If $n > m$, show that the probability that there are always more white than black balls in the urn (until, of course, the urn is empty) equals $(n - m)/(n + m)$. Explain why this probability is equal to the probability that the set of withdrawn balls always contains more white than black balls. [This latter probability is $(n - m)/(n + m)$ by the ballot problem.]

77. A coin that comes up heads with probability p is flipped n consecutive times. What is the probability that starting with the first flip there are always more heads than tails that have appeared?

78. Let $X_i, i \geq 1$, be independent uniform $(0, 1)$ random variables, and define N by

$$N = \min\{n: X_n < X_{n-1}\}$$

where $X_0 = x$. Let $f(x) = E[N]$.

(a) Derive an integral equation for $f(x)$ by conditioning on X_1.
(b) Differentiate both sides of the equation derived in part (a).
(c) Solve the resulting equation obtained in part (b).
(d) For a second approach to determining $f(x)$ argue that

$$P\{N \geq k\} = \frac{(1 - x)^{k-1}}{(k - 1)!}$$

(e) Use part (d) to obtain $f(x)$.

79. Let $X_1, X_2, \ldots$ be independent continuous random variables with a common distribution function F and density $f = F'$, and for $k \geq 1$ let

$$N_k = \min\{n \geq k: X_n = k\text{th largest of } X_1, \ldots, X_n\}$$

(a) Show that $P\{N_k = n\} = (k - 1)/n(n - 1), n \geq k$.
(b) Argue that

$$f_{X_{N_k}}(x) = f(x)(\bar{F}(x))^{k-1} \sum_{i=0}^{\infty} \binom{i + k - 2}{i}(F(x))^i$$

(c) Prove the following identity:

$$a^{1-k} = \sum_{i=0}^{\infty} \binom{i+k-2}{i} (1-a)^i, \qquad 0 < a < 1, k \geq 2$$

Hint: Use induction. First prove it when $k = 2$, and then assume it for k. To prove it for $k + 1$, use the fact that

$$\sum_{i=1}^{\infty} \binom{i+k-1}{i} (1-a)^i = \sum_{i=1}^{\infty} \binom{i+k-2}{i} (1-a)^i$$
$$+ \sum_{i=1}^{\infty} \binom{i+k-2}{i-1} (1-a)^i$$

where the preceding used the combinatorial identity

$$\binom{m}{i} = \binom{m-1}{i} + \binom{m-1}{i-1}$$

Now, use the induction hypothesis to evaluate the first term on the right side of the preceding equation.

(d) Conclude that X_{N_k} has distribution F.

80. An urn contains n balls, with ball i having weight w_i, $i = 1, \ldots, n$. The balls are withdrawn from the urn one at a time according to the following scheme: When S is the set of balls that remains, ball $i, i \in S$, is the next ball withdrawn with probability $w_i / \sum_{j \in S} w_j$. Find the expected number of balls that are withdrawn before ball i, $i = 1, \ldots, n$.

81. In the list example of Section 3.6.1 suppose that the initial ordering at time $t = 0$ is determined completely at random; that is, initially all $n!$ permutations are equally likely. Following the front-of-the-line rule, compute the expected position of the element requested at time t.

Hint: To compute $P\{e_j \text{ precedes } e_i \text{ at time } t\}$ condition on whether or not either e_i or e_j has ever been requested prior to t.

82. In the list problem, when the P_i are known, show that the best ordering (best in the sense of minimizing the expected position of the element requested) is to place the elements in decreasing order of their probabilities. That is, if $P_1 > P_2 > \cdots > P_n$, show that $1, 2, \ldots, n$ is the best ordering.

83. Consider the random graph of Section 3.6.2 when $n = 5$. Compute the probability distribution of the number of components and verify your solution by using it to compute $E[C]$ and then comparing your solution with

$$E[C] = \sum_{k=1}^{5} \binom{5}{k} \frac{(k-1)!}{5^k}$$

84. (a) From the results of Section 3.6.3 we can conclude that there are $\binom{n+m-1}{m-1}$ nonnegative integer valued solutions of the equation $x_1 + \cdots + x_m = n$. Prove this directly.

(b) How many positive integer valued solutions of $x_1 + \cdots + x_m = n$ are there?

Hint: Let $y_i = x_i - 1$.

(c) For the Bose–Einstein distribution, compute the probability that exactly k of the X_i are equal to 0.

85. In Section 3.6.3, we saw that if U is a random variable that is uniform on $(0, 1)$ and if, conditional on $U = p$, X is binomial with parameters n and p, then

$$P\{X = i\} = \frac{1}{n+1}, \qquad i = 0, 1, \ldots, n$$

For another way of showing this result, let $U, X_1, X_2, \ldots, X_n$ be independent uniform $(0, 1)$ random variables. Define X by

$$X = \#i : X_i < U$$

That is, if the $n + 1$ variables are ordered from smallest to largest, then U would be in position $X + 1$.

(a) What is $P\{X = i\}$?
(b) Explain how this proves the result of Exercise 85.

86. Let $I_1, \ldots, I_n$ be independent random variables, each of which is equally likely to be either 0 or 1. A well-known nonparametric statistical test (called the signed rank test) is concerned with determining $P_n(k)$ defined by

$$P_n(k) = P\left\{ \sum_{j=1}^{n} jI_j \leq k \right\}$$

Justify the following formula:

$$P_n(k) = \tfrac{1}{2}P_{n-1}(k) + \tfrac{1}{2}P_{n-1}(k - n)$$

Markov Chains

4

◆◆

4.1. Introduction

In this chapter, we consider a stochastic process $\{X_n, n = 0, 1, 2, \ldots\}$ that takes on a finite or countable number of possible values. Unless otherwise mentioned, this set of possible values of the process will be denoted by the set of nonnegative integers $\{0, 1, 2, \ldots\}$. If $X_n = i$, then the process is said to be in state i at time n. We suppose that whenever the process is in state i, there is a fixed probability P_{ij} that it will next be in state j. That is, we suppose that

$$P\{X_{n+1} = j \mid X_n = i, X_{n-1} = i_{n-1}, \ldots, X_1 = i_1, X_0 = i_0\} = P_{ij} \quad (4.1)$$

for all states $i_0, i_1, \ldots, i_{n-1}, i, j$ and all $n \geq 0$. Such a stochastic process is known as a *Markov chain*. Equation (4.1) may be interpreted as stating that, for a Markov chain, the conditional distribution of any future state X_{n+1} given the past states $X_0, X_1, \ldots, X_{n-1}$ and the present state X_n, is independent of the past states and depends only on the present state.

The value P_{ij} represents the probability that the process will, when in state i, next make a transition into state j. Since probabilities are nonnegative and since the process must make a transition into some state, we have that

$$P_{ij} \geq 0, \qquad i, j \geq 0; \qquad \sum_{j=0}^{\infty} P_{ij} = 1, \qquad i = 0, 1, \ldots.$$

Let $\mathbf{P}$ denote the matrix of one-step transition probabilities P_{ij}, so that

$$\mathbf{P} = \left\| \begin{array}{cccc} P_{00} & P_{01} & P_{02} & \cdots \\ P_{10} & P_{11} & P_{12} & \cdots \\ \vdots & & & \\ P_{i0} & P_{i1} & P_{i2} & \cdots \\ \vdots & \vdots & \vdots & \end{array} \right\|$$

Example 4.1   (Forecasting the Weather): Suppose that the chance of rain tomorrow depends on previous weather conditions only through whether or not it is raining today and not on past weather conditions. Suppose also that if it rains today, then it will rain tomorrow with probability α; and if it does not rain today, then it will rain tomorrow with probability β.

If we say that the process is in state 0 when it rains and state 1 when it does not rain, then the above is a two-state Markov chain whose transition probabilities are given by

$$\mathbf{P} = \left\|\begin{matrix} \alpha & 1 - \alpha \\ \beta & 1 - \beta \end{matrix}\right\| \quad \blacklozenge$$

Example 4.2   (A Communications System): Consider a communications system which transmits the digits 0 and 1. Each digit transmitted must pass through several stages, at each of which there is a probability p that the digit entered will be unchanged when it leaves. Letting X_n denote the digit entering the nth stage, then $\{X_n, n = 0, 1, \ldots\}$ is a two-state Markov chain having a transition probability matrix

$$\mathbf{P} = \left\|\begin{matrix} p & 1 - p \\ 1 - p & p \end{matrix}\right\| \quad \blacklozenge$$

Example 4.3   On any given day Gary is either cheerful (C), so-so (S), or glum (G). If he is cheerful today, then he will be C, S, or G tomorrow with respective probabilities 0.5, 0.4, 0.1. If he is feeling so-so today, then he will be C, S, or G tomorrow with probabilities 0.3, 0.4, 0.3. If he is glum today, then he will be C, S, or G tomorrow with probabilities 0.2, 0.3, 0.5.

Letting X_n denote Gary's mood on the nth day, then $\{X_n, n \geq 0\}$ is a three-state Markov chain (state $0 = C$, state $1 = S$, state $2 = G$) with transition probability matrix

$$\mathbf{P} = \left\|\begin{matrix} 0.5 & 0.4 & 0.1 \\ 0.3 & 0.4 & 0.3 \\ 0.2 & 0.3 & 0.5 \end{matrix}\right\| \quad \blacklozenge$$

Example 4.4   (Transforming a Process into a Markov Chain): Suppose that whether or not it rains today depends on previous weather conditions through the last two days. Specifically, suppose that if it has rained for the past two days, then it will rain tomorrow with probability 0.7; if it rained today but not yesterday, then it will rain tomorrow with probability 0.5; if it rained yesterday but not today, then it will rain tomorrow with probability 0.4; if it has not rained in the past two days, then it will rain tomorrow with probability 0.2.

If we let the state at time n depend only on whether or not it is raining at time n, then the above model is not a Markov chain (why not?). However, we can transform the above model into a Markov chain by saying that the state at any time is determined by the weather conditions during both that day and the previous day. In other words, we can say that the process is in

state 0   if it rained both today and yesterday,
state 1   if it rained today but not yesterday,
state 2   if it rained yesterday but not today,
state 3   if it did not rain either yesterday or today.

The preceding would then represent a four-state Markov chain having a transition probability matrix

$$\mathbf{P} = \begin{Vmatrix} 0.7 & 0 & 0.3 & 0 \\ 0.5 & 0 & 0.5 & 0 \\ 0 & 0.4 & 0 & 0.6 \\ 0 & 0.2 & 0 & 0.8 \end{Vmatrix}$$

The reader should carefully check the matrix $\mathbf{P}$, and make sure he or she understands how it was obtained.   ◆

Example 4.5 (A Random Walk Model): A Markov chain whose state space is given by the integers $i = 0, \pm 1, \pm 2, \ldots$ is said to be a random walk if, for some number $0 < p < 1$,

$$P_{i,i+1} = p = 1 - P_{i,i-1}, \qquad i = 0, \pm 1, \ldots$$

The preceding Markov chain is called a *random walk* for we may think of it as being a model for an individual walking on a straight line who at each point of time either takes one step to the right with probability p or one step to the left with probability $1 - p$.   ◆

Example 4.6 (A Gambling Model): Consider a gambler who, at each play of the game, either wins \$1 with probability p or loses \$1 with probability $1 - p$. If we suppose that our gambler quits playing either when he goes broke or he attains a fortune of \$N, then the gambler's fortune is a Markov chain having transition probabilities

$$P_{i,i+1} = p = 1 - P_{i,i-1}, \qquad i = 1, 2, \ldots, N - 1$$
$$P_{00} = P_{NN} = 1$$

States 0 and N are called *absorbing* states since once entered they are never left. Note that the above is a finite state random walk with absorbing barriers (states 0 and N).   ◆

4.2.   Chapman–Kolmogorov Equations

We have already defined the one-step transition probabilities P_{ij}. We now define the n-step transition probabilities P_{ij}^n to be the probability that a process in state i will be in state j after n additional transitions. That is,

$$P_{ij}^n = P\{X_{n+m} = j \mid X_m = i\}, \qquad n \geqslant 0, i, j \geqslant 0$$

Of course $P_{ij}^1 = P_{ij}$. The *Chapman–Kolmogorov equations* provide a method for computing these n-step transition probabilities. These equations are

$$P_{ij}^{n+m} = \sum_{k=0}^{\infty} P_{ik}^n P_{kj}^m \qquad \text{for all } n, m \geqslant 0, \text{ all } i, j \tag{4.2}$$

and are most easily understood by noting that $P_{ik}^n P_{kj}^m$ represents the probability that starting in i the process will go to state j in $n + m$ transitions through a path which takes it into state k at the nth transition. Hence, summing over all intermediate states k yields the probability that the process will be in state j after $n + m$ transitions. Formally, we have

$$
\begin{aligned}
P_{ij}^{n+m} &= P\{X_{n+m} = j \mid X_0 = i\} \\
&= \sum_{k=0}^{\infty} P\{X_{n+m} = j, X_n = k \mid X_0 = i\} \\
&= \sum_{k=0}^{\infty} P\{X_{n+m} = j \mid X_n = k, X_0 = i\} P\{X_n = k \mid X_0 = i\} \\
&= \sum_{k=0}^{\infty} P_{kj}^m P_{ik}^n
\end{aligned}
$$

If we let $\mathbf{P}^{(n)}$ denote the matrix of n-step transition probabilities P_{ij}^n, then Equation (4.2) asserts that

$$\mathbf{P}^{(n+m)} = \mathbf{P}^{(n)} \cdot \mathbf{P}^{(m)}$$

where the dot represents matrix multiplication.* Hence, in particular,

$$\mathbf{P}^{(2)} = \mathbf{P}^{(1+1)} = \mathbf{P} \cdot \mathbf{P} = \mathbf{P}^2$$

and by induction

$$\mathbf{P}^{(n)} = \mathbf{P}^{(n-1+1)} = \mathbf{P}^{n-1} \cdot \mathbf{P} = \mathbf{P}^n$$

That is, the n-step transition matrix may be obtained by multiplying the matrix $\mathbf{P}$ by itself n times.

---

*If $\mathbf{A}$ is an $N \times M$ matrix whose element in the ith row and jth column is a_{ij} and $\mathbf{B}$ is a $M \times K$ matrix whose element in the ith row and jth column is b_{ij}, then $\mathbf{A} \cdot \mathbf{B}$ is defined to be the $N \times K$ matrix whose element in the ith row and jth column is $\sum_{k=1}^{M} a_{ik} b_{kj}$.

Example 4.7 Consider Example 4.1 in which the weather is considered as a two-state Markov chain. If $\alpha = 0.7$ and $\beta = 0.4$, then calculate the probability that it will rain four days from today given that it is raining today.

Solution: The one-step transition probability matrix is given by

$$\mathbf{P} = \begin{Vmatrix} 0.7 & 0.3 \\ 0.4 & 0.6 \end{Vmatrix}$$

Hence,

$$\mathbf{P}^{(2)} = \mathbf{P}^2 = \begin{Vmatrix} 0.7 & 0.3 \\ 0.4 & 0.6 \end{Vmatrix} \cdot \begin{Vmatrix} 0.7 & 0.3 \\ 0.4 & 0.6 \end{Vmatrix}$$

$$= \begin{Vmatrix} 0.61 & 0.39 \\ 0.52 & 0.48 \end{Vmatrix},$$

$$\mathbf{P}^{(4)} = (\mathbf{P}^2)^2 = \begin{Vmatrix} 0.61 & 0.39 \\ 0.52 & 0.48 \end{Vmatrix} \cdot \begin{Vmatrix} 0.61 & 0.39 \\ 0.52 & 0.48 \end{Vmatrix}$$

$$= \begin{Vmatrix} 0.5749 & 0.4251 \\ 0.5668 & 0.4332 \end{Vmatrix}$$

and the desired probability P_{00}^4 equals 0.5749. ✦

Example 4.8 Consider Example 4.4. Given that it rained on Monday and Tuesday, what is the probability that it will rain on Thursday?

Solution: The two-step transition matrix is given by

$$\mathbf{P}^{(2)} = \mathbf{P}^2 = \begin{Vmatrix} 0.7 & 0 & 0.3 & 0 \\ 0.5 & 0 & 0.5 & 0 \\ 0 & 0.4 & 0 & 0.6 \\ 0 & 0.2 & 0 & 0.8 \end{Vmatrix} \cdot \begin{Vmatrix} 0.7 & 0 & 0.3 & 0 \\ 0.5 & 0 & 0.5 & 0 \\ 0 & 0.4 & 0 & 0.6 \\ 0 & 0.2 & 0 & 0.8 \end{Vmatrix}$$

$$= \begin{Vmatrix} 0.49 & 0.12 & 0.21 & 0.18 \\ 0.35 & 0.20 & 0.15 & 0.30 \\ 0.20 & 0.12 & 0.20 & 0.48 \\ 0.10 & 0.16 & 0.10 & 0.64 \end{Vmatrix}$$

Since rain on Thursday is equivalent to the process being in either state 0 or state 1 on Thursday, the desired probability is given by $P_{00}^2 + P_{01}^2 = 0.49 + 0.12 = 0.61$. ✦

So far, all of the probabilities we have considered are conditional probabilities. For instance, P^n_{ij} is the probability that the state at time n is j *given* that the initial state at time 0 is i. If the unconditional distribution of the state at time n is desired, it is necessary to specify the probability distribution of the initial state. Let us denote this by

$$\alpha_i \equiv P\{X_0 = i\}, \qquad i \geqslant 0 \left(\sum_{i=0}^{\infty} \alpha_i = 1 \right)$$

All unconditional probabilities may be computed by conditioning on the initial state. That is,

$$P\{X_n = j\} = \sum_{i=0}^{\infty} P\{X_n = j \mid X_0 = i\} P\{X_0 = i\}$$

$$= \sum_{i=0}^{\infty} P^n_{ij} \alpha_i$$

For instance, if $\alpha_0 = 0.4$, $\alpha_1 = 0.6$, in Example 4.7, then the (unconditional) probability that it will rain four days after we begin keeping weather records is

$$P\{X_4 = 0\} = 0.4 P^4_{00} + 0.6 P^4_{10}$$

$$= (0.4)(0.5749) + (0.6)(0.5668)$$

$$= 0.5700$$

4.3. Classification of States

State j is said to be *accessible* from state i if $P^n_{ij} > 0$ for some $n \geqslant 0$. Note that this implies that state j is accessible from state i if and only if, starting in i, it is possible that the process will ever enter state j. This is true since if j is not accessible from i, then

$$P\{\text{ever enter } j \mid \text{start in } i\} = P\left\{ \bigcup_{n=0}^{\infty} \{X_n = j\} \mid X_0 = i \right\}$$

$$\leqslant \sum_{n=0}^{\infty} P\{X_n = j \mid X_0 = i\}$$

$$= \sum_{n=0}^{\infty} P^n_{ij}$$

$$= 0$$

Two states i and j that are accessible to each other are said to *communicate*, and we write $i \leftrightarrow j$.

Note that any state communicates with itself since, by definition,

$$P_{ii}^0 = P\{X_0 = i \,|\, X_0 = i\} = 1$$

The relation of communication satisfies the following three properties:

(i) State i communicates with state i, all $i \geqslant 0$.
(ii) If state i communicates with state j, then state j communicates with state i.
(iii) If state i communicates with state j, and state j communicates with state k, then state i communicates with state k.

Properties (i) and (ii) follow immediately from the definition of communication. To prove (iii) suppose that i communicates with j, and j communicates with k. Thus, there exist integers n and m such that $P_{ij}^n > 0$, $P_{jk}^m > 0$. Now by the Chapman–Kolmogorov equations, we have that

$$P_{ik}^{n+m} = \sum_{r=0}^{\infty} P_{ir}^n P_{rk}^m \geqslant P_{ij}^n P_{jk}^m > 0$$

Hence, state k is accessible from state i. Similarly, we can show that state i is accessible from state k. Hence, states i and k communicate.

Two states that communicate are said to be in the same *class*. It is an easy consequence of (i), (ii), and (iii) that any two classes of states are either identical or disjoint. In other words, the concept of communication divides the state space up into a number of separate classes. The Markov chain is said to be *irreducible* if there is only one class, that is, if all states communicate with each other.

Example 4.9   Consider the Markov chain consisting of the three states 0, 1, 2 and having transition probability matrix

$$\mathbf{P} = \begin{Vmatrix} \frac{1}{2} & \frac{1}{2} & 0 \\ \frac{1}{2} & \frac{1}{4} & \frac{1}{4} \\ 0 & \frac{1}{3} & \frac{2}{3} \end{Vmatrix}$$

It is easy to verify that this Markov chain is irreducible. For example, it is possible to go from state 0 to state 2 since

$$0 \to 1 \to 2$$

That is, one way of getting from state 0 to state 2 is to go from state 0 to state 1 (with probability $\frac{1}{2}$) and then go from state 1 to state 2 (with probability $\frac{1}{4}$). ✦

Example 4.10 Consider a Markov chain consisting of the four states 0, 1, 2, 3 and having transition probability matrix

$$\mathbf{P} = \begin{Vmatrix} \frac{1}{2} & \frac{1}{2} & 0 & 0 \\ \frac{1}{2} & \frac{1}{2} & 0 & 0 \\ \frac{1}{4} & \frac{1}{4} & \frac{1}{4} & \frac{1}{4} \\ 0 & 0 & 0 & 1 \end{Vmatrix}$$

The class of this Markov chain are $\{0, 1\}$, $\{2\}$, and $\{3\}$. Note that while state 0 (or 1) is accessible from state 2, the reverse is not true. Since state 3 is an absorbing state, that is, $P_{33} = 1$, no other state is accessible from it. ◆

For any state i we let f_i denote the probability that, starting in state i, the process will ever reenter state i. State i is said to be *recurrent* if $f_i = 1$ and *transient* if $f_i < 1$.

Suppose that the process starts in state i and i is recurrent. Hence, with probability 1, the process will eventually reenter state i. However, by the definition of a Markov chain, it follows that the process will be starting over again when it reenters state i and, therefore, state i will eventually be visited again. Continual repetition of this argument leads to the conclusion that *if state i is recurrent then, starting in state i, the process will reenter state i again and again and again—in fact, infinitely often.*

On the other hand, suppose that state i is transient. Hence, each time the process enters state i there will be a positive probability, namely, $1 - f_i$, that it will never again enter that state. Therefore, starting in state i, the probability that the process will be in state i for exactly n time periods equals $f_i^{n-1}(1 - f_i)$, $n \geq 1$. In other words, *if state i is transient then, starting in state i, the number of time periods that the process will be in state i has a geometric distribution with finite mean $1/(1 - f_i)$.*

From the preceding two paragraphs, it follows that *state i is recurrent if and only if, starting in state i, the expected number of time periods that the process is in state i is infinite.* But, letting

$$I_n = \begin{cases} 1, & \text{if } X_n = i \\ 0, & \text{if } X_n \neq i \end{cases}$$

we have that $\sum_{n=0}^{\infty} I_n$ represents the number of periods that the process is in state i. Also,

$$E\left[\sum_{n=0}^{\infty} I_n \,\middle|\, X_0 = i\right] = \sum_{n=0}^{\infty} E[I_n | X_0 = i]$$
$$= \sum_{n=0}^{\infty} P\{X_n = i | X_0 = i\}$$
$$= \sum_{n=0}^{\infty} P_{ii}^n$$

We have thus proven the following.

Proposition 4.1 State i is

$$\text{recurrent if } \sum_{n=1}^{\infty} P_{ii}^n = \infty,$$

$$\text{transient if } \sum_{n=1}^{\infty} P_{ii}^n < \infty$$

The argument leading to the preceding proposition is doubly important because it also shows that a transient state will only be visited a finite number of times (hence the name transient). This leads to the conclusion that in a finite-state Markov chain not all states can be transient. To see this, suppose the states are $0, 1, \ldots, M$ and suppose that they are all transient. Then after a finite amount of time (say, after time T_0) state 0 will never be visited, and after a time (say, T_1) state 1 will never be visited, and after a time (say, T_2) state 2 will never be visited, etc. Thus, after a finite time $T = \max\{T_0, T_1, \ldots, T_M\}$ no states will be visited. But as the process must be in some state after time T we arrive at a contradiction, which shows that at least one of the states must be recurrent.

Another use of Proposition 4.1 is that it enables us to show that recurrence is a class property.

Corollary 4.2 If state i is recurrent, and state i communicates with state j, then state j is recurrent.

Proof To prove this we first note that, since state i communicates with state j, there exist integers k and m such that $P_{ij}^k > 0$, $P_{ji}^m > 0$. Now, for any integer n

$$P_{jj}^{m+n+k} \geqslant P_{ji}^m P_{ii}^n P_{ij}^k$$

This follows since the left side of the above is the probability of going from j to j in $m + n + k$ steps, while the right side is the probability of going from j to j in $m + n + k$ steps via a path that goes from j to i in m steps, then from i to i in an additional n steps, then from i to j in an additional k steps.

From the preceding we obtain, by summing over n, that

$$\sum_{n=1}^{\infty} P_{jj}^{m+n+k} \geqslant P_{ji}^m P_{ij}^k \sum_{n=1}^{\infty} P_{ii}^n = \infty$$

since $P_{ji}^m P_{ij}^k > 0$, and $\sum_{n=1}^{\infty} P_{ii}^n$ is infinite since state i is recurrent. Thus, by Proposition 4.1 it follows that state j is also recurrent. ✦

Remarks   (i) Corollary 4.2 also implies that transience is a class property. For if state i is transient and communicates with state j, then state j must also be transient. For if j were recurrent then, by Corollary 4.2, i would also be recurrent and hence could not be transient.

   (ii) Corollary 4.2 along with our previous result that not all states in a finite Markov chain can be transient leads to the conclusion that all states of a finite irreducible Markov chain are recurrent.

Example 4.11   Let the Markov chain consisting of the states 0, 1, 2, 3 have the transition probability matrix

$$\mathbf{P} = \begin{Vmatrix} 0 & 0 & \frac{1}{2} & \frac{1}{2} \\ 1 & 0 & 0 & 0 \\ 0 & 1 & 0 & 0 \\ 0 & 1 & 0 & 0 \end{Vmatrix}$$

Determine which states are transient and which are recurrent.

   Solution: It is a simple matter to check that all states communicate and hence, since this is a finite chain, all states must be recurrent.   ✦

Example 4.12   Consider the Markov chain having states 0, 1, 2, 3, 4 and

$$\mathbf{P} = \begin{Vmatrix} \frac{1}{2} & \frac{1}{2} & 0 & 0 & 0 \\ \frac{1}{2} & \frac{1}{2} & 0 & 0 & 0 \\ 0 & 0 & \frac{1}{2} & \frac{1}{2} & 0 \\ 0 & 0 & \frac{1}{2} & \frac{1}{2} & 0 \\ \frac{1}{4} & \frac{1}{4} & 0 & 0 & \frac{1}{2} \end{Vmatrix}$$

Determine the recurrent state.

   Solution: This chain consists of the three classes $\{0, 1\}$, $\{2, 3\}$, and $\{4\}$. The first two classes are recurrent and the third transient.   ✦

Example 4.13   (A Random Walk): Consider a Markov chain whose state space consists of the integers $i = 0, \pm1, \pm2, \ldots,$ and have transition probabilities given by

$$P_{i,i+1} = p = 1 - P_{i,i-1}, \qquad i = 0, \pm1, \pm2, \ldots$$

where $0 < p < 1$. In other words, on each transition the process either moves one step to the right (with probability p) or one step to the left (with probability $1 - p$). One colorful interpretation of this process is that it represents the wanderings of a drunken man as he walks along a straight line. Another is that it represents the winnings of a gambler who on each play of the game either wins or loses one dollar.

Since all states clearly communicate, it follows from Corollary 4.2 that they are either all transient or all recurrent. So let us consider state 0 and attempt to determine if $\sum_{n=1}^{\infty} P_{00}^n$ is finite or infinite.

Since it is impossible to be even (using the gambling model interpretation) after an odd number of plays we must, of course, have that

$$P_{00}^{2n-1} = 0, \qquad n = 1, 2, \ldots$$

On the other hand, we would be even after $2n$ trials if and only if we won n of these and lost n of these. Because each play of the game results in a win with probability p and a loss with probability $1 - p$, the desired probability is thus the binomial probability

$$P_{00}^{2n} = \binom{2n}{n} p^n (1-p)^n = \frac{(2n)!}{n!n!} (p(1-p))^n, \qquad n = 1, 2, 3, \ldots$$

By using an approximation, due to Stirling, which asserts that

$$n! \sim n^{n+1/2} e^{-n} \sqrt{2\pi} \tag{4.3}$$

where we say that $a_n \sim b_n$ when $\lim_{n \to \infty} a_n/b_n = 1$, we obtain

$$P_{00}^{2n} \sim \frac{(4p(1-p))^n}{\sqrt{\pi n}}$$

Now it is easy to verify that if $a_n \sim b_n$, then $\sum_n a_n < \infty$ if and only if $\sum_n b_n < \infty$. Hence, $\sum_{n=1}^{\infty} P_{00}^n$ will converge if and only if

$$\sum_{n=1}^{\infty} \frac{(4p(1-p))^n}{\sqrt{\pi n}}$$

does. However, $4p(1 - p) \leqslant 1$ with equality holding if and only if $p = \frac{1}{2}$. Hence, $\sum_{n=1}^{\infty} P_{00}^n = \infty$ if and only if $p = \frac{1}{2}$. Thus, the chain is recurrent when $p = \frac{1}{2}$ and transient if $p \neq \frac{1}{2}$.

When $p = \frac{1}{2}$, the above process is called a *symmetric random walk*. We could also look at symmetric random walks in more than one dimension. For instance, in the two-dimensional symmetric random walk the process would, at each transition, either take one step to the left, right, up, or down, each having probability $\frac{1}{4}$. That is, the state is the pair of integers (i, j) and the transition probabilities are given by

$$P_{(i,j),(i+1,j)} = P_{(i,j),(i-1,j)} = P_{(i,j),(i,j+1)} = P_{(i,j),(i,j-1)} = \frac{1}{4}$$

By using the same method as in the one-dimensional case, we now show that this Markov chain is also recurrent.

Since the preceding chain is irreducible, it follows that all states will be recurrent if state $\mathbf{0} = (0, 0)$ is recurrent. So consider P_{00}^{2n}. Now after $2n$ steps, the chain will be back in its original location if for some i, $0 \leqslant i \leqslant n$, the $2n$ steps consist of i steps to the left, i to the right, $n - i$ up, and $n - i$ down. Since each step will be either of these four types with probability $\frac{1}{4}$, it follows that the desired probability is a multinomial probability. That is,

$$
\begin{aligned}
P_{00}^{2n} &= \sum_{i=0}^{n} \frac{(2n)!}{i!i!(n-i)!(n-i)!} \left(\frac{1}{4}\right)^{2n} \\
&= \sum_{i=0}^{n} \frac{(2n)!}{n!n!} \frac{n!}{(n-i)!i!} \frac{n!}{(n-i)!i!} \left(\frac{1}{4}\right)^{2n} \\
&= \left(\frac{1}{4}\right)^{2n} \binom{2n}{n} \sum_{i=0}^{n} \binom{n}{i}\binom{n}{n-i} \\
&= \left(\frac{1}{4}\right)^{2n} \binom{2n}{n}\binom{2n}{n}
\end{aligned}
$$

(4.4)

where the last equality uses the combinatorial identity

$$
\binom{2n}{n} = \sum_{i=0}^{n} \binom{n}{i}\binom{n}{n-i}
$$

which follows upon noting that both sides represent the number of subgroups of size n one can select from a set of n white and n black objects. Now,

$$
\begin{aligned}
\binom{2n}{n} &= \frac{(2n)!}{n!n!} \\
&\sim \frac{(2n)^{2n+1/2} e^{-2n} \sqrt{2\pi}}{n^{2n+1} e^{-2n}(2\pi)} \qquad \text{by Stirling's approximation} \\
&= \frac{4^n}{\sqrt{\pi n}}
\end{aligned}
$$

Hence, from Equation (4.4) we see that

$$
P_{00}^{2n} \sim \frac{1}{\pi n}
$$

which shows that $\sum_n P_{00}^{2n} = \infty$, and thus all states are recurrent.

Interestingly enough, whereas the symmetric random walks in one and two dimensions are both recurrent, all high-dimensional symmetric random walks turn out to be transient. (For instance, the three-dimensional symmetric random walk is at each transition equally likely to move in any of six ways — either to the left, right, up, down, in, or out.) ◆

Remark We can compute the probability of whether the one-dimensional random walk of Example 4.13 ever returns to state 0 when $p \neq 1/2$ by conditioning on the initial transition:

$$P\{\text{ever return}\} = P\{\text{ever return} \mid X_1 = 1\}p$$
$$+ P\{\text{ever return} \mid X_1 = -1\}(1 - p)$$

Suppose that $p > 1/2$. Then it can be shown (see Exercise 13 at the end of this chapter) that $P\{\text{ever return} \mid X_1 = -1\} = 1$, and thus

$$P\{\text{ever return}\} = P\{\text{ever return} \mid X_1 = 1\}p + 1 - p \qquad (4.5)$$

Let $\alpha = P\{\text{ever return} \mid X_1 = 1\}$. Conditioning on the next transition gives

$$\alpha = P\{\text{ever return} \mid X_1 = 1, X_2 = 0\}(1 - p)$$
$$+ P\{\text{ever return} \mid X_1 = 1, X_2 = 2\}p$$
$$= 1 - p + P\{\text{ever enter } 0 \mid X_0 = 2\}p$$

Now, if the chain is at state 2 then in order for it to enter state 0 it must first enter state 1 and the probability that this ever occurs is α (why is that?). Also, if it does enter state 1 then the probability that it ever enters state 0 is also α. Thus, we see that the probability of ever entering state 0 starting at state 2 is α^2. Therefore, we have that

$$\alpha = 1 - p + p\alpha^2$$

The two roots of this equation are $\alpha = 1$ and $\alpha = (1 - p)/p$. The first is impossible since we know by transience that $\alpha < 1$. Hence, $\alpha = (1 - p)/p$, and we obtain from Equation (4.5) that

$$P\{\text{ever return}\} = 1 - p + 1 - p = 2(1 - p)$$

Similarly, when $p < 1/2$ we can show that $P\{\text{ever return}\} = 2p$. Thus, in general we have that

$$P\{\text{ever return}\} = 2 \min(p, 1 - p)$$

Example 4.14 (On the Ultimate Instability of the Aloha Protocol): Consider a communications facility in which the numbers of messages arriving during each of the time periods $n = 1, 2, \ldots$ are independent and identically distributed random variables. Let $a_i = P\{i \text{ arrivals}\}$, and suppose that $a_0 + a_1 < 1$. Each arriving message will transmit at the end of the period in which it arrives. If exactly one message is transmitted, then the transmission is successful and the message leaves the system. However, if at any time two or more messages simultaneously transmit, then a collision is deemed to occur and these messages remain in the system. Once a message is involved in a collision it will, independently of all else, transmit at the end of each additional period with probability p—the so-called Aloha protocol (because it was first instituted at the University of Hawaii). We will show that such a system is asymptotically unstable in the sense that the number of successful transmissions will, with probability 1, be finite.

To begin let X_n denote the number of messages in the facility at the beginning of the nth period, and note that $\{X_n, n \geq 0\}$ is a Markov chain. Now for $k \geq 0$ define the indicator variables I_k by

$$I_k = \begin{cases} 1, & \text{if the first time that the chain departs state } k \text{ it} \\ & \text{directly goes to state } k-1 \\ 0, & \text{otherwise} \end{cases}$$

and let it be 0 if the system is never in state k, $k \geq 0$. (For instance, if the succesive states are 0, 1, 3, 3, 4, ..., then $I_3 = 0$ since when the chain first departs state 3 it goes to state 4; whereas, if they are 0, 3, 3, 2, ..., then $I_3 = 1$ since this time it goes to state 2.) Now,

$$E\left[\sum_{k=0}^{\infty} I_k\right] = \sum_{k=0}^{\infty} E[I_k]$$

$$= \sum_{k=0}^{\infty} P\{I_k = 1\}$$

$$\leq \sum_{k=0}^{\infty} P\{I_k = 1 \mid k \text{ is ever visited}\} \quad (4.6)$$

Now, $P\{I_k = 1 \mid k \text{ is ever visited}\}$ is the probability that when state k is departed the next state is $k-1$. That is, it is the conditional probability that a transition from k is to $k-1$ given that it is not back into k, and so

$$P\{I_k = 1 \mid k \text{ is ever visited}\} = \frac{P_{k,k-1}}{1 - P_{kk}}$$

Because

$$P_{k,k-1} = a_0 k p (1-p)^{k-1}$$
$$P_{k,k} = a_0[1 - kp(1-p)^{k-1}] + a_1(1-p)^k$$

which is seen by noting that if there are k messages present on the beginning of a day, then (a) there will be $k-1$ at the beginning of the next day if there are no new messages that day and exactly one of the k messages transmits; and (b) there will be k at the beginning of the next day if either

(i) there are no new messages and it is not the case that exactly one of the existing k messages transmits, or
(ii) there is exactly one new message (which automatically transmits) and none of the other k messages transmits.

Substitution of the preceding into Equation (4.6) yields

$$E\left[\sum_{k=0}^{\infty} I_k\right] \leqslant \sum_{k=0}^{\infty} \frac{a_0 kp(1-p)^{k-1}}{1 - a_0[1 - kp(1-p)^{k-1}] - a_1(1-p)^k}$$

$$< \infty$$

where the convergence follows by noting that when k is large the denominator of the expression in the preceding sum converges to $1 - a_0$ and so the convergence or divergence of the sum is determined by whether or not the sum of the terms in the numerator converge and $\sum_{k=0}^{\infty} k(1-p)^{k-1} < \infty$.

Hence, $E[\sum_{k=0}^{\infty} I_k] < \infty$, which implies that $\sum_{k=0}^{\infty} I_k < \infty$ with probability 1 (for if there was a positive probability that $\sum_{k=0}^{\infty} I_k$ could be ∞, then its mean would be ∞). Hence, with probability 1, there will be only a finite number of states that are initially departed via a successful transmission; or equivalently, there will be some finite integer N such that whenever there are N or more messages in the system, there will never again be a successful transmission. From this (and the fact that such higher states will eventually be reached — why?) it follows that, with probability 1, there will only be a finite number of successful transmissions. ✦

Remark For a (slightly less than rigorous) probabilistic proof of Stirling's approximation, let $X_1 X_2, \ldots$ be independent Poisson random variables each having mean 1. Let $S_n = \sum_{i=1}^{n} X_i$, and note that both the mean and variance of S_n are equal to n. Now,

$$P\{S_n = n\} = P\{n - 1 < S_n \leqslant n\}$$

$$= P\{-1/\sqrt{n} < (S_n - n)/\sqrt{n} \leqslant 0$$

$$\approx \int_{-1/\sqrt{n}}^{0} (2\pi)^{-1/2} e^{-x^2/2} \, dx \qquad \begin{array}{l}\text{when } n \text{ is large, by the}\\ \text{central limit theorem}\end{array}$$

$$\approx (2\pi)^{-1/2}(1/\sqrt{n})$$

$$= (2\pi n)^{-1/2}$$

But S_n is Poisson with mean n, and so

$$P\{S_n = n\} = \frac{e^{-n}n^n}{n!}$$

Hence, for n large

$$\frac{e^{-n}n^n}{n!} \approx (2\pi n)^{-1/2}$$

or, equivalently

$$n! \approx n^{n+1/2}e^{-n}\sqrt{2\pi}$$

which is Stirling's approximation.

4.4. Limiting Probabilities

In Example 4.7, we calculated $\mathbf{P}^{(4)}$ for a two-state Markov chain; it turned out to be

$$\mathbf{P}^{(4)} = \begin{Vmatrix} 0.5749 & 0.4251 \\ 0.5668 & 0.4332 \end{Vmatrix}$$

From this it follows that $\mathbf{P}^{(8)} = \mathbf{P}^{(4)} \cdot \mathbf{P}^{(4)}$ is given (to three significant places) by

$$\mathbf{P}^{(8)} = \begin{Vmatrix} 0.572 & 0.428 \\ 0.570 & 0.430 \end{Vmatrix}$$

Note that the matrix $\mathbf{P}^{(8)}$ is almost identical to the matrix $\mathbf{P}^{(4)}$, and secondly, that each of the rows of $\mathbf{P}^{(8)}$ has almost identical entries. In fact it seems that P_{ij}^n is converging to some value (as $n \to \infty$) which is the same for all i. In other words, there seems to exist a limiting probability that the process will be in state j after a large number of transitions, and this value is independent of the initial state.

To make the above heuristics more precise, two additional properties of the states of a Markov chain need to be considered. State i is said to have *period d* if $P_{ii}^n = 0$ whenever n is not divisible by d, and d is the largest integer with this property. For instance, starting in i, it may be possible for the process to enter state i only at the times 2, 4, 6, 8, ..., in which case state i has period 2. A state with period 1 is said to be *aperiodic*. It can be shown that periodicity is a class property. That is, if state i has period d, and states i and j communicate, then state j also has period d.

If state i is recurrent, then it is said to be *positive recurrent* if, starting in i, the expected time until the process returns to state i is finite. It can be shown that positive recurrence is a class property. While there exist recurrent states that are not positive recurrent,* it can be shown that *in a finite-state Markov chain all recurrent states are positive recurrent*. Positive recurrent, aperiodic states are called *ergodic*.

We are now ready for the following important theorem which we state without proof.

---

* Such states are called *null recurrent*.

Theorem 4.1 For an irreducible ergodic Markov chain $\lim_{n \to \infty} P_{ij}^n$ exists and is independent of i. Furthermore, letting

$$\pi_j = \lim_{n \to \infty} P_{ij}^n, \quad j \geq 0$$

then π_j is the unique nonnegative solution of

$$\pi_j = \sum_{i=0}^{\infty} \pi_i P_{ij}, \quad j \geq 0$$

$$\sum_{j=0}^{\infty} \pi_j = 1 \tag{4.7}$$

Remarks (i) Given that $\pi_j = \lim_{n \to \infty} P_{ij}^n$ exists and is independent of the initial state i, it is not difficult to (heuristically) see that the πs must satisfy Equation (4.7). Let us derive an expression for $P\{X_{n+1} = j\}$ by conditioning on the state at time n. That is,

$$P\{X_{n+1} = j\} = \sum_{i=0}^{\infty} P\{X_{n+1} = j \mid X_n = i\} P\{X_n = i\}$$

$$= \sum_{i=0}^{\infty} P_{ij} P\{X_n = i\}$$

Letting $n \to \infty$, and assuming that we can bring the limit inside the summation, leads to

$$\pi_j = \sum_{i=0}^{\infty} P_{ij} \pi_i$$

(ii) It can be shown that π_j, the limiting probability that the process will be in state j at time n, also equals the long-run proportion of time that the process will be in state j.

(iii) In the irreducible, positive recurrent, *periodic* case we still have that the $\pi_j, j \geq 0$, are the unique nonnegative solution of

$$\pi_j = \sum_i \pi_i P_{ij},$$

$$\sum_j \pi_j = 1$$

But now π_j must be interpreted as the long-run proportion of time that the Markov chain is in state j.

Example 4.15 Consider Example 4.1, in which we assume that if it rains today, then it will rain tomorrow with probability α; and if it does not rain today, then it will rain tomorrow with probability β. If we say that the state is 0 when it rains and 1 when it does not rain, then by Equation (4.7) the limiting probabilities π_0 and π_1 are given by

$$\pi_0 = \alpha\pi_0 + \beta\pi_1,$$

$$\pi_1 = (1 - \alpha)\pi_0 + (1 - \beta)\pi_1,$$

$$\pi_0 + \pi_1 = 1$$

which yields that

$$\pi_0 = \frac{\beta}{1 + \beta - \alpha}, \qquad \pi_1 = \frac{1 - \alpha}{1 + \beta - \alpha}$$

For example if $\alpha = 0.7$ and $\beta = 0.4$, then the limiting probability of rain is $\pi_0 = \frac{4}{7} = 0.571$. ✦

Example 4.16 Consider Example 4.3 in which the mood of an individual is considered as a three-state Markov chain having a transition probability matrix

$$\mathbf{P} = \begin{Vmatrix} 0.5 & 0.4 & 0.1 \\ 0.3 & 0.4 & 0.3 \\ 0.2 & 0.3 & 0.5 \end{Vmatrix}$$

In the long run, what proportion of time is the process in each of the three states?

Solution: The limiting probabilities π_i, $i = 0, 1, 2$, are obtained by solving the set of equations in Equation (4.1). In this case these equations are

$$\pi_0 = 0.5\pi_0 + 0.3\pi_1 + 0.2\pi_2,$$

$$\pi_1 = 0.4\pi_0 + 0.4\pi_1 + 0.3\pi_2,$$

$$\pi_2 = 0.1\pi_0 + 0.3\pi_1 + 0.5\pi_2,$$

$$\pi_0 + \pi_1 + \pi_2 = 1$$

Solving yields

$$\pi_0 = \tfrac{21}{62}, \qquad \pi_1 = \tfrac{23}{62}, \qquad \pi_2 = \tfrac{18}{62} \quad ✦$$

Example 4.17 (A Model of Class Mobility): A problem of interest to sociologists is to determine the proportion of society that has an upper- or lower-class occupation. One possible mathematical model would be to assume that transitions between social classes of the successive generations in a family can be regarded as transitions of a Markov chain. That is, we assume that the occupation of a child depends only on his or her parent's occupation. Let us suppose that such a model is appropriate and that the transition probability matrix is given by

$$\mathbf{P} = \begin{Vmatrix} 0.45 & 0.48 & 0.07 \\ 0.05 & 0.70 & 0.25 \\ 0.01 & 0.50 & 0.49 \end{Vmatrix} \tag{4.8}$$

That is, for instance, we suppose that the child of a middle-class worker will attain an upper-, middle-, or lower-class occupation with respective probabilities 0.05, 0.70, 0.25.

The limiting probabilities π_i, thus satisfy

$$\pi_0 = 0.45\pi_0 + 0.05\pi_1 + 0.01\pi_2,$$

$$\pi_1 = 0.48\pi_0 + 0.70\pi_1 + 0.50\pi_2,$$

$$\pi_2 = 0.07\pi_0 + 0.25\pi_1 + 0.49\pi_2,$$

$$\pi_0 + \pi_1 + \pi_2 = 1$$

Hence,

$$\pi_0 = 0.07, \qquad \pi_1 = 0.62, \qquad \pi_2 = 0.31$$

In other words, a society in which social mobility between classes can be described by a Markov chain with transition probability matrix given by Equation (4.8) has, in the long run, 7 percent of its people in upper-class jobs, 62 percent of its people in middle-class jobs, and 31 percent in lower-class jobs. ◆

Example 4.18 (The Hardy–Weinberg Law and a Markov Chain in Genetics): Consider a large population of individuals each of whom possesses a particular pair of genes, of which each individual gene is classified as being of type A or type a. Assume that the proportions of individuals whose gene pairs are AA, aa, or Aa are respectively p_0, q_0, and r_0 ($p_0 + q_0 + r_0 = 1$). When two individuals mate, each contributes one of his or her genes, chosen at random, to the resultant offspring. Assuming that the mating occurs at random, in that each individual is equally likely to mate with any other individual, we are interested in determining the proportions of individuals in the next generation whose genes are AA, aa, or Aa. Calling these proportions p, q, and r, they are easily obtained by focusing attention on an individual of the next generation and then determining the probabilities for the gene pair of that individual.

To begin, note that randomly choosing a parent and then randomly choosing one of its genes is equivalent to just randomly choosing a gene from the total gene population. By conditioning on the gene pair of the parent, we see that a randomly chosen gene will be type A with probability

$$P\{A\} = P\{A \mid AA\}p_0 + P\{A \mid aa\}q_0 + P\{A \mid Aa\}r_0$$

$$= p_0 + r_0/2$$

Similarly, it will be type a with probability

$$P\{a\} = q_0 + r_0/2$$

Thus, under random mating a randomly chosen member of the next generation will be type AA with probability p, where

$$p = P\{A\}P\{A\} = (p_0 + r_0/2)^2$$

Similarly, the randomly chosen member will be type aa with probability

$$q = P\{a\}P\{a\} = (q_0 + r_0/2)^2$$

and will be type Aa with probability

$$r = 2P\{A\}P\{a\} = 2(p_0 + r_0/2)(q_0 + r_0/2)$$

Since each member of the next generation will independently be of each of the three gene types with probabilities p, q, r, it follows that the percentages of the members of the next generation that are type AA, aa, or Aa are respectively p, q, and r.

If we now consider the total gene pool of this next generation, then $p + r/2$, the fraction of its genes that are A, will be unchanged from the previous generation. This follows either by arguing that the total gene pool has not changed from generation to generation or by the following simple algebra:

$$p + r/2 = (p_0 + r_0/2)^2 + (p_0 + r_0/2)(q_0 + r_0/2)$$

$$= (p_0 + r_0/2)[p_0 + r_0/2 + q_0 + r_0/2]$$

$$= p_0 + r_0/2 \qquad \text{since } p_0 + r_0 + q_0 = 1$$

$$= P\{A\} \tag{4.9}$$

Thus, the fractions of the gene pool that are A and a are the same as in the initial generation. From this it follows that, under random mating, in all successive generations after the initial one the percentages of the population having gene pairs AA, aa, and Aa will remain fixed at the values p, q, and r. This is known as the *Hardy–Weinberg law*.

Suppose now that the gene pair population has stabilized in the percentages p, q, r, and let us follow the genetic history of a single individual and her descendants. (For simplicity, assume that each individual has exactly one offspring.) So, for a given individual, let X_n denote the genetic state of her descendant in the nth generation. The transition probability matrix of this Markov chain, namely,

$$
\begin{array}{c c c c}
 & AA & aa & Aa \\
AA & \left\| p + \dfrac{r}{2} \right. & 0 & \left. q + \dfrac{r}{2} \right\| \\
aa & \left\| 0 \right. & q + \dfrac{r}{2} & \left. p + \dfrac{r}{2} \right\| \\
Aa & \left\| \dfrac{p}{2} + \dfrac{r}{4} \right. & \dfrac{q}{2} + \dfrac{r}{4} & \left. \dfrac{p}{2} + \dfrac{q}{2} + \dfrac{r}{2} \right\|
\end{array}
$$

is easily verified by conditioning on the state of the randomly chosen mate. It is quite intuitive (why?) that the limiting probabilities for this Markov chain (which also equal the fractions of the individual's descendants that are in each of the three genetic states) should just be p, q, and r. To verify this we must show that they satisfy Equation (4.7). Because one of the equations in Equation (4.7) is redundant, it suffices to show that

$$
p = p\left(p + \frac{r}{2}\right) + r\left(\frac{p}{2} + \frac{r}{4}\right) = \left(p + \frac{r}{2}\right)^2,
$$

$$
q = q\left(q + \frac{r}{2}\right) + r\left(\frac{q}{2} + \frac{r}{4}\right) = \left(q + \frac{r}{2}\right)^2,
$$

$$
p + q + r = 1
$$

But this follows from Equation (4.9), and thus the result is established. ✦

Example 4.19 Suppose that a production process changes states in accordance with a Markov chain having transition probabilities $P_{ij}, i, j = 1, \ldots, n$, and suppose that certain of the states are considered acceptable and the remaining unacceptable. Let A denote the acceptable states and A^c the unacceptable ones. If the production process is said to be "up" when in an acceptable state and "down" when in an unacceptable state, determine

1. the rate at which the production process goes from up to down (that is, the rate of breakdowns);
2. the average length of time the process remains down when it goes down; and
3. the average length of time the process remains up when it goes up.

Solution: Let π_k, $k = 1, \ldots, n$, denote the limiting probabilities. Now for $i \in A$ and $j \in A^c$ the rate at which the process enters state j from state i is

$$\text{rate enter } j \text{ from } i = \pi_i P_{ij}$$

and so the rate at which the production process enters state j from an acceptable state is

$$\text{rate enter } j \text{ from } A = \sum_{i \in A} \pi_i P_{ij}$$

Hence, the rate at which it enters an unacceptable state from an acceptable one (which is the rate at which breakdowns occur) is

$$\text{rate breakdowns occur} = \sum_{j \in A^c} \sum_{i \in A} \pi_i P_{ij} \tag{4.10}$$

Now let $\bar{U}$ and $\bar{D}$ denote the average time the process remains up when it goes up and down when it goes down. Because there is a single breakdown every $\bar{U} + \bar{D}$ time units on the average, it follows heuristically that

$$\text{rate at which breakdowns occur} = \frac{1}{\bar{U} + \bar{D}}$$

and, so from Equation (4.10),

$$\frac{1}{\bar{U} + \bar{D}} = \sum_{j \in A^c} \sum_{i \in A} \pi_i P_{ij} \tag{4.11}$$

To obtain a second equation relating $\bar{U}$ and $\bar{D}$, consider the percentage of time the process is up, which, of course, is equal to $\sum_{i \in A} \pi_i$. However, since the process is up on the average $\bar{U}$ out of every $\bar{U} + \bar{D}$ time units, it follows (again somewhat heuristically) that the

$$\text{proportion of up time} = \frac{\bar{U}}{\bar{U} + \bar{D}}$$

and so

$$\frac{\bar{U}}{\bar{U} + \bar{D}} = \sum_{i \in A} \pi_i \tag{4.12}$$

Hence, from Equations (4.11) and (4.12) we obtain

$$\bar{U} = \frac{\sum_{i \in A} \pi_i}{\sum_{j \in A^c} \sum_{i \in A} \pi_i P_{ij}},$$

$$\bar{D} = \frac{1 - \sum_{i \in A} \pi_i}{\sum_{j \in A^c} \sum_{i \in A} \pi_i P_{ij}}$$

$$= \frac{\sum_{i \in A^c} \pi_i}{\sum_{j \in A^c} \sum_{i \in A} \pi_i P_{ij}}$$

For example, suppose the transition probability matrix is

$$\mathbf{P} = \begin{Vmatrix} \frac{1}{4} & \frac{1}{4} & \frac{1}{2} & 0 \\ 0 & \frac{1}{4} & \frac{1}{2} & \frac{1}{4} \\ \frac{1}{4} & \frac{1}{4} & \frac{1}{4} & \frac{1}{4} \\ \frac{1}{4} & \frac{1}{4} & 0 & \frac{1}{2} \end{Vmatrix}$$

where the acceptable (up) states are 1, 2 and the unacceptable (down) ones are 3, 4. The limiting probabilities satisfy

$$\pi_1 = \pi_1 \tfrac{1}{4} + \pi_3 \tfrac{1}{4} + \pi_4 \tfrac{1}{4},$$

$$\pi_2 = \pi_1 \tfrac{1}{4} + \pi_2 \tfrac{1}{4} + \pi_3 \tfrac{1}{4} + \pi_4 \tfrac{1}{4},$$

$$\pi_3 = \pi_1 \tfrac{1}{2} + \pi_2 \tfrac{1}{2} + \pi_3 \tfrac{1}{4},$$

$$\pi_1 + \pi_2 + \pi_3 + \pi_4 = 1$$

These solve to yield

$$\pi_1 = \tfrac{3}{16}, \qquad \pi_2 = \tfrac{1}{4}, \qquad \pi_3 = \tfrac{14}{48}, \qquad \pi_4 = \tfrac{13}{48}$$

and thus

$$\text{rate of breakdowns} = \pi_1(P_{13} + P_{14}) + \pi_2(P_{23} + P_{24})$$

$$= \tfrac{9}{32},$$

$$\bar{U} = \tfrac{14}{9} \qquad \text{and} \qquad \bar{D} = 2$$

Hence, on the average, breakdowns occur about $\tfrac{9}{32}$ (or 28 percent) of the time. They last, on the average, 2 time units, and then there follows a stretch of (on the average) $\tfrac{14}{9}$ time units when the system is up. ◆

Remarks (i) The long run proportions $\pi_j, j \geqslant 0$, are often called *station-ary* probabilities. The reason being that if the initial state is chosen according to the probabilities $\pi_j, j \geqslant 0$, then the probability of being in state j at any time n is also equal to π_j. That is, if

$$P\{X_0 = j\} = \pi_j, \qquad j \geqslant 0$$

then

$$P\{X_n = j\} = \pi_j \qquad \text{for all } n, j \geqslant 0$$

The preceding is easily proven by induction, for if we suppose it true for

$n - 1$, then writing

$$P\{X_n = j\} = \sum_i P\{X_n = j \mid X_{n-1} = i\} P\{X_{n-1} = i\}$$

$$= \sum_i P_{ij} \pi_i \qquad \text{by the induction hypothesis}$$

$$= \pi_j \qquad \text{by Equation (4.7)}$$

(ii) For state j, define m_{jj} to be the expected number of transitions until a Markov chain, starting in state j, returns to that state. Since, on the average, the chain will spend 1 unit of time in state j for every m_{jj} units of time, it follows that

$$\pi_j = \frac{1}{m_{jj}}$$

In words, the proportion of time in state j equals the inverse of the mean time between visits to j. (The above is a special case of a general result, sometimes called the strong law for renewal processes, which will be presented in Chapter 7.)

Example 4.20   Consider independent tosses of a coin that, on each toss, lands on heads (H) with probability p and on tails (T) with probability $q = 1 - p$. What is the expected number of tosses needed for the pattern HTHT to appear?

Solution: To answer the question, let us imagine that the coin tossing does not stop when the pattern appears, but rather it goes on indefinitely. If we define the state at time n to be the most recent 4 outcomes when $n \geqslant 4$, and the most recent n outcomes when $n < 4$, then it is easy to see that the successive states constitute a Markov chain. For instance, if the first 5 outcomes are TTTHH, then the successive states of the Markov chain are $X_1 = T$, $X_2 = TT$, $X_3 = TTT$, $X_4 = TTTH$, and $X_5 = TTHH$. It therefore follows from remark (ii) that π_{HTHT}, the limiting probability of state HTHT, is equal to the inverse of the mean time to go from state HTHT to HTHT. However, for any $n \geqslant 4$, the probability that the state at time n is HTHT is just the probability that the toss at n is T, the one at $n - 1$ is H, the one at $n - 2$ is T, and the one at $n - 3$ is H. Since the successive tosses are independent, it follows that

$$P\{X_n = HTHT\} = p^2 q^2, \qquad n \geqslant 4$$

and so

$$\pi_{HTHT} = \lim_{n \to \infty} P\{X_n = HTHT\} = p^2 q^2$$

Hence, $1/(p^2q^2)$ is the mean time to go from HTHT to HTHT. But this means that starting with HT the expected number of additional trials to obtain HTHT is $1/(p^2q^2)$. Therefore, since in order to obtain HTHT one must first obtain HT, it follows that

$$E[\text{time to pattern HTHT}] = E[\text{time to the pattern HT}] + \frac{1}{p^2q^2}$$

To determine the expected time to the pattern HT, we can reason in the same way and let the state be the most recent 2 tosses. By the same argument as used before, it follows that the expected time between appearances of HT is equal to $1/\pi_{\text{HT}} = 1/(pq)$. Because this is the same as the expected time until HT first appears, we finally obtain that

$$E[\text{time until HTHT appears}] = \frac{1}{pq} + \frac{1}{p^2q^2}$$

The same approach can be used to obtain the mean time until any given pattern appears. For instance, reasoning as before, we obtain that

$$E[\text{time until HTHHTHTHH}] = E[\text{time until HTHH}] + \frac{1}{p^6q^3}$$

$$= E[\text{time until H}] + \frac{1}{p^3q} + \frac{1}{p^6q^3}$$

$$= \frac{1}{p} + \frac{1}{p^3q} + \frac{1}{p^6q^3}$$

Also, it is not necessary for the basic experiment to have only two possible outcomes (which we designated as H and T). For instance, if the successive values are independently and identically distributed with p_j denoting the probability that any given value is equal to $j, j \geqslant 0$, then

$$E[\text{time until } 012301] = E[\text{time until } 01] + \frac{1}{p_0^2 p_1^2 p_2 p_3}$$

$$= \frac{1}{p_0 p_1} + \frac{1}{p_0^2 p_1^2 p_2 p_3} \quad \blacklozenge$$

The following result is quite useful.

Proposition 4.3. Let $\{X_n, n \geqslant 1\}$ be an irreducible Markov chain with stationary probabilities $\pi_j, j \geqslant 0$, and let r be a bounded function on the state space. Then, with probability 1,

$$\lim_{N \to \infty} \frac{\Sigma_{n=1}^{N} r(X_n)}{N} = \sum_{j=0}^{\infty} r(j)\pi_j$$

Proof   If we let $a_j(N)$ be the amount of time the Markov chain spends in state j during time periods $1, \ldots, N$, then

$$\sum_{n=1}^{N} r(X_n) = \sum_{j=0}^{\infty} a_j(N)r(j)$$

Since $a_j(N)/N \to \pi_j$ the result follows from the preceding upon dividing by N and then letting $N \to \infty$.   ◆

If we suppose that we earn a reward $r(j)$ whenever the chain is in state j, then Proposition 4.3 states that our average reward per unit time is $\Sigma_j r(j)\pi_j$.

4.5 Some Applications

4.5.1. The Gambler's Ruin Problem

Consider a gambler who at each play of the game has probability p of winning one unit and probability $q = 1 - p$ of losing one unit. Assuming that successive plays of the game are independent, what is the probability that, starting with i units, the gambler's fortune will reach N before reaching 0?

If we let X_n denote the player's fortune at time n, then the process $\{X_n, n = 0, 1, 2, \ldots\}$ is a Markov chain with transition probabilities

$$P_{00} = P_{NN} = 1$$

$$P_{i,i+1} = p = 1 - P_{i,i-1}, \qquad i = 1, 2, \ldots, N-1$$

This Markov chain has three classes, namely, $\{0\}$, $\{1, 2, \ldots, N-1\}$, and $\{N\}$; the first and third class being recurrent and the second transient. Since each transient state is visited only finitely often, it follows that, after some finite amount of time, the gambler will either attain his goal of N or go broke.

Let P_i, $i = 0, 1, \ldots, N$, denote the probability that, starting with i, the gambler's fortune will eventually reach N. By conditioning on the outcome of the initial play of the game we obtain

$$P_i = pP_{i+1} + qP_{i-1}, \qquad i = 1, 2, \ldots, N-1$$

or equivalently, since $p + q = 1$,

$$pP_i + qP_i = pP_{i+1} + qP_{i-1}$$

or

$$P_{i+1} - P_i = \frac{q}{p}(P_i - P_{i-1}), \qquad i = 1, 2, \ldots, N - 1$$

Hence, since $P_0 = 0$, we obtain from the preceding line that

$$P_2 - P_1 = \frac{q}{p}(P_1 - P_0) = \frac{q}{p}P_1,$$

$$P_3 - P_2 = \frac{q}{p}(P_2 - P_1) = \left(\frac{q}{p}\right)^2 P_1,$$

$$\vdots$$

$$P_i - P_{i-1} = \frac{q}{p}(P_{i-1} - P_{i-2}) = \left(\frac{q}{p}\right)^{i-1} P_1,$$

$$\vdots$$

$$P_N - P_{N-1} = \left(\frac{q}{p}\right)(P_{N-1} - P_{N-2}) = \left(\frac{q}{p}\right)^{N-1} P_1$$

Adding the first $i - 1$ of these equations yields

$$P_i - P_1 = P_1 \left[\left(\frac{q}{p}\right) + \left(\frac{q}{p}\right)^2 + \cdots + \left(\frac{q}{p}\right)^{i-1} \right]$$

or

$$P_i = \begin{cases} \dfrac{1 - (q/p)^i}{1 - (q/p)} P_1, & \text{if } \dfrac{q}{p} \neq 1 \\ iP_1, & \text{if } \dfrac{q}{p} = 1 \end{cases}$$

Now, using the fact that $P_N = 1$, we obtain that

$$P_1 = \begin{cases} \dfrac{1 - (q/p)}{1 - (q/p)^N}, & \text{if } p \neq \dfrac{1}{2} \\ \dfrac{1}{N}, & \text{if } p = \dfrac{1}{2} \end{cases}$$

and hence

$$P_i = \begin{cases} \dfrac{1 - (q/p)^i}{1 - (q/p)^N}, & \text{if } p \neq \dfrac{1}{2} \\ \dfrac{i}{N}, & \text{if } p = \dfrac{1}{2} \end{cases} \tag{4.13}$$

Note that, as $N \to \infty$,

$$P_i \to \begin{cases} 1 - \left(\dfrac{q}{p}\right)^i, & \text{if } p > \dfrac{1}{2} \\[2mm] 0, & \text{if } p \leqslant \dfrac{1}{2} \end{cases}$$

Thus, if $p > \frac{1}{2}$, there is a positive probability that the gambler's fortune will increase indefinitely; while if $p \leqslant \frac{1}{2}$, the gambler will, with probability 1, go broke against an infinitely rich adversary.

Example 4.21   Suppose Max and Patty decide to flip pennies; the one coming closest to the wall wins. Patty, being the better player, has a probability 0.6 of winning on each flip. (a) If Patty starts with five pennies and Max with ten, then what is the probability that Patty will wipe Max out? (b) What if Patty starts with ten and Max with 20?

Solution: (a) The desired probability is obtained from Equation (4.13) by letting $i = 5$, $N = 15$, and $p = 0.6$. Hence, the desired probability is

$$\frac{1 - \left(\frac{2}{3}\right)^5}{1 - \left(\frac{2}{3}\right)^{15}} \approx 0.87$$

(b) The desired probability is

$$\frac{1 - \left(\frac{2}{3}\right)^{10}}{1 - \left(\frac{2}{3}\right)^{30}} \approx 0.98 \quad \blacklozenge$$

For an application of the gambler's ruin problem to drug testing, suppose that two new drugs have been developed for treating a certain disease. Drug i has a cure rate P_i, $i = 1, 2$, in the sense that each patient treated with drug i will be cured with probability P_i. These cure rates are, however, not known, and suppose we are interested in a method for deciding whether $P_1 > P_2$ or $P_2 > P_1$. To decide upon one of these alternatives, consider the following test: Pairs of patients are treated sequentially wth one member of the pair receiving drug 1 and the other drug 2. The results for each pair are determined, and the testing stops when the cumulative number of cures using one of the drugs exceeds the cumulative number of cures when using the other by some fixed predetermined number. More formally, let

$$X_j = \begin{cases} 1, & \text{if the patient in the } j\text{th pair to receive drug number 1 is cured} \\ 0, & \text{otherwise} \end{cases}$$

$$Y_j = \begin{cases} 1, & \text{if the patient in the } j\text{th pair to receive drug number 2 is cured} \\ 0, & \text{otherwise} \end{cases}$$

For a predetermined positive integer M the test stops after pair N where N is the first value of n such that either

$$X_1 + \cdots + X_n - (Y_1 + \cdots + Y_n) = M$$

or

$$X_1 + \cdots + X_n - (Y_1 + \cdots + Y_n) = -M$$

In the former case we then assert that $P_1 > P_2$, and in the latter that $P_2 > P_1$.

In order to help ascertain whether the preceding is a good test, one thing we would like to know is the probability of it leading to an incorrect decision. That is, for given P_1 and P_2 where $P_1 > P_2$, what is the probability that the test will incorrectly assert that $P_2 > P_1$? To determine this probability, note that after each pair is checked the cumulative difference of cures using drug 1 versus drug 2 will either go up by 1 with probability $P_1(1 - P_2)$ — since this is the probability that drug 1 leads to a cure and drug 2 does not — or go down by 1 with probability $(1 - P_1)P_2$, or remain the same with probability $P_1 P_2 + (1 + P_1)(1 - P_2)$. Hence, if we only consider those pairs in which the cumulative difference changes, then the difference will go up 1 with probability

$$p = P\{\text{up } 1 \mid \text{up } 1 \text{ or down } 1\}$$

$$= \frac{P_1(1 - P_2)}{P_1(1 - P_2) + (1 - P_1)P_2}$$

and down 1 with probability

$$q = 1 - p = \frac{P_2(1 - P_1)}{P_1(1 - P_2) + (1 - P_1)P_2}$$

Hence, the probability that the test will assert that $P_2 > P_1$ is equal to the probability that a gambler who wins each (one unit) bet with probability p will go down M before going up M. But Equation (4.12) with $i = M$, $N = 2M$, shows that this probability is given by

$$P\{\text{test asserts that } P_2 > P_1\} = 1 - \frac{1 - (q/p)^M}{1 - (q/p)^{2M}}$$

$$= \frac{1}{1 + (p/q)^M}$$

Thus, for instance, if $P_1 = 0.6$ and $P_2 = 0.4$ then the probability of an incorrect decision is 0.017 when $M = 5$ and reduces to 0.0003 when $M = 10$.

4.5.2. A Model for Algorithmic Efficiency

The following optimization problem is called a linear program:

$$\text{minimize } \mathbf{cx},$$

$$\text{subject to } \mathbf{Ax} = \mathbf{b},$$

$$\mathbf{x} \geqslant 0$$

where $\mathbf{A}$ is an $m \times n$ matrix of fixed constants; $\mathbf{c} = (c_1, \ldots, c_n)$ and $\mathbf{b} = (b_1, \ldots, b_m)$ are vectors of fixed constants; and $\mathbf{x} = (x_1, \ldots, x_n)$ is the n-vector of nonnegative values that is to be chosen to minimize $\mathbf{cx} \equiv \sum_{i=1}^{n} c_i x_i$. Supposing that $n > m$, it can be shown that the optimal $\mathbf{x}$ can always be chosen to have at least $n - m$ components equal to 0—that is, it can always be taken to be one of the so-called extreme points of the feasibility region.

The simplex algorithm solves this linear program by moving from an extreme point of the feasibility region to a better (in terms of the objective function $\mathbf{cx}$) extreme point (via the pivot operation) until the optimal is reached. Because there can be as many as $N \equiv \binom{n}{m}$ such extreme points, it would seem that this method might take many iterations, but, surprisingly to some, this does not appear to be the case in practice.

To obtain a feel for whether or not the preceding statement is surprising, let us consider a simple probabilistic (Markov chain) model as to how the algorithm moves along the extreme points. Specifically, we will suppose that if at any time the algorithm is at the jth best extreme point then after the next pivot the resulting extreme point is equally likely to be any of the $j - 1$ best. Under this assumption, we show that the time to get from the Nth best to the best extreme point has approximately, for large N, a normal distribution with mean and variance equal to the logarithm (base e) of N.

Consider a Markov chain for which $P_{11} = 1$ and

$$P_{ij} = \frac{1}{i-1}, \qquad j = 1, \ldots, i-1, \ i > 1$$

and let T_i denote the number of transitions needed to go from state i to state 1. A recursive formula for $E[T_i]$ can be obtained by conditioning on the initial transition:

$$E[T_i] = 1 + \frac{1}{i-1} \sum_{j=1}^{i-1} E[T_j]$$

Starting with $E[T_1] = 0$, we successively see that

$$E[T_2] = 1,$$

$$E[T_3] = 1 + \tfrac{1}{2},$$

$$E[T_4] = 1 + \tfrac{1}{3}(1 + 1 + \tfrac{1}{2}) = 1 + \tfrac{1}{2} + \tfrac{1}{3}$$

and it is not difficult to guess and then prove inductively that

$$E[T_i] = \sum_{j=1}^{i-1} 1/j$$

However, to obtain a more complete description of T_N, we will use the representation

$$T_N = \sum_{j=1}^{N-1} I_j$$

where

$$I_j = \begin{cases} 1, & \text{if the process ever enters } j \\ 0, & \text{otherwise} \end{cases}$$

The importance of the preceding representation stems from the following:

Proposition 4.4 $I_1, \ldots, I_{N-1}$ are independent and

$$P\{I_j = 1\} = 1/j, \qquad 1 \leq j \leq N-1$$

Proof   Given $I_{j+1}, \ldots, I_N$, let $n = \min\{i : i > j, I_i = 1\}$ denote the lowest numbered state, greater than j, that is entered. Thus we know that the process enters state n and the next state entered is one of the states $1, 2, \ldots, j$. Hence, as the next state from state n is equally likely to be any of the lower number states $1, 2, \ldots, n-1$ we see that

$$P\{I_j = 1 \mid I_{j+1}, \ldots, I_N\} = \frac{1/(n-1)}{j/(n-1)} = 1/j$$

Hence, $P\{I_j = 1\} = 1/j$, and independence follows since the preceding conditional probability does not depend on $I_{j+1}, \ldots, I_N$. ◆

Corollary 4.5

(i) $E[T_N] = \sum_{j=1}^{N-1} 1/j$.
(ii) $\mathrm{Var}(T_N) = \sum_{j=1}^{N-1} (1/j)(1 - 1/j)$.
(iii) For N large, T_N has approximately a normal distribution with mean $\log N$ and variance $\log N$.

Proof   Parts (i) and (ii) follow from Proposition 4.4 and the representation $T_N = \sum_{j=1}^{N-1} I_j$. Part (iii) follows from the central limit theorem since

$$\int_1^N \frac{dx}{x} < \sum_1^{N-1} 1/j < 1 + \int_1^{N-1} \frac{dx}{x}$$

or

$$\log N < \sum_{1}^{N-1} 1/j < 1 + \log(N-1)$$

and so

$$\log N \approx \sum_{j=1}^{N-1} 1/j \quad \blacklozenge$$

Returning to the simplex algorithm, if we assume that n, m, and $n - m$ are all large, we have by Stirling's approximation that

$$N = \binom{n}{m} \sim \frac{n^{n+1/2}}{(n-m)^{n-m+1/2} m^{m+1/2} \sqrt{2\pi}}$$

and so, letting $c = n/m$,

$$\log N \sim (mc + \tfrac{1}{2}) \log(mc) - (m(c-1) + \tfrac{1}{2}) \log(m(c-1))$$
$$- (m + \tfrac{1}{2}) \log m - \tfrac{1}{2} \log(2\pi)$$

or

$$\log N \sim m \left[c \log \frac{c}{c-1} + \log(c-1) \right]$$

Now, as $\lim_{x \to \infty} x \log[x/(x-1)] = 1$, it follows that, when c is large,

$$\log N \sim m[1 + \log(c-1)]$$

Thus, for instance, if $n = 8000$, $m = 1000$, then the number of necessary transitions is approximately normally distributed with mean and variance equal to $1000(1 + \log 7) \approx 3000$. Hence, the number of necessary transitions would be roughly between

$$3000 \pm 2\sqrt{3000} \qquad \text{or roughly } 3000 \pm 110$$

95 percent of the time.

4.5.3. Using a Random Walk to Analyze a Probabilistic Algorithm for the Satisfiability Problem

Consider a Markov chain with states $0, 1, \ldots, n$ having

$$P_{0,1} = 1, \qquad P_{i,i+1} = p, \qquad P_{i,i-1} = q = 1 - p, \qquad 1 \leqslant i \leqslant n$$

and suppose that we are interested in studying the time that it takes for the chain to go from state 0 to state n. One approach to obtaining the mean time to reach state n would be to let m_i denote the mean time to go from state i to state $n, i = 0, \ldots, n - 1$. If we then condition on the initial transition, we obtain the following set of equations:

$$m_0 = 1 + m_1$$

$$m_i = E[\text{time to reach } n \,|\, \text{next state is } i + 1]p$$

$$+ E[\text{time to reach } n \,|\, \text{next state is } i - 1]q$$

$$= (1 + m_{i+1})p + (1 + m_{i-1})q$$

$$= 1 + pm_{i+1} + qm_{i-1}, \qquad i = 1, \ldots, n - 1$$

Whereas the preceding equations can be solved for $m_i, i = 0, \ldots, n - 1$, we do not pursue their solution; we instead make use of the special structure of the Markov chain to obtain a simpler set of equations. To start, let N_i denote the number of additional transitions that it takes the chain when it first enters state i until it enters state $i + 1$. By the Markovian property, it follows that these random variables $N_i, i = 0, \ldots, n - 1$ are independent. Also, we can express $N_{0,n}$, the number of transitions that it takes the chain to go from state 0 to state n, as

$$N_{0,n} = \sum_{i=0}^{n-1} N_i \tag{4.14}$$

Letting $\mu_i = E[N_i]$ we obtain, upon conditioning on the next transition after the chain enters state i, that for $i = 1, \ldots, n - 1$

$$\mu_i = 1 + E[\text{number of additional transitions to reach } i + 1 \,|\, \text{chain to } i - 1]q$$

Now, if the chain next enters state $i - 1$, then in order for it to reach $i + 1$ it must first return to state i and must then go from state $i + 1$. Hence, we have from the preceding that

$$\mu_i = 1 + E[N_{i-1}^* + N_i^*]q$$

where N_{i-1}^* and N_i^* are, respectively, the additional number of transitions to return to state i from $i - 1$ and the number to then go from i to $i + 1$. Now, it follows from the Markovian property that these random variables have, respectively, the same distributions as N_{i-1} and N_i. In addition, they are independent (although we will only use this when we compute the variance of $N_{0,n}$). Hence, we see that

$$\mu_i = 1 + q(\mu_{i-1} + \mu_i)$$

or

$$\mu_i = \frac{1}{p} + \frac{q}{p}\mu_{i-1}, \qquad i = 1, \ldots, n - 1$$

Starting with $\mu_0 = 1$, and letting $\alpha = q/p$, we obtain from the preceding recursion that

$$\mu_1 = 1/p + \alpha$$
$$\mu_2 = 1/p + \alpha(1/p + \alpha) = 1/p + \alpha/p + \alpha^2$$
$$\mu_3 = 1/p + \alpha(1/p + \alpha/p + \alpha^2)$$
$$= 1/p + \alpha/p + \alpha^2/p + \alpha^3$$

In general, we see that

$$\mu_i = \frac{1}{p} \sum_{j=0}^{i-1} \alpha^j + \alpha^i, \qquad i = 1, \ldots, n-1 \qquad (4.15)$$

Using Equation (4.14), we now get

$$E[N_{0,n}] = 1 + \frac{1}{p} \sum_{i=1}^{n-1} \sum_{j=0}^{i-1} \alpha^j + \sum_{i=1}^{n-1} \alpha^i$$

When $p = \frac{1}{2}$, and so $\alpha = 1$, we see from the preceding that

$$E[N_{0,n}] = 1 + (n-1)n + n - 1 = n^2$$

When $p \neq \frac{1}{2}$, we obtain that

$$E[N_{0,n}] = 1 + \frac{1}{p(1-\alpha)} \sum_{i=1}^{n-1} (1 - \alpha^i) + \frac{\alpha - \alpha^n}{1 - \alpha}$$
$$= 1 + \frac{1+\alpha}{1-\alpha}\left[n - 1 - \frac{(\alpha - \alpha^n)}{1-\alpha} \right] + \frac{\alpha - \alpha^n}{1-\alpha}$$
$$= 1 + \frac{2\alpha^{n+1} - (n+1)\alpha^2 + n - 1}{(1-\alpha)^2}$$

where the second equality used the fact that $p = 1/(1 + \alpha)$. Therefore, we see that when $\alpha > 1$, or equivalently when $p < \frac{1}{2}$, the expected number of transitions to reach n is an exponentially increasing function of n. On the other hand, when $p = \frac{1}{2}$, $E[N_{0,n}] = n^2$, and when $p > \frac{1}{2}$, $E[N_{0,n}]$ is, for large n, essentially linear in n.

Let us now compute $\text{Var}(N_{0,n})$. To do so, we will again make use of the representation given by Equation (4.14). Letting $v_i = \text{Var}(N_i)$, we start by determining the v_i recursively by using the conditional variance formula. Let $S_i = 1$ if the first transition out of state i is into state $i + 1$, and let $S_i = -1$ if the transition is into state $i - 1$, $i = 1, \ldots, n - 1$. Then,

given that $S_i = 1$: $N_i = 1$

given that $S_i = -1$: $N_i = 1 + N_{i-1}^* + N_i^*$

Hence,

$$E[N_i \,|\, S_i = 1] = 1$$

$$E[N_i \,|\, S_i = -1] = 1 + \mu_{i-1} + \mu_i$$

implying that

$$\begin{aligned}
\text{Var}(E[N_i \,|\, S_i]) &= \text{Var}(E[N_i \,|\, S_i] - 1) \\
&= (\mu_{i-1} + \mu_i)^2 q - (\mu_{i-1} + \mu_i)^2 q^2 \\
&= qp(\mu_{i-1} + \mu_i)^2
\end{aligned}$$

Also, since N_{i-1}^* and N_i^*, the numbers of transitions to return from state $i - 1$ to i and to then go from state i to state $i + 1$ are, by the Markovian property, independent random variables having the same distributions as N_{i-1} and N_i, respectively, we see that

$$\text{Var}(N_i \,|\, S_i = 1) = 0$$

$$\text{Var}(N_i \,|\, S_i = -1) = v_{i-1} + v_i$$

Hence,

$$E[\text{Var}(N_i \,|\, S_i)] = q(v_{i-1} + v_i)$$

From the conditional variance formula, we thus obtain that

$$v_i = pq(\mu_{i-1} + \mu_i)^2 + q(v_{i-1} + v_i)$$

or, equivalently,

$$v_i = q(\mu_{i-1} + \mu_i)^2 + \alpha v_{i-1} \qquad i = 1, \dots, n - 1$$

Starting with $v_0 = 0$, we obtain from the preceding recursion that

$$v_1 = q(\mu_0 + \mu_1)^2,$$

$$v_2 = q(\mu_1 + \mu_2)^2 + \alpha q(\mu_0 + \mu_1)^2,$$

$$v_3 = q(\mu_2 + \mu_3)^2 + \alpha q(\mu_1 + \mu_2)^2 + \alpha^2 q(\mu_0 + \mu_1)^2$$

In general, we have for $i > 0$,

$$v_i = q \sum_{j=1}^{i} \alpha^{i-j} (\mu_{j-1} + \mu_j)^2 \tag{4.16}$$

Therefore, we see that

$$\text{Var}(N_{0,n}) = \sum_{i=0}^{n-1} v_i = q \sum_{i=1}^{n-1} \sum_{j=1}^{i} \alpha^{i-j} (\mu_{j-1} + \mu_j)^2$$

where μ_j is given by Equation (4.15).

We see from Equations (4.15) and (4.16) that when $p \geqslant \frac{1}{2}$, and so $\alpha \leqslant 1$, that μ_i and v_i, the mean and variance of the number of transitions to go from state i to $i + 1$, do not increase too rapidly in i. For instance, when $p = \frac{1}{2}$ it follows from Equations (4.15) and (4.16) that

$$\mu_i = 2i + 1$$

and

$$v_i = \frac{1}{2} \sum_{j=1}^{i} (4j)^2 = 8 \sum_{j=1}^{i} j^2$$

Hence, since $N_{0,n}$ is the sum of independent random variables, which are of roughly similar magnitudes when $p \geqslant \frac{1}{2}$, it follows in this case from the central limit theorem that $N_{0,n}$ is, for large n, approximately normally distributed. In particular, when $p = \frac{1}{2}$, $N_{0,n}$ is approximately normal with mean n^2 and variance

$$\begin{aligned}
\text{Var}(N_{0,n}) &= 8 \sum_{i=1}^{n-1} \sum_{j=1}^{i} j^2 \\
&= 8 \sum_{j=1}^{n-1} \sum_{i=j}^{n-1} j^2 \\
&= 8 \sum_{j=1}^{n-1} (n - j)j^2 \\
&\approx 8 \int_{1}^{n-1} (n - x)x^2 \, dx \\
&\approx \tfrac{2}{3} n^4
\end{aligned}$$

Example 4.22 (The Satisfiability Problem): A Boolean variable x is one that takes on either of two values: TRUE or FALSE. If $x_i, i \geqslant 1$ are Boolean variables, then a Boolean clause of the form

$$x_1 + \bar{x}_2 + x_3$$

is TRUE if x_1 is TRUE, or if x_2 is FALSE, or if x_3 is TRUE. That is, the symbol "$+$" means "or" and $\bar{x}$ is TRUE if x is FALSE and vice versa. A Boolean formula is a combination of clauses such as

$$(x_1 + \bar{x}_2) * (x_1 + x_3) * (x_2 + \bar{x}_3) * (\bar{x}_1 + \bar{x}_2) * (x_1 + x_2)$$

In the preceding, the terms between the parentheses represent clauses, and the formula is TRUE if all the clauses are TRUE, and is FALSE otherwise. For a given Boolean formula, the *satisfiability problem* is to either determine values for the variables that result in the formula being TRUE, or to determine that the formula is never true. For instance, one set of values that makes the preceding formula TRUE is to set $x_1 = $ TRUE, $x_2 = $ FALSE, and $x_3 = $ FALSE.

Consider a formula of the n Boolean variables $x_1, \ldots, x_n$ and suppose that each clause in this formula refers to exactly two variables. We will now present a *probabilistic algorithm* that will either find values that satisfy the formula or determine *to a high probability* that it is not possible to satisfy it. To begin, start with an arbitrary setting of values. Then, at each stage choose a clause whose value is FALSE, and randomly choose one of the Boolean variables in that clause and change its value. That is, if the variable has value TRUE then change its value to FALSE, and vice versa. If this new setting makes the formula TRUE then stop, otherwise continue in the same fashion. If you have not stopped after $n^2(1 + 4\sqrt{\frac{2}{3}})$ repetitions, then declare that the formula cannot be satisfied. We will now argue that if there is a satisfiable assignment then this algorithm will find such an assignment with a probability very close to 1.

Let us start by assuming that there is a satisfiable assignment of truth values and let $\mathscr{A}$ be such an assignment. At each stage of the algorithm there is a certain assignment of values. Let Y_j denote the number of the n variables whose values at the jth stage of the algorithm agree with their values in $\mathscr{A}$. For instance, suppose that $n = 3$ and $\mathscr{A}$ consists of the settings $x_1 = x_2 = x_3 = $ TRUE. If the assignment of values at the jth step of the algorithm is $x_1 = $ TRUE, $x_2 = x_3 = $ FALSE, then $Y_j = 1$. Now, at each stage, the algorithm considers a clause that is not satisfied, thus implying that at least one of the values of the two variables in this clause does not agree with its value in $\mathscr{A}$. As a result, when we randomly choose one of the variables in this clause then there is a probability of at least $\frac{1}{2}$ that $Y_{j+1} = Y_j + 1$ and at most $\frac{1}{2}$ that $Y_{j+1} = Y_j - 1$. That is, independent of what has previously transpired in the algorithm, at each stage the number of settings in agreement with those in $\mathscr{A}$ will either increase or decrease by 1 and the probability of an increase is at least $\frac{1}{2}$ (it is 1 if both variables have values different from their values in $\mathscr{A}$). Thus, even though the process $Y_j, j \geq 0$ is not itself a Markov chain (why not?) it is intuitively clear that both the expectation and the variance of the number of stages of the algorithm needed to obtain the values of $\mathscr{A}$ will be less than or equal to the expectation and variance of the number of transitions to go from state 0 to state n in the Markov chain of Section 4.5.2. Hence, if the algorithm has not yet terminated because it found a set of satisfiable values different from that of $\mathscr{A}$, it will do so within an expected time of at most n^2 and with a standard deviation of at most $n^2\sqrt{\frac{2}{3}}$. In addition, since the time for the Markov chain to go from 0 to n is approximately normal when n is large we can be quite certain that a satisfiable assignment will be reached by $n^2 + 4(n^2\sqrt{\frac{2}{3}})$ stages, and thus if one has not been found by this number of stages of the algorithm we can be quite certain that there is no satisfiable assignment.

Our analysis also makes it clear why we assumed that there are only two variables in each clause. For if there were $k, k > 2$, variables in a clause then as any clause that is not presently satisfied may only have 1 incorrect setting, a randomly chosen variable whose value is changed might only increase the number of values in agreement with $\mathscr{A}$ with probability $1/k$ and so we could only conclude from our prior Markov chain results that the mean time to obtain the values in $\mathscr{A}$ is an exponential function of n, which is not an efficient algorithm when n is large. ✦

4.6. Mean Time Spent in Transient States

Consider now a finite state Markov chain and suppose that the states are numbered so that $\mathbf{T} = \{1, 2, \ldots, t\}$ denotes the set of transient states. Let

$$\mathbf{P_T} = \begin{bmatrix} P_{11} & P_{12} & \cdots & P_{1t} \\ \cdots & \cdots & \cdots & \cdots \\ P_{t1} & P_{t2} & \cdots & P_{tt} \end{bmatrix}$$

and note that since $\mathbf{P_T}$ specifies only the transition probabilities from transient states into transient states, some of its row sums are less than 1 (otherwise, T would be a closed class of states).

For transient states i and j, let s_{ij} denote the expected number of time periods that the Markov chain is in state j, given that it starts in state i. Let $\delta_{i,j} = 1$ when $i = j$ and let it be 0 otherwise. Condition on the initial transition to obtain

$$s_{ij} = \delta_{i,j} + \sum_k P_{ik} s_{kj}$$

$$= \delta_{i,j} + \sum_{k=1}^{t} P_{ik} s_{kj} \tag{4.17}$$

where the final equality follows since it is impossible to go from a recurrent to a transient state, implying that $s_{kj} = 0$ when k is a recurrent state.

Let $\mathbf{S}$ denote the matrix of values $s_{ij}, i, j = 1, \ldots, t$. That is,

$$\mathbf{S} = \begin{bmatrix} s_{11} & s_{12} & \cdots & s_{1t} \\ \cdots & \cdots & \cdots & \cdots \\ s_{t1} & s_{t2} & \cdots & s_{tt} \end{bmatrix}$$

In matrix notation, Equation (4.17) can be written as

$$\mathbf{S} = \mathbf{I} + \mathbf{P_T} \mathbf{S}$$

where $\mathbf{I}$ is the identity matrix of size t. Because the preceding equation is equivalent to

$$(\mathbf{I} - \mathbf{P}_T)\mathbf{S} = \mathbf{I}$$

we obtain, upon multiplying both sides by $(\mathbf{I} - \mathbf{P}_T)^{-1}$,

$$\mathbf{S} = (\mathbf{I} - \mathbf{P}_T)^{-1}$$

That is, the quantities s_{ij}, $i \in T$, $j \in T$, can be obtained by inverting the matrix $\mathbf{I} - \mathbf{P}_T$. (The existence of the inverse is easily established.)

Example 4.23 Consider the gambler's ruin problem with $p = 0.4$ and $N = 7$. Starting with 3 units, determine

(a) the expected amount of time the gambler has 5 units,
(b) the expected amount of time the gambler has 2 units.

Solution: The matrix $\mathbf{P}_T$, which specifies P_{ij}, $i, j \in \{1, 2, 3, 4, 5, 6\}$, is as follows:

		1	2	3	4	5	6
	1	0	0.4	0	0	0	0
	2	0.6	0	0.4	0	0	0
$\mathbf{P}_T =$	3	0	0.6	0	0.4	0	0
	4	0	0	0.6	0	0.4	0
	5	0	0	0	0.6	0	0.4
	6	0	0	0	0	0.6	0

Inverting $\mathbf{I} - \mathbf{P}_T$ gives

$$\mathbf{S} = (\mathbf{I} - \mathbf{P}_T)^{-1} = \begin{bmatrix} 1.6149 & 1.0248 & 0.6314 & 0.3691 & 0.1943 & 0.0777 \\ 1.5372 & 2.5619 & 1.5784 & 0.9228 & 0.4857 & 0.1943 \\ 1.4206 & 2.3677 & 2.9990 & 1.7533 & 0.9228 & 0.3691 \\ 1.2458 & 2.0763 & 2.6299 & 2.9990 & 1.5784 & 0.6314 \\ 0.9835 & 1.6391 & 2.0763 & 2.3677 & 2.5619 & 1.0248 \\ 0.5901 & 0.9835 & 1.2458 & 1.4206 & 1.5372 & 1.6149 \end{bmatrix}$$

Hence,

$$s_{3,5} = 0.9228, \qquad s_{3,2} = 2.3677 \quad \blacklozenge$$

For $i \in T$, $j \in T$, the quantity f_{ij}, equal to the probability that the Markov chain ever makes a transition into state j given that it starts in state i, is easily determined from $\mathbf{P}_T$. To determine the relationship, let us start by deriving an expression for s_{ij} by conditioning on whether state j is ever entered. This yields

$$s_{ij} = E[\text{time in } j \,|\, \text{start in } i, \text{ ever transit to } j] f_{ij}$$

$$+ E[\text{time in } j \,|\, \text{start in } i, \text{ never transit to } j](1 - f_{ij})$$

$$= (\delta_{i,j} + s_{jj}) f_{ij} + \delta_{i,j}(1 - f_{i,j})$$

$$= \delta_{i,j} + f_{ij} s_{jj}$$

since s_{jj} is the expected number of additional time periods spent in state j given that it is eventually entered from state i. Solving the preceding equation yields

$$f_{ij} = \frac{s_{ij} - \delta_{i,j}}{s_{jj}}$$

Example 4.24 In Example 4.23, what is the probability that the gambler ever has a fortune of 1?

Solution: Since $s_{3,1} = 1.4206$ and $s_{1,1} = 1.6149$, then

$$f_{3,1} = \frac{s_{3,1}}{s_{1,1}} = 0.8797$$

As a check, note that $f_{3,1}$ is just the probability that a gambler starting with 3 reaches 1 before 7. That is, it is the probability that the gambler's fortune will go down 2 before going up 4; which is the probability that a gambler starting with 2 will go broke before reaching 6. Therefore,

$$f_{3,1} = 1 - \frac{1 - (0.6/0.4)^2}{1 - (0.6/0.4)^6} = 0.8797$$

which checks with our earlier answer. ✦

4.7. Branching Processes

In this section we consider a class of Markov chains, known as *branching processes*, which have a wide variety of applications in the biological, sociological, and engineering sciences.

Consider a population consisting of individuals able to produce off-spring of the same kind. Suppose that each individual will, by the end of its lifetime, have produced j new offspring with probability P_j, $j \geqslant 0$, independently of the number produced by any other individual. We suppose that $P_j < 1$ for all $j \geqslant 0$. The number of individuals initially present, denoted by X_0, is called the size of the zeroth generation. All offspring of the zeroth generation constitute the first generation and their number is denoted by

X_1. In general, let X_n denote the size of the nth generation. It follows that $\{X_n, n = 0, 1, \ldots\}$ is a Markov chain having as its state space the set of nonnegative integers.

Note that state 0 is a recurrent state, since clearly $P_{00} = 1$. Also, if $P_0 > 0$, all other states are transient. This follows since $P_{i0} = P_0^i$, which implies that starting with i individuals there is a positive probability of at least P_0^i that no later generation will ever consist of i individuals. Moreover, since any finite set of transient states $\{1, 2, \ldots, n\}$ will be visited only finitely often, this leads to the important conclusion that, *if $P_0 > 0$, then the population will either die out or its size will converge to infinity.*

Let

$$\mu = \sum_{j=0}^{\infty} jP_j$$

denote the mean number of offspring of a single individual, and let

$$\sigma^2 = \sum_{j=0}^{\infty} (j - \mu)^2 P_j$$

be the variance of the number of offspring produced by a single individual.

Let us suppose that $X_0 = 1$, that is, initially there is a single individual present. We calculate $E[X_n]$ and $\text{Var}(X_n)$ by first noting that we may write

$$X_n = \sum_{i=1}^{X_{n-1}} Z_i$$

where Z_i represents the number of offspring of the ith individual of the $(n - 1)$st generation. By conditioning on X_{n-1}, we obtain

$$E[X_n] = E[E[X_n \mid X_{n-1}]]$$

$$= E\left[E\left[\sum_{i=1}^{X_{n-1}} Z_i \mid X_{n-1}\right]\right]$$

$$= E[X_{n-1}\mu]$$

$$= \mu E[X_{n-1}] \tag{4.18}$$

where we have used the fact that $E[Z_i] = \mu$. Since $E[X_0] = 1$, Equation (4.18) yields

$$E[X_1] = \mu,$$

$$E[X_2] = \mu E[X_1] = \mu^2,$$

$$\vdots$$

$$E[X_n] = \mu E[X_{n-1}] = \mu^n$$

Similarly, $\mathrm{Var}(X_n)$ may be obtained by using the conditional variance formula

$$\mathrm{Var}(X_n) = E[\mathrm{Var}(X_n \mid X_{n-1})] + \mathrm{Var}(E[X_n \mid X_{n-1}])$$

Now, given X_{n-1}, X_n is just the sum of X_{n-1} independent random variables each having the distribution $\{P_j, j \geqslant 0\}$. Hence,

$$\mathrm{Var}(X_n \mid X_{n-1}) = X_{n-1}\sigma^2$$

Thus, the conditional variance formula yields

$$\mathrm{Var}(X_n) = E[X_{n-1}\sigma^2] + \mathrm{Var}(X_{n-1}\mu)$$
$$= \sigma^2 \mu^{n-1} + \mu^2 \, \mathrm{Var}(X_{n-1})$$

Using the fact that $\mathrm{Var}(X_0) = 0$ we can show by mathematical induction that the preceding implies

$$\mathrm{Var}(X_n) = \begin{cases} \sigma^2 \mu^{n-1}\left(\dfrac{\mu^n - 1}{\mu - 1}\right), & \text{if } \mu \neq 1 \\[2mm] n\sigma^2, & \text{if } \mu = 1 \end{cases} \tag{4.19}$$

Let π_0 denote the probability that the population will eventually die out (under the assumption that $X_0 = 1$). More formally,

$$\pi_0 = \lim_{n \to \infty} P\{X_n = 0 \mid X_0 = 1\}$$

The problem of determining the value of π_0 was first raised in connection with the extinction of family surnames by Galton in 1889.

We first note that $\pi_0 = 1$ if $\mu < 1$. This follows since

$$\mu^n = E[X_n] = \sum_{j=1}^{\infty} j P\{X_n = j\}$$

$$\geqslant \sum_{j=1}^{\infty} 1 \cdot P\{X_n = j\}$$

$$= P\{X_n \geqslant 1\}$$

Since $\mu^n \to 0$ when $\mu < 1$, it follows that $P\{X_n \geqslant 1\} \to 0$, and hence $P\{X_n = 0\} \to 1$.

In fact, it can be shown that $\pi_0 = 1$ even when $\mu = 1$. When $\mu > 1$, it turns out that $\pi_0 < 1$, and an equation determining π_0 may be derived by conditioning on the number of offspring of the initial individual, as follows:

$$\pi_0 = P\{\text{population dies out}\}$$

$$= \sum_{j=0}^{\infty} P\{\text{population dies out} \mid X_1 = j\}P_j$$

Now, given that $X_1 = j$, the population will eventually die out if and only if each of the j families started by the members of the first generation eventually dies out. Since each family is assumed to act independently, and since the probability that any particular family dies out is just π_0, this yields

$$P\{\text{population dies out} \,|\, X_1 = j\} = \pi_0^j$$

and thus π_0 satisfies

$$\pi_0 = \sum_{j=0}^{\infty} \pi_0^j P_j \qquad (4.20)$$

in fact when $\mu > 1$, it can be shown that π_0 is the smallest positive number satisfying Equation (4.20).

Example 4.25   If $P_0 = \frac{1}{2}$, $P_1 = \frac{1}{4}$, $P_2 = \frac{1}{4}$, then determine π_0.

Solution: Since $\mu = \frac{3}{4} \leqslant 1$, it follows that $\pi_0 = 1$.   ✦

Example 4.26   If $P_0 = \frac{1}{4}$, $P_1 = \frac{1}{4}$, $P_2 = \frac{1}{2}$, then determine π_0.

Solution: π_0 satisfies

$$\pi_0 = \tfrac{1}{4} + \tfrac{1}{4}\pi_0 + \tfrac{1}{2}\pi_0^2$$

or

$$2\pi_0^2 - 3\pi_0 + 1 = 0$$

The smallest positive solution of this quadratic equation is $\pi_0 = \frac{1}{2}$.   ✦

Example 4.27   In Examples 4.25 and 4.26, what is the probability that the population will die out if it initially consists of n individuals?

Solution: Since the population will die out if and only if the families of each of the members of the initial generation die out, the desired probability is π_0^n. For Example 4.25 this yields $\pi_0^n = 1$, and for Example 4.26, $\pi_0^n = (\frac{1}{2})^n$.   ✦

4.8. Time Reversible Markov Chains

Consider a stationary ergodic Markov chain (that is, an ergodic Markov chain that has been in operation for a long time) having transition probabilities P_{ij} and stationary probabilities π_i, and suppose that starting at some time we trace the sequence of states going backward in time. That is, starting at time n, consider the sequence of states $X_n, X_{n-1}, X_{n-2}, \ldots$. It turns out that this sequence of states is itself a Markov chain with transition probabilities Q_{ij} defined by

$$
\begin{aligned}
Q_{ij} &= P\{X_m = j \mid X_{m+1} = i\} \\
&= \frac{P\{X_m = j, X_{m+1} = i\}}{P\{X_{m+1} = i\}} \\
&= \frac{P\{X_m = j\} P\{X_{m+1} = i \mid X_m = j\}}{P\{X_{m+1} = i\}} \\
&= \frac{\pi_j P_{ji}}{\pi_i}
\end{aligned}
$$

To prove that the reversed process is indeed a Markov chain, we must verify that

$$
P\{X_m = j \mid X_{m+1} = i, X_{m+2}, X_{m+3}, \ldots\} = P\{X_m = j \mid X_{m+1} = i\}
$$

To see that this is so, suppose that the present time is $m + 1$. Now, since $X_0, X_1, X_2, \ldots$ is a Markov chain, it follows that the conditional distribution of the future $X_{m+2}, X_{m+3}, \ldots$ given the present state X_{m+1} is independent of the past state X_m. However, independence is a symmetric relationship (that is, if A is independent of B, then B is independent of A), and so this means that given X_{m+1}, X_m is independent of $X_{m+2}, X_{m+3}, \ldots$. But this is exactly what we had to verify.

Thus, the reversed process is also a Markov chain with transition probabilities given by

$$
Q_{ij} = \frac{\pi_j P_{ji}}{\pi_i}
$$

If $Q_{ij} = P_{ij}$ for all i, j, then the Markov chain is said to be *time reversible*. The condition for time reversibility, namely, $Q_{ij} = P_{ij}$, can also be expressed as

$$
\pi_i P_{ij} = \pi_j P_{ji} \qquad \text{for all } i, j \tag{4.21}
$$

The condition in Equation (4.21) can be stated that, for all states i and j, the rate at which the process goes from i to j (namely, $\pi_i P_{ij}$) is equal to the rate at which it goes from j to i (namely, $\pi_j P_{ji}$). It is worth noting that this is an obvious necessary condition for time reversibility since a transition from i to j going backward in time is equivalent to a transition from j to i going forward in time; i.e., if $X_m = i$ and $X_{m-1} = j$, then a transition from i to j is observed if we are looking backward, and one from j to i if we are looking forward in time. Thus, the rate at which the forward process makes a transition from j to i is always equal to the rate at which the reverse process makes a transition from i to j; if time reversible, this must equal the rate at which the forward process makes a transition from i to j.

If we can find nonnegative numbers, summing to one, that satisfy Equation (4.21), then it follows that the Markov chain is time reversible and the numbers represent the limiting probabilities. This is so since if

$$x_i P_{ij} = x_j P_{ji} \qquad \text{for all } i, j, \quad \sum_i x_i = 1 \tag{4.22}$$

then summing over i yields

$$\sum_i x_i P_{ij} = x_j \sum_i P_{ji} = x_j, \qquad \sum_i x_i = 1$$

and, because the limiting probabilities π_i are the unique solution of the above, it follows that $x_i = \pi_i$ for all i.

Example 4.28 Consider a random walk with states $0, 1, \ldots, M$ and transition probabilities

$$P_{i,i+1} = \alpha_i = 1 - P_{i,i-1}, \qquad i = 1, \ldots, M-1,$$

$$P_{0,1} = \alpha_0 = 1 - P_{0,0},$$

$$P_{M,M} = \alpha_M = 1 - P_{M,M-1}$$

Without the need for any computations, it is possible to argue that this Markov chain, which can only make transitions from a state to one of its two nearest neighbors, is time reversible. This follows by noting that the number of transitions from i to $i+1$ must at all times be within 1 of the number from $i+1$ to i. This is so since between any two transitions from i to $i+1$ there must be one from $i+1$ to i (and conversely) since the only way to reenter i from a higher state is via state $i+1$. Hence, it follows that the rate of transitions from i to $i+1$ equals the rate from $i+1$ to i, and so the process is time reversible.

We can easily obtain the limiting probabilities by equating for each state $i = 0, 1, \ldots, M-1$ the rate at which the process goes from i to $i+1$ with the rate at which it goes from $i+1$ to i. This yields

$$\pi_0 \alpha_0 = \pi_1(1 - \alpha_1),$$

$$\pi_1 \alpha_1 = \pi_2(1 - \alpha_2),$$

$$\vdots$$

$$\pi_i \alpha_i = \pi_{i+1}(1 - \alpha_{i+1}), \qquad i = 0, 1, \ldots, M-1$$

Solving in terms of π_0 yields

$$\pi_1 = \frac{\alpha_0}{1 - \alpha_1} \pi_0,$$

$$\pi_2 = \frac{\alpha_1}{1 - \alpha_2} \pi_1 = \frac{\alpha_1 \alpha_0}{(1 - \alpha_2)(1 - \alpha_1)} \pi_0$$

and, in general,

$$\pi_i = \frac{\alpha_{i-1} \cdots \alpha_0}{(1 - \alpha_i) \cdots (1 - \alpha_1)} \pi_0, \qquad i = 1, 2, \ldots, M$$

Since $\Sigma_0^M \pi_i = 1$, we obtain

$$\pi_0 \left[1 + \sum_{j=1}^{M} \frac{\alpha_{j-1} \cdots \alpha_0}{(1 - \alpha_j) \cdots (1 - \alpha_1)} \right] = 1$$

or

$$\pi_0 = \left[1 + \sum_{j=1}^{M} \frac{\alpha_{j-1} \cdots \alpha_0}{(1 - \alpha_j) \cdots (1 - \alpha_1)} \right]^{-1} \qquad (4.23)$$

and

$$\pi_i = \frac{\alpha_{i-1} \cdots \alpha_0}{(1 - \alpha_i) \cdots (1 - \alpha_1)} \pi_0, \qquad i = 1, \ldots, M \qquad (4.24)$$

For instance, if $\alpha_i \equiv \alpha$, then

$$\pi_0 = \left[1 + \sum_{j=1}^{M} \left(\frac{\alpha}{1 - \alpha} \right)^j \right]^{-1}$$

$$= \frac{1 - \beta}{1 - \beta^{M+1}}$$

and, in general,

$$\pi_i = \frac{\beta^i (1 - \beta)}{1 - \beta^{M+1}}, \qquad i = 0, 1, \ldots, M$$

where

$$\beta = \frac{\alpha}{1 - \alpha} \quad \blacklozenge$$

Another special case of Example 4.28 is the following urn model, proposed by the physicists P. and T. Ehrenfest to describe the movements of molecules. Suppose that M molecules are distributed among two urns; and at each time point one of the molecules is chosen at random, removed from its urn, and placed in the other one. The number of molecules in urn I is a special case of the Markov chain of Example 4.28 having

$$\alpha_i = \frac{M - i}{M}, \qquad i = 0, 1, \ldots, M$$

Hence, using Equations (4.23) and (4.24) the limiting probabilities in this case are

$$\pi_0 = \left[1 + \sum_{j=1}^{M} \frac{(M - j + 1) \cdots (M - 1)M}{j(j-1)\cdots 1} \right]^{-1}$$

$$= \left[\sum_{j=0}^{M} \binom{M}{j} \right]^{-1}$$

$$= \left(\frac{1}{2} \right)^{M}$$

where we have used the identity

$$1 = \left(\frac{1}{2} + \frac{1}{2} \right)^{M}$$

$$= \sum_{j=0}^{M} \binom{M}{j} \left(\frac{1}{2} \right)^{M}$$

Hence, from Equation (4.24)

$$\pi_i = \binom{M}{i} \left(\frac{1}{2} \right)^{M}, \qquad i = 0, 1, \ldots, M$$

Because the preceding are just the binomial probabilities, it follows that in the long run, the positions of each of the M balls are independent and each one is equally likely to be in either urn. This, however, is quite intuitive, for if we focus on any one ball, it becomes quite clear that its position will be independent of the positions of the other balls (since no matter where the other $M - 1$ balls are, the ball under consideration at each stage will be moved with probability $1/M$) and by symmetry, it is equally likely to be in either urn.

Example 4.29   Consider an arbitrary connected graph (see Section 3.6 for definitions) having a number w_{ij} associated with arc (i, j) for each arc. One instance of such a graph is given by Figure 4.1. Now consider a particle moving from node to node in this manner: If at any time the particle resides

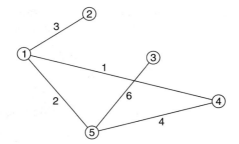

Figure 4.1. A connected graph with arc weights.

at node i, then it will next move to node j with probability P_{ij} where

$$P_{ij} = \frac{w_{ij}}{\Sigma_j w_{ij}}$$

and where w_{ij} is 0 if (i, j) is not an arc. For instance, for the graph of Figure 4.1, $P_{12} = 3/(3 + 1 + 2) = \frac{1}{2}$.

The time reversibility equations

$$\pi_i P_{ij} = \pi_j P_{ji}$$

reduce to

$$\pi_i \frac{w_{ij}}{\Sigma_j w_{ij}} = \pi_j \frac{w_{ji}}{\Sigma_i w_{ji}}$$

or, equivalently, since $w_{ij} = w_{ji}$

$$\frac{\pi_i}{\Sigma_j w_{ij}} = \frac{\pi_j}{\Sigma_i w_{ji}}$$

which is equivalent to

$$\frac{\pi_i}{\Sigma_j w_{ij}} = c$$

or

$$\pi_i = c \sum_j w_{ij}$$

or, since $1 = \Sigma_i \pi_i$

$$\pi_i = \frac{\Sigma_j w_{ij}}{\Sigma_i \Sigma_j w_{ij}}$$

Because the π_is given by this equation satisfy the time reversibility equations, it follows that the process is time reversible with these limiting probabilities.

For the graph of Figure 4.1 we have that

$$\pi_1 = \tfrac{6}{32}, \qquad \pi_2 = \tfrac{3}{32}, \qquad \pi_3 = \tfrac{6}{32}, \qquad \pi_4 = \tfrac{5}{32}, \qquad \pi_5 = \tfrac{12}{32} \quad \blacklozenge$$

If we try to solve Equation (4.22) for an arbitrary Markov chain with states $0, 1, \ldots, M$, it will usually turn out that no solution exists. For example, from Equation (4.22),

$$x_i P_{ij} = x_j P_{ji},$$
$$x_k P_{kj} = x_j P_{jk}$$

implying (if $P_{ij} P_{jk} > 0$) that

$$\frac{x_i}{x_k} = \frac{P_{ji} P_{kj}}{P_{ij} P_{jk}}$$

which in general need not equal P_{ki}/P_{ik}. Thus, we see that a necessary condition for time reversibility is that

$$P_{ik} P_{kj} P_{ji} = P_{ij} P_{jk} P_{ki} \qquad \text{for all } i, j, k \tag{4.25}$$

which is equivalent to the statement that, starting in state i, the path $i \to k \to j \to i$ has the same probability as the reversed path $i \to j \to k \to i$. To understand the necessity of this, note that time reversibility implies that the rate at which a sequence of transitions from i to k to j to i occurs must equal the rate of ones from i to j to k to i (why?), and so we must have

$$\pi_i P_{ik} P_{kj} P_{ji} = \pi_i P_{ij} P_{jk} P_{ki}$$

implying Equation (4.25) when $\pi_i > 0$.

In fact, we can show the following.

Theorem 4.2 An ergodic Markov chain for which $P_{ij} = 0$ whenever $P_{ji} = 0$ is time reversible if and only if starting in state i, any path back to i has the same probability as the reversed path. That is, if

$$P_{i,i_1} P_{i_1,i_2} \cdots P_{i_k,i} = P_{i,i_k} P_{i_k,i_{k-1}} \cdots P_{i_1,i} \tag{4.26}$$

for all states $i, i_1, \ldots, i_k$.

Proof We have already proven necessity. To prove sufficiency, fix states i and j and rewrite (4.26) as

$$P_{i,i_1} P_{i_1,i_2} \cdots P_{i_k,j} P_{ji} = P_{ij} P_{j,i_k} \cdots P_{i_1,i}$$

Summing the above over all states $i_1, \ldots, i_k$ yields

$$P_{ij}^{k+1} P_{ji} = P_{ij} P_{ji}^{k+1}$$

Letting $k \to \infty$ yields

$$\pi_j P_{ji} = P_{ij} \pi_i$$

which proves the theorem.   ◆

Example 4.30 Suppose we are given a set of n elements, numbered 1 through n, which are to be arranged in some ordered list. At each unit of time a request is made to retrieve one of these elements, element i being requested (independently of the past) with probability P_i. After being requested, the element then is put back but not necessarily in the same position. In fact, let us suppose that the element requested is moved one closer to the front of the list; for instance, if the present list ordering is 1, 3, 4, 2, 5 and element 2 is requested, then the new ordering becomes 1, 3, 2, 4, 5. We are interested in the long-run average position of the element requested.

For any given probability vector $P = (P_1, \ldots, P_n)$, the preceding can be modeled as a Markov chain with $n!$ states, with the state at any time being the list order at that time. We shall show that this Markov chain is time reversible and then use this to show that the average position of the element requested when this one-closer rule is in effect is less than when the rule of always moving the requested element to the front of the line is used. The time reversibility of the resulting Markov chain when the one-closer reordering rule is in effect easily follows from Theorem 4.2. For instance, suppose $n = 3$ and consider the following path from state $(1, 2, 3)$ to itself:

$$(1, 2, 3) \to (2, 1, 3) \to (2, 3, 1) \to (3, 2, 1) \to (3, 1, 2) \to (1, 3, 2) \to (1, 2, 3)$$

The product of the transition probabilities in the forward direction is

$$P_2 P_3 P_3 P_1 P_1 P_2 = P_1^2 P_2^2 P_3^2$$

whereas in the reverse direction, it is

$$P_3 P_3 P_2 P_2 P_1 P_1 = P_1^2 P_2^2 P_3^2$$

Because the general result follows in much the same manner, the Markov chain is indeed time reversible. [For a formal argument note that if f_i denotes the number of times element i moves forward in the path, then as the path goes from a fixed state back to itself, it follows that element i will also move backward f_i times. Therefore, since the backward moves of element i are precisely the times that it moves forward in the reverse path, it follows that the product of the transition probabilities for both the path and its reversal will equal

$$\prod_i P_i^{f_i + r_i}$$

where r_i is equal to the number of times that element i is in the first position and the path (or the reverse path) does not change states.]

For any permutation $i_1, i_2, \ldots, i_n$ of $1, 2, \ldots, n$, let $\pi(i_1 i_2, \ldots, i_n)$ denote the limiting probability under the one-closer rule. By time reversibility we have

$$P_{i_{j+1}} \pi(i_1, \ldots, i_j, i_{j+1}, \ldots, i_n) = P_{i_j} \pi(i_1, \ldots, i_{j+1}, i_j, \ldots, i_n) \qquad (4.27)$$

for all permutations.

Now the average position of the element requested can be expressed (as in Section 3.6.1) as

Average position $= \sum_i P_i E[\text{Position of element } i]$

$$= \sum_i P_i \left[1 + \sum_{j \neq i} P\{\text{element } j \text{ precedes element } i\} \right]$$

$$= 1 + \sum_i \sum_{j \neq i} P_i P\{e_j \text{ precedes } e_i\}$$

$$= 1 + \sum_{i<j} [P_i P\{e_j \text{ precedes } e_i\} + P_j P\{e_i \text{ precedes } e_j\}]$$

$$= 1 + \sum_{i<j} [P_i P\{e_j \text{ precedes } e_i\} + P_j(1 - P\{e_j \text{ precedes } e_i\})]$$

$$= 1 + \sum_{i<j} (P_i - P_j) P\{e_j \text{ precedes } e_i\} + \sum_{i<j} P_j$$

Hence, to minimize the average position of the element requested, we would want to make $P\{e_j \text{ precedes } e_i\}$ as large as possible when $P_j > P_i$ and as small as possible when $P_i > P_j$. Now under the front-of-the-line rule we showed in Section 3.6.1 that

$$P\{e_j \text{ precedes } e_i\} = \frac{P_j}{P_j + P_i}$$

(since under the front-of-the-line rule element j will precede element i if and only if the last request for either i or j was for j).

Therefore, to show that the one-closer rule is better than the front-of-the-line rule, it suffices to show that under the one-closer rule

$$P\{e_j \text{ precedes } e_i\} > \frac{P_j}{P_j + P_i} \qquad \text{when } P_j > P_i$$

Now consider any state where element i precedes element j, say, $(\ldots, i, i_1, \ldots, i_k, j, \ldots)$. By successive transpositions using Equation (4.27), we have

$$\pi(\ldots, i, i_1, \ldots, i_k, j, \ldots) = \left(\frac{P_i}{P_j}\right)^{k+1} \pi(\ldots, j, i_1, \ldots, i_k, i, \ldots) \qquad (4.28)$$

For instance,

$$\pi(1, 2, 3) = \frac{P_2}{P_3} \pi(1, 3, 2) = \frac{P_2}{P_3} \frac{P_1}{P_3} \pi(3, 1, 2)$$

$$= \frac{P_2}{P_3} \frac{P_1}{P_3} \frac{P_1}{P_2} \pi(3, 2, 1) = \left(\frac{P_1}{P_3}\right)^2 \pi(3, 2, 1)$$

Now when $P_j > P_i$, Equation (4.28) implies that

$$\pi(\ldots, i, i_1, \ldots, i_k, j, \ldots) < \frac{P_i}{P_j} \pi(\ldots, j, i_1, \ldots, i_k, i, \ldots)$$

Letting $\alpha(i, j) = P\{e_i \text{ precedes } e_j\}$, we see by summing over all states for which i precedes j and by using the preceding that

$$\alpha(i, j) < \frac{P_i}{P_j} \alpha(j, i)$$

which, since $\alpha(i, j) = 1 - \alpha(j, i)$, yields

$$\alpha(j, i) > \frac{P_j}{P_j + P_i}$$

Hence, the average position of the element requested is indeed smaller under the one-closer rule than under the front-of-the-line rule. ◆

The concept of the reversed chain is useful even when the process is not time reversible. To illustrate this, we start with the following proposition whose proof is left as an exercise.

Proposition 4.6 Consider an irreducible Markov chain with transition probabilities P_{ij}. If one can find positive numbers π_i, $i \geq 0$, summing to one, and a transition probability matrix $\mathbf{Q} = [Q_{ij}]$ such that

$$\pi_i P_{ij} = \pi_j Q_{ji} \qquad (4.29)$$

then the Q_{ij} are the transition probabilities of the reversed chain and the π_i are the stationary probabilities both for the original and reversed chain.

The importance of the preceding proposition is that, by thinking backward, we can sometimes guess at the nature of the reversed chain and then use the set of equations (4.29) to obtain both the stationary probabilities and the Q_{ij}.

Example 4.31 A single bulb is necessary to light a given room. When the bulb in use fails, it is replaced by a new one at the beginning of the next day. Let X_n equal i if the bulb in use at the beginning of day n is in its ith day of use (that is, if its present age is i). For instance, if a bulb fails on day $n - 1$, then a new bulb will be put in use at the beginning of day n and so $X_n = 1$. If we suppose that each bulb, independently, fails on its ith day of use with probability p_i, $i \geq 1$, then it is easy to see that $\{X_n, n \geq 1\}$ is a Markov chain whose transition probabilities are as follows:

$$P_{i,1} = P\{\text{bulb, on its } i\text{th day of use, fails}\}$$

$$= P\{\text{life of bulb} = i \,|\, \text{life of bulb} \geq i\}$$

$$= \frac{P\{L = i\}}{P\{L \geq i\}}$$

where L, a random variable representing the lifetime of a bulb, is such that $P\{L = i\} = p_i$. Also,

$$P_{i,i+1} = 1 - P_{i,1}$$

Suppose now that this chain has been in operation for a long (in theory, an infinite) time and consider the sequence of states going backward in time. Since, in the forward direction, the state is always increasing by 1 until it reaches the age at which the item fails, it is easy to see that the reverse chain will always decrease by 1 until it reaches 1 and then it will jump to a random value representing the lifetime of the (in real time) previous bulb. Thus, it seems that the reverse chain should have transition probabilities given by

$$Q_{i,i-1} = 1 \qquad i > 1$$

$$Q_{1,i} = p_i, \qquad i \geq 1$$

To check this, and at the same time determine the stationary probabilities, we must see if we can find, with the $Q_{i,j}$ as given above, positive numbers $\{\pi_i\}$ such that

$$\pi_i P_{i,j} = \pi_j Q_{j,i}$$

To begin, let $j = 1$ and consider the resulting equations:

$$\pi_i P_{i,1} = \pi_1 Q_{1,i}$$

This is equivalent to

$$\pi_i \frac{P\{L = i\}}{P\{L \geq i\}} = \pi_1 P\{L = i\}$$

or

$$\pi_i = \pi_1 P\{L \geq i\}$$

Summing over all i yields

$$1 = \sum_{i=1}^{\infty} \pi_i = \pi_1 \sum_{i=1}^{\infty} P\{L \geq i\} = \pi_1 E[L]$$

and so, for the Q_{ij} above to represent the reverse transition probabilities, it is necessary for the stationary probabilities to be

$$\pi_i = \frac{P\{L \geq i\}}{E[L]}, \qquad i \geq 1$$

To finish the proof that the reverse transition probabilities and stationary probabilities are as given, all that remains is to show that they satisfy

$$\pi_i P_{i,i+1} = \pi_{i+1} Q_{i+1,i}$$

which is equivalent to

$$\frac{P\{L \geq i\}}{E[L]} \left(1 - \frac{P\{L = i\}}{P\{L \geq i\}} \right) = \frac{P\{L \geq i+1\}}{E[L]}$$

and which is true since $P\{L \geq i\} - P\{L = i\} = P\{L \geq i+1\}$. ✦

4.9. Markov Chain Monte Carlo Methods

Let $\mathbf{X}$ be a discrete random vector whose set of possible values is $\mathbf{x}_j, j \geq 1$. Let the probability mass function of $\mathbf{X}$ be given by $P\{\mathbf{X} = \mathbf{x}_j\}, j \geq 1$, and suppose that we are interested in calculating

$$\theta = E[h(\mathbf{X})] = \sum_{j=1}^{\infty} h(\mathbf{x}_j) P\{\mathbf{X} = \mathbf{x}_j\}$$

for some specified function h. In situations where it is computationally difficult to evaluate the function $h(\mathbf{x}_j), j \geq 1$, we often turn to simulation to approximate θ. The usual approach, called *Monte Carlo simulation*, is to use random numbers to generate a partial sequence of independent and identically distributed random vectors $\mathbf{X}_1, \mathbf{X}_2, \ldots, \mathbf{X}_n$ having the mass function $P\{\mathbf{X} = \mathbf{x}_j\}, j \geq 1$ (see Chapter 11 for a discussion as to how this can be accomplished). Since the strong law of large numbers yields

$$\lim_{n \to \infty} \sum_{i=1}^{n} \frac{h(\mathbf{X}_i)}{n} = \theta \tag{4.30}$$

it follows that we can estimate θ by letting n be large and using the average of the values of $h(\mathbf{X}_i)$, $i = 1, \ldots, n$ as the estimator.

It often, however, turns out that it is difficult to generate a random vector having the specified probability mass function, particularly if $\mathbf{X}$ is a vector of dependent random variables. In addition, its probability mass function is sometimes given in the form $P\{\mathbf{X} = \mathbf{x}_j\} = Cb_j$, $j \geqslant 1$, where the b_j are specified, but C must be computed, and in many applications it is not computationally feasible to sum the b_j so as to determine C. Fortunately, however, there is another way of using simulation to estimate θ in these situations. It works by generating a sequence, not of independent random vectors, but of the successive states of a vector-valued Markov chain $\mathbf{X}_1, \mathbf{X}_2, \ldots$ whose stationary probabilities are $P\{\mathbf{X} = \mathbf{x}_j\}, j \geqslant 1$. If this can be accomplished, then it would follow from Proposition 4.3 that Equation (4.30) remains valid, implying that we can then use $\sum_{i=1}^{n} h(\mathbf{X}_i)/n$ as an estimator of θ.

We now show how to generate a Markov chain with arbitrary stationary probabilities that may only be specified up to a multiplicative constant. Let $b(j)$, $j = 1, \ldots$ be positive numbers whose sum $B = \sum_{j=1}^{\infty} b(j)$ is finite. The following, known as the *Hastings–Metropolis algorithm*, can be used to generate a time reversible Markov chain whose stationary probabilities are

$$\pi(j) = b(j)/B, \qquad j = 1, \ldots$$

To begin, let $\mathbf{Q}$ be any specified irreducible Markov transition probability matrix on the integers, with $q(i, j)$ representing the row i column j element of $\mathbf{Q}$. Now define a Markov chain $\{X_n, n \geqslant 0\}$ as follows. When $X_n = i$, generate a random variable Y such that $P\{Y = j\} = q(i, j)$, $j = 1, \ldots$. If $Y = j$, then set X_{n+1} equal to j with probability $\alpha(i, j)$, and set it equal to i with probability $1 - \alpha(i, j)$. Under these conditions, it is easy to see that the sequence of states constitutes a Markov chain with transition probabilities $P_{i,j}$ given by

$$P_{i,j} = q(i, j)\alpha(i, j), \qquad \text{if } j \neq i$$

$$P_{i,i} = q(i, i) + \sum_{k \neq i} q(i, k)(1 - \alpha(i, k))$$

This Markov chain will be time reversible and have stationary probabilities $\pi(j)$ if

$$\pi(i)P_{i,j} = \pi(j)P_{j,i} \qquad \text{for } j \neq i$$

which is equivalent to

$$\pi(i)q(i, j)\alpha(i, j) = \pi(j)q(j, i)\alpha(j, i) \tag{4.31}$$

But if we take $\pi_j = b(j)/B$ and set

$$\alpha(i, j) = \min\left(\frac{\pi(j)q(j, i)}{\pi(i)q(i, j)}, 1\right) \tag{4.32}$$

then Equation (4.31) is easily seen to be satisfied. For if

$$\alpha(i, j) = \frac{\pi(j)q(j, i)}{\pi(i)q(i, j)}$$

then $\alpha(j, i) = 1$ and Equation (4.31) follows, and if $\alpha(i, j) = 1$ then

$$\alpha(j, i) = \frac{\pi(i)q(i, j)}{\pi(j)q(j, i)}$$

and again Equation (4.31) holds, thus showing that the Markov chain is time reversible with stationary probabilities $\pi(j)$. Also, since $\pi(j) = b(j)/B$, we see from (4.32) that

$$\alpha(i, j) = \min\left(\frac{b(j)q(j, i)}{b(i)q(i, j)}, 1\right)$$

which shows that the value of B is not needed to define the Markov chain, because the values $b(j)$ suffice. Also, it is almost always the case that $\pi(j)$, $j \geqslant 1$ will not only be stationary probabilities but will also be limiting probabilities. (Indeed, a sufficient condition is that $P_{i,i} > 0$ for some i.)

Example 4.32 Suppose that we want to generate a uniformly distributed element in $\mathscr{S}$, the set of all permutations $(x_1, \ldots, x_n)$ of the numbers $(1, \ldots, n)$ for which $\sum_{j=1}^{n} jx_j > a$ for a given constant a. To utilize the Hastings–Metropolis algorithm we need to define an irreducible Markov transition probability matrix on the state space $\mathscr{S}$. To accomplish this, we first define a concept of "neighboring" elements of $\mathscr{S}$, and then construct a graph whose vertex set is $\mathscr{S}$. We start by putting an arc between each pair of neighboring elements in $\mathscr{S}$, where any two permutations in $\mathscr{S}$ are said to be neighbors if one results from an interchange of two of the positions of the other. That is, $(1, 2, 3, 4)$ and $(1, 2, 4, 3)$ are neighbors whereas $(1, 2, 3, 4)$ and $(1, 3, 4, 2)$ are not. Now, define the q transition probability function as follows. With $N(\mathbf{s})$ defined as the set of neighbors of $\mathbf{s}$, and $|N(\mathbf{s})|$ equal to the number of elements in the set $N(\mathbf{s})$, let

$$q(\mathbf{s}, \mathbf{t}) = \frac{1}{|N(\mathbf{s})|} \qquad \text{if } \mathbf{t} \in N(\mathbf{s})$$

That is, the candidate next state from $\mathbf{s}$ is equally likely to be any of its neighbors. Since the desired limiting probabilities of the Markov chain are $\pi(\mathbf{s}) = C$, it follows that $\pi(\mathbf{s}) = \pi(\mathbf{t})$, and so

$$\alpha(\mathbf{s}, \mathbf{t}) = \min(|N(\mathbf{s})|/|N(\mathbf{t})|, 1)$$

That is, if the present state of the Markov chain is $\mathbf{s}$ then one of its neighbors is randomly chosen, say, $\mathbf{t}$. If $\mathbf{t}$ is a state with fewer neighbors than $\mathbf{s}$ (in graph theory language, if the degree of vertex $\mathbf{t}$ is less than that of vertex $\mathbf{s}$), then the next state is $\mathbf{t}$. If not, a uniform $(0, 1)$ random number U is generated and the next state is $\mathbf{t}$ if $U < |N(\mathbf{s})|/|N(\mathbf{t})|$ and is $\mathbf{s}$ otherwise. The limiting probabilities of this Markov chain are $\pi(\mathbf{s}) = 1/|\mathscr{S}|$, where $|\mathscr{S}|$ is the (unknown) number of permutations in $\mathscr{S}$.

The preceding does not specify how to randomly choose a neighbor permutation of $\mathbf{s}$. One possibility, which is efficient when n is small enough so that we can easily keep track of all the neighbors of $\mathbf{s}$, is to just randomly choose one, call it $\mathbf{t}$, as the target next state. The number of the neighbors of $\mathbf{t}$ would then have to be determined, and the next state of the Markov chain would then either be $\mathbf{t}$ with probability $\min(1, |N(\mathbf{s})|/|N(\mathbf{t})|)$ or it would remain $\mathbf{s}$ otherwise. However, if n is large this may be impractical, and a better approach might be to expand the state space to consist of all $n!$ permutations. The desired limiting probability mass function is then

$$\pi(\mathbf{s}) = \begin{cases} C, & \mathbf{s} \in \mathscr{S} \\ 0, & \mathbf{s} \notin \mathscr{S} \end{cases}$$

With this setup, each permutation $\mathbf{s}$ has $\binom{n}{2}$ neighbors, and one can be randomly chosen by generating a random subset of size two from the set $1, \ldots, n$ and if i and j are chosen then the candidate next state $\mathbf{t}$ is obtained by interchanging the values of the ith and jth coordinates of $\mathbf{s}$. If $\mathbf{t} \in \mathscr{S}$ then $\mathbf{t}$ becomes the next state of the chain, and if not then the next state remains $\mathbf{s}$.   ✦

The most widely used version of the Hastings–Metropolis algorithm is the Gibbs sampler. Let $\mathbf{X} = (X_1, \ldots, X_n)$ be a random vector with probability mass function $p(\mathbf{x})$, which may only be specified up to a multiplicative constant, and suppose that we want to generate a random vector whose distribution is that of the conditional distribution of $\mathbf{X}$ given that $\mathbf{X} \in \mathscr{A}$ for some set $\mathscr{A}$. That is, we want to generate a random vector having mass function

$$f(\mathbf{x}) = \frac{p(\mathbf{x})}{P\{\mathbf{X} \in \mathscr{A}\}} \qquad \text{for } \mathbf{x} \in \mathscr{A}$$

The Gibbs sampler assumes that for any $i, i = 1, \ldots, n$ and values $x_j, j \neq i$, we can generate a random variable X having the probability mass function

$$P\{X = x\} = P\{X_i = x \mid X_j = x_j, j \neq i\}$$

It operates by considering a Markov chain with states

$$\mathbf{x} = (x_1, \ldots, x_i, \ldots, x_n) \in \mathscr{A}$$

and then uses the Hastings–Metropolis algorithm with Markov transition probabilities defined as follows. Whenever the present state is $\mathbf{x}$, a coordinate that is equally likely to be any of $1, \ldots, n$ is generated. If coordinate i is the one chosen, then a random variable X having probability mass function $P\{X = x\} = P\{X_i = x \mid X_j = x_j, j \neq i\}$ is generated, and if $X = x$ then the state $\mathbf{y} = (x_1, \ldots, x_{i-1}, x, x_{i+1}, \ldots, x_n)$ is considered as a candidate for transition. In other words, the Gibbs sampler uses the Hastings–Metropolis algorithm with

$$q(\mathbf{x}, \mathbf{y}) = \frac{1}{n} P\{X_i = x \mid X_j = x_j, j \neq i\} = \frac{1}{n} \frac{p(\mathbf{y})}{P\{X_j = x_j, j \neq i\}}$$

Since we want the limiting mass function to be f, we have from Equation (4.32) that the vector $\mathbf{y}$ is then accepted as the new state with probability

$$\alpha(\mathbf{x}, \mathbf{y}) = \min\left(\frac{f(\mathbf{y})q(\mathbf{y}, \mathbf{x})}{f(\mathbf{x})q(\mathbf{x}, \mathbf{y})}, 1\right)$$

Now, for $\mathbf{x} \in \mathscr{A}$ and $\mathbf{y} \in \mathscr{A}$

$$\frac{f(\mathbf{y})q(\mathbf{y}, \mathbf{x})}{f(\mathbf{x})q(\mathbf{x}, \mathbf{y})} = \frac{f(\mathbf{y})p(\mathbf{x})}{f(\mathbf{x})p(\mathbf{y})} = 1$$

whereas for $\mathbf{x} \in \mathscr{A}$ and $\mathbf{y} \notin \mathscr{A}$ we have [since $f(\mathbf{y}) = 0$]

$$\frac{f(\mathbf{y})q(\mathbf{y}, \mathbf{x})}{f(\mathbf{x})q(\mathbf{x}, \mathbf{y})} = 0$$

Hence, the next state is either $\mathbf{y}$ if $\mathbf{y} \in \mathscr{A}$ or it remains $\mathbf{x}$ if $\mathbf{y} \notin \mathscr{A}$.

Example 4.33 Suppose we want to generate n uniformly distributed points in the circle of radius 1 centered at the origin, conditional on the event that no two points are within a distance d of each other, where

$$\beta = P\{\text{no two points are within } d \text{ of each other}\}$$

is assumed to be a small positive number. (If β were not small, then we could just continue to generate sets of n uniformly distributed points in the circle, stopping the first time that no two points in the set are within d of each other.) This can be accomplished by the Gibbs sampler by starting with any n points in the circle, $\mathbf{x}_1, \ldots, \mathbf{x}_n$, for which no two are within a distance d of each other. Then generate the value of a random variable I that is equally likely to be any of the values $1, \ldots, n$. Also generate a random point in the circle (see Chapter 11 for details of these generations). If this point is not within d of any of the other $n - 1$ points excluding $\mathbf{x}_I$ then replace $\mathbf{x}_I$ by this generated point, otherwise do not make a change. After a large number of iterations the set of n points will approximately have the desired distribution. ✦

The Gibbs sampler for generating a random vector $\mathbf{X}$ conditional on the event that $\mathbf{X} \in \mathscr{A}$ moves from state to state by choosing a coordinate I at random and then generating a random variable from the conditional distribution of X_I given the values of the other random variables, $X_j, j \neq I$. If the vector obtained by replacing the old value of X_I by this generated value remains in $\mathscr{A}$ then it becomes the next state, and if not then the next state remains unchanged from the previous one. However, if we can easily generate X_I conditional both on the values of $X_j, j \neq I$ and on the condition that $\mathbf{X} \in A$, then the Gibbs sampler may be performed by doing this generation and then obtaining the next state of the Markov chain by replacing the old value of X_I by the value generated. This is illustrated by our next example.

Example 4.34 Let X_i, $i = 1, \ldots, n$ be independent random variables with X_i having an exponential distribution with rate λ_i, $i = 1, \ldots, n$. Let $S = \sum_{i=1}^{n} X_i$ and suppose that we want to generate the random vector $\mathbf{X} = (X_1, \ldots, X_n)$ conditional on the event that $S > c$ for some large positive constant c. That is, we want to generate the value of a random vector whose density function is given by

$$f(x_1, \ldots, x_n) = \frac{1}{P\{S > c\}} \prod_{i=1}^{n} \lambda_i e^{-\lambda_i x_i} \qquad \text{if } \sum_{i=1}^{n} x_i > c$$

This is easily accomplished by starting with an initial vector $\mathbf{x} = (x_1, \ldots, x_n)$ satisfying $x_i > 0$, $i = 1, \ldots, n$ and $\sum_{i=1}^{n} x_i > c$. Then generate a variable I that is equally likely to be any of $1, \ldots, n$. Now, we want to generate an exponential random variable X with rate λ_I, conditioned on the event that $X + \sum_{j \neq I} x_j > c$. That is, we want to generate the value of X conditional on the event that it exceeds $c - \sum_{j \neq I} x_j$. Hence, using the fact that an exponential conditioned to be greater than a positive constant is distributed as the constant plus the exponential, we see that we should generate an exponential random variable Y with rate λ_I, and set

$$X = Y + \left(c - \sum_{j \neq I} x_j \right)^+$$

where b^+ is equal to b when $b > 0$ and is 0 otherwise. The value of x_I should then be reset to equal X and a new iteration of the algorithm begun. ✦

Remark   As can be seen by Examples 4.33 and 4.34, although the theory
for the Gibbs sampler was presented under the assumption that the
distribution to be generated was discrete, it also holds when this distribution
is continuous.

4.10. Markov Decision Processes

Consider a process that is observed at discrete time points to be in any one
of M possible states, which we number by $1, 2, \ldots, M$. After observing the
state of the process, an action must be chosen, and we let A, assumed finite,
denote the set of all possible actions.

If the process is in state i at time n and action a is chosen, then the next
state of the system is determined according to the transition probabilities
$P_{ij}(a)$. If we let X_n denote the state of the process at time n and a_n the action
chosen at time n, then the above is equivalent to stating that

$$P\{X_{n+1} = j \,|\, X_0, a_0, X_1, a_1, \ldots, X_n = i, a_n = a\} = P_{ij}(a)$$

Thus, the transition probabilities are functions only of the present state and
the subsequent action.

By a policy, we mean a rule for choosing actions. We shall restrict
ourselves to policies which are of the form that the action they prescribe at
any time depends only on the state of the process at that time (and not on
any information concerning prior states and actions). However, we shall
allow the policy to be "randomized" in that its instructions may be to
choose actions according to a probability distribution. In other words, a
policy β is a set of numbers $\beta = \{\beta_i(a), a \in A, i = 1, \ldots, M\}$ with the interpretation that if the process is in state i, then action a is to be chosen with
probability $\beta_i(a)$. Of course, we need have that

$$0 \leqslant \beta_i(a) \leqslant 1, \qquad \text{for all } i, a$$

$$\sum_a \beta_i(a) = 1, \qquad \text{for all } i$$

Under any given policy β, the sequence of states $\{X_n, n = 0, 1, \ldots\}$
constitutes a Markov chain with transition probabilities $P_{ij}(\beta)$ given by

$$P_{ij}(\beta) = P_\beta\{X_{n+1} = j \,|\, X_n = i\}^*$$

$$= \sum_a P_{ij}(a)\beta_i(a)$$

where the last equality follows by conditioning on the action chosen when

---

* We use the notation P_β to signify that the probability is conditional on the fact that policy
β is used.

in state i. Let us suppose that for every choice of a policy β, the resultant Markov chain $\{X_n, n = 0, 1, \ldots\}$ is ergodic.

For any policy β, let π_{ia} denote the limiting (or steady-state) probability that the process will be in state i and action a will be chosen if policy β is employed. That is,

$$\pi_{ia} = \lim_{n \to \infty} P_\beta\{X_n = i, a_n = a\}$$

The vector $\pi = (\pi_{ia})$ must satisfy

(i) $\pi_{ia} \geq 0$ for all i, a
(ii) $\Sigma_i \Sigma_a \pi_{ia} = 1$
(iii) $\Sigma_a \pi_{ja} = \Sigma_i \Sigma_a \pi_{ia} P_{ij}(a)$ for all j (4.33)

Equations (i) and (ii) are obvious, and Equation (iii) which is an analogue of Equation (4.7) follows as the left-hand side equals the steady-state probability of being in state j and the right-hand side is the same probability computed by conditioning on the state and action chosen one stage earlier.

Thus for any policy β, there is a vector $\pi = (\pi_{ia})$ which satisfies (i)–(iii) and with the interpretation that π_{ia} is equal to the steady-state probability of being in state i and choosing action a when policy β is employed. Moreover, it turns out that the reverse is also true. Namely, for any vector $\pi = (\pi_{ia})$ which satisfies (i)–(iii), there exists a policy β such that if β is used, then the steady-state probability of being in i and choosing action a equals π_{ia}. To verify this last statement, suppose that $\pi = (\pi_{ia})$ is a vector which satisfies (i)–(iii). Then, let the policy $\beta = (\beta_i(a))$ be

$$\beta_i(a) = P\{\beta \text{ chooses } a \,|\, \text{state is } i\}$$

$$= \frac{\pi_{ia}}{\Sigma_a \pi_{ia}}$$

Now let P_{ia} denote the limiting probability of being in i and choosing a when policy β is employed. We need to show that $P_{ia} = \pi_{ia}$. To do so, first note that $\{P_{ia}, i = 1, \ldots, M, \ a \in A\}$ are the limiting probabilities of the two-dimensional Markov chain $\{(X_n, a_n), n \geq 0\}$. Hence, by the fundamental Theorem 4.1, they are the unique solution of

(i') $P_{ia} \geq 0$
(ii') $\Sigma_i \Sigma_a P_{ia} = 1$
(iii') $P_{ja} = \Sigma_i \Sigma_{a'} P_{ia'} P_{ij}(a') \beta_j(a)$

where (iii') follows since

$$P\{X_{n+1} = j, a_{n+1} = a \,|\, X_n = i, a_n = a'\} = P_{ij}(a') \beta_j(a)$$

Since

$$\beta_j(a) = \frac{\pi_{ja}}{\sum_a \pi_{ja}}$$

we see that (P_{ia}) is the unique solution of

$$P_{ia} \geqslant 0,$$

$$\sum_i \sum_a P_{ia} = 1,$$

$$P_{ja} = \sum_i \sum_{a'} P_{ia'} P_{ij}(a') \frac{\pi_{ja}}{\sum_a \pi_{ja}}$$

Hence, to show that $P_{ia} = \pi_{ia}$, we need show that

$$\pi_{ia} \geqslant 0,$$

$$\sum_i \sum_a \pi_{ia} = 1,$$

$$\pi_{ja} = \sum_i \sum_{a'} \pi_{ia'} P_{ij}(a') \frac{\pi_{ja}}{\sum_a \pi_{ja}}$$

The top two equations follow from (i) and (ii) of Equation (4.33), and the third, which is equivalent to,

$$\sum_a \pi_{ja} = \sum_i \sum_{a'} \pi_{ia'} P_{ij}(a')$$

follows from condition (iii) of Equation (4.33).

Thus we have shown that a vector $\boldsymbol{\pi} = (\pi_{ia})$ will satisfy (i), (ii), and (iii) of Equation (4.33) if and only if there exists a policy $\boldsymbol{\beta}$ such that π_{ia} is equal to the steady-state probability of being in state i and choosing action a when $\boldsymbol{\beta}$ is used. In fact, the policy $\boldsymbol{\beta}$ is defined by $\beta_i(a) = \pi_{ia}/\sum_a \pi_{ia}$.

The preceding is quite important in the determination of "optimal" policies. For instance, suppose that a reward $R(i, a)$ is earned whenever action a is chosen in state i. Since $R(X_i, a_i)$ would then represent the reward earned at time i, the expected average reward per unit time under policy $\boldsymbol{\beta}$ can be expressed as

$$\text{expected average reward under } \boldsymbol{\beta} = \lim_{n \to \infty} E_{\boldsymbol{\beta}} \left[\frac{\sum_{i=1}^{n} R(X_i, a_i)}{n} \right]$$

Now, if π_{ia} denotes the steady-state probability of being in state i and choosing action a, it follows that the limiting expected reward at time n equals

$$\lim_{n \to \infty} E[R(X_n, a_n)] = \sum_i \sum_a \pi_{ia} R(i, a)$$

which implies that

$$\text{expected average reward under } \boldsymbol{\beta} = \sum_i \sum_a \pi_{ia} R(i, a)$$

Hence, the problem of determining the policy that maximizes the expected average reward is

$$\underset{\pi = (\pi_{ia})}{\text{maximize}} \sum_i \sum_a \pi_{ia} R(i, a)$$

$$\text{subject to } \pi_{ia} \geqslant 0, \qquad \text{for all } i, a,$$

$$\sum_i \sum_a \pi_{ia} = 1,$$

$$\sum_a \pi_{ja} = \sum_i \sum_a \pi_{ia} P_{ij}(a), \qquad \text{for all } j \qquad (4.34)$$

However, the above maximization problem is a special case of what is known as a *linear program** and can thus be solved by a standard linear programming algorithm known as the *simplex algorithm*. If $\pi^* = (\pi_{ia}^*)$ maximizes the preceding, then the optimal policy will be given by $\boldsymbol{\beta}^*$ where

$$\beta_i^*(a) = \frac{\pi_{ia}^*}{\sum_a \pi_{ia}^*}$$

Remarks   (i) It can be shown that there is a π^* maximizing Equation (4.34) that has the property that for each i, π_{ia}^* is zero for all but one value of a, which implies that the optimal policy is nonrandomized. That is, the action it prescribes when in state i is a deterministic function of i.

(ii) The linear programming formulation also often works when there are restrictions placed on the class of allowable policies. For instance, suppose there is a restriction on the fraction of time the process spends in some state, say, state 1. Specifically, suppose that we are only allowed to consider policies having the property that their use results in the process being in state 1 less than 100α percent of time. To determine the optimal policy subject to this requirement, we add to the linear programming problem the additional constraint

$$\sum_a \pi_{1a} \leqslant \alpha$$

since $\sum_a \pi_{1a}$ represents the proportion of time that the process is in state 1.

---

* It is called a linear program since the objective function $\sum_i \sum_a R(i, a)\pi_{ia}$ and the constraints are all linear functions of the π_{ia}.

Exercises

*1.   Three white and three black balls are distributed in two urns in such a way that each contains three balls. We say that the system is in state i, $i = 0, 1, 2, 3$, if the first urn contains i white balls. At each step, we draw one ball from each urn and place the ball drawn from the first urn into the second, and conversely with the ball from the second urn. Let X_n denote the state of the system after the nth step. Explain why $\{X_n, n = 0, 1, 2, \ldots\}$ is a Markov chain and calculate its transition probability matrix.

2.   Suppose that whether or not it rains today depends on previous weather conditions through the last three days. Show how this system may be analyzed by using a Markov chain. How many states are needed?

3.   In Exercise 2, suppose that if it has rained for the past three days, then it will rain today with probability 0.8; if it did not rain for any of the past three days, then it will rain today with probability 0.2; and in any other case the weather today will, with probability 0.6, be the same as the weather yesterday. Determine **P** for this Markov chain.

*4.   Consider a process $\{X_n, n = 0, 1, \ldots\}$ which takes on the values 0, 1, or 2. Suppose

$$P\{X_{n+1} = j \mid X_n = i, X_{n-1} = i_{n-1}, \ldots, X_0 = i_0\} = \begin{cases} P_{ij}^{\text{I}}, & \text{when } n \text{ is even} \\ P_{ij}^{\text{II}}, & \text{when } n \text{ is odd} \end{cases}$$

where $\sum_{j=0}^{2} P_{ij}^{\text{I}} = \sum_{j=0}^{2} P_{ij}^{\text{II}} = 1$, $i = 0, 1, 2$. Is $\{X_n, n \geq 0\}$ a Markov chain? If not, then show how, by enlarging the state space, we may transform it into a Markov chain.

5.   Consider the Markov chain $\{X_n, n \geq 0\}$ with states $0, 1, 2$, whose transition probability matrix is

$$\mathbf{P} = \begin{bmatrix} 0 & \frac{1}{2} & \frac{1}{2} \\ \frac{1}{2} & \frac{1}{2} & 0 \\ 1 & 0 & 0 \end{bmatrix}$$

Let $f(0) = 0$, $f(1) = f(2) = 1$. If $Y_n = f(X_n)$, is $\{Y_n, n \geq 0\}$ a Markov chain?

6.   Let the transition probability matrix of a two-state Markov chain be given, as in Example 4.2, by

$$\mathbf{P} = \left\| \begin{matrix} p & 1 - p \\ 1 - p & p \end{matrix} \right\|$$

Show by mathematical induction that

$$\mathbf{P}^{(n)} = \begin{Vmatrix} \frac{1}{2} + \frac{1}{2}(2p-1)^n & \frac{1}{2} - \frac{1}{2}(2p-1)^n \\ \frac{1}{2} - \frac{1}{2}(2p-1)^n & \frac{1}{2} + \frac{1}{2}(2p-1)^n \end{Vmatrix}$$

7. In Example 4.4 suppose that it has rained neither yesterday nor the day before yesterday. What is the probability that it will rain tomorrow?

8. Suppose that coin 1 has probability 0.7 of coming up heads, and coin 2 has probability 0.6 of coming up heads. If the coin flipped today comes up heads, then we select coin 1 to flip tomorrow, and if it comes up tails, then we select coin 2 to flip tomorrow. If the coin initially flipped is equally likely to be coin 1 or coin 2, then what is the probability that the coin flipped on the third day after the initial flip is coin 1?

9. Let $\mathbf{P}$ be the transition probability matrix of a Markov chain. Argue that if for some positive integer r, $\mathbf{P}^r$ has all positive entries, then so does $\mathbf{P}^n$, for all integers $n \geqslant r$.

10. Specify the classes of the following Markov chains, and determine whether they are transient or recurrent:

$$\mathbf{P}_1 = \begin{Vmatrix} 0 & \frac{1}{2} & \frac{1}{2} \\ \frac{1}{2} & 0 & \frac{1}{2} \\ \frac{1}{2} & \frac{1}{2} & 0 \end{Vmatrix} \qquad \mathbf{P}_2 = \begin{Vmatrix} 0 & 0 & 0 & 1 \\ 0 & 0 & 0 & 1 \\ \frac{1}{2} & \frac{1}{2} & 0 & 0 \\ 0 & 0 & 1 & 0 \end{Vmatrix}$$

$$\mathbf{P}_3 = \begin{Vmatrix} \frac{1}{2} & 0 & \frac{1}{2} & 0 & 0 \\ \frac{1}{4} & \frac{1}{2} & \frac{1}{4} & 0 & 0 \\ \frac{1}{2} & 0 & \frac{1}{2} & 0 & 0 \\ 0 & 0 & 0 & \frac{1}{2} & \frac{1}{2} \\ 0 & 0 & 0 & \frac{1}{2} & \frac{1}{2} \end{Vmatrix} \qquad \mathbf{P}_4 = \begin{Vmatrix} \frac{1}{4} & \frac{3}{4} & 0 & 0 & 0 \\ \frac{1}{2} & \frac{1}{2} & 0 & 0 & 0 \\ 0 & 0 & 1 & 0 & 0 \\ 0 & 0 & \frac{1}{3} & \frac{2}{3} & 0 \\ 1 & 0 & 0 & 0 & 0 \end{Vmatrix}$$

11. Prove that if the number of states in a Markov chain is M, and if state j can be reached from state i, then it can be reached in M steps or less.

*12. Show that if state i is recurrent and state i does not communicate with state j, then $P_{ij} = 0$. This implies that once a process enters a recurrent class of states it can never leave that class. For this reason, a recurrent class is often referred to as a *closed* class.

13. For the random walk of Example 4.13 use the strong law of large numbers to give another proof that the Markov chain is transient when $p \neq \frac{1}{2}$.

Hint: Note that the state at time n can be written as $\sum_{i=1}^{n} Y_i$ where the Y_is are independent and $P\{Y_i = 1\} = p = 1 - P\{Y_i = -1\}$. Argue that if

$p > \frac{1}{2}$, then, by the strong law of large numbers, $\Sigma_1^n Y_i \to \infty$ as $n \to \infty$ and hence the initial state 0 can be visited only finitely often, and hence must be transient. A similar argument holds when $p < \frac{1}{2}$.

14. Coin 1 comes up heads with probability 0.6 and coin 2 with probability 0.5. A coin is continually flipped until it comes up tails, at which time that coin is put aside and we start flipping the other one.

(a) What proportion of flips use coin 1?
(b) If we start the process with coin 1 what is the probability that coin 2 is used on the fifth flip?

15. For Example 4.4, calculate the proportion of days that it rains.

16. A transition probability matrix **P** is said to be doubly stochastic if the sum over each column equals one; that is,

$$\sum_i P_{ij} = 1, \qquad \text{for all } j$$

If such a chain is irreducible and aperiodic and consists of $M + 1$ states $0, 1, \ldots, M$, show that the limiting probabilities are given by

$$\pi_j = \frac{1}{M + 1}, \qquad j = 0, 1, \ldots, M$$

*17. A particle moves on a circle through points which have been marked $0, 1, 2, 3, 4$ (in a clockwise order). At each step it has a probability p of moving to the right (clockwise) and $1 - p$ to the left (counterclockwise). Let X_n denote its location on the circle after the nth step. The process $\{X_n, n \geq 0\}$ is a Markov chain.

(a) Find the transition probability matrix.
(b) Calculate the limiting probabilities.

18. Let Y_n be the sum of n independent rolls of a fair die. Find

$$\lim_{n \to \infty} P\{Y_n \text{ is a multiple of } 13\}$$

Hint: Define an appropriate Markov chain and apply the results of Exercise 16.

19. Trials are performed in sequence. If the last two trials were successes, then the next trial is a success with probability 0.8; otherwise the next trial is a success with probability 0.5. In the long run, what proportion of trials are successes?

20. Consider three urns, one colored red, one white, and one blue. The red urn contains 1 red and 4 blue balls; the white urn contains 3 white balls, 2 red balls, and 2 blue balls; the blue urn contains 4 white balls, 3 red balls, and 2 blue balls. At the initial stage, a ball is randomly selected from the red urn and then returned to that urn. At every subsequent stage, a ball is randomly selected from the urn whose color is the same as that of the ball previously selected and is then returned to that urn. In the long run, what proportion of the selected balls are red? What proportion are white? What proportion are blue?

21. Each morning an individual leaves his house and goes for a run. He is equally likely to leave either from his front or back door. Upon leaving the house, he chooses a pair of running shoes (or goes running barefoot if there are no shoes at the door from which he departed). On his return he is equally likely to enter, and leave his running shoes, either by the front or back door. If he owns a total of k pairs of running shoes, what proportion of the time does he run barefooted?

22. Consider the following approach to shuffling a deck of n cards. Starting with any initial ordering of the cards, one of the numbers $1, 2, \ldots, n$ is randomly chosen in such a manner that each one is equally likely to be selected. If number i is chosen, then we take the card that is in position i and put it on top of the deck — that is, we put that card in position 1. We then repeatedly perform the same operation. Show that, in the limit, the deck is perfectly shuffled in the sense that the resultant ordering is equally likely to be any of the $n!$ possible orderings.

*23. Determine the limiting probabilities π_j for the model presented in Exercise 1. Give an intuitive explanation of your answer.

24. For a series of dependent trials the probability of success on any trial is $(k + 1)/(k + 2)$ where k is equal to the number of successes on the previous two trials. Compute $\lim_{n \to \infty} P\{\text{success on the } n\text{th trial}\}$.

25. An organization has N employees where N is a large number. Each employee has one of three possible job classifications and changes classifications (independently) according to a Markov chain with transition probabilities

$$\begin{bmatrix} 0.7 & 0.2 & 0.1 \\ 0.2 & 0.6 & 0.2 \\ 0.1 & 0.4 & 0.5 \end{bmatrix}$$

What percentage of employees are in each classification?

26. Three out of every four trucks on the road are followed by a car, while only one out of every five cars is followed by a truck. What fraction of vehicles on the road are trucks?

27. A certain town never has two sunny days in a row. Each day is classified as being either sunny, cloudy (but dry), or rainy. If it is sunny one day, then it is equally likely to be either cloudy or rainy the next day. If it is rainy or cloudy one day, then there is one chance in two that it will be the same the next day, and if it changes then it is equally likely to be either of the other two possibilities. In the long run, what proportion of days are sunny? What proportion are cloudy?

*28. Each of two switches is either on or off during a day. On day n, each switch will independently be on with probability

$$[1 + \text{number of on switches during day } n - 1]/4$$

For instance, if both switches are on during day $n - 1$, then each will independently be on during day n with probability $3/4$. What fraction of days are both switches on? What fraction are both off?

29. A professor continually gives exams to her students. She can give three possible types of exams, and her class is graded as either having done well or badly. Let p_i denote the probability that the class does well on a type i exam, and suppose that $p_1 = 0.3$, $p_2 = 0.6$, and $p_3 = 0.9$. If the class does well on an exam, then the next exam is equally likely to be any of the three types. If the class does badly, then the next exam is always type 1. What proportion of exams are type i, $i = 1, 2, 3$?

30. A flea moves around the vertices of a triangle in the following manner: Whenever it is at vertex i it moves to its clockwise neighbor vertex with probability p_i and to the counterclockwise neighbor with probability $q_i = 1 - p_i$, $i = 1, 2, 3$.

   (a) Find the proportion of time that the flea is at each of the vertices.
   (b) How often does the flea make a counterclockwise move which is then followed by 5 consecutive clockwise moves?

31. Consider a Markov chain with states 0, 1, 2, 3, 4. Suppose $P_{0,4} = 1$; and suppose that when the chain is in state i, $i > 0$, the next state is equally likely to be any of the states $0, 1, \ldots, i - 1$. Find the limiting probabilities of this Markov chain.

*32. Let π_i denote the long-run proportion of time a given Markov chain is in state i.

   (a) Explain why π_i is also the proportion of transitions that are into state i as well as being the proportion of transitions that are from state i.
   (b) $\pi_i P_{ij}$ represents the proportion of transitions that satisfy what property?

(c) $\sum_i \pi_i P_{ij}$ represent the proportion of transitions that satisfy what property?

(d) Using the preceding explain why

$$\pi_j = \sum_i \pi_i P_{ij}$$

33. Let A be a set of states, and let A^c be the remaining states.

(a) What is the interpretation of

$$\sum_{i \in A} \sum_{j \in A^c} \pi_i P_{ij}?$$

(b) What is the interpretation of

$$\sum_{i \in A^c} \sum_{j \in A} \pi_i P_{ij}?$$

(c) Explain the identity

$$\sum_{i \in A} \sum_{j \in A^c} \pi_i P_{ij} = \sum_{i \in A^c} \sum_{j \in A} \pi_i P_{ij}$$

34. Each day, one of n possible elements is requested, the ith one with probability P_i, $i \geq 1$, $\sum_1^n P_i = 1$. These elements are at all times arranged in an ordered list which is revised as follows: The element selected is moved to the front of the list with the relative positions of all the other elements remaining unchanged. Define the state at any time to be the list ordering at that time and note that there are $n!$ possible states.

(a) Argue that the preceding is a Markov chain.

(b) For any state $i_1, \ldots, i_n$ (which is a permutation of $1, 2, \ldots, n$), let $\pi(i_1, \ldots, i_n)$ denote the limiting probability. In order for the state to be $i_1, \ldots, i_n$, it is necessary for the last request to be for i_1, the last non-i_1 request for i_2, the last non-i_1 or i_2 request for i_3, and so on. Hence, it appears intuitive that

$$\pi(i_1, \ldots, i_n) = P_{i_1} \frac{P_{i_2}}{1 - P_{i_1}} \frac{P_{i_3}}{1 - P_{i_1} - P_{i_2}} \cdots \frac{P_{i_{n-1}}}{1 - P_{i_1} - \cdots - P_{i_{n-2}}}$$

Verify when $n = 3$ that the above are indeed the limiting probabilities.

35. Suppose that a population consists of a fixed number, say, m, of genes in any generation. Each gene is one of two possible genetic types. If any generation has exactly i (of its m) genes being type 1, then the next generation will have j type 1 (and $m - j$ type 2) genes with probability

$$\binom{m}{j}\left(\frac{i}{m}\right)^j\left(\frac{m - i}{m}\right)^{m-j}, \qquad j = 0, 1, \ldots, m$$

Let X_n denote the number of type 1 genes in the nth generation, and assume that $X_0 = i$.

(a) Find $E[X_n]$.
(b) What is the probability that eventually all the genes will be type 1?

36. Consider an irreducible finite Markov chain with states $0, 1, \ldots, N$.

(a) Starting in state i, what is the probability the process will ever visit state j? Explain!
(b) Let $x_i = P\{\text{visit state } N \text{ before state } 0 \,|\, \text{start in } i\}$. Compute a set of linear equations which the x_i satisfy, $i = 0, 1, \ldots, N$.
(c) If $\sum_j j p_{ij} = i$ for $i = 1, \ldots, N - 1$, show that $x_i = i/N$ is a solution to the equations in part (b).

37. An individual possesses r umbrellas which he employs in going from his home to office, and vice versa. If he is at home (the office) at the beginning (end) of a day and it is raining, then he will take an umbrella with him to the office (home), provided there is one to be taken. If it is not raining, then he never takes an umbrella. Assume that, independent of the past, it rains at the beginning (end) of a day with probability p.

(i) Define a Markov chain with $r + 1$ states which will help us to determine the proportion of time that our man gets wet. (*Note:* He gets wet if it is raining, and all umbrellas are at his other location.)
(ii) Show that the limiting probabilities are given by

$$\pi_i = \begin{cases} \dfrac{q}{r + q}, & \text{if } i = 0 \\[2mm] \dfrac{1}{r + q}, & \text{if } i = 1, \ldots, r \end{cases} \qquad \text{where } q = 1 - p$$

(iii) What fraction of time does our man get wet?
(iv) When $r = 3$, what value of p maximizes the fraction of time he gets wet?

***38.** Let $\{X_n, n \geq 0\}$ denote an ergodic Markov chain with limiting probabilities π. Define the process $\{Y_n, n \geq 1\}$ by $Y_n = (X_{n-1}, X_n)$. That is, Y_n keeps track of the last two states of the original chain. Is $\{Y_n, n \geq 1\}$ a Markov chain? If so, determine its transition probabilities and find

$$\lim_{n \to \infty} P\{Y_n = (i, j)\}$$

39. Verify the transition probability matrix given in Example 4.18.

40. Let $P^{(1)}$ and $P^{(2)}$ denote transition probability matrices for ergodic Markov chains having the same state space. Let π^1 and π^2 denote the stationary (limiting) probability vectors for the two chains. Consider a process defined as follows:

(i) $X_0 = 1$. A coin is then flipped and if it comes up heads, then the remaining states $X_1, \ldots$ are obtained from the transition probability matrix $P^{(1)}$ and if tails from the matrix $P^{(2)}$. Is $\{X_n, n \geq 0\}$ a Markov chain? If $p = P\{\text{coin comes up heads}\}$, what is $\lim_{n \to \infty} P(X_n = i)$?

(ii) $X_0 = 1$. At each stage the coin is flipped and if it comes up heads, then the next state is chosen according to $P^{(1)}$ and if tails comes up, then it is chosen according to $P^{(2)}$. In this case do the successive states constitute a Markov chain? If so, determine the transition probabilities. Show by a counterexample that the limiting probabilities are not the same as in part (i).

41. A fair coin is continually flipped. Compute the expected number of flips until the following patterns appear:

(a) HHTTHT
*(b) HHTTHH
(c) HHTHHT

42. Consider an irreducible Markov chain with transition probabilities $P_{i,j}$ and stationary probabilities π_j. Starting in state 1 find the expected number of transitions until the pattern 0, 0, 0, 1 occurs.

43. Consider the Ehrenfest urn model in which M molecules are distributed among two urns, and at each time point one of the molecules is chosen at random and is then removed from its urn and placed in the other one. Let X_n denote the number of molecules in urn 1 after the nth switch and let $\mu_n = E[X_n]$. Show that

(i) $\mu_{n+1} = 1 + (1 - 2/M)\mu_n$
(ii) Use (i) to prove that

$$\mu_n = \frac{M}{2} + \left(\frac{M-2}{M}\right)^n \left(E[X_0] - \frac{M}{2}\right)$$

44. Consider a population of individuals each of whom possesses two genes which can be either type A or type a. Suppose that in outward appearance type A is dominant and type a is recessive. (That is, an individual will only have the outward characteristics of the recessive gene if its pair is aa.) Suppose that the population has stabilized, and the percentages of individuals having respective gene pairs AA, aa, and Aa are p, q, and r. Call an individual dominant or recessive depending on the outward characteristics it exhibits. Let S_{11} denote the probability that an offspring of two dominant parents will be recessive; and let S_{10} denote the probability that the offspring of one dominant and one recessive parent will be recessive. Compute S_{11} and S_{10} to show that $S_{11} = S_{10}^2$. (The quantities S_{10} and S_{11} are known in the genetics literature as *Snyder's ratios*.)

45.   Suppose that on each play of the game a gambler either wins 1 with probability p or loses 1 with probability $1 - p$. The gambler continues betting until she or he is either winning n or losing m. What is the probability that the gambler quits a winner?

46.   A particle moves among $n + 1$ vertices that are situated on a circle in the following manner. At each step it moves one step either in the clockwise direction with probability p or the counterclockwise direction with probability $q = 1 - p$. Starting at a specified state, call it state 0, let T be the time of the first return to state 0. Find the probability that all states have been visited by time T.

Hint: Condition on the initial transition and then use results from the gambler's ruin problem.

47.   In the gambler's ruin problem of Section 4.5.1, suppose the gambler's fortune is presently i, and suppose that we know that the gambler's fortune will eventually reach N (before it goes to 0). Given this information, show that the probability he wins the next gamble is

$$\frac{p[1 - (q/p)^{i+1}]}{1 - (q/p)^i}, \qquad \text{if } p \neq \tfrac{1}{2}$$

$$\frac{i+1}{2i}, \qquad \text{if } p = \tfrac{1}{2}$$

Hint: The probability we want is

$$P\{X_{n+1} = i + 1 \mid X_n = i, \lim_{m \to \infty} X_m = N\}$$

$$= \frac{P\{X_{n+1} = i + 1, \lim_m X_m = N \mid X_n = i\}}{P\{\lim_m X_m = N \mid X_n = i\}}$$

48.   For the gambler's ruin model of Section 4.5.1, let M_i denote the mean number of games that must be played until the gambler either goes broke or reaches a fortune of N, given that he starts with $i, i = 0, 1, \ldots, N$. Show that M_i satisfies

$$M_0 = M_N = 0; \qquad M_i = 1 + pM_{i+1} + qM_{i-1}, \qquad i = 1, \ldots, N-1$$

49. Solve the equations given in Exercise 48 to obtain

$$M_i = i(N - i), \qquad\qquad\qquad\qquad\text{if } p = \tfrac{1}{2}$$

$$= \frac{i}{q - p} - \frac{N}{q - p}\frac{1 - (q/p)^i}{1 - (q/p)^N}, \qquad \text{if } p \neq \tfrac{1}{2}$$

50. Suppose in the gambler's ruin problem that the probability of winning a bet depends on the gambler's present fortune. Specifically, suppose that α_i is the probability that the gambler wins a bet when his or her fortune is i. Given that the gambler's initial fortune is i, let $P(i)$ denote the probability that the gambler's fortune reaches N before 0.

(a) Derive a formula that relates $P(i)$ to $P(i - 1)$ and $P(i + 1)$.
(b) Using the same approach as in the gambler's ruin problem, solve the equation of part (a) for $P(i)$.
(c) Suppose that i balls are initially in urn 1 and $N - i$ are in urn 2, and suppose that at each stage one of the N balls is randomly chosen, taken from whichever urn it is in, and placed in the other urn. Find the probability that the first urn becomes empty before the second.

*51. In Exercise 17,

(a) what is the expected number of steps the particle takes to return to the starting position?
(b) what is the probability that all other positions are visited before the particle returns to its starting state?

52. For the Markov chain with states 1, 2, 3, 4 whose transition probability matrix $\mathbf{P}$ is as specified below find f_{i3} and s_{i3} for $i = 1, 2, 3$.

$$\mathbf{P} = \begin{bmatrix} 0.4 & 0.2 & 0.1 & 0.3 \\ 0.1 & 0.5 & 0.2 & 0.2 \\ 0.3 & 0.4 & 0.2 & 0.1 \\ 0 & 0 & 0 & 1 \end{bmatrix}$$

53. Consider a branching process having $\mu < 1$. Show that if $X_0 = 1$, then the expected number of individuals that ever exist in this population is given by $1/(1 - \mu)$. What if $X_0 = n$?

54. In a branching process having $X_0 = 1$ and $\mu > 1$, prove that π_0 is the *smallest* positive number satisfying Equation (4.15).

Hint: Let π be any solution of $\pi = \sum_{j=0}^{\infty} \pi^j P_j$. Show by mathematical induction that $\pi \geqslant P\{X_n = 0\}$ for all n, and let $n \to \infty$. In using the induction argue that

$$P\{X_n = 0\} = \sum_{j=0}^{\infty} (P\{X_{n-1} = 0\})^j P_j$$

55. For a branching process, calculate π_0 when

(a) $P_0 = \frac{1}{4}$, $P_2 = \frac{3}{4}$
(b) $P_0 = \frac{1}{4}$, $P_1 = \frac{1}{2}$, $P_2 = \frac{1}{4}$
(c) $P_0 = \frac{1}{6}$, $P_1 = \frac{1}{2}$, $P_3 = \frac{1}{3}$

56. At all times, an urn contains N balls — some white balls and some black balls. At each stage, a coin having probability p, $0 < p < 1$, of landing heads is flipped. If heads appears, then a ball is chosen at random from the urn and is replaced by a white ball; if tails appears, then a ball is chosen from the urn and is replaced by a black ball. Let X_n denote the number of white balls in the urn after the nth stage.

(a) Is $\{X_n, n \geqslant 0\}$ a Markov chain? If so, explain why.
(b) What are its classes? What are their periods? Are they transient or recurrent?
(c) Compute the transition probabilities P_{ij}.
(d) Let $N = 2$. Find the proportion of time in each state.
(e) Based on your answer in part (d) and your intuition, guess the answer for the limiting probability in the general case.
(f) Prove your guess in part (e) either by showing that Equation (4.7) is satisfied or by using the results of Example 4.28.
(g) If $p = 1$, what is the expected time until there are only white balls in the urn if initially there are i white and $N - i$ black?

*57. (a) Show that the limiting probabilities of the reversed Markov chain are the same as for the forward chain by showing that they satisfy the equations

$$\pi_j = \sum_i \pi_i Q_{ij}$$

(b) Give an intuitive explanation for the result of part (a).

58. M balls are initially distributed among m urns. At each stage one of the balls is selected at random, taken from whichever urn it is in, and then placed, at random, in one of the other $M - 1$ urns. Consider the Markov chain whose state at any time is the vector $(n_1, \ldots, n_m)$ where n_i denotes the number of balls in urn i. Guess at the limiting probabilities for this Markov chain and then verify your guess and show at the same time that the Markov chain is time reversible.

59. A total of m white and m black balls are distributed among two urns, with each urn containing m balls. At each stage, a ball is randomly selected from each urn and the two selected balls are interchanged. Let X_n denote the number of black balls in urn 1 after the nth interchange.

(a) Give the transition probabilities of the Markov chain $X_n, n \geq 0$.

(b) Without any computations, what do you think are the limiting probabilities of this chain?

(c) Find the limiting probabilities and show that the stationary chain is time reversible.

60. It follows from Theorem 4.2 that for a time reversible Markov chain

$$P_{ij}P_{jk}P_{ki} = P_{ik}P_{kj}P_{ji}, \qquad \text{for all } i, j, k$$

It turns out that if the state space is finite and $P_{ij} > 0$ for all i, j, then the preceding is also a sufficient condition for time reversibility. [That is, in this case, we need only check Equation (4.26) for paths from i to i that have only two intermediate states.] Prove this.

Hint: Fix i and show that the equations

$$\pi_j P_{jk} = \pi_k P_{kj}$$

are satisfied by $\pi_j = c P_{ij}/P_{ji}$, where c is chosen so that $\Sigma_j \pi_j = 1$.

61. For a time reversible Markov chain, argue that the rate at which transitions from i to j to k occur must equal the rate at which transitions from k to j to i occur.

62. Show that the Markov chain of Exercise 27 is time reversible.

63. A group of n processors is arranged in an ordered list. When a job arrives, the first processor in line attempts it; if it is unsuccessful, then the next in line tries it; if it too is unsuccessful, then the next in line tries it, and so on. When the job is successfully processed or after all processors have been unsuccessful, the job leaves the system. At this point we are allowed to reorder the processors, and a new job appears. Suppose that we use the one-closer reordering rule, which moves the processor that was successful one closer to the front of the line by interchanging its position with the one in front of it. If all processors were unsuccessful (or if the processor in the first position was successful), then the ordering remains the same. Suppose that each time processor i attempts a job then, independently of anything else, it is successful with probability p_i.

(a) Define an appropriate Markov chain to analyze this model.

(b) Show that this Markov chain is time reversible.

(c) Find the long-run probabilities.

64. A Markov chain is said to be a tree process if

(i) $P_{ij} > 0$ whenever $P_{ji} > 0$.
(ii) for every pair of states i and j, $i \neq j$, there is a unique sequence of distinct states $i = i_0, i_1, \ldots, i_{n-1}, i_n = j$ such that

$$P_{i_k, i_{k+1}} > 0, \qquad k = 0, 1, \ldots, n-1$$

In other words, a Markov chain is a tree process if for every pair of distinct states i and j there is a unique way for the process to go from i to j without reentering a state (and this path is the reverse of the unique path from j to i). Argue that an ergodic tree process is time reversible.

65. On a chessboard compute the expected number of plays it takes a knight, starting in one of the four corners of the chessboard, to return to its initial position if we assume that at each play it is equally likely to choose any of its legal moves. (No other pieces are on the board.)

Hint: Make use of Example 4.29.

66. In a Markov decision problem, another criterion often used, different than the expected average return per unit time, is that of the expected discounted return. In this criterion we choose a number α, $0 < \alpha < 1$, and try to choose a policy so as to maximize $E[\sum_{i=0}^{\infty} \alpha^i R(X_i, a_i)]$, (That is, rewards at time n are discounted at rate α^n.) Suppose that the initial state is chosen according to the probabilities b_i. That is,

$$P\{X_0 = i\} = b_i, \qquad i = 1, \ldots, n$$

For a given policy β let y_{ja} denote the expected discounted time that the process is in state j and action a is chosen. That is,

$$y_{ja} = E_\beta \left[\sum_{n=0}^{\infty} \alpha^n I_{\{X_n = j, a_n = a\}} \right]$$

where for any event A the indicator variable I_A is defined by

$$I_A = \begin{cases} 1, & \text{if } A \text{ occurs} \\ 0, & \text{otherwise} \end{cases}$$

(a) Show that

$$\sum_a y_{ja} = E \left[\sum_{n=0}^{\infty} \alpha^n I_{\{X_n = j\}} \right]$$

or, in other words, $\sum_a y_{ja}$ is the expected discounted time in state j under β.
(b) Show that

$$\sum_j \sum_a y_{ja} = \frac{1}{1 - \alpha},$$

$$\sum_a y_{ja} = b_j + \alpha \sum_i \sum_a y_{ia} P_{ij}(a)$$

Hint: For the second equation, use the identity

$$I_{\{X_{n+1}=j\}} = \sum_i \sum_a I_{\{X_n=i,\, a_n=a\}} I_{\{x_{n+1}=j\}}$$

Take expectations of the preceding to obtain

$$E[I_{X_{n+1}=j}] = \sum_i \sum_a E[I_{\{X_n=i,\, a_n=a\}}] P_{ij}(a)$$

(c) Let $\{y_{ja}\}$ be a set of numbers satisfying

$$\sum_j \sum_a y_{ja} = \frac{1}{1-\alpha}$$

$$\sum_a y_{ja} = b_j + \alpha \sum_i \sum_a y_{ia} P_{ij}(a) \tag{4.35}$$

Argue that y_{ja} can be interpreted as the expected discounted time that the process is in state j and action a is chosen when the initial state is chosen according to the probabilities b_j and the policy β, given by

$$\beta_i(a) = \frac{y_{ia}}{\sum_a y_{ia}}$$

is employed.

Hint: Derive a set of equations for the expected discounted times when policy β is used and show that they are equivalent to Equation (4.35).

(d) Argue that an optimal policy with respect to the expected discounted return criterion can be obtained by first solving the linear program

$$\text{maximize} \quad \sum_j \sum_a y_{ja} R(j,a),$$

$$\text{such that} \quad \sum_j \sum_a y_{ja} = \frac{1}{1-\alpha},$$

$$\sum_a y_{ja} = b_j + \alpha \sum_i \sum_a y_{ia} P_{ij}(a),$$

$$y_{ja} \geqslant 0, \qquad \text{all } j,a;$$

and then defining the policy β^* by

$$\beta_i^*(a) = \frac{y_{ia}^*}{\sum_a y_{ia}^*}$$

where the y_{ja}^* are the solutions of the linear program.

References

1. K. L. Chung, "Markov Chains with Stationary Transition Probabilities," Springer, Berlin, 1960.
2. S. Karlin and H. Taylor, "A First Course in Stochastic Processes," Second Edition, Academic Press, New York, 1975.
3. J. G. Kemeny and J. L. Snell, "Finite Markov Chains," Van Nostrand Reinhold, Princeton, New Jersey, 1960.
4. S. M. Ross, "Stochastic Processes," Second Edition, John Wiley, New York, 1996.

The Exponential Distribution and the Poisson Process **5**

5.1. Introduction

In making a mathematical model for a real-world phenomenon it is always necessary to make certain simplifying assumptions so as to render the mathematics tractable. On the other hand, however, we cannot make too many simplifying assumptions, for then our conclusions, obtained from the mathematical model, would not be applicable to the real-world situation. Thus, in short, we must make enough simplifying assumptions to enable us to handle the mathematics but not so many that the mathematical model no longer resembles the real-world phenomenon. One simplifying assumption that is often made is to assume that certain random variables are exponentially distributed. The reason for this is that the exponential distribution is both relatively easy to work with and is often a good approximation to the actual distribution.

The property of the exponential distribution that makes it easy to analyze is that it does not deteriorate with time. By this we mean that if the lifetime of an item is exponentially distributed, then an item that has been in use for ten (or any number of) hours is as good as a new item in regards to the amount of time remaining until the item fails. This will be formally defined in Section 5.2 where it will be shown that the exponential is the only distribution which possesses this property.

In Section 5.3 we shall study counting processes with an emphasis on a kind of counting process known as the Poisson process. Among other things we shall discover about this process is its intimate connection with the exponential distribution.

5.2. The Exponential Distribution

5.2.1. Definition

A continuous random variable X is said to have an *exponential distribution* with parameter $\lambda, \lambda > 0$, if its probability density function is given by

$$f(x) = \begin{cases} \lambda e^{-\lambda x} & x \geq 0 \\ 0, & x < 0 \end{cases}$$

or, equivalently, if its cdf is given by

$$F(x) = \int_{-\infty}^{x} f(y)\, dy = \begin{cases} 1 - e^{-\lambda x}, & x \geq 0 \\ 0, & x < 0 \end{cases}$$

The mean of the exponential distribution, $E[X]$, is given by

$$E[X] = \int_{-\infty}^{\infty} xf(x)\, dx$$

$$= \int_{0}^{\infty} \lambda x e^{-\lambda x}\, dx$$

Integrating by parts ($u = x, dv = \lambda e^{-\lambda x}\, dx$) yields

$$E[X] = -xe^{-\lambda x}\Big|_{0}^{\infty} + \int_{0}^{\infty} e^{-\lambda x}\, dx = \frac{1}{\lambda}$$

The moment generating function $\phi(t)$ of the exponential distribution is given by

$$\phi(t) = E[e^{tX}]$$

$$= \int_{0}^{\infty} e^{tx} \lambda e^{-\lambda x}\, dx$$

$$= \frac{\lambda}{\lambda - t} \qquad \text{for } t < \lambda \tag{5.1}$$

All the moments of X can now be obtained by differentiating Equation (5.1). For example,

$$E[X^2] = \frac{d^2}{dt^2}\phi(t)\Big|_{t=0}$$

$$= \frac{2\lambda}{(\lambda - t)^3}\Big|_{t=0}$$

$$= \frac{2}{\lambda^2}$$

Also, from the preceding, we obtain

$$\text{Var}(X) = E[X^2] - (E[X])^2$$

$$= \frac{2}{\lambda^2} - \frac{1}{\lambda^2}$$

$$= \frac{1}{\lambda^2}$$

5.2.2. Properties of the Exponential Distribution

A random variable X is said to be without memory, or *memoryless*, if

$$P\{X > s + t \mid X > t\} = P\{X > s\} \qquad \text{for all } s, t \geq 0 \qquad (5.2)$$

If we think of X as being the lifetime of some instrument, then Equation (5.2) states that the probability that the instrument lives for at least $s + t$ hours given that it has survived t hours is the same as the initial probability that it lives for at least s hours. In other words, if the instrument is alive at time t, then the distribution of the remaining amount of time that it survives is the same as the original lifetime distribution, that is; the instrument does not remember that it has already been in use for a time t.

The condition in Equation (5.2) is equivalent to

$$\frac{P\{X > s + t, X > t\}}{P\{X > t\}} = P\{X > s\}$$

or

$$P\{X > s + t\} = P\{X > s\}P\{X > t\} \qquad (5.3)$$

Since Equation (5.3) is satisfied when X is exponentially distributed (for $e^{-\lambda(s+t)} = e^{-\lambda s}e^{-\lambda t}$), it follows that exponentially distributed random variables are memoryless.

Example 5.1   Suppose that the amount of time one spends in a bank is exponentially distributed with mean ten minutes, that is, $\lambda = \frac{1}{10}$. What is the probability that a customer will spend more than fifteen minutes in the bank? What is the probability that a customer will spend more than fifteen minutes in the bank given that she is still in the bank after ten minutes?

Solution: If X represents the amount of time that the customer spends in the bank, then the first probability is just

$$P\{X > 15\} = e^{-15\lambda} = e^{-3/2} \approx 0.220$$

The second question asks for the probability that a customer who has spent ten minutes in the bank will have to spend at least five more minutes. However, since the exponential distribution does not "remember" that the customer has already spent ten minutes in the bank, this must equal the probability that an entering customer spends at least five minutes in the bank. That is, the desired probability is just

$$P\{X > 5\} = e^{-5\lambda} = e^{-1/2} \approx 0.604 \quad \blacklozenge$$

Example 5.2   Consider a post office that is run by two clerks. Suppose that when Mr. Smith enters the system he discovers that Mr. Jones is being served by one of the clerks and Mr. Brown by the other. Suppose also that Mr. Smith is told that his service will begin as soon as either Jones or Brown leaves. If the amount of time that a clerk spends with a customer is exponentially distributed with mean $1/\lambda$, what is the probability that, of the three customers, Mr. Smith is the last to leave the post office?

Solution: The answer is obtained by this reasoning: Consider the time at which Mr. Smith first finds a free clerk. At this point either Mr. Jones or Mr. Brown would have just left and the other one would still be in service. However, by the lack of memory of the exponential, it follows that the amount of time that this other man (either Jones or Brown) would still have to spend in the post office is exponentially distributed with mean $1/\lambda$. That is, it is the same as if he were just starting his service at this point. Hence, by symmetry, the probability that he finishes before Smith must equal $\frac{1}{2}$. $\blacklozenge$

It turns out that not only is the exponential distribution "memoryless," but it is the unique distribution possessing this property. To see this, suppose that X is memoryless and let $\bar{F}(x) = P\{X > x\}$. Then by Equation (5.3) it follows that

$$\bar{F}(s + t) = \bar{F}(s)\bar{F}(t)$$

That is, $\bar{F}(x)$ satisfies the functional equation

$$g(s + t) = g(s)g(t)$$

However, it turns out that the only right continuous solution of this functional equation is

$$g(x) = e^{-\lambda x}*$$

and since a distribution function is always right continuous we must have

$$\bar{F}(x) = e^{-\lambda x}$$

or

$$F(x) = P\{X \leqslant x\} = 1 - e^{-\lambda x}$$

which shows that X is exponentially distributed.

Example 5.3 Suppose that the amount of time that a lightbulb works before burning itself out is exponentially distributed with mean ten hours. Suppose that a person enters a room in which a lightbulb is burning. If this person desires to work for five hours, then what is the probability that she will be able to complete her work without the bulb burning out? What can be said about this probability when the distribution is not exponential?

Solution: Since the bulb is burning when the person enters the room it follows, by the memoryless property of the exponential, that its remaining lifetime is exponential with mean ten. Hence the desired probability is

$$P\{\text{remaining lifetime} > 5\} = 1 - F(5) = e^{-5\lambda} = e^{-1/2}$$

However, if the lifetime distribution F is not exponential, then the relevant probability is

$$P\{\text{lifetime} > t + 5 \,|\, \text{lifetime} > t\} = \frac{1 - F(t + 5)}{1 - F(t)}$$

where t is the amount of time that the bulb had been in use prior to the person entering the room. That is, if the distribution is not exponential then additional information is needed (namely, t) before the desired

---

*This is proven as follows: If $g(s + t) = g(s)g(t)$, then

$$g\left(\frac{2}{n}\right) = g\left(\frac{1}{n} + \frac{1}{n}\right) = g^2\left(\frac{1}{n}\right)$$

and repeating this yields $g(m/n) = g^m(1/n)$. Also

$$g(1) = g\left(\frac{1}{n} + \frac{1}{n} + \cdots + \frac{1}{n}\right) = g^n\left(\frac{1}{n}\right) \qquad \text{or} \qquad g\left(\frac{1}{n}\right) = (g(1))^{1/n}$$

Hence $g(m/n) = (g(1))^{m/n}$, which implies, since g is right continuous, that $g(x) = (g(1))^x$. Since $g(1) = (g(\frac{1}{2}))^2 \geqslant 0$ we obtain $g(x) = e^{-\lambda x}$, where $\lambda = -\log(g(1))$.

probability can be calculated. In fact, it is for this reason, namely, that the distribution of the remaining lifetime is independent of the amount of time that the object has already survived, that the assumption of an exponential distribution is so often made. ✦

The memoryless property is further illustrated by the failure rate function (also called the hazard rate function) of the exponential distribution.

Consider a continuous positive random variable X having distribution function F and density f. The *failure* (or *hazard*) *rate* function $r(t)$ is defined by

$$r(t) = \frac{f(t)}{1 - F(t)} \tag{5.4}$$

To interpret $r(t)$, suppose that an item, having lifetime X, has survived for t hours, and we desire the probability that it does not survive for an additional time dt. That is, consider $P\{X \in (t, t + dt) \mid X > t\}$. Now

$$P\{X \in (t, t + dt) \mid X > t\} = \frac{P\{X \in (t, t + dt), X > t\}}{P\{X > t\}}$$

$$= \frac{P\{X \in (t, t + dt)\}}{P\{X > t\}}$$

$$\approx \frac{f(t)\, dt}{1 - F(t)} = r(t)\, dt$$

That is, $r(t)$ represents the conditional probability density that a t-year-old item will fail.

Suppose now that the lifetime distribution is exponential. Then, by the memoryless property, it follows that the distribution of remaining life for a t-year-old item is the same as for a new item. Hence $r(t)$ should be constant. This checks out since

$$r(t) = \frac{f(t)}{1 - F(t)}$$

$$= \frac{\lambda e^{-\lambda t}}{e^{-\lambda t}} = \lambda$$

Thus, the failure rate function for the exponential distribution is constant. The parameter λ is often referred to as the *rate* of the distribution. (Note that the rate is the reciprocal of the mean, and vice versa.)

It turns out that the failure rate function $r(t)$ uniquely determines the distribution F. To prove this, we note by Equation (5.4) that

$$r(t) = \frac{d/dt\, F(t)}{1 - F(t)}$$

Integrating both sides yields

$$\log(1 - F(t)) = -\int_0^t r(t)\, dt + k$$

or

$$1 - F(t) = e^k \exp\left\{ -\int_0^t r(t)\, dt \right\}$$

Letting $t = 0$ shows that $k = 0$ and thus

$$F(t) = 1 - \exp\left\{ -\int_0^t r(t)\, dt \right\}$$

Example 5.4   Let $X_1, \ldots, X_n$ be independent exponential random variables with respective rates $\lambda_1, \ldots, \lambda_n$, where $\lambda_i \neq \lambda_j$ when $i \neq j$. Let N be independent of these random variables and suppose that

$$\sum_{j=1}^n P_j = 1 \qquad \text{where } P_j = P\{N = j\}$$

The random variable X_N is said to be a *hyperexponential* random variable. By conditioning on the value of N, we obtain that its density function is

$$f(t) = f_{X_N}(t) = \sum_{j=1}^n f_{X_N}(t \mid N = j) P_j$$

$$= \sum_{j=1}^n f_{X_j}(t) P_j$$

$$= \sum_{j=1}^n P_j \lambda_j e^{-\lambda_j t}$$

where the next to last equality used the fact that N is independent of X_j.

To see how such a random variable might originate, imagine that a bin contains n different types of batteries, with a type j battery lasting for an exponential distributed time with rate λ_j, $j = 1, \ldots, n$. Suppose further that P_j is the proportion of batteries in the bin that are type j for each $j = 1, \ldots, n$. If a battery is randomly chosen, in the sense that it is equally likely to be any of the batteries in the bin, then the lifetime of the battery selected will have the hyperexponential distribution specified in the preceding.

Since

$$1 - F(t) = \int_t^\infty f(t)\, dt = \sum_{j=1}^n P_j e^{-\lambda_j t}$$

we see that the failure rate function of a hyperexponential random variable is

$$r(t) = \frac{\sum_{j=1}^{n} P_j \lambda_j e^{-\lambda_j t}}{\sum_{j=1}^{n} P_j e^{-\lambda_j t}}$$

If we let $\lambda_i = \min(\lambda_1, \ldots, \lambda_n)$ then, upon multiplying the numerator and denominator of $r(t)$ by $e^{\lambda_i t}$, we have

$$r(t) = \frac{\sum_{j=1}^{n} P_j \lambda_j e^{-(\lambda_j - \lambda_i)t}}{\sum_{j=1}^{n} P_j e^{-(\lambda_j - \lambda_i)t}} = \frac{P_i \lambda_i + \sum_{j \neq i} P_j \lambda_j e^{-(\lambda_j - \lambda_i)t}}{P_i + \sum_{j \neq i} P_j e^{-(\lambda_j - \lambda_i)t}}$$

Hence, since for $j \neq i$, $\lambda_j - \lambda_i > 0$, we see that

$$\lim_{t \to \infty} r(t) = \frac{P_i \lambda_i}{P_i} = \lambda_i$$

That is, as a randomly chosen battery ages its failure rate converges to the failure rate of the exponential type having the smallest failure rate, which is intuitive since the longer the battery lasts, the most likely it seems that it is a battery type with the smallest failure rate. ✦

5.2.3. Further Properties of the Exponential Distribution

Let $X_1, \ldots, X_n$ be independent and identically distributed exponential random variables having mean $1/\lambda$. It follows from the results of Example 2.38 that $X_1 + \cdots + X_n$ has a gamma distribution with parameters n and λ. Let us now give a second verification of this result by using mathematical induction. Because there is nothing to prove when $n = 1$, let us start by assuming that $X_1 + \cdots + X_{n-1}$ has density given by

$$f_{X_1 + \cdots + X_{n-1}}(t) = \lambda e^{-\lambda t} \frac{(\lambda t)^{n-2}}{(n-2)!}$$

Hence,

$$f_{X_1 + \cdots + X_{n-1} + X_n}(t) = \int_0^{\infty} f_{X_n}(t - s) f_{X_1 + \cdots + X_{n-1}}(s) \, ds$$

$$= \int_0^t \lambda e^{-\lambda(t-s)} \lambda e^{-\lambda s} \frac{(\lambda s)^{n-2}}{(n-2)!} \, ds$$

$$= \lambda e^{-\lambda t} \frac{(\lambda t)^{n-1}}{(n-1)!}$$

which proves the result.

Another useful calculation is to determine the probability that one exponential random variable is smaller than another. That is, suppose that

X_1 and X_2 are independent exponential random variables with respective means $1/\lambda_1$ and $1/\lambda_2$; what is $P\{X_1 < X_2\}$? This probability is easily calculated by conditioning on X_1:

$$P\{X_1 < X_2\} = \int_0^\infty P\{X_1 < X_2 \mid X_1 = x\}\lambda_1 e^{-\lambda_1 x}\,dx$$

$$= \int_0^\infty P\{x < X_2\}\lambda_1 e^{-\lambda_1 x}\,dx$$

$$= \int_0^\infty e^{-\lambda_2 x}\lambda_1 e^{-\lambda_1 x}\,dx$$

$$= \int_0^\infty \lambda_1 e^{-(\lambda_1 + \lambda_2)x}\,dx$$

$$= \frac{\lambda_1}{\lambda_1 + \lambda_2} \tag{5.5}$$

Example 5.5 Suppose one has a stereo system consisting of two main parts, a radio and a speaker. If the lifetime of the radio is exponential with mean 1000 hours and the lifetime of the speaker is exponential with mean 500 hours independent of the radio's lifetime, then what is the probability that the system's failure (when it occurs) will be caused by the radio failing?

Solution: From Equation (5.5) (with $\lambda_1 = 1/1000$, $\lambda_2 = 1/500$) we see that the answer is

$$\frac{1/1000}{1/1000 + 1/500} = \frac{1}{3} \quad \blacklozenge$$

Suppose that $X_1, X_2, \ldots, X_n$ are independent exponential random variables, with X_i having rate μ_i, $i = 1, \ldots, n$. It turns out that the smallest of the X_i is exponential with a rate equal to the sum of the μ_i. This is shown as follows:

$$P\{\text{minimum}(X_1, \ldots, X_n) > x\} = P\{X_i > x \text{ for each } i = 1, \ldots, n\}$$

$$= \prod_{i=1}^n P\{X_i > x\} \quad \text{(by independence)}$$

$$= \prod_{i=1}^n e^{-\mu_i x}$$

$$= \exp\left\{-\left(\sum_{i=1}^n \mu_i\right)x\right\} \tag{5.6}$$

Example 5.6 (Analyzing Greedy Algorithms for the Assignment Problem): A group of n people is to be assigned to a set of n jobs, with one person assigned to each job. For a given set of n^2 values C_{ij}, $i, j = 1, \ldots, n$, a cost C_{ij} is incurred when person i is assigned to job j. The classical assignment problem is to determine the set of assignments that minimizes the sum of the n costs incurred.

Rather than trying to determine the optimal assignment, let us consider two heuristic algorithms for solving this problem. The first heuristic is as follows. Assign person 1 to the job that results in the least cost. That is, person 1 is assigned to job j_1 where $C(1, j_1) = \text{minimum}_j\, C(1, j)$. Now eliminate that job from consideration and assign person 2 to the job that results in the least cost. That is, person 2 is assigned to job j_2 where $C(2, j_2) = \text{minimum}_{j \neq j_1}\, C(2, j)$. This procedure is then continued until all n persons are assigned. Since this procedure always selects the best job for the person under consideration, we will call it Greedy Algorithm A.

The second algorithm, which we call Greedy Algorithm B, is a more "global" version of the first greedy algorithm. It considers all n^2 cost values and chooses the pair i_1, j_1 for which $C(i, j)$ is minimal. It then assigns person i_1 to job j_1. It then eliminates all cost values involving either person i_1 or job j_1 [so that $(n - 1)^2$ values remain] and continues in the same fashion. That is, at each stage it chooses the person and job that have the smallest cost among all the unassigned people and jobs.

Under the assumption that the C_{ij} constitute a set of n^2 independent exponential random variables each having mean 1, which of the two algorithms results in a smaller expected total cost?

Solution: Suppose first that Greedy Algorithm A is employed. Let C_i denote the cost associated with person $i, i = 1, \ldots, n$. Now C_1 is the minimum of n independent exponentials each having rate 1; so by Equation (5.6) it will be exponential with rate n. Similarly, C_2 is the minimum of $n - 1$ independent exponentials with rate 1, and so is exponential with rate $n - 1$. Indeed, by the same reasoning C_i will be exponential with rate $n - i + 1, i = 1, \ldots, n$. Thus, the expected total cost under Greedy Algorithm A is

$$E_A[\text{total cost}] = E[C_1 + \cdots + C_n]$$

$$= \sum_{i=1}^{n} 1/i$$

Let us now analyze Greedy Algorithm B. Let C_i be the cost of the ith person–job pair assigned by this algorithm. Since C_1 is the minimum of all the n^2 values C_{ij}, it follows from Equation (5.6) that C_1 is exponential with rate n^2. Now, it follows from the lack of memory property of the exponential that the amounts by which the other C_{ij} exceed C_1 will be independent exponentials with rates 1. As a result, C_2 is equal to C_1 plus the minimum of $(n - 1)^2$ independent exponentials with rate 1. Similarly, C_3 is equal to C_2 plus the minimum of $(n - 2)^2$ independent exponentials with rate 1, and so on. Therefore, we see that

$$E[C_1] = 1/n^2,$$

$$E[C_2] = E[C_1] + 1/(n - 1)^2,$$

$$E[C_3] = E[C_2] + 1/(n - 2)^2,$$

$$\vdots$$

$$E[C_j] = E[C_{j-1}] + 1/(n - j + 1)^2,$$

$$\vdots$$

$$E[C_n] = E[C_{n-1}] + 1$$

Therefore,

$$E[C_1] = 1/n^2,$$

$$E[C_2] = 1/n^2 + 1/(n - 1)^2,$$

$$E[C_3] = 1/n^2 + 1/(n - 1)^2 + 1/(n - 2)^2,$$

$$\vdots$$

$$E[C_n] = 1/n^2 + 1/(n - 1)^2 + 1/(n - 2)^2 + \cdots + 1$$

Adding up all the $E[C_i]$ yields that

$$E_B[\text{total cost}] = n/n^2 + (n - 1)/(n - 1)^2 + (n - 2)/(n - 2)^2 + \cdots + 1$$

$$= \sum_{i=1}^{n} \frac{1}{i}$$

The expected cost is thus the same for both greedy algorithms.   ✦

Let $X_1, \ldots, X_n$ be independent exponential random variables, with respective rates $\lambda_1, \ldots, \lambda_n$. A useful result, generalizing Equation (5.5), is that X_i is the smallest of these with probability $\lambda_i/\Sigma_j \lambda_j$. This is shown as follows:

$$P\left\{X_i = \min_j X_j\right\} = P\left\{X_i < \min_{j \neq i} X_j\right\}$$

$$= \frac{\lambda_i}{\Sigma_{j=1}^n \lambda_j}$$

where the final equality uses Equation (5.5) along with the fact that $\min_{j \neq i} X_j$ is exponential with rate $\Sigma_{j \neq i} \lambda_j$.

Another useful result is that the minimal value is independent of the rank ordering of these exponentials. To verify this, let $i_1, \ldots, i_n$ be a permutation of $1, \ldots, n$. Then,

$$P\{\min X_i > x \mid X_{i_1} < \cdots < X_{i_n}\} = \frac{P\{\min X_i > x, X_{i_1} < \cdots < X_{i_n}\}}{P\{X_{i_1} < \cdots < X_{i_n}\}}$$

$$= \frac{P\{\min X_i > x\} P\{X_{i_1} < \cdots < X_{i_n} \mid \min X_i > x\}}{P\{X_{i_1} < \cdots < X_{i_n}\}}$$

$$= P\{\min X_i > x\}$$

where the final equality used the memoryless property of exponentials to conclude that $P\{X_{i_1} < \cdots < X_{i_n} \mid \min X_i > x\} = P\{X_{i_1} < \cdots < X_{i_n}\}$.

Example 5.7 Suppose that customers are in line to receive service that is provided sequentially by a server; whenever a service is completed, the next person in line enters the service facility. However, each waiting customer will only wait an exponentially distributed time with rate θ; if its service has not yet begun by this time then it will immediately depart the system. These exponential times, one for each waiting customer, are independent. In addition, the service times are independent exponential random variables with rate μ. Suppose that someone is presently being served and consider the person who is nth in line.

(a) Find P_n, the probability that this customer is eventually served.
(b) Find W_n, the conditional expected amount of time this person spends waiting in line given that she is eventually served.

Solution: Consider the $n + 1$ random variables consisting of the remaining service time of the person in service along with the n additional exponential departure times with rate θ of the first n in line.

(a) Given that the smallest of these $n + 1$ independent exponentials is the departure time of the nth person in line, the conditional probability that this person will be served is 0; on the other hand, given that this person's departure time is not the smallest, the conditional probability that this person will be served is the same as if it were initially in position $n - 1$. Since the probability that a given departure time is the smallest of the $n + 1$ exponentials is $\theta/(n\theta + \mu)$, we obtain that

$$P_n = \frac{(n - 1)\theta + \mu}{n\theta + \mu} P_{n-1}$$

Using the preceding with $n - 1$ replacing n gives

$$P_n = \frac{(n - 1)\theta + \mu}{n\theta + \mu} \frac{(n - 2)\theta + \mu}{(n - 1)\theta + \mu} P_{n-2} = \frac{(n - 2)\theta + \mu}{n\theta + \mu} P_{n-2}$$

Continuing in this fashion yields the result

$$P_n = \frac{\theta + \mu}{n\theta + \mu} P_1 = \frac{\mu}{n\theta + \mu}$$

(b) To determine an expression for W_n, we use the fact that the minimum of independent exponentials is, independent of their rank ordering, exponential with a rate equal to the sum of the rates. Since the time until the nth person in line enters service is the minimum of these $n + 1$ random variables plus the additional time thereafter, we see, upon using the lack of memory property of exponential random variables, that

$$W_n = \frac{1}{n\theta + \mu} + W_{n-1}$$

Repeating the preceding argument with successively smaller values of n yields the solution:

$$W_n = \sum_{i=1}^{n} \frac{1}{i\theta + \mu} \quad \blacklozenge$$

5.2.4. Convolutions of Exponential Random Variables

Let X_i, $i = 1, \ldots, n$, be independent exponential random variables with respective rates λ_i, $i = 1, \ldots, n$, and suppose that $\lambda_i \neq \lambda_j$ for $i \neq j$. The random variable $\sum_{i=1}^{n} X_i$ is said to be a *hypoexponential* random variable. To compute its probability density function, let us start with the case $n = 2$. Now,

$$
\begin{aligned}
f_{X_1 + X_2}(t) &= \int_0^t f_{X_1}(s) f_{X_2}(t - s)\, ds \\
&= \int_0^t \lambda_1 e^{-\lambda_1 s} \lambda_2 e^{-\lambda_2(t-s)}\, ds \\
&= \lambda_1 \lambda_2 e^{-\lambda_2 t} \int_0^t e^{-(\lambda_1 - \lambda_2)s}\, ds \\
&= \frac{\lambda_1}{\lambda_1 - \lambda_2} \lambda_2 e^{-\lambda_2 t} (1 - e^{-(\lambda_1 - \lambda_2)t}) \\
&= \frac{\lambda_1}{\lambda_1 - \lambda_2} \lambda_2 e^{-\lambda_2 t} + \frac{\lambda_2}{\lambda_2 - \lambda_1} \lambda_1 e^{-\lambda_1 t}
\end{aligned}
$$

Using the preceding, a similar computation yields, when $n = 3$,

$$f_{X_1 + X_2 + X_3}(t) = \sum_{i=1}^{3} \lambda_i e^{-\lambda_i t} \left(\prod_{j \neq i} \frac{\lambda_j}{\lambda_j - \lambda_i} \right)$$

which suggests the general result:

$$f_{X_1 + \cdots + X_n}(t) = \sum_{i=1}^{n} C_{i,n} \lambda_i e^{-\lambda_i t}$$

where

$$C_{i,n} = \prod_{j \neq i} \frac{\lambda_j}{\lambda_j - \lambda_i}$$

We will now prove the preceding formula by induction on n. Since we have already established it for $n = 2$, assume it for n and consider $n + 1$ arbitrary independent exponentials X_i with distinct rates λ_i, $i = 1, \ldots, n + 1$. If necessary, renumber X_1 and X_{n+1} so that $\lambda_{n+1} < \lambda_1$. Now,

$$
\begin{aligned}
f_{X_1 + \cdots + X_{n+1}}(t) &= \int_0^t f_{X_1 + \cdots + X_n}(s) \lambda_{n+1} e^{-\lambda_{n+1}(t-s)} \, ds \\
&= \sum_{i=1}^{n} C_{i,n} \int_0^t \lambda_i e^{-\lambda_i s} \lambda_{n+1} e^{-\lambda_{n+1}(t-s)} \, ds \\
&= \sum_{i=1}^{n} C_{i,n} \left(\frac{\lambda_i}{\lambda_i - \lambda_{n+1}} \lambda_{n+1} e^{-\lambda_{n+1} t} + \frac{\lambda_{n+1}}{\lambda_{n+1} - \lambda_i} \lambda_i e^{-\lambda_i t} \right) \\
&= K_{n+1} \lambda_{n+1} e^{-\lambda_{n+1} t} + \sum_{i=1}^{n} C_{i,n+1} \lambda_i e^{-\lambda_i t} \qquad (5.7)
\end{aligned}
$$

where $K_{n+1} = \sum_{i=1}^{n} C_{i,n} \lambda_i / (\lambda_i - \lambda_{n+1})$ is a constant that does not depend on t. But, we also have that

$$f_{X_1 + \cdots + X_{n+1}}(t) = \int_0^t f_{X_2 + \cdots + X_{n+1}}(s) \lambda_1 e^{-\lambda_1(t-s)} \, ds$$

which implies, by the same argument that resulted in Equation (5.7), that for a constant K_1

$$f_{X_1 + \cdots + X_{n+1}}(t) = K_1 \lambda_1 e^{-\lambda_1 t} + \sum_{i=2}^{n+1} C_{i,n+1} \lambda_i e^{-\lambda_i t}$$

Equating these two expressions for $f_{X_1 + \cdots + X_{n+1}}(t)$ yields

$$K_{n+1} \lambda_{n+1} e^{-\lambda_{n+1} t} + C_{1,n+1} \lambda_1 e^{-\lambda_1 t} = K_1 \lambda_1 e^{-\lambda_1 t} + C_{n+1,n+1} \lambda_{n+1} e^{-\lambda_{n+1} t}$$

Multiplying both sides of the preceding equation by $e^{\lambda_{n+1} t}$ and then letting $t \to \infty$ yields [since $e^{-(\lambda_1 \lambda_{n+1}) t} \to 0$ as $t \to \infty$]

$$K_{n+1} = C_{n+1,n+1}$$

and this, using Equation (5.7), completes the induction proof. Thus, we have shown that if $S = \sum_{i=1}^{n} X_i$, then

$$f_S(t) = \sum_{i=1}^{n} C_{i,n} \lambda_i e^{-\lambda_i t} \qquad (5.8)$$

where

$$C_{i,n} = \prod_{j \neq i} \frac{\lambda_j}{\lambda_j - \lambda_i}$$

Integrating both sides of the expression for f_S from t to ∞ yields that the tail distribution function of S is given by

$$P\{S > t\} = \sum_{i=1}^{n} C_{i,n} e^{-\lambda_i t} \qquad (5.9)$$

Hence, we obtain from Equations (5.8) and (5.9) that $r_S(t)$, the failure rate function of S, is as follows:

$$r_S(t) = \frac{\sum_{i=1}^{n} C_{i,n} \lambda_i e^{-\lambda_i t}}{\sum_{i=1}^{n} C_{i,n} e^{-\lambda_i t}}$$

If we let $\lambda_j = \min(\lambda_1, \ldots, \lambda_n)$, then it follows, upon multiplying the numerator and denominator of $r_S(t)$ by $e^{\lambda_j t}$, that

$$\lim_{t \to \infty} r_S(t) = \lambda_j$$

From the preceding, we can conclude that the remaining lifetime of a hypoexponentially distributed item that has survived to age t is, for t large, approximately that of an exponentially distributed random variable with a rate equal to the minimum of the rates of the random variables whose sums make up the hypoexponential.

Remark   Although

$$1 = \int_0^{\infty} f_S(t)\, dt = \sum_{i=1}^{n} C_{i,n} = \sum_{i=1}^{n} \prod_{j \neq i} \frac{\lambda_j}{\lambda_j - \lambda_i}$$

it should not be thought that the $C_{i,n}$, $i = 1, \ldots, n$ are probabilities, because some of them will be negative. Thus, while the form of the hypoexponential density is similar to that of the hyperexponential density (see Example 5.4) these two random variables are very different.

Example 5.8   Let $X_1, \ldots, X_m$ be independent exponential random variables with respective rates $\lambda_1, \ldots, \lambda_m$, where $\lambda_i \neq \lambda_j$ when $i \neq j$. Let N be independent of these random variables and suppose that $\sum_{n=1}^{m} P_n = 1$, where

$P_n = P\{N = n\}$. The random variable

$$Y = \sum_{j=1}^{N} X_j$$

is said to be a *Coxian* random variable. Conditioning on N gives its density function:

$$f_Y(t) = \sum_{n=1}^{m} f_Y(t \mid N = n)P_n$$

$$= \sum_{n=1}^{m} f_{X_1 + \cdots + X_n}(t \mid N = n)P_n$$

$$= \sum_{n=1}^{m} f_{X_1 + \cdots + X_n}(t)P_n$$

$$= \sum_{n=1}^{m} P_n \sum_{i=1}^{n} C_{i,n} \lambda_i e^{-\lambda_i t}$$

Let

$$r(n) = P\{N = n \mid N \geqslant n\}$$

If we interpret N as a lifetime measured in discrete time periods, then $r(n)$ denotes the probability that an item will die in its nth period of use given that it has survived up to that time. Thus, $r(n)$ is the discrete time analog of the failure rate function $r(t)$, and is correspondingly referred to as the discrete time *failure* (or *hazard*) *rate* function.

Coxian random variables often arise in the following manner. Suppose that an item must go through m stages of treatment to be cured. However, suppose that after each stage there is a probability that the item will quit the program. If we suppose that the amounts of time that it takes the item to pass through the successive stages are independent exponential random variables, and that the probability that an item that has just completed stage n quits the program is (independent of how long it took to go through the n stages) equal to $r(n)$, then the total time that an item spends in the program is a Coxian random variable. ✦

5.3. The Poisson Process

5.3.1. Counting Processes

A stochastic process $\{N(t), t \geqslant 0\}$ is said to be a *counting process* if $N(t)$ represents the total number of "events" that have occurred up to time t. Some examples of counting processes are the following:

(a) If we let $N(t)$ equal the number of persons who have entered a particular store at or prior to time t, then $\{N(t), t \geq 0\}$ is a counting process in which an event corresponds to a person entering the store. Note that if we had let $N(t)$ equal the number of persons in the store at time t, then $\{N(t), t \geq 0\}$ would *not* be a counting process (why not?).

(b) If we say that an event occurs whenever a child is born, then $\{N(t), t \geq 0\}$ is a counting process when $N(t)$ equals the total number of people who were born by time t. [Does $N(t)$ include persons who have died by time t? Explain why it must.]

(c) If $N(t)$ equals the number of goals that a given soccer player has scored by time t, then $\{N(t), t \geq 0\}$ is a counting process. An event of this process will occur whenever the soccer player scores a goal.

From its definition we see that for a counting process $N(t)$ must satisfy:

(i) $N(t) \geq 0$.
(ii) $N(t)$ is integer valued.
(iii) If $s < t$, then $N(s) \leq N(t)$.
(iv) For $s < t$, $N(t) - N(s)$ equals the number of events that have occurred in the interval $(s, t]$.

A counting process is said to possess *independent increments* if the numbers of events that occur in disjoint time intervals are independent. For example, this means that the number of events that occur by time 10 [that is, $N(10)$] must be independent of the number of events that occur between times 10 and 15 [that is, $N(15) - N(10)$].

The assumption of independent increments might be reasonable for example (a), but it probably would be unreasonable for example (b). The reason for this is that if in example (b) $N(t)$ is very large, then it is probable that there are many people alive at time t; this would lead us to believe that the number of new births between time t and time $t + s$ would also tend to be large [that is, it does not seem reasonable that $N(t)$ is independent of $N(t + s) - N(t)$, and so $\{N(t), t \geq 0\}$ would not have independent increments in example (b)]. The assumption of independent increments in example (c) would be justified if we believed that the soccer player's chances of scoring a goal today do not depend on "how he's been going." It would not be justified if we believed in "hot streaks" or "slumps."

A counting process is said to possess *stationary increments* if the distribution of the number of events that occur in any interval of time depends only on the length of the time interval. In other words, the process has stationary increments if the number of events in the interval $(t_1 + s, t_2 + s)$ [that is, $N(t_2 + s) - N(t_1 + s)$] has the same distribution as the number of events in the interval (t_1, t_2) [that is, $N(t_2) - N(t_1)$] for all $t_1 < t_2$, and $s > 0$.

The assumption of stationary increments would only be reasonable in example (a) if there were no times or day at which people were more likely to enter the store. Thus, for instance, if there was a rush hour (say, between 12 P.M. and 1 P.M.) each day, then the stationarity assumption would not be justified. If we believed that the earth's population is basically constant (a belief not held at present by most scientists), then the assumption of stationary increments might be reasonable in example (b). Stationary increments do not seem to be a reasonable assumption in example (c) since, for one thing, most people would agree that the soccer player would probably score more goals while in the age bracket 25–30 than he would while in the age bracket 35–40.

5.3.2. Definition of the Poisson Process

One of the most important counting processes is the Poisson process which is defined as follows:

Definition 5.1   The counting process $\{N(t), t \geq 0\}$ is said to be a *Poisson process having rate* $\lambda, \lambda > 0$, if

(i) $N(0) = 0$.
(ii) The process has independent increments.
(iii) The number of events in any interval of length t is Poisson distributed with mean λt. That is, for all $s, t \geq 0$

$$P\{N(t + s) - N(s) = n\} = e^{-\lambda t} \frac{(\lambda t)^n}{n!}, \qquad n = 0, 1, \ldots$$

Note that it follows from condition (iii) that a Poisson process has stationary increments and also that

$$E[N(t)] = \lambda t$$

which explains why λ is called the rate of the process.

To determine if an arbitrary counting process is actually a Poisson process, we must show that conditions (i), (ii), and (iii) are satisfied. Condition (i), which simply states that the counting of events begins at time $t = 0$, and condition (ii) can usually be directly verified from our knowledge of the process. However, it is not at all clear how we would determine that condition (iii) is satisfied, and for this reason an equivalent definition of a Poisson process would be useful.

As a prelude to giving a second definition of a Poisson process we shall define the concept of a function $f(\cdot)$ being $o(h)$.

Definition 5.2   The function $f(\cdot)$ is said to be $o(h)$ if

$$\lim_{h \to 0} \frac{f(h)}{h} = 0$$

Example 5.9

(i) The function $f(x) = x^2$ is $o(h)$ since

$$\lim_{h \to 0} \frac{f(h)}{h} = \lim_{h \to 0} \frac{h^2}{h} = \lim_{h \to 0} h = 0$$

(ii) The function $f(x) = x$ is not $o(h)$ since

$$\lim_{h \to 0} \frac{f(h)}{h} = \lim_{h \to 0} \frac{h}{h} = \lim_{h \to 0} 1 = 1 \neq 0$$

(iii) If $f(\cdot)$ is $o(h)$ and $g(\cdot)$ is $o(h)$, then so is $f(\cdot) + g(\cdot)$. This follows since

$$\lim_{h \to 0} \frac{f(h) + g(h)}{h} = \lim_{h \to 0} \frac{f(h)}{h} + \lim_{h \to 0} \frac{g(h)}{h} = 0 + 0 = 0$$

(iv) If $f(\cdot)$ is $o(h)$, then so is $g(\cdot) = cf(\cdot)$. This follows since

$$\lim_{h \to 0} \frac{cf(h)}{h} = c \lim \frac{f(h)}{h} = c \cdot 0 = 0$$

(v) From (iii) and (iv) it follows that any finite linear combination of functions, each of which is $o(h)$, is $o(h)$. ◆

In order for the function $f(\cdot)$ to be $o(h)$ it is necessary that $f(h)/h$ go to zero as h goes to zero. But if h goes to zero, the only way for $f(h)/h$ to go to zero is for $f(h)$ to go to zero faster than h does. That is, for h small, $f(h)$ must be small compared with h.

We are now in a position to give an alternate definition of a Poisson process.

Definition 5.3   The counting process $\{N(t), t \geq 0\}$ is said to be a Poisson process having rate λ, $\lambda > 0$, if

(i) $N(0) = 0$.
(ii) The process has stationary and independent increments.
(iii) $P\{N(h) = 1\} = \lambda h + o(h)$.
(iv) $P\{N(h) \geq 2\} = o(h)$.

Theorem 5.1 Definitions 5.1 and 5.3 are equivalent.

Proof We show that Definition 5.3 implies Definition 5.1, and leave it to the reader to prove the reverse. To start, fix $u \geq 0$ and let

$$g(t) = E[\exp\{-uN(t)\}]$$

We derive a differential equation for $g(t)$ as follows:

$$
\begin{aligned}
g(t + h) &= E[\exp\{-uN(t + h)\}] \\
&= E[\exp\{-uN(t)\} \exp\{-u(N(t + h) - N(t))\}] \\
&= E[\exp\{-uN(t)\}] E[\exp\{-u(N(t + h) - N(t))\}] \\
&\qquad \text{by independent increments} \\
&= g(t) E[\exp\{-uN(h)\}] \quad \text{by stationary increments} \quad (5.10)
\end{aligned}
$$

Now, assumptions (iii) and (iv) imply that

$$P\{N(h) = 0\} = 1 - \lambda h + o(h)$$

Hence, conditioning on whether $N(h) = 0$ or $N(h) = 1$ or $N(h) \geq 2$ yields

$$
\begin{aligned}
E[\exp\{-uN(h)\}] &= 1 - \lambda h + o(h) + e^{-u}(\lambda h + o(h)) + o(h) \\
&= 1 - \lambda h + e^{-u}\lambda h + o(h) \quad (5.11)
\end{aligned}
$$

Therefore, from Equations (5.10) and (5.11) we obtain that

$$g(t + h) = g(t)(1 - \lambda h + e^{-u}\lambda h) + o(h)$$

implying that

$$\frac{g(t + h) - g(t)}{h} - = g(t) \lambda (e^{-u} - 1) + \frac{o(h)}{h}$$

Letting $h \to 0$ gives

$$g'(t) = g(t) \lambda (e^{-u} - 1)$$

or, equivalently,

$$\frac{g'(t)}{g(t)} = \lambda(e^{-u} - 1)$$

Integrating, and using $g(0) = 1$, shows that

$$\log(g(t)) = \lambda t(e^{-u} - 1)$$

or

$$g(t) = \exp\{\lambda t(e^{-u} - 1)\}$$

That is, the Laplace transform of $N(t)$ evaluated at u is $e^{\lambda t(e^{-u}-1)}$. Since that is also the Laplace transform of a Poisson random variable with mean λt, the result follows from the fact that the distribution of a nonnegative random variable is uniquely determined by its Laplace transform. ✦

Remarks (i) The result that $N(t)$ has a Poisson distribution is a consequence of the Poisson approximation to the binomial distribution (see Section 2.2.4). To see this, subdivide the interval $[0, t]$ into k equal parts where k is very large (Figure 5.1). Now it can be shown using axiom (iv) of Definition 5.3 that as k increases to ∞ the probability of having two or more events in any of the k subintervals goes to 0. Hence, $N(t)$ will (with a probability going to 1) just equal the number of subintervals in which an event occurs. However, by stationary and independent increments this number will have a binomial distribution with parameters k and $p = \lambda t/k + o(t/k)$. Hence, by the Poisson approximation to the binomial we see by letting k approach ∞ that $N(t)$ will have a Poisson distribution with mean equal to

$$\lim_{k \to \infty} k\left[\lambda \frac{t}{k} + o\left(\frac{t}{k}\right)\right] = \lambda t + \lim_{k \to \infty} \frac{t\, o(t/k)}{t/k}$$

$$= \lambda t$$

by using the definition of $o(h)$ and the fact that $t/k \to 0$ as $k \to \infty$.

(ii) The explicit assumption that the process has stationary increments can be eliminated from Definition 5.3 provided that we change assumptions (iii) and (iv) to require that for any t the probability of one event in the interval $(t, t + h)$ is $\lambda h + o(h)$ and the probability of two or more events in that interval is $o(h)$. That is, assumptions (ii), (iii), and (iv) of Definition 5.3 can be replaced by

(ii) The process has independent increments.
(iii) $P\{N(t + h) - N(t) = 1\} = \lambda h + o(h)$.
(iv) $P\{N(t + h) - N(t) \geq 2\} = o(h)$.

5.3.3. Interarrival and Waiting Time Distributions

Consider a Poisson process, and let us denote the time of the first event by

Figure 5.1.

T_1. Further, for $n > 1$, let T_n denote the elapsed time between the $(n - 1)$st and the nth event. The sequence $\{T_n, n = 1, 2, \dots\}$ is called the *sequence of interarrival times*. For instance, if $T_1 = 5$ and $T_2 = 10$, then the first event of the Poisson process would have occurred at time 5 and the second at time 15.

We shall now determine the distribution of the T_n. To do so, we first note that the event $\{T_1 > t\}$ takes place if and only if no events of the Poisson process occur in the interval $[0, t]$ and thus,

$$P\{T_1 > t\} = P\{N(t) = 0\} = e^{-\lambda t}$$

Hence, T_1 has an exponential distribution with mean $1/\lambda$. Now,

$$P\{T_2 > t\} = E[P\{T_2 > t \mid T_1\}]$$

However,

$$P\{T_2 > t \mid T_1 = s\} = P\{0 \text{ events in } (s, s + t] \mid T_1 = s\}$$
$$= P\{0 \text{ events in } (s, s + t]\}$$
$$= e^{-\lambda t} \qquad (5.12)$$

where the last two equations followed from independent and stationary increments. Therefore, from Equation (5.12) we conclude that T_2 is also an exponential random variable with mean $1/\lambda$, and furthermore, that T_2 is independent of T_1. Repeating the same argument yields the following.

Proposition 5.1 $T_n, n = 1, 2, \dots$, are independent identically distributed exponential random variables having mean $1/\lambda$.

Remarks   The proposition should not surprise us. The assumption of stationary and independent increments is basically equivalent to asserting that, at any point in time, the process *probabilistically* restarts itself. That is, the process from any point on is independent of all that has previously occurred (by independent increments), and also has the same distribution as the original process (by stationary increments). In other words, the process has no *memory*, and hence exponential interarrival times are to be expected.

Another quantity of interest is S_n, the arrival time of the nth event, also called the *waiting time* until the nth event. It is easily seen that

$$S_n = \sum_{i=1}^{n} T_i, \qquad n \geq 1$$

and hence from Proposition 5.1 and the results of Section 2.2 it follows that

S_n has a gamma distribution with parameters n and λ. That is, the probability density of S_n is given by

$$f_{S_n}(t) = \lambda e^{-\lambda t} \frac{(\lambda t)^{n-1}}{(n-1)!}, \qquad t \geq 0 \tag{5.13}$$

Equation (5.13) may also have been derived by noting that the nth event will occur prior to or at time t if and only if the number of events occurring by time t is at least n. That is,

$$N(t) \geq n \leftrightarrow S_n \leq t$$

and hence,

$$F_{S_n}(t) = P\{S_n \leq t\} = P\{N(t) \geq n\} = \sum_{j=n}^{\infty} e^{-\lambda t} \frac{(\lambda t)^j}{j!}$$

which, upon differentiation, yields

$$f_{S_n}(t) = -\sum_{j=n}^{\infty} \lambda e^{-\lambda t} \frac{(\lambda t)^j}{j!} + \sum_{j=n}^{\infty} \lambda e^{-\lambda t} \frac{(\lambda t)^{j-1}}{(j-1)!}$$

$$= \lambda e^{-\lambda t} \frac{(\lambda t)^{n-1}}{(n-1)!} + \sum_{j=n+1}^{\infty} \lambda e^{-\lambda t} \frac{(\lambda t)^{j-1}}{(j-1)!} - \sum_{j=n}^{\infty} \lambda e^{-\lambda t} \frac{(\lambda t)^j}{j!}$$

$$= \lambda e^{-\lambda t} \frac{(\lambda t)^{n-1}}{(n-1)!}$$

Example 5.10 Suppose that people immigrate into a territory at a Poisson rate $\lambda = 1$ per day.

(a) What is the expected time until the tenth immigrant arrives?
(b) What is the probability that the elapsed time between the tenth and the eleventh arrival exceeds two days?

Solution:

(a) $E[S_{10}] = 10/\lambda = 10$ days.
(b) $P\{T_{11} > 2\} = e^{-2\lambda} = e^{-2} \approx 0.133.$ ✦

Proposition 5.1 also gives us another way of defining a Poisson process. Suppose we start with a sequence $\{T_n, n \geq 1\}$ of independent identically distributed exponential random variables each having mean $1/\lambda$. Now let us define a counting process by saying that the nth event of this process occurs at time

$$S_n \equiv T_1 + T_2 + \cdots + T_n$$

The resultant counting process $\{N(t), t \geq 0\}$* will be Poisson with rate λ.

Remark   Another way of obtaining the density function of S_n is to note that since S_n is the time of the nth event, it follows that

$$P\{t < S_n < t + h\} = P\{N(t) = n - 1, \text{one event in } (t, t + h)\} + o(h)$$

$$= P\{N(t) = n - 1\}P\{\text{one event in } (t, t + h)\} + o(h)$$

$$= e^{-\lambda t} \frac{(\lambda t)^{n-1}}{(n-1)!} [\lambda h + o(h)] + o(h)$$

$$= \lambda e^{-\lambda t} \frac{(\lambda t)^{n-1}}{(n-1)!} h + o(h)$$

where the first equality uses the fact that the probability of 2 or more events in $(t, t + h)$ is $o(h)$. If we now divide both sides of the preceding equation by h and then let $h \to 0$, we obtain

$$f_{S_n}(t) = \lambda e^{-\lambda t} \frac{(\lambda t)^{n-1}}{(n-1)!}$$

5.3.4. Further Properties of Poisson Processes

Consider a Poisson process $\{N(t), t \geq 0\}$ having rate λ, and suppose that each time an event occurs it is classified as either a type I or a type II event. Suppose further that each event is classified as a type I event with probability p or a type II event with probability $1 - p$, independently of all other events. For example, suppose that customers arrive at a store in accordance with a Poisson process having rate λ; and suppose that each arrival is male with probability $\frac{1}{2}$ and female with probability $\frac{1}{2}$. Then a type I event would correspond to a male arrival and a type II event to a female arrival.

Let $N_1(t)$ and $N_2(t)$ denote respectively the number of type I and type II events occurring in $[0, t]$. Note that $N(t) = N_1(t) + N_2(t)$.

Proposition 5.2 $\{N_1(t), t \geq 0\}$ and $\{N_2(t), t \geq 0\}$ are both Poisson processes having respective rates λp and $\lambda(1 - p)$. Furthermore, the two processes are independent.

* A formal definition of $N(t)$ is given by $N(t) \equiv \max\{n : S_n \leq t\}$ where $S_0 \equiv 0$.

Proof   Let us calculate the joint probability $P\{N_1(t) = n, N_2(t) = m\}$. To do this, we first condition on $N(t)$ to obtain

$$P\{N_1(t) = n, N_2(t) = m\}$$

$$= \sum_{k=0}^{\infty} P\{N_1(t) = n, N_2(t) = m \,|\, N(t) = k\} P\{N(t) = k\}$$

Now, in order for there to have been n type I events and m type II events there must have been a total of $n + m$ events occurring in $[0, t]$. That is,

$$P\{N_1(t) = n, N_2(t) = m \,|\, N(t) = k\} = 0 \qquad \text{when } k \neq n + m$$

Hence,

$$P\{N_1(t) = n, N_2(t) = m\}$$

$$= P\{N_1(t) = n, N_2(t) = m \,|\, N(t) = n + m\} P\{N(t) = n + m\}$$

$$= P\{N_1(t) = n, N_2(t) = m \,|\, N(t) = n + m\} e^{-\lambda t} \frac{(\lambda t)^{n+m}}{(n + m)!}$$

However, given that $n + m$ events occurred, since each event has probability p of being a type I event and probability $1 - p$ of being a type II event, it follows that the probability that n of them will be type I and m of them type II events is just the binomial probability

$$\binom{n + m}{n} p^n (1 - p)^m$$

Thus,

$$P\{N_1(t) = n, N_2(t) = m\} = \binom{n + m}{n} p^n (1 - p)^m e^{-\lambda t} \frac{(\lambda t)^{n+m}}{(n + m)!}$$

$$= e^{-\lambda t p} \frac{(\lambda t p)^n}{n!} e^{-\lambda t (1 - p)} \frac{(\lambda t (1 - p))^m}{m!} \qquad (5.14)$$

Hence,

$$P\{N_1(t) = n\} = \sum_{m=0}^{\infty} P\{N_1(t) = n, N_2(t) = m\}$$

$$= e^{-\lambda t p} \frac{(\lambda t p)^n}{n!} \sum_{m=0}^{\infty} e^{-\lambda t (1 - p)} \frac{(\lambda t (1 - p))^m}{m!}$$

$$= e^{-\lambda t p} \frac{(\lambda t p)^n}{n!}$$

That is, $\{N_1(t),\ t \geqslant 0\}$ is a Poisson process having rate λp. (How do we know that the other conditions of Definition 5.1 are satisfied? Argue it out!)
    Similarly,

$$P\{N_2(t) = m\} = e^{-\lambda t(1-p)} \frac{(\lambda t(1-p))^m}{m!}$$

and so $\{N_2(t),\ t \geqslant 0\}$ is a Poisson process having rate $\lambda(1-p)$. Finally, it follows from Equation (5.14) that the two processes are independent (since the joint distribution factors). ✦

Remark   It is not surprising that $\{N_1(t),\ t \geqslant 0\}$ and $\{N_2(t),\ t \geqslant 0\}$ are Poisson processes. What is somewhat surprising is the fact that they are independent. Assume that customers arrive at a bank at a Poisson rate of $\lambda = 1$ per hour and suppose that each customer is a man with probability $\frac{1}{2}$ and a woman with probability $\frac{1}{2}$. Now suppose that 100 men arrived in the first 10 hours. Then how many women would we expect to have arrived in that time? One might argue that because the number of male arrivals is 100 and because each arrival is male with probability $\frac{1}{2}$, then the expected number of total arrivals should be 200 and hence the expected number of female arrivals should also be 100. But, as shown by the previous proposition, such reasoning is spurious and the expected number of female arrivals in the first 10 hours is five, independent of the number of male arrivals in that period.
    To obtain an intuition as to why Proposition 5.2 is true reason as follows: If we divide the interval $(0, t)$ into n subintervals of equal length t/n, where n is very large, then (neglecting events having probability "little o") each subinterval will have a small probability $\lambda t/n$ of containing a single event. Because each event has probability p of being of type I, it follows that each of the n subintervals will have either no events, a type I event, a type II event with respective probabilities

$$1 - \frac{\lambda t}{n}, \qquad \frac{\lambda t}{n} p, \qquad \frac{\lambda t}{n}(1 - p)$$

Hence from the result which gives the Poisson as the limit of binomials, we can immediately conclude that $N_1(t)$ and $N_2(t)$ are Poisson distributed with respective means $\lambda t p$ and $\lambda t(1 - p)$. To see that they are independent, suppose, for instance, that $N_1(t) = k$. Then of the n subintervals, k will contain a type I event, and thus the other $n - k$ will each contain a type II event with probability

$$P\{\text{type II} \mid \text{no type I}\} = \frac{(\lambda t/n)(1 - p)}{1 - (\lambda t/n)p}$$

$$= \frac{\lambda t}{n}(1 - p) + o\left(\frac{t}{n}\right)$$

Hence, because $n - k$ will still be a very large number, we see again from the Poisson limit theorem that, given $N_1(t) = k$, $N_2(t)$ will be Poisson with mean $\lim_{n \to \infty} [(n-k)\lambda t(1-p)/n] = \lambda t(1-p)$, and so independence is established.

Example 5.11 If immigrants to area A arrive at a Poisson rate of ten per week, and if each immigrant is of English descent with probability $\frac{1}{12}$, then what is the probability that no people of English descent will emigrate to area A during the month of February?

Solution: By the previous proposition it follows that the number of Englishmen emigrating to area A during the month of February is Poisson distributed with mean $4 \cdot 10 \cdot \frac{1}{12} = \frac{10}{3}$. Hence the desired probability is $e^{-10/3}$. ◆

It follows from Proposition 5.2 that if each of a Poisson number of individuals is independently classified into one of two possible groups with respective probabilities p and $1 - p$, then the number of individuals in each of the two groups will be independent Poisson random variables. Because this result easily generalizes to the case where the classification is into any one of r possible groups, we have the following application to a model of employees moving about in an organization.

Example 5.12 Consider a system in which individuals at any time are classified as being in one of r possible states, and assume that an individual changes states in accordance with a Markov chain having transition probabilities $P_{ij}, i, j = 1, \ldots, r$. That is, if an individual is in state i during a time period then, independently of its previous states, it will be in state j during the next time period with probability P_{ij}. The individuals are assumed to move through the system independently of each other. Suppose that the numbers of people initially in states $1, 2, \ldots, r$ are independent Poisson random variables with respective means $\lambda_1, \lambda_2, \ldots, \lambda_r$. We are interested in determining the joint distribution of the numbers of individuals in states $1, 2, \ldots, r$ at some time n.

Solution: For fixed i, let $N_j(i)$, $j = 1, \ldots, r$, denote the number of those individuals, initially in state i, that are in state j at time n. Now each of the (Poisson distributed) number of people initially in state i will, independently of each other, be in state j at time n with probability P_{ij}^n, where P_{ij}^n is the n-stage transition probability for the Markov chain having transition probabilities P_{ij}. Hence, the $N_j(i)$, $j = 1, \ldots, r$, will be independent Poisson random variables with respective means $\lambda_i P_{ij}^n$, $j = 1, \ldots, r$. Because the sum of independent Poisson random variables is itself a Poisson random variable, it follows that the number of individuals in state j at time n — namely $\sum_{i=1}^r N_j(i)$ — will be independent Poisson random variables with respective means $\sum_i \lambda_i P_{ij}^n$, for $j = 1, \ldots, r$.   ◆

Example 5.13   (The Coupon Collecting Problem): There are m different types of coupons. Each time a person collects a coupon it is, independently of ones previously obtained, a type j coupon with probability p_j, $\sum_{j=1}^m p_j = 1$. Let N denote the number of coupons one needs to collect in order to have a complete collection of at least one of each type. Find $E[N]$.

Solution: If we let N_j denote the number one must collect to obtain a type j coupon, then we can express N as

$$N = \max_{1 \leqslant j \leqslant m} N_j$$

However, even though each N_j is geometric with parameter p_j, the foregoing representation of N is not that useful, because the random variables N_j are not independent.

We can, however, transform the problem into one of determining the expected value of the maximum of *independent* random variables. To do so, suppose that coupons are collected at times chosen according to a Poisson process with rate $\lambda = 1$. Say that an event of this Poisson process is of type j, $1 \leqslant j \leqslant m$, if the coupon obtained at that time is a type j coupon. If we now let $N_j(t)$ denote the number of type j coupons collected by time t, then it follows from Proposition 5.2 that $\{N_j(t), t \geqslant 0\}$, $j = 1, \ldots, m$, are independent Poisson processes with respective rates $\lambda p_j = p_j$. Let X_j denote the time of the first event of the jth process, and let

$$X = \max_{1 \leqslant j \leqslant m} X_j$$

denote the time at which a complete collection is amassed. Since the X_j are independent exponential random variables with respective rates p_j, it follows that

$$P\{X < t\} = P\{\max X_j < t\}$$
$$= P\{X_j < t, \text{ for } j = 1, \ldots, m\}$$
$$= \prod_{j=1}^{m} (1 - e^{-p_j t})$$

Therefore,

$$E[X] = \int_0^\infty P\{X > t\}\, dt$$
$$= \int_0^\infty \left\{ 1 - \prod_{j=1}^{m} (1 - e^{-p_j t}) \right\} dt \qquad (5.15)$$

It remains to relate $E[X]$, the expected time until one has a complete set, to $E[N]$, the expected number of coupons it takes. This can be done by letting T_i denote the ith interarrival time of the Poisson process that counts the number of coupons obtained. Then it is easy to see that

$$X = \sum_{i=1}^{N} T_i$$

Since the T_i are independent exponentials with rate 1, and N is independent of the T_i, we see that

$$E[X \mid N] = NE[T_i] = N$$

Therefore,

$$E[X] = E[N]$$

and so $E[N]$ is as given in Equation (5.15). ✦

The next probability calculation related to Poisson processes that we shall determine is the probability that n events occur in one Poisson process before m events have occurred in a second and independent Poisson process. More formally let $\{N_1(t), t \geq 0\}$ and $\{N_2(t), t \geq 0\}$ be two independent Poisson processes having respective rates λ_1 and λ_2. Also, let S_n^1 denote the time of the nth event of the first process, and S_m^2 the time of the mth event of the second process. We seek

$$P\{S_n^1 < S_m^2\}$$

Before attempting to calculate this for general n and m, let us consider the special case $n = m = 1$. Since S_1^1, the time of the first event of the $N_1(t)$ process, and S_1^2, the time of the first event of the $N_2(t)$ process, are both exponentially distributed random variables (by Proposition 5.1) with respective means $1/\lambda_1$ and $1/\lambda_2$, it follows from Section 5.2.3 that

$$P\{S_1^1 < S_1^2\} = \frac{\lambda_1}{\lambda_1 + \lambda_2} \tag{5.16}$$

Let us now consider the probability that two events occur in the $N_1(t)$ process before a single event has occurred in the $N_2(t)$ process. That is, $P\{S_2^1 < S_1^2\}$. To calculate this we reason as follows: In order for the $N_1(t)$ process to have two events before a single event occurs in the $N_2(t)$ process, it is first necessary for the initial event that occurs to be an event of the $N_1(t)$ process [and this occurs, by Equation (5.16), with probability $\lambda_1/(\lambda_1 + \lambda_2)$]. Now given that the initial event is from the $N_1(t)$ process, the next thing that must occur for S_2^1 to be less than S_1^2 is for the second event also to be an event of the $N_1(t)$ process. However, when the first event occurs both processes start all over again (by the memoryless property of Poisson processes) and hence this conditional probability is also $\lambda_1/(\lambda_1 + \lambda_2)$, and hence the desired probability is given by

$$P\{S_2^1 < S_1^2\} = \left(\frac{\lambda_1}{\lambda_1 + \lambda_2}\right)^2$$

In fact this reasoning shows that *each event that occurs is going to be an event of the $N_1(t)$ process with probability $\lambda_1/(\lambda_1 + \lambda_2)$ and an event of the $N_2(t)$ process with probability $\lambda_2/(\lambda_1 + \lambda_2)$, independent of all that has previously occurred.* In other words, the probability that the $N_1(t)$ process reaches n before the $N_2(t)$ process reaches m is just the probability that n heads will appear before m tails if one flips a coin having probability $p = \lambda_1/(\lambda_1 + \lambda_2)$ of a head appearing. But by noting that this event will occur if and only if the first $n + m - 1$ tosses result in n or more heads, we see that our desired probability is given by

$$P\{S_n^1 < S_m^2\} = \sum_{k=n}^{n+m-1} \binom{n + m - 1}{k} \left(\frac{\lambda_1}{\lambda_1 + \lambda_2}\right)^k \left(\frac{\lambda_2}{\lambda_1 + \lambda_2}\right)^{n+m-1-k}$$

5.3.5. Conditional Distribution of the Arrival Times

Suppose we are told that exactly one event of a Poisson process has taken place by time t, and we are asked to determine the distribution of the time at which the event occurred. Now, since a Poisson process possesses stationary and independent increments it seems reasonable that each interval in $[0, t]$ of equal length should have the same probability of containing the event. In other words, the time of the event should be uniformly distributed over $[0, t]$. This is easily checked since, for $s \leqslant t$,

$$P\{T_1 < s \mid N(t) = 1\} = \frac{P\{T_1 < s, N(t) = 1\}}{P\{N(t) = 1\}}$$

$$= \frac{P\{1 \text{ event in } [0, s), 0 \text{ events in } [s, t]\}}{P\{N(t) = 1\}}$$

$$= \frac{P\{1 \text{ event in } [0, s)\} P\{0 \text{ events in } [s, t]\}}{P\{N(t) = 1\}}$$

$$= \frac{\lambda s e^{-\lambda s} e^{-\lambda(t-s)}}{\lambda t e^{-\lambda t}}$$

$$= \frac{s}{t}$$

This result may be generalized, but before doing so we need to introduce the concept of order statistics.

Let $Y_1, Y_2, \ldots, Y_n$ be n random variables. We say that $Y_{(1)}, Y_{(2)}, \ldots, Y_{(n)}$ are the order statistics corresponding to $Y_1, Y_2, \ldots, Y_n$ if $Y_{(k)}$ is the kth smallest value among $Y_1, \ldots, Y_n$, $k = 1, 2, \ldots, n$. For instance if $n = 3$ and $Y_1 = 4$, $Y_2 = 5$, $Y_3 = 1$ then $Y_{(1)} = 1$, $Y_{(2)} = 4$, $Y_{(3)} = 5$. If the Y_i, $i = 1, \ldots, n$, are independent identically distributed continuous random variables with probability density f, then the joint density of the order statistics $Y_{(1)}, Y_{(2)}, \ldots, Y_{(n)}$ is given by

$$f(y_1, y_2, \ldots, y_n) = n! \prod_{i=1}^{n} f(y_i), \qquad y_1 < y_2 < \cdots < y_n$$

The preceding follows since

(i) $(Y_{(1)}, Y_{(2)}, \ldots, Y_{(n)})$ will equal $(y_1, y_2, \ldots, y_n)$ if $(Y_1, Y_2, \ldots, Y_n)$ is equal to any of the $n!$ permutations of $(y_1, y_2, \ldots, y_n)$;

and

(ii) the probability density that $(Y_1, Y_2, \ldots, Y_n)$ is equal to $y_{i_1}, \ldots, y_{i_n}$ is $\Pi_{j=1}^{n} f(y_{i_j}) = \Pi_{j=1}^{n} f(y_j)$ when $i_1, \ldots, i_n$ is a permutation of $1, 2, \ldots, n$.

If the Y_i, $i = 1, \ldots, n$, are uniformly distributed over $(0, t)$, then we obtain from the preceding that the joint density function of the order statistics $Y_{(1)}, Y_{(2)}, \ldots, Y_{(n)}$ is

$$f(y_1, y_2, \ldots, y_n) = \frac{n!}{t^n}, \qquad 0 < y_1 < y_2 < \cdots < y_n < t$$

We are now ready for the following useful theorem.

Theorem 5.2   Given that $N(t) = n$, the n arrival times $S_1, \ldots, S_n$ have the same distribution as the order statistics corresponding to n independent random variables uniformly distributed on the interval $(0, t)$.

Proof   To obtain the conditional density of $S_1, \ldots, S_n$ given that $N(t) = n$ note that for $0 < s_1 < \cdots < t$ the event that $S_1 = s_1, S_2 = s_2, \ldots, S_n = s_n$, $N(t) = n$ is equivalent to the event that the first $n + 1$ interarrival times satisfy $T_1 = s_1, T_2 = s_2 - s_1, \ldots, T_n = s_n - s_{n-1}, T_{n+1} > t - s_n$. Hence, using Proposition 5.1, we have that the conditional joint density of $S_1, \ldots, S_n$ given that $N(t) = n$ is as follows:

$$f(s_1, \ldots, s_n \mid n) = \frac{f(s_1, \ldots, s_n, n)}{P\{N(t) = n\}}$$

$$= \frac{\lambda e^{-\lambda s_1} \lambda e^{-\lambda(s_2 - s_1)} \cdots \lambda e^{-\lambda(s_n - s_{n-1})} e^{-\lambda(t - s_n)}}{e^{-\lambda t}(\lambda t)^n / n!}$$

$$= \frac{n!}{t^n}, \qquad 0 < s_1 < \cdots < s_n < t$$

which proves the result.   ◆

Remark   The preceding result is usually paraphrased as stating that, under the condition that n events have occurred in $(0, t)$, the times $S_1, \ldots, S_n$ at which events occur, considered as unordered random variables, are distributed independently and uniformly in the interval $(0, t)$.

Application of Theorem 5.2 (Sampling a Poisson Process)

In Proposition 5.2 we showed that if each event of a Poisson process is independently classified as a type I event with probability p and as a type II event with probability $1 - p$ then the counting processes of type I and type II events are independent Poisson processes with respective rates λp and $\lambda(1 - p)$. Suppose now, however, that there are k possible types of events and that the probability that an event is classified as a type i event, $i = 1, \ldots, k$, depends on the time the event occurs. Specifically, suppose that if an event occurs at time y then it will be classified as a type i event, independently of anything that has previously occurred, with probability $P_i(y)$, $i = 1, \ldots, k$ where $\sum_{i=1}^{k} P_i(y) = 1$. Upon using Theorem 5.2 we can prove the following useful proposition.

Proposition 5.3 If $N_i(t)$, $i = 1, \ldots, k$, represents the number of type i events occurring by time t then $N_i(t)$, $i = 1, \ldots, k$, are independent Poisson random variables having means

$$E[N_i(t)] = \lambda \int_0^t P_i(s)\,ds$$

Before proving this proposition, let us first illustrate its use.

Example 5.14 (An Infinite Server Queue): Suppose that customers arrive at a service station in accordance with a Poisson process with rate λ. Upon arrival the customer is immediately served by one of an infinite number of possible servers, and the service times are assumed to be independent with a common distribution G. What is the distribution of $X(t)$, the number of customers that have completed service by time t? What is the distribution of $Y(t)$, the number of customers that are being served at time t?

To answer the preceding questions let us agree to call an entering customer a type I customer if he completes his service by time t and a type II customer if he does not complete his service by time t. Now, if the customer enters at time s, $s \leqslant t$, then he will be a type I customer if his service time is less than $t - s$. Since the service time distribution is G, the probability of this will be $G(t - s)$. Similarly, a customer entering at time s, $s \leqslant t$, will be a type II customer with probability $\bar{G}(t - s) = 1 - G(t - s)$. Hence, from Proposition 5.3 we obtain that the distribution of $X(t)$, the number of customers that have completed service by time t, is Poisson distributed with mean

$$E[X(t)] = \lambda \int_0^t G(t - s)\,ds = \lambda \int_0^t G(y)\,dy \qquad (5.17)$$

Similarly, the distribution of $Y(t)$, the number of customers being served at time t is Poisson with mean

$$E[Y(t)] = \lambda \int_0^t \bar{G}(t - s)\,ds = \lambda \int_0^t \bar{G}(y)\,dy \qquad (5.18)$$

Furthermore, $X(t)$ and $Y(t)$ are independent.

Suppose now that we are interested in computing the joint distribution of $Y(t)$ and $Y(t + s)$ — that is, the joint distribution of the number in the system at time t and at time $t + s$. To accomplish this, say that an arrival is

type 1: if he arrives before time t and completes service between t and $t + s$,
type 2: if he arrives before t and completes service after $t + s$,
type 3: if he arrives between t and $t + s$ and completes service after $t + s$,
type 4: otherwise.

Hence an arrival at time y will be type i with probability $P_i(y)$ given by

$$P_1(y) = \begin{cases} G(t + s - y) - G(t - y), & \text{if } y < t \\ 0, & \text{otherwise} \end{cases}$$

$$P_2(y) = \begin{cases} \bar{G}(t + s - y), & \text{if } y < t \\ 0, & \text{otherwise} \end{cases}$$

$$P_3(y) = \begin{cases} \bar{G}(t + s - y), & \text{if } t < y < t + s \\ 0, & \text{otherwise} \end{cases}$$

$$P_4(y) = 1 - P_1(y) - P_2(y) - P_3(y)$$

Hence, if $N_i = N_i(s + t)$, $i = 1, 2, 3$, denotes the number of type i events that occur, then from Proposition 5.3, N_i, $i = 1, 2, 3$, are independent Poisson random variables with respective means

$$E[N_i] = \lambda \int_0^{t+s} P_i(y) \, dy, \qquad i = 1, 2, 3$$

Because

$$Y(t) = N_1 + N_2,$$

$$Y(t + s) = N_2 + N_3$$

it is now an easy matter to compute the joint distribution of $Y(t)$ and $Y(t + s)$. For instance,

$$\begin{aligned}
\text{Cov}[Y(t),\ Y(t + s)] \\
&= \text{Cov}(N_1 + N_2, N_2 + N_3) \\
&= \text{Cov}(N_2, N_2) \qquad \text{by independence of } N_1, N_2, N_3 \\
&= \text{Var}(N_2) \\
&= \lambda \int_0^t \bar{G}(t + s - y) \, dy = \lambda \int_0^t \bar{G}(u + s) \, du
\end{aligned}$$

where the last equality follows since the variance of a Poisson random variable equals its mean, and from the substitution $u = t - y$. Also, the joint distribution of $Y(t)$ and $Y(t + s)$ is as follows:

$$P\{Y(t) = i, Y(t + s) = j\} = P\{N_1 + N_2 = i, N_2 + N_3 = j\}$$

$$= \sum_{l=0}^{\min(i,j)} P\{N_2 = l, N_1 = i - l, N_3 = j - l\}$$

$$= \sum_{l=0}^{\min(i,j)} P\{N_2 = l\} P\{N_1 = i - l\} P\{N_3 = j - l\} \quad \blacklozenge$$

Example 5.15 (Minimizing the Number of Encounters): Suppose that cars enter a one-way highway in accordance with a Poisson process with rate λ. The cars enter at point a and depart at point b (see Figure 5.2). Each car travels at a constant speed that is randomly determined, independently from car to car, from the distribution G. When a faster car encounters a slower one, it passes it with no time being lost. If your car enters the highway at time s and you are able to choose your speed, what speed minimizes the expected number of encounters you will have with other cars, where we say that an encounter occurs each time your car either passes or is passed by another car?

Solution: We will show that for large s the speed that minimizes the expected number of encounters is the median of the speed distribution G. To see this, suppose that the speed x is chosen. Let $d = b - a$ denote the length of the road. Upon choosing the speed x, it follows that your car will enter the road at time s and will depart at time $s + t_0$, where $t_0 = d/x$ is the travel time.

Now, the other cars enter the road according to a Poisson process with rate λ. Each of them chooses a speed X according to the distribution G, and this results in a travel time $T = d/X$. Let F denote the distribution of travel time T. That is,

$$F(t) = P\{T \leqslant t\} = P\{d/X \leqslant t\} = P\{X \geqslant d/t\} = \bar{G}(d/t)$$

Let us say that an event occurs at time t if a car enters the highway at time t. Also, say that the event is a type 1 event if it results in an encounter with your car. Now, your car will enter the road at time s and will exit at time $s + t_0$. Hence, a car will encounter your car if it enters before s and exits after $s + t_0$ (in which case your car will pass it on the road) or if it enters after s but exits before $s + t_0$ (in which case it will pass your car). As a result, a car that enters the road at time t will encounter your car if its travel time T is such that

$$
\begin{aligned}
t + T > s + t_0, && \text{if } t < s \\
t + T < s + t_0, && \text{if } s < t < s + t_0
\end{aligned}
$$

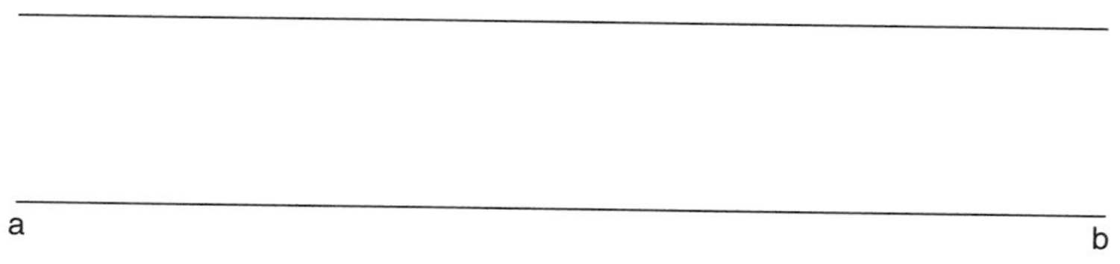

a
                                                                                      b

Figure 5.2. Cars enter at point a and depart at point b.

From the preceding, we see that an event at time t will, independently of other events, be a type 1 event with probability $p(t)$ given by

$$p(t) = \begin{cases} P\{t + T > s + t_0\} = \bar{F}(s + t_0 - t), & \text{if } t < s \\ P\{t + T < s + t_0\} = F(s + t_0 - t), & \text{if } s < t < s + t_0 \\ 0, & \text{if } t > s + t_0 \end{cases}$$

Since events (that is, cars entering the road) are occurring according to a Poisson process it follows, upon applying Proposition 5.3, that the total number of type 1 events that ever occurs is Poisson with mean

$$\lambda \int_0^\infty p(t)\, dt = \lambda \int_0^s \bar{F}(s + t_0 - t)\, dt + \lambda \int_s^{s+t_0} F(s + t_0 - t)\, dt$$

$$= \lambda \int_{t_0}^{s+t_0} \bar{F}(y)\, dy + \lambda \int_0^{t_0} F(y)\, dy$$

To choose the value of t_0 that minimizes the preceding quantity, we differentiate. This gives

$$\frac{d}{dt_0}\left\{ \lambda \int_0^\infty p(t)\, dt \right\} = \lambda\{\bar{F}(s + t_0) - \bar{F}(t_0) + F(t_0)\}$$

Setting this equal to 0, and using that $\bar{F}(s + t_0) \approx 0$ when s is large, we see that the optimal travel time t_0 is such that

$$F(t_0) - \bar{F}(t_0) = 0$$

or

$$F(t_0) - [1 - F(t_0)] = 0$$

or

$$F(t_0) = \tfrac{1}{2}$$

That is, the optimal travel time is the median of the travel time distribution. Since the speed X is equal to the distance d divided by the travel time T, it follows that the optimal speed $x_0 = d/t_0$ is such that

$$F(d/x_0) = \tfrac{1}{2}$$

Since

$$F(d/x_0) = \bar{G}(x_0)$$

we see that $\bar{G}(x_0) = \tfrac{1}{2}$, and so the optimal speed is the median of the distribution of speeds.

Summing up, we have shown that for any speed x the number of encounters with other cars will be a Poisson random variable, and the mean of this Poisson will be smallest when the speed x is taken to be the median of the distribution G. ◆

Example 5.16 (Tracking the Number of HIV Infections): There is a relatively long incubation period from the time when an individual becomes infected with the HIV virus, which causes AIDS, until the symptoms of the disease appear. As a result, it is difficult for public health officials to be certain of the number of members of the population that are infected at any given time. We will now present a first approximation model for this phenomenon, which can be used to obtain a rough estimate of the number of infected individuals.

Let us suppose that individuals contract the HIV virus in accordance with a Poisson process whose rate λ is unknown. Suppose that the time from when an individual becomes infected until symptoms of the disease appear is a random variable having a known distribution G. Suppose also that the incubation times of different infected individuals are independent.

Let $N_1(t)$ denote the number of individuals who have shown symptoms of the disease by time t. Also, let $N_2(t)$ denote the number who are HIV positive but have not yet shown any symptoms by time t. Now, since an individual that contracts the virus at time s will have symptoms by time t with probability $G(t - s)$ and will not with probability $\bar{G}(t - s)$, it follows from Proposition 5.3 that $N_1(t)$ and $N_2(t)$ are independent Poisson random variables with respective means

$$E[N_1(t)] = \lambda \int_0^t G(t - s)\, ds = \lambda \int_0^t G(y)\, dy$$

and

$$E[N_2(t)] = \lambda \int_0^t \bar{G}(t - s)\, ds = \lambda \int_0^t \bar{G}(y)\, dy$$

Now, if we knew λ, then we could use it to estimate $N_2(t)$, the number of individuals infected but without any outward symptoms at time t, by its mean value $E[N_2(t)]$. However, since λ is unknown, we must first estimate it. Now, we will presumably know the value of $N_1(t)$, and so we can use its known value as an estimate of its mean $E[N_1(t)]$. That is, if the number of individuals that have exhibited symptoms by time t is n_1, then we can estimate that

$$n_1 \approx E[N_1(t)] = \lambda \int_0^t G(y)\, dy$$

Therefore, we can estimate λ by the quantity $\hat{\lambda}$ given by

$$\hat{\lambda} = n_1 \Big/ \int_0^t G(y)\,dy$$

Using this estimate of λ, we can estimate the number of infected but symptomless individuals at time t by

$$\text{estimate of } N_2(t) = \hat{\lambda} \int_0^t \bar{G}(y)\,dy$$

$$= \frac{n_1 \int_0^t \bar{G}(y)\,dy}{\int_0^t G(y)\,dy}$$

For example, suppose that G is exponential with mean μ. Then $\bar{G}(y) = e^{-y/\mu}$, and a simple integration gives that

$$\text{estimate of } N_2(t) = \frac{n_1 \mu(1 - e^{-t/\mu})}{t - \mu(1 - e^{-t/\mu})}$$

If we suppose that $t = 16$ years, $\mu = 10$ years, and $n_1 = 220$ thousand, then the estimate of the number of infected but symptomless individuals at time 16 is

$$\text{estimate} = \frac{2200(1 - e^{-1.6})}{16 - 10(1 - e^{-1.6})} = 218.96$$

That is, if we suppose that the foregoing model is approximately correct (and we should be aware that the assumption of a constant infection rate λ that is unchanging over time is almost certainly a weak point of the model), then if the incubation period is exponential with mean 10 years and if the total number of individuals who have exhibited AIDS symptoms during the first 16 years of the epidemic is 220 thousand, then we can expect that approximately 219 thousand individuals are HIV positive though symptomless at time 16.   ✦

Proof of Proposition 5.3   Let us compute the joint probability $P\{N_i(t) = n_i, i = 1, \ldots, k\}$. To do so note first that in order for there to have been n_i type i events for $i = 1, \ldots, k$ there must have been a total of $\Sigma_{i=1}^k n_i$ events. Hence, conditioning on $N(t)$ yields

$$P\{N_1(t) = n_1, \ldots, N_k(t) = n_k\}$$

$$= P\left\{N_1(t) = n_1, \ldots, N_k(t) = n_k \,\middle|\, N(t) = \sum_{i=1}^k n_i\right\}$$

$$\times P\left\{N(t) = \sum_{i=1}^k n_i\right\}$$

Now consider an arbitrary event that occurred in the interval $[0, t]$. If it had occurred at time s, then the probability that it would be a type i event would be $P_i(s)$. Hence, since by Theorem 5.2 this event will have occurred at some time uniformly distributed on $(0, t)$, it follows that the probability that this event will be a type i event is

$$P_i = \frac{1}{t} \int_0^t P_i(s)\, ds$$

independently of the other events. Hence,

$$P\left\{ N_i(t) = n_i, i = 1, \ldots, k \,|\, N(t) = \sum_{i=1}^k n_i \right\}$$

will just equal the multinomial probability of n_i type i outcomes for $i = 1, \ldots, k$ when each of $\sum_{i=1}^k n_i$ independent trials results in outcome i with probability $P_i, i = 1, \ldots, k$. That is,

$$P\left\{ N_1(t) = n_1, \ldots, N_k(t) = n_k \,|\, N(t) = \sum_{i=1}^k n_i \right\} = \frac{(\sum_{i=1}^k n_i)!}{n_1! \cdots n_k!} P_1^{n_1} \cdots P_k^{n_k}$$

Consequently,

$$P\{N_1(t) = n_1, \ldots, N_k(t) = n_k\}$$

$$= \frac{(\sum_i n_i)!}{n_1! \cdots n_k!} P_1^{n_1} \cdots P_k^{n_k} \, e^{-\lambda t} \frac{(\lambda t)^{\sum_i n_i}}{(\sum_i n_i)!}$$

$$= \prod_{i=1}^k e^{-\lambda t P_i} (\lambda t P_i)^{n_i} / n_i!$$

and the proof is complete ✦

We now present some additional examples of the usefulness of Theorem 5.2.

Example 5.17 (An Electronic Counter): Suppose that electrical pulses having random amplitude arrive at a counter in accordance with a Poisson process with rate λ. The amplitude of a pulse is assumed to decrease with time at an exponential rate. That is, we suppose that if a pulse has an amplitude of A units upon arrival, then its amplitude a time t units later will be $Ae^{-\alpha t}$. We further suppose that the initial amplitudes of the pulses are independent and have a common distribution F.

Let $S_1, S_2, \ldots$ be the arrival times of the pulses and let $A_1, A_2, \ldots$ be their respective amplitudes. Then

$$A(t) = \sum_{i=1}^{N(t)} A_i e^{-\alpha(t - S_i)}$$

represents the total amplitude at time t. We can determine the expected value of $A(t)$ by conditioning on $N(t)$, the number of pulses to arrive by time t. This yields

$$E[A(t)] = \sum_{n=0}^{\infty} E[A(t) \mid N(t) = n] e^{-\lambda t} \frac{(\lambda t)^n}{n!}$$

Now, conditioned on $N(t) = n$, the unordered arrival times $(S_1, \ldots, S_n)$ are distributed as independent uniform $(0, t)$ random variables. Hence, given $N(t) = n$, $A(t)$ has the same distribution as $\sum_{j=1}^{n} A_j e^{-\alpha(t-Y_j)}$, where Y_j, $j = 1, \ldots, n$, are independent and uniformly distributed on $(0, t)$. Thus,

$$E[A(t) \mid N(t) = n] = E\left[\sum_{j=1}^{n} A_j e^{-\alpha(t-Y_j)} \right]$$

$$= nE[A]E[e^{-\alpha(t-Y)}]$$

where $E[A]$ is the mean initial amplitude of a pulse, and Y is a uniform $(0, t)$ random variable. Hence,

$$E[e^{-\alpha(t-Y)}] = \int_0^t e^{-\alpha(t-y)} \frac{dy}{t}$$

$$= \frac{e^{-\alpha t}}{\alpha t} e^{\alpha y} \bigg|_{y=0}^{y=t}$$

$$= \frac{1 - e^{-\alpha t}}{\alpha t}$$

and thus,

$$E[A(t) \mid N(t) = n] = nE[A] \frac{(1 - e^{-\alpha t})}{\alpha t}$$

or

$$E[A(t) \mid N(t)] = N(t)E[A] \frac{(1 - e^{-\alpha t})}{\alpha t}$$

Taking expectations and using the fact that $E[N(t)] = \lambda t$, we have

$$E[A(t)] = \frac{\lambda E[A]}{\alpha} (1 - e^{-\alpha t}) \quad \blacklozenge$$

Example 5.18 (An Optimization Example): Suppose that items arrive at a processing plant in accordance with a Poisson process with rate λ. At a fixed time T, all items are dispatched from the system. The problem is to choose an intermediate time, $t \in (0, T)$, at which all items in the system are dispatched, so as to minimize the total expected wait of all items.

If we dispatch at time t, $0 < t < T$, then the expected total wait of all items will be

$$\frac{\lambda t^2}{2} + \frac{\lambda(T - t)^2}{2}$$

To see why the above is true, we reason as follows: The expected number of arrivals in $(0, t)$ is λt, and each arrival is uniformly distributed on $(0, t)$, and hence has expected wait $t/2$. Thus, the expected total wait of items arriving in $(0, t)$ is $\lambda t^2/2$. Similar reasoning holds for arrivals in (t, T), and the above follows. To minimize this quantity, we differentiate with respect to t to obtain

$$\frac{d}{dt}\left[\lambda\frac{t^2}{2} + \lambda\frac{(T - t)^2}{2}\right] = \lambda t - \lambda(T - t)$$

and equating to 0 shows that the dispatch time that minimizes the expected total wait is $t = T/2$. ✦

We end this section with a result, quite similar in spirit to Theorem 5.2, which states that given S_n, the time of the nth event, then the first $n - 1$ event times are distributed as the ordered values of a set of $n - 1$ random variables uniformly distributed on $(0, S_n)$.

Proposition 5.4   Given that $S_n = t$, the set $S_1, \ldots, S_{n-1}$ has the distribution of a set of $n - 1$ independent uniform $(0, t)$ random variables.

Proof   We can prove the above in the same manner as we did Theorem 5.2, or we can argue more loosely as follows:

$$S_1, \ldots, S_{n-1} \mid S_n = t \sim S_1, \ldots, S_{n-1} \mid S_n = t, N(t^-) = n - 1$$

$$\sim S_1, \ldots, S_{n-1} \mid N(t^-) = n - 1$$

where $\sim$ means "has the same distribution as" and t^- is infinitesimally smaller than t. The result now follows from Theorem 5.2. ✦

5.3.6. Estimating Software Reliability

When a new computer software package is developed, a testing procedure is often put into effect to eliminate the faults, or bugs, in the package. One common procedure is to try the package on a set of well-known problems to see if any errors result. This goes on for some fixed time, with all resulting errors being noted. Then the testing stops and the package is carefully checked to determine the specific bugs that were responsible for the observed errors. The package is then altered to remove these bugs. Because we cannot be certain that all the bugs in the package have been eliminated, however, a problem of great importance is the estimation of the error rate of the revised software package.

To model the preceding, let us suppose that initially the package contains an unknown number, m, of bugs, which we will refer to as bug 1, bug 2, ..., bug m. Suppose also that bug i will cause errors to occur in accordance with a Poisson process having an unknown rate λ_i, $i = 1, ..., m$. Then, for instance, the number of errors due to bug i that occur in any s units of operating time is Poisson distributed with mean $\lambda_i s$. Also suppose that these Poisson processes caused by bugs $i, i = 1, ..., m$ are independent. In addition, suppose that the package is to be run for t time units with all resulting errors being noted. At the end of this time a careful check of the package is made to determine the specific bugs that caused the errors (that is, a *debugging*, takes place). These bugs are removed, and the problem is then to determine the error rate for the revised package.

If we let

$$\psi_i(t) = \begin{cases} 1, & \text{if bug } i \text{ has not caused an error by } t \\ 0, & \text{otherwise} \end{cases}$$

then the quantity we wish to estimate is

$$\Lambda(t) = \sum_i \lambda_i \psi_i(t)$$

the error rate of the final package. To start, note that

$$E[\Lambda(t)] = \sum_i \lambda_i E[\psi_i(t)]$$

$$= \sum_i \lambda_i e^{-\lambda_i t} \tag{5.19}$$

Now each of the bugs that is discovered would have been responsible for a certain number of errors. Let us denote by $M_j(t)$ the number of bugs that were responsible for j errors, $j \geqslant 1$. That is, $M_1(t)$ is the number of bugs that caused exactly 1 error, $M_2(t)$ is the number that caused 2 errors, and so on, with $\sum_j j M_j(t)$ equaling the total number of errors that resulted. To compute $E[M_1(t)]$, let us define the indicator variables, $I_i(t)$, $i \geqslant 1$, by

$$I_i(t) = \begin{cases} 1, & \text{bug } i \text{ causes exactly 1 error} \\ 0, & \text{otherwise} \end{cases}$$

Then,

$$M_1(t) = \sum_i I_i(t)$$

and so

$$E[M_1(t)] = \sum_i E[I_i(t)] = \sum_i \lambda_i t e^{-\lambda_i t} \qquad (5.20)$$

Thus, from (5.19) and (5.20) we obtain the intriguing result that

$$E\left[\Lambda(t) - \frac{M_1(t)}{t}\right] = 0 \qquad (5.21)$$

Thus suggests the possible use of $M_1(t)/t$ as an estimate of $\Lambda(t)$. To determine whether or not $M_1(t)/t$ constitutes a "good" estimate of $\Lambda(t)$ we shall look at how far apart these two quantities tend to be. That is, we will compute

$$E\left[\left(\Lambda(t) - \frac{M_1(t)}{t}\right)^2\right] = \mathrm{Var}\left(\Lambda(t) - \frac{M_1(t)}{t}\right) \qquad \text{from (5.21)}$$

$$= \mathrm{Var}(\Lambda(t)) - \frac{2}{t}\mathrm{Cov}(\Lambda(t), M_1(t)) + \frac{1}{t^2}\mathrm{Var}(M_1(t))$$

Now,

$$\mathrm{Var}(\Lambda(t)) = \sum_i \lambda_i^2\,\mathrm{Var}(\psi_i(t)) = \sum_i \lambda_i^2 e^{-\lambda_i t}(1 - e^{-\lambda_i t}),$$

$$\mathrm{Var}(M_1(t)) = \sum_i \mathrm{Var}(I_i(t)) = \sum_i \lambda_i t e^{-\lambda_i t}(1 - \lambda_i t e^{-\lambda_i t}),$$

$$\mathrm{Cov}(\Lambda(t), M_1(t)) = \mathrm{Cov}\left(\sum_i \lambda_i \psi_i(t), \sum_j I_j(t)\right)$$

$$= \sum_i \sum_j \mathrm{Cov}(\lambda_i \psi_i(t), I_j(t))$$

$$= \sum_i \lambda_i \,\mathrm{Cov}(\psi_i(t), I_i(t))$$

$$= -\sum_i \lambda_i e^{-\lambda_i t} \lambda_i t e^{-\lambda_i t}$$

where the last two equalities follow since $\psi_i(t)$ and $I_j(t)$ are independent when $i \neq j$ because they refer to different Poisson processes and $\psi_i(t)I_i(t) = 0$. Hence we obtain that

$$E\left[\left(\Lambda(t) - \frac{M_1(t)}{t}\right)^2\right] = \sum_i \lambda_i^2 e^{-\lambda_i t} + \frac{1}{t}\sum_i \lambda_i e^{-\lambda_i t}$$

$$= \frac{E[M_1(t) + 2M_2(t)]}{t^2}$$

where the last equality follows from (5.20) and the identity (which we leave as an exercise)

$$E[M_2(t)] = \tfrac{1}{2}\sum_i (\lambda_i t)^2 e^{-\lambda_i t} \tag{5.22}$$

Thus, we can estimate the average square of the difference between $\Lambda(t)$ and $M_1(t)/t$ by the observed value of $M_1(t) + 2M_2(t)$ divided by t^2.

Example 5.19   Suppose that in 100 units of operating time 20 bugs are discovered of which two resulted in exactly one, and three resulted in exactly two, errors. Then we would estimate that $\Lambda(100)$ is something akin to the value of a random variable whose mean is equal to 1/50 and whose variance is equal to 8/10,000.   ✦

5.4.   Generalizations of the Poisson Process

5.4.1.   Nonhomogeneous Poisson Process

In this section we consider two generalizations of the Poisson process. The first of these is the nonhomogeneous, also called the nonstationary, Poisson process, which is obtained by allowing the arrival rate at time t to be a function of t.

Definition 5.4   The counting process $\{N(t), t \geqslant 0\}$ is said to be a *nonhomogeneous Poisson process with intensity function* $\lambda(t)$, $t \geqslant 0$, if

  (i) $N(0) = 0$.
  (ii) $\{N(t), t \geqslant 0\}$ has independent increments.
  (iii) $P\{N(t + h) - N(t) \geqslant 2\} = o(h)$.
  (iv) $P\{N(t + h) - N(t) = 1\} = \lambda(t)h + o(h)$.

If we let

$$m(t) = \int_0^t \lambda(y)\, dy$$

then it can be shown that

$$P\{N(s + t) - N(s) = n\} = e^{-[m(s+t) - m(s)]} \frac{[m(s + t) - m(s)]^n}{n!}, \qquad n \geqslant 0 \tag{5.23}$$

That is, $N(s+t) - N(s)$ is a Poisson random variable with mean $m(s+t) - m(s)$. Since this implies that $N(t)$ is Poisson with mean $m(t)$, we call $m(t)$ the *mean value function* of the nonhomogeneous Poisson process.

The proof of Equation (5.23) follows along the lines of the proof of Theorem 5.1, with a slight modification. Fix nonnegative values s and u, let

$$N_s(t) = N(s + t) - N(s)$$

and define

$$g(t) = E[\exp\{-uN_s(t)\}]$$

Then,

$$
\begin{aligned}
g(t + h) &= E[\exp\{-uN_s(t + h)\}] \\
&= E[\exp\{-uN_s(t)\} \exp\{-u(N_s(t + h) - N_s(t))\}] \\
&= g(t) \, E[\exp\{-u(N_s(t + h) - N_s(t))\}] \quad \text{by independent increments}
\end{aligned}
$$

Conditioning on $N(s + t + h) - N(s + t)$ yields

$$
\begin{aligned}
E[\exp\{-u(N_s(t + h) - N_s(t))\}] &= E[\exp\{-u(N(s + t + h) - N(s + t))\}] \\
&= 1 - \lambda(s + t)h + e^{-u}\lambda(s + t)h + o(h)
\end{aligned}
$$

Hence,

$$g(t + h) = g(t) (1 - \lambda(s + t)h + e^{-u} \lambda(s + t) h) + o(h)$$

or

$$\frac{g(t + h) - g(t)}{h} = g(t) \lambda(s + t) (e^{-u} - 1) + \frac{o(h)}{h}$$

Letting $h \to 0$ gives

$$g'(t) = g(t) \lambda(s + t) (e^{-u} - 1)$$

or, equivalently,

$$\frac{g'(t)}{g(t)} = \lambda(s + t) (e^{-u} - 1)$$

Consequently,

$$\int_0^t \frac{g'(y)}{g(y)} \, dy = (e^{-u} - 1) \int_0^t \lambda(s + y) \, dy$$

Since $g(0) = 1$, the preceding yields that

$$\log g(t) = (e^{-u} - 1) \int_0^t \lambda(s + y) \, dy$$

or

$$g(t) = \exp\left\{ \int_0^t \lambda(s + y) \, dy \, (e^{-u} - 1) \right\}$$

Since the right side is the Laplace transform function of a Poisson variable with mean $\int_0^t \lambda(s + y)\, dy = m(s + t) - m(s)$, the result follows from the fact that the Laplace transform uniquely determines the distribution. ✦

The importance of the nonhomogeneous Poisson process resides in the fact that we no longer require the condition of stationary increments. Thus we now allow for the possibility that events may be more likely to occur at certain times than during other times.

Example 5.20   Siegbert runs a hot dog stand that opens at 8 A.M. From 8 until 11 A.M. customers seem to arrive, on the average, at a steadily increasing rate that starts with an initial rate of 5 customers per hour at 8 A.M. and reaches a maximum of 20 customers per hour at 11 A.M. From 11 A.M. until 1 P.M. the (average) rate seems to remain constant at 20 customers per hour. However, the (average) arrival rate then drops steadily from 1 P.M. until closing time at 5 P.M. at which time it has the value of 12 customers per hour. If we assume that the numbers of customers arriving at Siegbert's stand during disjoint time periods are independent, then what is a good probability model for the above? What is the probability that no customers arrive between 8:30 A.M. and 9:30 A.M. on Monday morning? What is the expected number of arrivals in this period?

Solution: A good model for the above would be to assume that arrivals constitute a nonhomogeneous Poisson process with intensity function $\lambda(t)$ given by

$$\lambda(t) = \begin{cases} 5 + 5t, & 0 \leqslant t \leqslant 3 \\ 20, & 3 \leqslant t \leqslant 5 \\ 20 - 2(t - 5), & 5 \leqslant t \leqslant 9 \end{cases}$$

and

$$\lambda(t) = \lambda(t - 9) \qquad \text{for } t > 9$$

Note that $N(t)$ represents the number of arrivals during the first t hours that the store is open. That is, we do not count the hours between 5 P.M. and 8 A.M. If for some reasons we wanted $N(t)$ to represent the number of arrivals during the first t hours regardless of whether the store was open or not, then, assuming that the process begins at midnight we would let

$$\lambda(t) = \begin{cases} 0, & 0 \leqslant t < 8 \\ 5 + 5(t-8), & 8 \leqslant t \leqslant 11 \\ 20, & 11 \leqslant t \leqslant 13 \\ 20 - 2(t-13), & 13 \leqslant t \leqslant 17 \\ 0, & 17 < t \leqslant 24 \end{cases}$$

and

$$\lambda(t) = \lambda(t-24) \qquad \text{for } t > 24$$

As the number of arrivals between 8:30 A.M. and 9:30 A.M. will be Poisson with mean $m(\frac{3}{2}) - m(\frac{1}{2})$ in the first representation [and $m(\frac{19}{2}) - m(\frac{17}{2})$ in the second representation], we have that the probability that this number is zero is

$$\exp\left\{ -\int_{1/2}^{3/2} (5 + 5t)\, dt \right\} = e^{-10}$$

and the mean number of arrivals is

$$\int_{1/2}^{3/2} (5 + 5t)\, dt = 10 \quad \blacklozenge$$

When the intensity function $\lambda(t)$ is bounded, we can think of the nonhomogeneous process as being a random sample from a homogeneous Poisson process. Specifically, let λ be such that

$$\lambda(t) \leqslant \lambda \qquad \text{for all } t \geqslant 0$$

and consider a Poisson process with rate λ. Now if we suppose that an event of the Poisson process that occurs at time t is counted with probability $\lambda(t)/\lambda$, then the process of counted events is a nonhomogeneous Poisson process with intensity function $\lambda(t)$. This last statement easily follows from Definition 5.4. For instance (i), (ii), and (iii) follow since they are also true for the homogeneous Poisson process. Axiom (iv) follows since

$$P\{\text{one counted event in } (t, t+h)\} = P\{\text{one event in } (t, t+h)\} \frac{\lambda(t)}{\lambda} + o(h)$$

$$= \lambda h \frac{\lambda(t)}{\lambda} + o(h)$$

$$= \lambda(t)h + o(h)$$

Example 5.21   The Output Process of an Infinite Server Poisson Queue: It turns out that the output process of the $M/G/\infty$ queue — that is, of the infinite server queue having Poisson arrivals and general service distribution G — is a nonhomogeneous Poisson process having intensity function $\lambda(t) = \lambda G(t)$. To prove this claim, note first that the (joint) probability (density) that a customer arrives at time s and departs at time t is equal to λ, the probability (intensity) of an arrival at time s, multiplied by $g(t - s)$, the probability (density) that its service time is $t - s$. Hence,

$$\lambda(t) = \int_0^t \lambda\, g(t - s)\, ds = \lambda G(t)$$

is the probability intensity of a departure at time t. Now suppose we are told that a departure occurs at time t — how does this affect the probabilities of other departure times? Well, even if we knew the arrival time of the customer who departed at time t, this would not affect the arrival times of other customers (because of the independent increment assumption of the Poisson arrival process). Hence, because there are always servers available for arrivals, this information cannot affect the probabilities of other departure times. Thus, the departure process has independent increments, and departures occur at t with intensity $\lambda(t)$, which verifies the claim. ◆

If we let S_n denote the time of the nth event of the nonhomogeneous Poisson process, then we can obtain its density as follows:

$$P\{t < S_n < t + h\} = P\{N(t) = n - 1, \text{one event in } (t, t + h)\} + o(h)$$

$$= P\{N(t) = n - 1\}P\{\text{one event in } (t, t + h)\} + o(h)$$

$$= e^{-m(t)} \frac{[m(t)]^{n-1}}{(n-1)!} [\lambda(t)h + o(h)] + o(h)$$

$$= \lambda(t)\, e^{-m(t)} \frac{[m(t)]^{n-1}}{(n-1)!} h + o(h)$$

which implies that

$$f_{S_n}(t) = \lambda(t)\, e^{-m(t)} \frac{[m(t)]^{n-1}}{(n-1)!}$$

where

$$m(t) = \int_0^t \lambda(s)\, ds$$

5.4.2. Compound Poisson Process

A stochastic process $\{X(t), t \geq 0\}$ is said to be a *compound Poisson process* if it can be represented as

$$X(t) = \sum_{i=1}^{N(t)} Y_i, \qquad t \geq 0 \tag{5.24}$$

where $\{N(t), t \geq 0\}$ is a Poisson process, and $\{Y_i, i \geq t\}$ is a family of independent and identically distributed random variables which are also independent of $\{N(t), t \geq 0\}$. The random variable $X(t)$ is said to be a compound Poisson random variable.

Examples of Compound Poisson Processes

(i) If $Y_i \equiv 1$, then $X(t) = N(t)$, and so we have the usual Poisson process.
(ii) Suppose that buses arrive at a sporting event in accordance with a Poisson process, and suppose that the numbers of customers in each bus are assumed to be independent and identically distributed. Then $\{X(t), t \geq 0\}$ is a compound Poisson process where $X(t)$ denotes the number of customers who have arrived by t. In Equation (5.24) Y_i represents the number of customers in the ith bus.
(iii) Suppose customers leave a supermarket in accordance with a Poisson process. If Y_i, the amount spent by the ith customer, $i = 1, 2, \ldots$, are independent and identically distributed, then $\{X(t), t \geq 0\}$ is a compound Poisson process when $X(t)$ denotes the total amount of money spent by time t. ✦

Let us calculate the mean and variance of $X(t)$. To calculate $E[X(t)]$, we first condition on $N(t)$ to obtain

$$E[X(t)] = E[E[X(t) \mid N(t)]]$$

Now

$$E[X(t) \mid N(t) = n] = E\left[\sum_{i=1}^{N(t)} Y_i \mid N(t) = n\right]$$

$$= E\left[\sum_{i=1}^{n} Y_i \mid N(t) = n\right]$$

$$= E\left[\sum_{i=1}^{n} Y_i\right]$$

$$= nE[Y_1]$$

where we have used the assumed independence of the Y_i's and $N(t)$. Hence,

$$E[X(t) \mid N(t)] = N(t)E[Y_1] \tag{5.25}$$

and therefore

$$E[X(t)] = \lambda t E[Y_1] \tag{5.26}$$

To calculate $\text{Var}[X(t)]$ we use the conditional variance formula (see Exercise 43 of Chapter 3)

$$\text{Var}[X(t)] = E[\text{Var}(X(t) \mid N(t))] + \text{Var}(E[X(t) \mid N(t)]) \tag{5.27}$$

Now,

$$\text{Var}[X(t) \mid N(t) = n] = \text{Var}\left[\sum_{i=1}^{N(t)} Y_i \mid N(t) = n \right]$$

$$= \text{Var}\left[\sum_{i=1}^{n} Y_i \right]$$

$$= n \, \text{Var}(Y_1)$$

and thus

$$\text{Var}[X(t) \mid N(t)] = N(t) \, \text{Var}(Y_1) \tag{5.28}$$

Therefore, using Equations (5.25) and (5.28), we obtain by Equation (5.27) that

$$\text{Var}[X(t)] = E[N(t) \, \text{Var}(Y_1)] + \text{Var}(N(t)E[Y_1])$$

$$= \lambda t \, \text{Var}(Y_1) + (EY_1)^2 \lambda t$$

$$= \lambda t[\text{Var}(Y_1) + (EY_1)^2]$$

$$= \lambda t E[Y_1^2] \tag{5.29}$$

where we have used the fact that $\text{Var}[N(t)] = \lambda t$ since $N(t)$ is Poisson distributed with mean λt.*

Example 5.22 Suppose that families migrate to an area at a Poisson rate $\lambda = 2$ per week. If the number of people in each family is independent and takes on the values 1, 2, 3, 4 with respective probabilities $\frac{1}{6}, \frac{1}{3}, \frac{1}{3}, \frac{1}{6}$, then what is the expected value and variance of the number of individuals migrating to this area during a fixed five-week period?

---

* This derivation of the variance can be compared with the one given in Example 3.16.

Solution: Letting Y_i denote the number of people in the ith family, we have that

$$E[Y_i] = 1 \cdot \tfrac{1}{6} + 2 \cdot \tfrac{1}{3} + 3 \cdot \tfrac{1}{3} + 4 \cdot \tfrac{1}{6} = \tfrac{5}{2},$$

$$E[Y_i^2] = 1^2 \cdot \tfrac{1}{6} + 2^2 \cdot \tfrac{1}{3} + 3^2 \cdot \tfrac{1}{3} + 4^2 \cdot \tfrac{1}{6} = \tfrac{43}{6}$$

Hence, letting $X(5)$ denote the number of immigrants during a five-week period, we obtain from Equations (5.26) and (5.29) that

$$E[X(5)] = 2 \cdot 5 \cdot \tfrac{5}{2} = 25$$

and

$$\mathrm{Var}[X(5)] = 2 \cdot 5 \cdot \tfrac{43}{6} = \tfrac{215}{3} \quad \blacklozenge$$

Example 5.23 (Busy Periods in Single-Server Poisson Arrival Queues): Consider a single-server service station in which customers arrive according to a Poisson process having rate λ. An arriving customer is immediately served if the server is free; if not, the customer waits in line (that is, he or she joins the queue). The successive service times are independent with a common distribution.

Such a system will alternate between idle periods when there are no customers in the system, so the server is idle, and busy periods when there are customers in the system, so the server is busy. A busy period will begin when an arrival finds the system empty, and because of the memoryless property of the Poisson arrivals it follows that the distribution of the length of a busy period will be the same for each such period. Let B denote the length of a busy period. We will compute its mean and variance.

To begin, let S denote the service time of the first customer in the busy period and let $N(S)$ denote the number of arrivals during that time. Now, if $N(S) = 0$ then the busy period will end when the initial customer completes his service, and so B will equal S in this case. Now, suppose that one customer arrives during the service time of the initial customer. Then, at time S there will be a single customer in the system who is just about to enter service. Because the arrival stream from time S on will still be a Poisson process with rate λ, it thus follows that the additional time from S until the system becomes empty will have the same distribution as a busy period. That is, if $N(S) = 1$ then

$$B = S + B_1$$

where B_1 is independent of S and has the same distribution as B.

Now, consider the general case where $N(S) = n$, so there will be n customers waiting when the server finishes his initial service. To determine the distribution of remaining time in the busy period note that the order in which customers are served will not affect the remaining time. Hence, let us suppose that the n arrivals, call them $C_1, \ldots, C_n$, during the initial service period are served as follows. Customer C_1 is served first, but C_2 is not served until the only customers in the system are $C_2, \ldots, C_n$. For instance, any customers arriving during C_1's service time will be served before C_2. Similarly, C_3 is not served until the system is free of all customers but $C_3, \ldots, C_n$, and so on. A little thought reveals that the times between the beginnings of service of customers C_i and C_{i+1}, $i = 1, \ldots, n-1$, and the time from the beginning of service of C_n until there are no customers in the system, are independent random variables, each distributed as a busy period.

It follows from the preceding that if we let $B_1, B_2, \ldots$ be a sequence of independent random variables, each distributed as a busy period, then we can express B as

$$B = S + \sum_{i=1}^{N(S)} B_i$$

Hence,

$$E[B \mid S] = S + E\left[\sum_{i=1}^{N(S)} B_i \mid S \right]$$

and

$$\mathrm{Var}(B \mid S) = \mathrm{Var}\left(\sum_{i=1}^{N(S)} B_i \mid S \right)$$

However, given S, $\sum_{i=1}^{N(S)} B_i$ is a compound Poisson random variable, and thus from Equations (5.26) and (5.29) we obtain

$$E[B \mid S] = S + \lambda S E[B] = (1 + \lambda E[B])S$$
$$\mathrm{Var}(B \mid S) = \lambda S E[B^2]$$

Hence,

$$E[B] = E[E[B \mid S]] = (1 + \lambda E[B])E[S]$$

implying that

$$E[B] = \frac{E[S]}{1 - \lambda E[S]}$$

Also, by the conditional variance formula

$$\mathrm{Var}(B) = \mathrm{Var}(E[B \mid S]) + E[\mathrm{Var}(B \mid S)]$$
$$= (1 + \lambda E[B])^2 \, \mathrm{Var}(S) + \lambda E[S] E[B^2]$$
$$= (1 + \lambda E[B])^2 \, \mathrm{Var}(S) + \lambda E[S](\mathrm{Var}(B) + E^2[B])$$

yielding

$$\text{Var}(B) = \frac{\text{Var}(S)\,(1 + \lambda E[B])^2 + \lambda E[S]E^2[B]}{1 - \lambda E[S]}$$

Using $E[B] = E[S]/(1 - \lambda E[S])$, we obtain

$$\text{Var}(B) = \frac{\text{Var}(S) + \lambda E^3[S]}{(1 - \lambda E[S])^3} \quad \blacklozenge$$

There is a very nice representation of the compound Poisson process when the set of possible values of the Y_i is finite or countably infinite. So let us suppose that there are numbers $\alpha_j, j \geq 1$, such that

$$P\{Y_1 = \alpha_j\} = p_j, \qquad \sum_j p_j = 1$$

Now, a compound Poisson process arises when events occur according to a Poisson process and each event results in a random amount Y being added to the cumulative sum. Let us say that the event is a type j event whenever it results in adding the amount $\alpha_j, j \geq 1$. That is, the ith event of the Poisson process is a type j event if $Y_i = \alpha_j$. If we let $N_j(t)$ denote the number of type j events by time t, then it follows from Proposition 5.2 that the random variables $N_j(t)$, $j \geq 1$, are independent Poisson random variables with respective means

$$E[N_j(t)] = \lambda p_j t$$

Since, for each j, the amount α_j is added to the cumulative sum a total of $N_j(t)$ times by time t, it follows that the cumulative sum at time t can be expressed as

$$X(t) = \sum_j \alpha_j N_j(t) \qquad (5.30)$$

As a check of Equation (5.30), let us use it to compute the mean and variance of $X(t)$. This yields

$$E[X(t)] = E\left[\sum_j \alpha_j N_j(t)\right]$$

$$= \sum_j \alpha_j E[N_j(t)]$$

$$= \sum_j \alpha_j \lambda p_j t$$

$$= \lambda t E[Y_1]$$

Also,

$$
\begin{aligned}
\mathrm{Var}[X(t)] &= \mathrm{Var}\left[\sum_{j} \alpha_j N_j(t)\right] \\
&= \sum_{j} \alpha_j^2 \, \mathrm{Var}[N_j(t)] \qquad \text{by the independence of the } N_j(t),\ j \geqslant 1 \\
&= \sum_{j} \alpha_j^2 \lambda p_j t \\
&= \lambda t E[Y_1^2]
\end{aligned}
$$

where the next to last equality follows since the variance of the Poisson random variable $N_j(t)$ is equal to its mean.

Thus, we see that the representation (5.30) results in the same expressions for the mean and variance of $X(t)$ as were previously derived.

One of the uses of the representation (5.30) is that it enables us to conclude that as t grows large, the distribution of $X(t)$ converges to the normal distribution. To see why, note first that it follows by the central limit theorem that the distribution of a Poisson random variable converges to a normal distribution as its mean increases. (Why is this?) Therefore, each of the random variables $N_j(t)$ converges to a normal random variable as t increases. Because they are independent, and because the sum of independent normal random variables is also normal, it follows that $X(t)$ also approaches a normal distribution as t increases.

Example 5.24   In Example 5.22, find the approximate probability that at least 240 people migrate to the area within the next 50 weeks.

Solution: Since $\lambda = 2$, $E[Y_i] = 5/2$, $E[Y_i^2] = 43/6$, we see that

$$
E[X(50)] = 250, \qquad \mathrm{Var}[X(50)] = 4300/6
$$

Now, the desired probability is

$$
\begin{aligned}
P\{X(50) \geqslant 240\} &= P\{X(50) \geqslant 239.5\} \\
&= P\left\{\frac{X(50) - 250}{\sqrt{4300/6}} \geqslant \frac{239.5 - 250}{\sqrt{4300/6}}\right\} \\
&= 1 - \phi(-0.3922) \\
&= \phi(0.3922) \\
&= 0.6525
\end{aligned}
$$

where Table 2.3 was used to determine $\phi(0.3922)$, the probability that a standard normal is less than 0.3922.   ◆

Exercises

1. The time T required to repair a machine is an exponentially distributed random variable with mean $\frac{1}{2}$ (hours).

(a) What is the probability that a repair time exceeds $\frac{1}{2}$ hour?
(b) What is the probability that a repair takes at least $12\frac{1}{2}$ hours given that its duration exceeds 12 hours?

2. Suppose that you arrive at a single-teller bank to find five other customers in the bank, one being served and the other four waiting in line. You join the end of the line. If the service times are all exponential with rate μ, what is the expected amount of time you will spend in the bank?

3. Let X be an exponential random variable. Without any computations, tell which one of the following is correct. Explain your answer.

(a) $E[X^2 \mid X > 1] = E[(X + 1)^2]$
(b) $E[X^2 \mid X > 1] = E[X^2] + 1$
(c) $E[X^2 \mid X > 1] = (1 + E[X])^2$

4. Consider a post office with two clerks. Three people, A, B, and C, enter simultaneously. A and B go directly to the clerks, and C waits until either A or B leaves before he begins service. What is the probability that A is still in the post office after the other two have left when

(a) the service time for each clerk is exactly (nonrandom) ten minutes?
(b) the service times are i with probability $\frac{1}{3}$, $i = 1, 2, 3$?
(c) the service times are exponential with mean $1/\mu$?

5. The lifetime of a radio is exponentially distributed with a mean of ten years. If Jones buys a ten-year-old radio, what is the probability that it will be working after an additional ten years?

6. In Example 5.2 if server i serves at an exponential rate λ_i, $i = 1, 2$, show that

$$P\{\text{Smith is not last}\} = \left(\frac{\lambda_1}{\lambda_1 + \lambda_2}\right)^2 + \left(\frac{\lambda_2}{\lambda_1 + \lambda_2}\right)^2$$

*7. If X_1 and X_2 are independent nonnegative continuous random variables, show that

$$P\{X_1 < X_2 \mid \min(X_1, X_2) = t\} = \frac{r_1(t)}{r_1(t) + r_2(t)}$$

where $r_i(t)$ is the failure rate function of X_i.

8. Show that the failure rate function of a gamma distribution with parameters n and λ is increasing when $n \geqslant 1$.

9. Norb and Diana enter a barbershop simultaneously — Norb to get a shave and Diana a haircut. If the amount of time it takes to receive a haircut (shave) is exponentially distributed with mean 20 (15) minutes, and if Norb and Diana are immediately served, what is the probability that Diana finishes before Norb?

*10. If X and Y are independent exponential random variables with respective means $1/\lambda_1$ and $1/\lambda_2$, then

(a) use the lack of memory property of the exponential to intuitively explain why $Z = \min(X, Y)$ is exponential.
(b) what is the conditional distribution of Z given that $Z = X$?
(c) give a heuristic argument that the conditional distribution of $Y - Z$, given that $Z = X$, is exponential with mean $1/\lambda_2$.

11. Let $X, Y_1, \ldots, Y_n$ be independent exponential random variables; X having rate λ, and the Y_i having rate μ. Let A_j be the event that the jth smallest of these $n + 1$ random variables is one of the Y_i. Find $p = P\{X > \max_i Y_i\}$, by using the identity

$$p = P(A_1 \cdots A_n) = P(A_1)P(A_2 \mid A_1) \cdots P(A_n \mid A_1 \cdots A_{n-1})$$

Verify your answer when $n = 2$ by conditioning on X to obtain p.

12. If X_i, $i = 1, 2, 3$, are independent exponential random variables with rates $\lambda_i, i = 1, 2, 3$, find

(a) $P\{X_1 < X_2 < X_3\}$
(b) $P\{X_1 < X_2 \mid \max(X_1, X_2, X_3) = X_3\}$

13. Find, in Example 5.7, the expected time until the nth person on line leaves the line (either by entering service or departing without service).

14. Let X be an exponential random variable with rate λ.

(a) Use the definition of conditional expectation to determine $E[X \mid X < c]$.
(b) Now determine $E[X \mid X < c]$ by using the following identity:

$$E[X] = E[X \mid X < c]P\{X < c\} + E[X \mid X > c]P\{X > c\}$$

15. In Example 5.2, what is the expected time until all three customers have left the post office?

16. Suppose in Example 5.2 that the time it takes server i to serve customers is exponentially distributed with mean $1/\lambda_i$, $i = 1, 2$. What is the expected time until all three customers have left the post office?

17. A set of n cities is to be connected via communication links. The cost to construct a link between cities i and j is C_{ij}, $i \neq j$. Enough links should be constructed so that for each pair of cities there is path of links that connects them. As a result, only $n - 1$ links need be constructed. A minimal cost algorithm for solving this problem (known as the minimal spanning tree problem) first constructs the cheapest of all the $\binom{n}{2}$ links. Then, at each additional stage it chooses the cheapest link that connects a city without any links to one with links. That is, if the first link is between cities 1 and 2, then the second link will either be between 1 and one of the links $3, \ldots, n$ or between 2 and one of the links $3, \ldots, n$. Suppose that all of the $\binom{n}{2}$ costs C_{ij} are independent exponential random variables with mean 1. Find the expected cost of the preceding algorithm if

(a) $n = 3$
(b) $n = 4$

*18. Let X_1 and X_2 be independent exponential random variables, each having rate μ. Let

$$X_{(1)} = \text{minimum}(X_1, X_2) \quad \text{and} \quad X_{(2)} = \text{maximum}(X_1, X_2)$$

Find

(a) $E[X_{(1)}]$
(b) $\text{Var}[X_{(1)}]$
(c) $E[X_{(2)}]$
(d) $\text{Var}[X_{(2)}]$

19. Repeat Exercise 18, but this time suppose that the X_i are independent exponentials with respective rates μ_i, $i = 1, 2$.

20. Consider a two-server system in which a customer is served first by server 1, then by server 2, and then departs. The service times at server i are exponential random variables with rates μ_i, $i = 1, 2$. When you arrive, you find server 1 free and two customers at server 2—customer A in service and customer B waiting in line.

(a) Find P_A, the probability that A is still in service when you move over to server 2.
(b) Find P_B, the probability that B is still in the system when you move over to 2.
(c) Find $E[T]$, where T is the time that you spend in the system.

Hint: Write

$$T = S_1 + S_2 + W_A + W_B$$

where S_i is your service time at server i, W_A is the amount of time you wait in queue while A is being served, and W_B is the amount of time you wait in queue while B is being served.

21. In a certain system, a customer must first be served by server 1 and then by server 2. The service times at server i are exponential with rate μ_i, $i = 1, 2$. An arrival finding server 1 busy waits in line for that server. Upon completion of service at server 1, a customer either enters service with server 2 if that server is free or else remains with server 1 (blocking any other customer from entering service) until server 2 is free. Customers depart the system after being served by server 2. Suppose that when you arrive there is one customer in the system and that customer is being served by server 1. What is the expected total time you spend in the system?

22. Suppose in Exercise 21 you arrive to find two others in the system, one being served by server 1 and one by server 2. What is the expected time you spend in the system? Recall that if server 1 finishes before server 2, then server 1's customer will remain with him (thus blocking your entrance) until server 2 becomes free.

*23. A flashlight needs two batteries to be operational. Consider such a flashlight along with a set of n functional batteries — battery 1, battery 2,..., battery n. Initially, battery 1 and 2 are installed. Whenever a battery fails, it is immediately replaced by the lowest numbered functional battery that has not yet been put in use. Suppose that the lifetimes of the different batteries are independent exponential random variables each having rate μ. At a random time, call it T, a battery will fail and our stockpile will be empty. At that moment exactly one of the batteries — which we call battery X — will not yet have failed.

    (a) What is $P\{X = n\}$?
    (b) What is $P\{X = 1\}$?
    (c) What is $P\{X = i\}$?
    (d) Find $E[T]$.
    (e) What is the distribution of T?

24. Let X and Y be independent exponential random variables having respective rates λ and μ. Let I, independent of X, Y, be such that

$$I = \begin{cases} 1, & \text{with probability } \dfrac{\mu}{\lambda + \mu} \\[2mm] 0, & \text{with probability } \dfrac{\lambda}{\lambda + \mu} \end{cases}$$

and define Z by

$$Z = \begin{cases} X, & \text{if } I = 1 \\ -Y, & \text{if } I = 0 \end{cases}$$

(a) Show, by computing their moment generating functions, that $X - Y$ and Z have the same distribution.

(b) Using the lack of memory property of exponential random variables, give a simple explanation of the result of part (a).

25. Two individuals, A and B, both require kidney transplants. If she does not receive a new kidney, then A will die after an exponential time with rate μ_A, and B after an exponential time with rate μ_B. New kidneys arrive in accordance with a Poisson process having rate λ. It has been decided that the first kidney will go to A (or to B if B is alive and A is not at that time) and the next one to B (if still living).

(a) What is the probability that A obtains a new kidney?

(b) What is the probability that B obtains a new kidney?

26. The lifetimes of A's dog and cat are independent exponential random variables with respective rates λ_d and λ_c. One of them has just died. Find the expected additional lifetime of the other pet.

27. A doctor has scheduled two appointments, one at 1 P.M. and the other at 1:30 P.M. The amounts of time that appointments last are independent exponential random variables with mean 30 minutes. Assuming that both patients are on time, find the expected amount of time that the 1:30 appointment spends at the doctor's office.

28. Let X_i be independent exponential random variables with rates i, $i = 1, 2, 3$. Find $E[\max_i X_i]$ by conditioning on all the possible orderings of X_1, X_2, X_3.

29. There are three jobs and a single worker who works first on job 1, then on job 2, and finally on job 3. The amounts of time that he spends on each job are independent exponential random variables with mean 1. Let C_i be the time at which job i is completed, $i = 1, 2, 3$, and let $X = \sum_{i=1}^{3} C_i$ be the sum of these completion times. Find (a) $E[X]$, (b) Var(X).

30. Let X and Y be independent exponential random variables with respective rates λ and μ, where $\lambda > \mu$. Let $c > 0$.

(a) Show that the conditional density function of X, given that $X + Y = c$, is

$$f_{X|X+Y}(x \mid c) = \frac{(\lambda - \mu)e^{-(\lambda - \mu)x}}{1 - e^{-(\lambda - \mu)x}}, \qquad 0 < x < c$$

(b) Use part (a) and the result of Exercise 14 to find $E[X | X + Y = c]$.
(c) Find $E[Y | X + Y = c]$.

31. Let $X_1, \ldots, X_n$ be independent exponential random variables, each having rate λ. Also, let $X_{(i)}$ be the ith smallest of these values, $i = 1, \ldots, n$. Find

(a) $E[X_{(1)} X_{(2)}]$
(b) $E[X_{(i)} X_{(i+1)}]$
(c) $E[X_{(i)} X_{(j)}], i < j$

32. Argue that if λ_i, $i = 1, \ldots, n$ are distinct positive numbers then

$$\sum_{i=1}^{n} \frac{1}{\lambda_i} \prod_{j \neq i} \frac{\lambda_j}{\lambda_j - \lambda_i} = \sum_{i=1}^{n} \frac{1}{\lambda_i}$$

Hint: Relate this problem to Section 5.2.4.

33. Show that Definition 5.1 of a Poisson process implies Definition 5.3.

*34. Show that assumption (iv) of Definition 5.3 follows from assumptions (ii) and (iii).

Hint: Derive a functional equation for $g(t) = P\{N(t) = 0\}$.

35. Cars cross a certain point in the highway in accordance with a Poisson process with rate $\lambda = 3$ per minute. If Reb blindly runs across the highway, then what is the probability that she will be uninjured if the amount of time that it takes her to cross the road is s seconds? (Assume that if she is on the highway when a car passes by, then she will be injured.) Do it for $s = 2, 5, 10, 20$.

36. Suppose in Exercise 35 that Reb is agile enough to escape from a single car, but if she encounters two or more cars while attempting to cross the road, then she will be injured. What is the probability that she will be unhurt if it takes her s seconds to cross? Do it for $s = 5, 10, 20, 30$.

37. A certain scientific theory supposes that mistakes in cell division occur according to a Poisson process with rate 2.5 per year, and that an individual dies when 196 such mistakes have occurred. Assuming this theory, find

(a) the mean lifetime of an individual,
(b) the variance of the lifetime of an individual.

Also approximate

(c) the probability that an individual dies before age 67.2.
(d) the probability that an individual reaches age 90.
(e) the probability that an individual reaches age 100.

*38.   Show that if $\{N_i(t), t \geq 0\}$ are independent Poisson processes with rate λ_i, $i = 1, 2$, then $\{N(t), t \geq 0\}$ is a Poisson process with rate $\lambda_1 + \lambda_2$ where $N(t) = N_1(t) + N_2(t)$.

39.   In Exercise 38 what is the probability that the first event of the combined process is from the N_1 process?

40.   Let $\{N(t), t \geq 0\}$ be a Poisson process with rate λ. Let S_n denote the time of the nth event. Find

(a) $E[S_4]$
(b) $E[S_4 \mid N(1) = 2]$
(c) $E[N(4) - N(2) \mid N(1) = 3]$

41.   Customers arrive at a two-server service station according to a Poisson process with rate λ. Whenever a new customer arrives, any customer that is in the system immediately departs. A new arrival enters service first with server 1 and then with server 2. If the service times at the servers are independent exponentials with respective rates μ_1 and μ_2, what proportion of entering customers completes their service with server 2?

42.   Cars pass a certain street location according to a Poisson process with rate λ. A woman who want to cross the street at that location waits until she can see that no cars will come by in the next T time units.

(a) Find the probability that her waiting time is 0.
(b) Find her expected waiting time.

Hint: Condition on the time of the first car.

43.   Let $\{N(t), t \geq 0\}$ be a Poisson process with rate λ, that is independent of the nonnegative random variable T with mean μ and variance σ^2. Find

(a) $\text{Cov}(T, N(T))$
(b) $\text{Var}(N(T))$

44.   Let $\{N(t), t \geq 0\}$ be a Poisson process with rate λ, that is independent of the sequence $X_1, X_2, \ldots$ of independent and identically distributed random variables with mean μ and variance σ^2. Find

$$\text{Cov}\left(N(t), \sum_{i=1}^{N(t)} X_i\right)$$

45.   Consider a two-server parallel queueing system where customers arrive according to a Poisson process with rate λ, and where the service times are exponential with rate μ. Moreover suppose that arrivals finding both servers busy immediately depart without receiving any service (such a customer is said to be lost), whereas those finding at least one free server immediately enter service and then depart when their service is completed.

(a) If both servers are presently busy, find the expected time until the next customer enters the system.
(b) Starting empty, find the expected time until both servers are busy.
(c) Find the expected time between two successive lost customers.

46. Consider an n-server parallel queueing system where customers arrive according to a Poisson process with rate λ, where the service times are exponential random variables with rate μ, and where any arrival finding all servers busy immediately departs without receiving any service. If an arrival finds all servers busy, find

(a) the expected number of busy servers found by the next arrival,
(b) the probability that the next arrival finds all servers free,
(c) the probability that the next arrival finds exactly i of the servers free.

47. Events occur according to a Poisson process with rate λ. Each time an event occurs, we must decide whether or not to stop, with our objective being to stop at the last event to occur prior to some specified time T, where $T > 1/\lambda$. That is, if an event occurs at time $t, 0 \leqslant t \leqslant T$, and we decide to stop, then we win if there are no additional events by time T, and we lose otherwise. If we do not stop when an event occurs and no additional events occur by time T, then we lose. Also, if no events occur by time T, then we lose. Consider the strategy that stops at the first event to occur after some fixed time $s, 0 \leqslant s \leqslant T$.

(a) Using this strategy, what is the probability of winning?
(b) What value of s maximizes the probability of winning?
(c) Show that one's probability of winning when using the preceding strategy with the value of s specified in part (b) is $1/e$.

48. The number of hours between successive train arrivals at the station is uniformly distributed on $(0, 1)$. Passengers arrive according to a Poisson process with rate 7 per hour. Suppose a train has just left the station. Let X denote the number of people that get on the next train. Find

(a) $E[X]$
(b) $\text{Var}(X)$

49. For a Poisson process show, for $s < t$, that

$$P\{N(s) = k \mid N(t) = n\} = \binom{n}{k}\left(\frac{s}{t}\right)^k\left(1 - \frac{s}{t}\right)^{n-k}, \qquad k = 0, 1, \ldots, n$$

50. Men and women enter a supermarket according to independent Poisson processes having respective rates two and four per minute. Starting at an arbitrary time, compute the probability that at least two men arrive before three women arrive.

51. A viral linear DNA molecule of length, say, 1 is often known to contain a certain "marked position," with the exact location of this mark being unknown. One approach to locating the marked position is to cut the molecule by agents that break it at points chosen according to a Poisson process with rate λ. It is then possible to determine the fragment that contains the marked position. For instance, letting m denote the location on the line of the marked position, then if L_1 denotes the last Poisson event time before m (or 0 if there are no Poisson events in $[0, m]$), and R_1 denotes the first Poisson event time after m (or 1 if there are no Poisson events in $[m, 1]$), then it would be learned that the marked position lies between L_1 and R_1. Find

(a) $P\{L_1 = 0\}$
(b) $P\{L_1 < x\}, 0 < x < m$
(c) $P\{R_1 = 1\}$
(d) $P\{R_1 > x\}, m < x < 1$

By repeating the preceding process on identical copies of the DNA molecule one is able to zero in on the location of the marked position. If the cutting procedure is utilized on n identical copies of the molecule, yielding the data $L_i, R_i, i = 1, \ldots, n$, then it follows that the marked position lies between L and R, where

$$L = \max_i L_i, \qquad R = \min_i R_i$$

(e) Find $E[R - L]$, and in doing so, show that $E[R - L] \approx \dfrac{2}{n\lambda}$.

52. Consider a single server queueing system where customers arrive according to a Poisson process with rate λ, service times are exponential with rate μ, and customers are served in the order of their arrival. Suppose that a customer arrives and finds $n - 1$ others in the system. Let X denote the number in the system at the moment that customer departs. Find the probability mass function of X.

Hint: Relate this to a negative binomial random variable.

53. An event independently occurs on each day with probability p. Let $N(n)$ denote the total number of events that occur on the first n days, and let T_r denote the day on which the rth event occurs.

(a) What is the distribution of $N(n)$?
(b) What is the distribution of T_1?

(c) What is the distribution of T_r?

(d) Given that $N(n) = r$, show that the unordered set of r days on which events occurred has the same distribution as a random selection (without replacement) of r of the values $1, 2, \ldots, n$.

*54. Events occur according to a Poisson process with rate $\lambda = 2$ per hour.

(a) What is the probability that no event occurs between 8 P.M. and 9 P.M.?

(b) Starting at noon, what is the expected time at which the fourth event occurs?

(c) What is the probability that two or more events occur between 6 P.M. and 8 P.M.?

55. Pulses arrive at a Geiger counter in accordance with a Poisson process at a rate of three arrivals per minute. Each particle arriving at the counter has a probability $\frac{2}{3}$ of being recorded. Let $X(t)$ denote the number of pulses recorded by time t minutes.

(a) $P\{X(t) = 0\} = ?$

(b) $E[X(t)] = ?$

56. Cars pass a point on the highway at a Poisson rate of one per minute. If 5 percent of the cars on the road are vans, then

(a) what is the probability that at least one van passes by during an hour?

(b) given that ten vans have passed by in an hour, what is the expected number of cars to have passed by in that time?

(c) if 50 cars have passed by in an hour, what is the probability that five of them were vans?

*57. Customers arrive at a bank at a Poisson rate λ. Suppose two customers arrived during the first hour. What is the probability that

(a) both arrived during the first 20 minutes?

(b) at least one arrived during the first 20 minutes?

58. A system has a random number of flaws that we will suppose is Poisson distributed with mean c. Each of these flaws will, independently, cause the system to fail at a random time having distribution G. When a system failure occurs, suppose that the flaw causing the failure is immediately located and fixed.

(a) What is the distribution of the number of failures by time t?

(b) What is the distribution of the number of flaws that remain in the system at time t?

(c) Are the random variables in parts (a) and (b) dependent or independent?

59. Suppose that the number of typographical errors in a new text is Poisson distributed with mean λ. Two proofreaders independently read the text. Suppose that each error is independently found by proofreader i with probability p_i, $i = 1, 2$. Let X_1 denote the number of errors that are found by proofreader 1 but not by proofreader 2. Let X_2 denote the number of errors that are found by proofreader 2 but not by proofreader 1. Let X_3 denote the number of errors that are found by both proofreaders. Finally, let X_4 denote the number of errors found by neither proofreader.

(a) Describe the joint probability distribution of X_1, X_2, X_3, X_4.
(b) Show that

$$\frac{E[X_1]}{E[X_3]} = \frac{1 - p_2}{p_2} \quad \text{and} \quad \frac{E[X_2]}{E[X_3]} = \frac{1 - p_1}{p_1}$$

Suppose now that λ, p_1, and p_2 are all unknown.

(c) By using X_i as an estimator of $E[X_i]$, $i = 1, 2, 3$, present estimators of p_1, p_2, and λ.
(d) Give an estimator of X_4, the number of errors not found by either proofreader.

60. Consider an infinite server queueing system in which customers arrive in accordance with a Poisson process and where the service distribution is exponential with rate μ. Let $X(t)$ denote the number of customers in the system at time t. Find

(a) $E[X(t + s) \,|\, X(s) = n]$
(b) $\text{Var}[X(t + s) \,|\, X(s) = n]$

Hint: Divide the customers in the system at time $t + s$ into two groups, one consisting of "old" customers and the other of "new" customers.

*61. Suppose that people arrive at a bus stop in accordance with a Poisson process with rate λ. The bus departs at time t. Let X denote the total amount of waiting time of all those that get on the bus at time t. We want to determine $\text{Var}(X)$. Let $N(t)$ denote the number of arrivals by time t.

(a) What is $E[X \,|\, N(t)]$?
(b) Argue that $\text{Var}[X \,|\, N(t)] = N(t)t^2/12$.
(c) What is $\text{Var}(X)$?

62. An average of 500 people pass the California bar exam each year. A California lawyer practices law, on average, for 30 years. Assuming these numbers remain steady, how many lawyers would you expect California to have in 2050?

63. Let S_n denote the time of the nth event of the Poisson process $\{N(t), t \geq 0\}$ having rate λ. Show, for an arbitrary function g, that the random variable $\sum_{i=1}^{N(t)} g(S_i)$ has the same distribution as the compound Poisson random variable $\sum_{i=1}^{N} g(U_i)$, where $U_1, U_2, \ldots$ is a sequence of independent and identically distributed uniform $(0, t)$ random variables that is independent of N, a Poisson random variable with mean λt. Consequently, conclude that

$$E\left[\sum_{i=1}^{N(t)} g(S_i)\right] = \lambda \int_0^t g(x)\,dx \qquad \mathrm{Var}\left(\sum_{i=1}^{N(t)} g(S_i)\right) = \lambda \int_0^t g^2(x)\,dx$$

64. A cable car starts off with n riders. The times between successive stops of the car are independent exponential random variables with rate λ. At each stop one rider gets off. This takes no time, and no additional riders get on. After a rider gets off the car, he or she walks home. Independently of all else, the walk takes an exponential time with rate μ.

(a) What is the distribution of the time at which the last rider departs the car?

(b) Suppose the last rider departs the car at time t. What is the probability that all the other riders are home at that time?

65. Shocks occur according to a Poisson process with rate λ, and each shock independently causes a certain system to fail with probability p. Let T denote the time at which the system fails and let N denote the number of shocks that it takes.

(a) Find the conditional distribution of T given that $N = n$.

(b) Calculate the conditional distribution of N, given that $T = t$, and notice that it is distributed as 1 plus a Poisson random variable with mean $\lambda(1 - p)t$.

(c) Explain how the result in part (b) could have been obtained without any calculations.

66. The number of missing items in a certain location, call it X, is a Poisson random variable with mean λ. When searching the location, each item will independently be found after an exponentially distributed time with rate μ. A reward of R is received for each item found, and a searching cost of C per unit of search time is incurred. Suppose that you search for a fixed time t and then stop.

(a) Find your total expected return.

(b) Find the value of t that maximizes the total expected return.

(c) The policy of searching for a fixed time is a static policy. Would a dynamic policy which allows the decision as to whether to stop at each time t depend on the number already found by t be beneficial?

Hint: How does the distribution of the number of items not yet found by time t depend on the number already found by that time?

67. Suppose that the times between successive arrivals of customers at a single-server station are independent random variables having a common distribution F. Suppose that when a customer arrives, he or she either immediately enters service if the server is free or else joins the end of the waiting line if the server is busy with another customer. When the server completes work on a customer, that customer leaves the system and the next waiting customer, if there are any, enters service. Let X_n denote the number of customers in the system immediately before the nth arrival, and let Y_n denote the number of customers that remain in the system when the nth customer departs. The successive service times of customers are independent random variables (which are also independent of the interarrival times) having a common distribution G.

(a) If F is the exponential distribution with rate λ, which, if any, of the processes $\{X_n\}$, $\{Y_n\}$ is a Markov chain?
(b) If G is the exponential distribution with rate μ, which, if any, of the processes $\{X_n\}$, $\{Y_n\}$ is a Markov chain?
(c) Give the transition probabilities of any Markov chains in parts (a) and (b).

68. For the model of Example 5.23, find the mean and variance of the number of customers served in a busy period.

69. Events occur according to a nonhomogeneous Poisson process whose mean value function is given by

$$m(t) = t^2 + 2t, \qquad t \geqslant 0$$

What is the probability that n events occur between times $t = 4$ and $t = 5$?

70. A store opens at 8 A.M. From 8 until 10 customers arrive at a Poisson rate of four an hour. Between 10 and 12 they arrive at a Poisson rate of eight an hour. From 12 to 2 the arrival rate increases steadily from eight per hour at 12 to ten per hour at 2; and from 2 to 5 the arrival rate drops steadily from ten per hour at 2 to four per hour at 5. Determine the probability distribution of the number of customers that enter the store on a given day.

*71. Consider a nonhomogeneous Poisson process whose intensity function $\lambda(t)$ is bounded and continuous. Show that such a process is equivalent to a process of counted events from a (homogeneous) Poisson process having rate λ, where an event at time t is counted (independent of the past) with probability $\lambda(t)/\lambda$; and where λ is chosen so that $\lambda(s) < \lambda$ for all s.

72. Let $T_1, T_2, \ldots$ denote the interarrival times of events of a nonhomogeneous Poisson process having intensity function $\lambda(t)$.

(a) Are the T_i independent?
(b) Are the T_i identically distributed?
(c) Find the distribution of T_1.

73. (a) Let $\{N(t), t \geq 0\}$ be a nonhomogeneous Poisson process with mean value function $m(t)$. Given $N(t) = n$, show that the unordered set of arrival times has the same distribution as n independent and identically distributed random variables having distribution function

$$F(x) = \begin{cases} \dfrac{m(x)}{m(t)}, & x \leq t \\ 1, & x > t \end{cases}$$

(b) Suppose that workmen incur accidents in accordance with a nonhomogeneous Poisson process with mean value function $m(t)$. Suppose further that each injured man is out of work for a random amount of time having distribution F. Let $X(t)$ be the number of workers who are out of work at time t. By using part (a), find $E[X(t)]$.

74. Suppose that events occur according to a nonhomogeneous Poisson process with intensity function $\lambda(t)$, $t \geq 0$. Suppose that, independently of anything that has previously occurred, an event at time s will be counted with probability $p(s)$, $s \geq 0$. Let $N_c(t)$ denote the number of counted events by time t.

(a) What type of process if $\{N_c(t), t \geq 0\}$?
(b) Prove your answer to part (a).

75. Suppose that $\{N_0(t), t \geq 0\}$ is a Poisson process with rate $\lambda = 1$. Let $\lambda(t)$ denote a nonnegative function of t, and let

$$m(t) = \int_0^t \lambda(s) \, ds$$

Define $N(t)$ by

$$N(t) = N_0(m(t))$$

Argue that $\{N(t), t \geq 0\}$ is a nonhomogeneous Poisson process with intensity function $\lambda(t)$, $t \geq 0$.

Hint: Make use of the identity

$$m(t + h) - m(t) = m'(t)h + o(h)$$

*76. Let $X_1, X_2, \ldots$ be independent and identically distributed non-negative continuous random variables having density function $f(x)$. We say that record occurs at time n if X_n is larger than each of the previous values $X_1, \ldots, X_{n-1}$. (A record automatically occurs at time 1.) If a record occurs at time n, then X_n is called a *record value*. In other words, a record occurs whenever a new high is reached, and that new high is called the record value. Let $N(t)$ denote the number of record values that are less than or equal to t. Characterize the process $\{N(t), t \geqslant 0\}$ when

(a) f is an arbitrary continuous density function.
(b) $f(x) = \lambda e^{-\lambda x}$.

Hint: Finish the following sentence: There will be a record whose value is between t and $t + dt$ if the first X_i that is greater than t lies between....

77. An insurance company pays out claims on its life insurance policies in accordance with a Poisson process having rate $\lambda = 5$ per week. If the amount of money paid on each policy is exponentially distributed with mean $2000, what is the mean and variance of the amount of money paid by the insurance company in a four-week span?

78. In good years, storms occur according to a Poisson process with rate 3 per unit time, while in other years they occur according to a Poisson process with rate 5 per unit time. Suppose next year will be a good year with probability 0.3. Let $N(t)$ denote the number of storms during the first t time units of next year.

(a) Find $P\{N(t) = n\}$.
(b) Is $\{N(t)\}$ a Poisson process?
(c) Does $\{N(t)\}$ have stationary increments? Why or why not?
(d) Does it have independent increments? Why or why not?
(e) If next year starts off with 3 storms by time $t = 1$, what is the conditional probability it is a good year?

79. Determine

$$\text{Cov}[X(t), X(t + s)]$$

when $\{X(t), t \geqslant 0\}$ is a compound Poisson process.

80. Customers arrive at the automatic teller machine in accordance with a Poisson process with rate 12 per hour. The amount of money withdrawn on each transaction is a random variable with mean $30 and standard deviation $50. (A negative withdrawal means that money was deposited.) The machine is in use for 15 hours daily. Approximate the probability that the total daily withdrawal is less than $6000.

81. Some components of a two-component system fail after receiving a shock. Shocks of three types arrive independently and in accordance with Poisson processes. Shocks of the first type arrive at a Poisson rate λ_1 and cause the first component to fail. Those of the second type arrive at a Poisson rate λ_2 and cause the second component to fail. The third type of shock arrives at a Poisson rate λ_3 and causes both components to fail. Let X_1 and X_2 denote the survival times for the two components. Show that the joint distribution of X_1 and X_2 is given by

$$P\{X_1 > s, X_2 > t\} = \exp\{-\lambda_1 s - \lambda_2 t - \lambda_3 \max(s, t)\}$$

This distribution is known as the *bivariate exponential distribution.*

82. In Exercise 81 show that X_1 and X_2 both have exponential distributions.

*83. Let $X_1, X_2, \ldots, X_n$ be independent and identically distributed exponential random variables. Show that the probability that the largest of them is greater than the sum of the others is $n/2^{n-1}$. That is, if

$$M = \max_j X_j$$

then show

$$P\left\{M > \sum_{i=1}^{n} X_i - M\right\} = \frac{n}{2^{n-1}}$$

Hint: What is $P\{X_1 > \sum_{i=2}^{n} X_i\}$?

84. Prove Equation (5.22).

85. Prove that

(a) $\max(X_1, X_2) = X_1 + X_2 - \min(X_1, X_2)$ and, in general,

(b) $$\max(X_1, \ldots, X_n) = \sum_{1}^{n} X_i - \sum \sum_{i<j} \min(X_i, X_j)$$
$$+ \sum \sum \sum_{i<j<k} \min(X_i, X_j, X_k) + \cdots$$
$$+ (-1)^{n-1} \min(X_1, X_2, \ldots, X_n)$$

Show by defining appropriate random variables X_i, $i = 1, \ldots, n$, and by taking expectations in part (b) how to obtain the well-known formula

$$P\left(\bigcup_{1}^{n} A_i\right) = \sum_i P(A_i) - \sum \sum_{i<j} P(A_i A_j) + \cdots + (-1)^{n-1} P(A_1 \cdots A_n)$$

(c) Consider n independent Poisson processes — the ith having rate λ_i. Derive an expression for the expected time until an event has occurred in all n processes.

86.   A two-dimensional Poisson process is a process of randomly occurring events in the plane such that

  (i) for any region of area A the number of events in that region has a Poisson distribution with mean λA and
  (ii) the number of events in nonoverlapping regions are independent.

For such a process, consider an arbitrary point in the plane and let X denote its distance from its nearest event (where distance is measured in the usual Euclidean manner). Show that

(a)
$$P\{X > t\} = e^{-\lambda \pi t^2}$$

(b)
$$E[X] = \frac{1}{2\sqrt{\lambda}}$$

87.   Show, in Example 5.6, that the distributions of the total cost are the same for the two algorithms.

References

1. H. Cramér and M. Leadbetter, "Stationary and Related Stochastic Processes," John Wiley, New York, 1966.
2. S. Ross, "Stochastic Processes," Second Edition, John Wiley, New York, 1996.

Continuous-Time Markov Chains

6

◆◆◆

6.1. Introduction

In this chapter we consider a class of probability models that has a wide variety of applications in the real world. The members of this class are the continuous-time analogs of the Markov chains of Chapter 4 and as such are characterized by the Markovian property that, given the present state, the future is independent of the past.

One example of a continuous-time Markov chain has already been met. This is, of course, the Poisson process of Chapter 5. For if we let the total number of arrivals by time t [that is, $N(t)$] be the state of the process at time t, then the Poisson process is a continuous-time Markov chain having states $0, 1, 2, \ldots$ and which always proceeds from state n to state $n + 1$, where $n \geqslant 0$. Such a process is known as a *pure birth process* since when a transition occurs the state of the system is always increased by one. More generally, an exponential model which can go (in one transition) only from state n to either state $n - 1$ or state $n + 1$ is called a *birth and death model*. For such a model, transitions from state n to state $n + 1$ are designated as births, and those from n to $n - 1$ as deaths. Birth and death models have wide applicability in the study of biological systems and in the study of waiting line systems in which the state represents the number of customers in the system. These models will be studied extensively in this chapter.

In Section 6.2 we define continuous-time Markov chains and then relate them to the discrete-time Markov chains of Chapter 4. In Section 6.3 we consider birth and death processes and in Section 6.4 we derive two sets of differential equations—the forward and backward equations—which describe the probability laws for the system. The material in Section 6.5 is concerned with determining the limiting (or long-run) probabilities connected with a continuous-time Markov chain. In Section 6.6 we consider the topic of time reversibility. We show that all birth and death processes are time reversible, and then illustrate the importance of this observation to queueing systems. In the final section we show how to "uniformize" Markov chains, a technique useful for numerical computations.

6.2. Continuous-Time Markov Chains

Suppose we have a continuous-time stochastic process $\{X(t), t \geq 0\}$ taking on values in the set of nonnegative integers. In analogy with the definition of a discrete-time Markov chain, given in Chapter 4, we say that the process $\{X(t), t \geq 0\}$ is a *continuous-time Markov chain* if for all s, $t \geq 0$ and nonnegative integers $i, j, x(u), 0 \leq u < s$

$$P\{X(t+s) = j \,|\, X(s) = i, X(u) = x(u), 0 \leq u < s\} = P\{X(t+s) = j \,|\, X(s) = i\}$$

In other words, a continuous-time Markov chain is a stochastic process having the Markovian property that the conditional distribution of the future $X(t + s)$ given the present $X(s)$ and the past $X(u)$, $0 \leq u < s$, depends only on the present and is independent of the past. If, in addition,

$$P\{X(t + s) = j \,|\, X(s) = i\}$$

is independent of s, then the continuous-time Markov chain is said to have stationary or homogeneous transition probabilities.

All Markov chains considered in this text will be assumed to have stationary transition probabilities.

Suppose that a continuous-time Markov chain enters state i at some time, say, time 0, and suppose that the process does not leave state i (that is, a transition does not occur) during the next ten minutes. What is the probability that the process will not leave state i during the following five minutes? Now since the process is in state i at time 10 it follows, by the Markovian property, that the probability that it remains in that state during the interval $[10, 15]$ is just the (unconditional) probability that it stays in state i for at least five minutes. That is, if we let T_i denote the amount of time that the process stays in state i before making a transition into a different state, then

$$P\{T_i > 15 \,|\, T_i > 10\} = P\{T_i > 5\}$$

or, in general, by the same reasoning,

$$P\{T_i > s + t \,|\, T_i > s\} = P\{T_i > t\}$$

for all s, $t \geq 0$. Hence, the random variable T_i is *memoryless* and must thus (see Section 5.2.2) be *exponentially* distributed.

In fact, the preceding gives us another way of defining a continuous-time Markov chain. Namely, it is a stochastic process having the properties that each time it enters state i

(i) the amount of time it spends in that state before making a transition into a different state is exponentially distributed with mean, say, $1/v_i$, and

(ii) when the process leaves state i, it next enters state j with some probability, say, P_{ij}. Of course, the P_{ij} must satisfy

$$P_{ii} = 0, \quad \text{all } i$$

$$\sum_j P_{ij} = 1, \quad \text{all } i$$

In other words, a continuous-time Markov chain is a stochastic process that moves from state to state in accordance with a (discrete-time) Markov chain, but is such that the amount of time it spends in each state, before proceeding to the next state, is exponentially distributed. In addition, the amount of time the process spends in state i, and the next state visited, must be independent random variables. For if the next state visited were dependent on T_i, then information as to how long the process has already been in state i would be relevant to the prediction of the next state — and this contradicts the Markovian assumption.

Example 6.1 (A Shoeshine Shop): Consider a shoeshine establishment consisting of two chairs — chair 1 and chair 2. A customer upon arrival goes initially to chair 1 where his shoes are cleaned and polish is applied. After this is done the customer moves on to chair 2 where the polish is buffed. The service times at the two chairs are assumed to be independent random variables that are exponentially distributed with respective rates μ_1 and μ_2. Suppose that potential customers arrive in accordance with a Poisson process having rate λ, and that a potential customer will only enter the system if both chairs are empty.

The preceding model can be analyzed as a continuous-time Markov chain, but first we must decide upon an appropriate state space. Since a potential customer will enter the system only if there are no other customers present, it follows that there will always either be 0 or 1 customers in the system. However, if there is 1 customer in the system, then we would also need to know which chair he was presently in. Hence, an appropriate state space might consist of the three states 0, 1, and 2 where the states have the following interpretation:

$$\textit{State} \qquad\qquad \textit{Interpretation}$$

$$
\begin{array}{ll}
0 & \text{system is empty} \\
1 & \text{a customer is in chair 1} \\
2 & \text{a customer is in chair 2}
\end{array}
$$

We leave it as an exercise for the reader to verify that

$$v_0 = \lambda, \qquad v_1 = \mu_1, \qquad v_2 = \mu_2,$$

$$P_{01} = P_{12} = P_{20} = 1 \quad \blacklozenge$$

6.3.   Birth and Death Processes

Consider a system whose state at any time is represented by the number of people in the system at that time. Suppose that whenever there are n people in the system, then (i) new arrivals enter the system at an exponential rate λ_n, and (ii) people leave the system at an exponential rate μ_n. That is, whenever there are n persons in the system, then the time until the next arrival is exponentially distributed with mean $1/\lambda_n$ and is independent of the time until the next departure which is itself exponentially distributed with mean $1/\mu_n$. Such a system is called a birth and death process. The parameters $\{\lambda_n\}_{n=0}^{\infty}$ and $\{\mu_n\}_{n=1}^{\infty}$ are called respectively the arrival (or birth) and departure (or death) rates.

Thus, a birth and death process is a continuous-time Markov chain with states $\{0, 1, \ldots\}$ for which transitions from state n may go only to either state $n - 1$ or state $n + 1$. The relationships between the birth and death rates and the state transition rates and probabilities are

$$v_0 = \lambda_0,$$

$$v_i = \lambda_i + \mu_i, \qquad i > 0$$

$$P_{01} = 1,$$

$$P_{i,i+1} = \frac{\lambda_i}{\lambda_i + \mu_i}, \qquad i > 0$$

$$P_{i,i-1} = \frac{\mu_i}{\lambda_i + \mu_i}, \qquad i > 0$$

The preceding follows, since when there are i in the system, then the next state will be $i + 1$ if a birth occurs before a death; and the probability that an exponential random variable with rate λ_i will occur earlier than an (independent) exponential with rate μ_i is $\lambda_i/(\lambda_i + \mu_i)$ (and so, $P_{i,i+1} = \lambda_i/(\lambda_i + \mu_i)$), and the time until either occurs is exponentially distributed with rate $\lambda_i + \mu_i$ (and so, $v_i = \lambda_i + \mu_i$).

Example 6.2   (The Poisson Process): Consider a birth and death process for which

$$\mu_n = 0, \qquad \text{for all } n \geqslant 0$$

$$\lambda_n = \lambda, \qquad \text{for all } n \geqslant 0$$

This is a process in which departures never occur, and the time between successive arrivals is exponential with mean $1/\lambda$. Hence, this is just the Poisson process. ◆

A birth and death process for which $\mu_n = 0$ for all n is called a pure birth process. Another pure birth process is given by the next example.

Example 6.3   (A Birth Process with Linear Birth Rate): Consider a population whose members can give birth to new members but cannot die. If each member acts independently of the others and takes an exponentially distributed amount of time, with mean $1/\lambda$, to give birth, then if $X(t)$ is the population size at time t, then $\{X(t), t \geqslant 0\}$ is a pure birth process with $\lambda_n = n\lambda, n \geqslant 0$. This follows since if the population consists of n persons and each gives birth at an exponential rate λ, then the total rate at which births occur is $n\lambda$. This pure birth process is known as a Yule process after G. Yule who used it in his mathematical theory of evolution. ◆

Example 6.4   (A Linear Growth Model with Immigration): A model in which

$$\mu_n = n\mu, \qquad n \geqslant 1$$

$$\lambda_n = n\lambda + \theta, \qquad n \geqslant 0$$

is called a linear growth process with immigration. Such processes occur naturally in the study of biological reproduction and population growth. Each individual in the population is assumed to give birth at an exponential rate λ; in addition, there is an exponential rate of increase θ of the population due to an external source such as immigration. Hence, the total birth rate where there are n persons in the system is $n\lambda + \theta$. Deaths are assumed to occur at an exponential rate μ for each member of the population, and hence $\mu_n = n\mu$.

Let $X(t)$ denote the population size at time t. Suppose that $X(0) = i$ and let

$$M(t) = E[X(t)]$$

We will determine $M(t)$ by deriving and then solving a differential equation that it satisfies.

We start by deriving an equation for $M(t + h)$ by conditioning on $X(t)$. This yields

$$M(t + h) = E[X(t + h)]$$
$$= E[E[X(t + h) \mid X(t)]]$$

Now, given the size of the population at time t then, ignoring events whose probability is $o(h)$, the population at time $t + h$ will either increase in size by 1 if a birth or an immigration occurs in $(t, t + h)$, or decrease by 1 if a death occurs in this interval, or remain the same if neither of these two possibilities occurs. That is, given $X(t)$,

$$X(t+h) = \begin{cases} X(t)+1, & \text{with probability } [\theta + X(t)\lambda]h + o(h) \\ X(t)-1, & \text{with probability } X(t)\mu h + o(h) \\ X(t), & \text{with probability } 1 - [\theta + X(t)\lambda + X(t)\mu]h + o(h) \end{cases}$$

Therefore,

$$E[X(t + h) \mid X(t)] = X(t) + [\theta + X(t)\lambda - X(t)\mu]h + o(h)$$

Taking expectations yields

$$M(t + h) = M(t) + (\lambda - \mu)M(t)h + \theta h + o(h)$$

or, equivalently,

$$\frac{M(t + h) - M(t)}{h} = (\lambda - \mu)M(t) + \theta + \frac{o(h)}{h}$$

Taking the limit as $h \to 0$ yields the differential equation

$$M'(t) = (\lambda - \mu)M(t) + \theta \qquad (6.1)$$

If we now define the function $h(t)$ by

$$h(t) = (\lambda - \mu)M(t) + \theta$$

then

$$h'(t) = (\lambda - \mu)M'(t)$$

Therefore, the differential equation (6.1) can be rewritten as

$$\frac{h'(t)}{\lambda - \mu} = h(t)$$

or

$$\frac{h'(t)}{h(t)} = \lambda - \mu$$

Integration yields

$$\log[h(t)] = (\lambda - \mu)t + c$$

or

$$h(t) = Ke^{(\lambda - \mu)t}$$

Putting this back in terms of $M(t)$ gives

$$\theta + (\lambda - \mu)M(t) = Ke^{(\lambda - \mu)t}$$

To determine the value of the constant K, we use the fact that $M(0) = i$ and evaluate the preceding at $t = 0$. This gives

$$\theta + (\lambda - \mu)i = K$$

Substituting this back in the preceding equation for $M(t)$ yields the following solution for $M(t)$:

$$M(t) = \frac{\theta}{\lambda - \mu}[e^{(\lambda - \mu)t} - 1] + ie^{(\lambda - \mu)t}$$

Note that we have implicitly assumed that $\lambda \neq \mu$. If $\lambda = \mu$, then the differential equation (6.1) reduces to

$$M'(t) = \theta \tag{6.2}$$

Integrating (6.2) and using that $M(0) = i$ gives the solution

$$M(t) = \theta t + i \quad \blacklozenge$$

Example 6.5 (The Queueing System $M/M/1$): Suppose that customers arrive at a single-server service station in accordance with a Poisson process having rate λ. That is, the times between successive arrivals are independent exponential random variables having mean $1/\lambda$. Upon arrival, each customer goes directly into service if the server is free; if not, then the customer joins the queue (that is, he waits in line). When the server finishes serving a customer, the customer leaves the system and the next customer in line, if there are any waiting, enters the service. The successive service times are assumed to be independent exponential random variables having mean $1/\mu$.

The preceding is known as the $M/M/1$ queueing system. The first M refers to the fact that the interarrival process is Markovian (since it is a Poisson process) and the second to the fact that the service distribution is exponential (and, hence, Markovian). The 1 refers to the fact that there is a single server.

If we let $X(t)$ denote the number in the system at time t then $\{X(t), t \geqslant 0\}$ is a birth and death process with

$$\mu_n = \mu, \qquad n \geqslant 1$$
$$\lambda_n = \lambda, \qquad n \geqslant 0 \quad \blacklozenge$$

Example 6.6 (A Multiserver Exponential Queueing System): Consider an exponential queueing system in which there are s servers available. An entering customer first waits in line and then goes to the first free server. This is a birth and death process with parameters

$$\mu_n = \begin{cases} n\mu, & 1 \leqslant n \leqslant s \\ s\mu, & n > s \end{cases}$$

$$\lambda_n = \lambda, \qquad n \geqslant 0$$

To see why this is true, reason as follows: If there are n customers in the system, where $n \leqslant s$, then n servers will be busy. Since each of these servers works at a rate μ, the total departure rate will be $n\mu$. On the other hand, if there are n customers in the system, where $n > s$, then all s of the servers will be busy, and thus the total departure rate will be $s\mu$. This is known as an $M/M/s$ queueing model (why?). $\blacklozenge$

Consider now a general birth and death process with birth rates $\{\lambda_n\}$ and death rates $\{\mu_n\}$, where $\mu_0 = 0$, and let T_i denote the time, starting from state i, it takes for the process to enter state $i + 1$, $i \geqslant 0$. We will recursively compute $E[T_i]$, $i \geqslant 0$, by starting with $i = 0$. Since T_0 is exponential with rate λ_0, we have that

$$E[T_0] = \frac{1}{\lambda_0}$$

For $i > 0$, we condition whether the first transition takes the process into state $i - 1$ or $i + 1$. That is, let

$$I_i = \begin{cases} 1, & \text{if the first transition from } i \text{ is to } i + 1 \\ 0, & \text{if the first transition from } i \text{ is to } i - 1 \end{cases}$$

and note that

$$E[T_i \mid I_i = 1] = \frac{1}{\lambda_i + \mu_i},$$

$$E[T_i \mid I_i = 0] = \frac{1}{\lambda_i + \mu_i} + E[T_{i-1}] + E[T_i]$$

(6.3)

This follows since, independent of whether the first transition is from a birth or death, the time until it occurs is exponential with rate $\lambda_i + \mu_i$; now if this first transition is a birth, then the population size is at $i + 1$, so no additional time is needed; whereas if it is death, then the population size becomes $i - 1$ and the additional time needed to reach $i + 1$ is equal to the time it takes to return to state i (and this has mean $E[T_{i-1}]$) plus the additional time it then takes to reach $i + 1$ (and this has mean $E[T_i]$). Hence, since the probability that the first transition is a birth is $\lambda_i/(\lambda_i + \mu_i)$, we see that

$$E[T_i] = \frac{1}{\lambda_i + \mu_i} + \frac{\mu_i}{\lambda_i + \mu_i}(E[T_{i-1}] + E[T_i])$$

or, equivalently,

$$E[T_i] = \frac{1}{\lambda_i} + \frac{\mu_i}{\lambda_i} E[T_{i-1}], \qquad i \geqslant 1$$

Starting with $E[T_0] = 1/\lambda_0$, the preceding yields an efficient method to successively compute $E[T_1]$, $E[T_2]$, and so on.

Suppose now that we wanted to determine the expected time to go from state i to state j where $i < j$. This can be accomplished using the preceding by noting that this quantity will equal $E[T_i] + E[T_{i+1}] + \cdots + E[T_{j-1}]$.

Example 6.7 For the birth and death process having parameters $\lambda_i \equiv \lambda$, $\mu_i \equiv \mu$,

$$E[T_i] = \frac{1}{\lambda} + \frac{\mu}{\lambda} E[T_{i-1}]$$

$$= \frac{1}{\lambda}(1 + \mu E[T_{i-1}])$$

Starting with $E[T_0] = 1/\lambda$, we see that

$$E[T_1] = \frac{1}{\lambda}\left(1 + \frac{\mu}{\lambda}\right),$$

$$E[T_2] = \frac{1}{\lambda}\left[1 + \frac{\mu}{\lambda} + \left(\frac{\mu}{\lambda}\right)^2\right]$$

and, in general,

$$E[T_i] = \frac{1}{\lambda}\left[1 + \frac{\mu}{\lambda} + \left(\frac{\mu}{\lambda}\right)^2 + \cdots + \left(\frac{\mu}{\lambda}\right)^i\right]$$

$$= \frac{1 - (\mu/\lambda)^{i+1}}{\lambda - \mu}, \qquad i \geqslant 0$$

The expected time to reach state j, starting at state k, $k < j$, is

$$E[\text{time to go from } k \text{ to } j] = \sum_{i=k}^{j-1} E[T_i]$$

$$= \frac{j-k}{\lambda - \mu} - \frac{(\mu/\lambda)^{k+1}}{\lambda - \mu} \frac{[1 - (\mu/\lambda)^{j-k}]}{1 - \mu/\lambda}$$

The foregoing assumes that $\lambda \neq \mu$. If $\lambda = \mu$, then

$$E[T_i] = \frac{i+1}{\lambda},$$

$$E[\text{time to go from } k \text{ to } j] = \frac{j(j+1) - k(k+1)}{2\lambda} \quad \blacklozenge$$

We can also compute the variance of the time to go from 0 to $i+1$ by utilizing the conditional varince formula (see Exercise 43 of Chapter 3). First note that Equation (6.3) can be written as

$$E[T_i | I_i] = \frac{1}{\lambda_i + \mu_i} + (1 - I_i)(E[T_{i-1}] + E[T_i])$$

so

$$\text{Var}(E[T_i | I_i]) = (E[T_{i-1}] + E[T_i])^2 \, \text{Var}(I_i)$$

$$= (E[T_{i-1}] + E[T_i])^2 \frac{\mu_i \lambda_i}{(\mu_i + \lambda_i)^2} \qquad (6.4)$$

where $\text{Var}(I_i)$ is as shown since I_i is a Bernoulli random variable with parameter $p = \lambda_i/(\lambda_i + \mu_i)$. Also, note that if we let X_i denote the time until the transition from i occurs, then

$$\text{Var}(T_i | I_i = 1) = \text{Var}(X_i | I_i = 1)$$

$$= \text{Var}(X_i)$$

$$= \frac{1}{(\lambda_i + \mu_i)^2} \qquad (6.5)$$

where the preceding uses the fact that the time until transition is independent of the next state visited. Also,

$$\text{Var}(T_i | I_i = 0)$$

$$= \text{Var}(X_i + \text{time to get back to } i + \text{time to then reach } i + 1)$$

$$= \text{Var}(X_i) + \text{Var}(T_{i-1}) + \text{Var}(T_i) \qquad (6.6)$$

where the foregoing uses the fact that the three random variables are independent. We can rewrite Equations (6.5) and (6.6) as

$$\text{Var}(T_i \mid I_i) = \text{Var}(X_i) + (1 - I_i)[\text{Var}(T_{i-1}) + \text{Var}(T_i)]$$

so

$$E[\text{Var}(T_i \mid I_i)] = \frac{1}{(\mu_i + \lambda_i)^2} + \frac{\mu_i}{\mu_i + \lambda_i}[\text{Var}(T_{i-1}) + \text{Var}(T_i)] \quad (6.7)$$

Hence, using the conditional variance formula, which states that $\text{Var}(T_i)$ is the sum of Equations (6.7) and (6.4), we obtain

$$\text{Var}(T_i) = \frac{1}{(\mu_i + \lambda_i)^2} + \frac{\mu_i}{\mu_i + \lambda_i}[\text{Var}(T_{i-1}) + \text{Var}(T_i)]$$

$$+ \frac{\mu_i \lambda_i}{(\mu_i + \lambda_i)^2}(E[T_{i-1}] + E[T_i])^2$$

or, equivalently,

$$\text{Var}(T_i) = \frac{1}{\lambda_i(\lambda_i + \mu_i)} + \frac{\mu_i}{\lambda_i}\text{Var}(T_{i-1}) + \frac{\mu_i}{\mu_i + \lambda_i}(E[T_{i-1}] + E[T_i])^2$$

Starting with $\text{Var}(T_0) = 1/\lambda_0^2$ and using the former recursion to obtain the expectations, we can recursively compute $\text{Var}(T_i)$. In addition, if we want the variance of the time to reach state j, starting from state k, $k < j$, then this can be expressed as the time to go from k to $k + 1$ plus the additional time to go from $k + 1$ to $k + 2$, and so on. Since, by the Markovian property, these successive random variables are independent, it follows that

$$\text{Var(time to go from } k \text{ to } j) = \sum_{i=k}^{j-1} \text{Var}(T_i)$$

6.4. The Transition Probability Function $P_{ij}(t)$

Let

$$P_{ij}(t) = P\{X(t + s) = j \mid X(s) = i\}$$

denote the probability that a process presently in state i will be in state j a time t later. These quantities are often called the *transition probabilities* of the continuous-time Markov chain.

We can explicitly determine $P_{ij}(t)$ in the case of a pure birth process having distinct birth rates. For such a process, let X_k denote the time the process spends in state k before making a transition into state $k + 1$, $k \geq 1$. Suppose that the process is presently in state i, and let $j > i$. Then, as X_i is the time it spends in state i before moving to state $i + 1$, and X_{i+1} is the time it then spends in state $i + 1$ before moving to state $i + 2$, and so on, it follows that $\sum_{k=i}^{j-1} X_k$ is the time it takes until the process enters state j. Now, if the process has not yet entered state j by time t, then its state at time t is smaller than j, and vice versa. That is,

$$X(t) < j \Leftrightarrow X_i + \cdots + X_{j-1} > t$$

Therefore, for $i < j$, we have for a pure birth process that

$$P\{X(t) < j \mid X(0) = i\} = P\left\{\sum_{k=i}^{j-1} X_k > t\right\}$$

However, since $X_i, \ldots, X_{j-1}$ are independent exponential random variables with respective rates $\lambda_i, \ldots, \lambda_{j-1}$, we obtain from the preceding and Equation (5.9), which gives the tail distribution function of $\sum_{k=i}^{j-1} X_k$, that

$$P\{X(t) < j \mid X(0) = i\} = \sum_{k=i}^{j-1} e^{-\lambda_k t} \prod_{r \neq k, r=i}^{j-1} \frac{\lambda_r}{\lambda_r - \lambda_k}$$

Replacing j by $j + 1$ in the preceding gives that

$$P\{X(t) < j + 1 \mid X(0) = i\} = \sum_{k=i}^{j} e^{-\lambda_k t} \prod_{r \neq k, r=i}^{j} \frac{\lambda_r}{\lambda_r - \lambda_k}$$

Since

$$P\{X(t) = j \mid X(0) = i\} = P\{X(t) < j + 1 \mid X(0) = i\} - P\{X(t) < j \mid X(0) = i\}$$

and since $P_{ii}(t) = P\{X_i > t\} = e^{-\lambda_i t}$, we have shown the following.

Proposition 6.1   For a pure birth process having $\lambda_i \neq \lambda_j$ when $i \neq j$

$$P_{ij}(t) = \sum_{k=i}^{j} e^{-\lambda_k t} \prod_{r \neq k, r=i}^{j} \frac{\lambda_r}{\lambda_r - \lambda_k} - \sum_{k=i}^{j-1} e^{-\lambda_k t} \prod_{r \neq k, r=i}^{j-1} \frac{\lambda_r}{\lambda_r - \lambda_k}, \qquad i < j$$

$$P_{ii}(t) = e^{-\lambda_i t}$$

Example 6.8   Consider the Yule process, which is a pure birth process in which each individual in the population independently gives birth at rate λ, and so $\lambda_n = n\lambda$, $n \geq 1$. Letting $i = 1$, we obtain from Proposition 6.1

$$P_{1j}(t) = \sum_{k=1}^{j} e^{-k\lambda t} \prod_{r \neq k, r=1}^{j} \frac{r}{r - k} - \sum_{k=1}^{j-1} e^{-k\lambda t} \prod_{r \neq k, r=1}^{j-1} \frac{r}{r - k}$$

$$= e^{-j\lambda t} \prod_{r=1}^{j-1} \frac{r}{r - j} + \sum_{k=1}^{j-1} e^{-k\lambda t} \left(\prod_{r \neq k, r=1}^{j} \frac{r}{r - k} - \prod_{r \neq k, r=1}^{j-1} \frac{r}{r - k} \right)$$

$$= e^{-j\lambda t}(-1)^{j-1} + \sum_{k=1}^{j-1} e^{-k\lambda t} \left(\frac{j}{j - k} - 1 \right) \prod_{r \neq k, r=1}^{j-1} \frac{r}{r - k}$$

Now,

$$\frac{k}{j-k} \prod_{r \neq k, r=1}^{j-1} \frac{r}{r-k} = \frac{(j-1)!}{(1-k)(2-k)\cdots(j-1-k)(j-k)!}$$

$$= (-1)^{k-1} \binom{j-1}{k-1}$$

so

$$P_{1j}(t) = \sum_{k=1}^{j} \binom{j-1}{k-1} e^{-k\lambda t}(-1)^{k-1}$$

$$= e^{-\lambda t} \sum_{i=0}^{j-1} \binom{j-1}{i} e^{-i\lambda t}(-1)^{i}$$

$$= e^{-\lambda t}(1 - e^{-\lambda t})^{j-1}$$

Thus, starting with a single individual, the population size at time t has a geometric distribution with mean $e^{\lambda t}$. If the population starts with i individuals, then we can regard each of these individuals as starting her own independent Yule process, and so the population at time t will be the sum of i independent and identically distributed geometric random variables with rate $e^{-\lambda t}$. But this means that the conditional distribution of $X(t)$, given that $X(0) = i$, is the same as the distribution of the number of times that a coin that lands heads on each flip with probability $e^{-\lambda t}$ must be flipped to amass a total of i heads. Hence, the population size at time t has a negative binomial distribution with parameters i and $e^{-\lambda t}$, so

$$P_{ij}(t) = \binom{j-1}{i-1} e^{-i\lambda t}(1 - e^{-\lambda t})^{j-i}, \qquad j \geq i \geq 1$$

[We could, of course, have used Proposition 6.1 to immediately obtain an equation for $P_{ij}(t)$, rather than just using it for $P_{1j}(t)$, but the algebra that would have then been needed to show the equivalence of the resulting expression to the preceding result is somewhat involved.] ✦

We shall now derive a set of differential equations that the transition probabilities $P_{ij}(t)$ satisfy in a general continuous-time Markov chain. However, first we need a definition and a pair of lemmas.
For any pair of states i and j, let

$$q_{ij} = v_i P_{ij}$$

Since v_i is the rate at which the process makes a transition when in state i and P_{ij} is the probability that this transition is into state j, it follows that q_{ij} is the rate, when in state i, at which the process makes a transition into state j. The quantities q_{ij} are called the *instantaneous transition rates*. Since

$$v_i = \sum_j v_i P_{ij} = \sum_j q_{ij}$$

and

$$P_{ij} = \frac{q_{ij}}{v_i} = \frac{q_{ij}}{\sum_j q_{ij}}$$

it follows that specifying the instantaneous transition rates determines the parameters of the continuous-time Markov chain.

Lemma 6.1

(a)

$$\lim_{h \to 0} \frac{1 - P_{ii}(h)}{h} = v_i$$

(b)

$$\lim_{h \to 0} \frac{P_{ij}(h)}{h} = q_{ij} \qquad \text{when } i \neq j$$

Proof   We first note that since the amount of time until a transition occurs is exponentially distributed it follows that the probability of two or more transitions in a time h is $o(h)$. Thus, $1 - P_{ii}(h)$, the probability that a process in state i at time 0 will not be in state i at time h, equals the probability that a transition occurs within time h plus something small compared to h. Therefore,

$$1 - P_{ii}(h) = v_i h + o(h)$$

and part (a) is proven. To prove part (b), we note that $P_{ij}(h)$, the probability that the process goes from state i to state j in a time h, equals the probability that a transition occurs in this time multiplied by the probability that the transition is into state j, plus something small compared to h. That is,

$$P_{ij}(h) = h v_i P_{ij} + o(h)$$

and part (b) is proven.   ✦

Lemma 6.2   For all $s \geqslant 0$, $t \geqslant 0$,

$$P_{ij}(t + s) = \sum_{k=0}^{\infty} P_{ik}(t) P_{kj}(s) \qquad (6.8)$$

Proof   In order for the process to go from state i to state j in time $t + s$, it must be somewhere at time t and thus

$$P_{ij}(t + s) \equiv P\{X(t + s) = j \mid X(0) = i\}$$

$$= \sum_{k=0}^{\infty} P\{X(t + s) = j, X(t) = k \mid X(0) = i\}$$

$$= \sum_{k=0}^{\infty} P\{X(t + s) = j \mid X(t) = k, X(0) = i\} \cdot P\{X(t) = k \mid X(0) = i\}$$

$$= \sum_{k=0}^{\infty} P\{X(t + s) = j \mid X(t) = k\} \cdot P\{X(t) = k \mid X(0) = i\}$$

$$= \sum_{k=0}^{\infty} P_{kj}(s)P_{ik}(t)$$

and the proof is completed. ✦

The set of equations (6.8) is known as the *Chapman–Kolmogorov* equations. From Lemma 6.2, we obtain

$$P_{ij}(h + t) - P_{ij}(t) = \sum_{k=0}^{\infty} P_{ik}(h)P_{kj}(t) - P_{ij}(t)$$

$$= \sum_{k \ne i} P_{ik}(h)P_{kj}(t) - [1 - P_{ii}(h)]P_{ij}(t)$$

and thus

$$\lim_{h \to 0} \frac{P_{ij}(t + h) - P_{ij}(t)}{h} = \lim_{h \to 0} \left\{ \sum_{k \ne i} \frac{P_{ik}(h)}{h} P_{kj}(t) - \left[\frac{1 - P_{ii}(h)}{h} \right] P_{ij}(t) \right\}$$

Now assuming that we can interchange the limit and the summation in the preceding and applying Lemma 6.1, we obtain

$$P'_{ij}(t) = \sum_{k \ne i} q_{ik} P_{kj}(t) - v_i P_{ij}(t)$$

It turns out that this interchange can indeed be justified and, hence, we have the following theorem.

Theorem 6.1   (Kolmogorov's Backward Equations): For all states i, j, and times $t \geqslant 0$,

$$P'_{ij}(t) = \sum_{k \ne i} q_{ik} P_{kj}(t) - v_i P_{ij}(t)$$

Example 6.9   The backward equations for the pure birth process become

$$P'_{ij}(t) = \lambda_i P_{i+1, j}(t) - \lambda_i P_{ij}(t) \quad ✦$$

Example 6.10 The backward equations for the birth and death process become

$$P'_{0j}(t) = \lambda_0 P_{1j}(t) - \lambda_0 P_{0j}(t),$$

$$P'_{ij}(t) = (\lambda_i + \mu_i)\left[\frac{\lambda_i}{\lambda_i + \mu_i} P_{i+1,j}(t) + \frac{\mu_i}{\lambda_i + \mu_i} P_{i-1,j}(t)\right] - (\lambda_i + \mu_i)P_{ij}(t)$$

or equivalently

$$P'_{0j}(t) = \lambda_0[P_{1j}(t) - P_{0j}(t)],$$

$$P'_{ij}(t) = \lambda_i P_{i+1,j}(t) + \mu_i P_{i-1,j}(t) - (\lambda_i + \mu_i)P_{ij}(t), \qquad i > 0 \qquad \blacklozenge \qquad (6.9)$$

Example 6.11 (A Continuous-Time Markov Chain Consisting of Two States): Consider a machine that works for an exponential amount of time having mean $1/\lambda$ before breaking down; and suppose that it takes an exponential amount of time having mean $1/\mu$ to repair the machine. If the machine is in working condition at time 0, then what is the probability that it will be working at time $t = 10$?

To answer this question, we note that the process is a birth and death process (with state 0 meaning that the machine is working and state 1 that it is being repaired) having parameters

$$\lambda_0 = \lambda \qquad\qquad \mu_1 = \mu$$
$$\lambda_i = 0, i \neq 0, \qquad \mu_i = 0, i \neq 1$$

We shall derive the desired probability, namely, $P_{00}(10)$ by solving the set of differential equations given in Example 6.10. From Equation (6.9), we obtain

$$P'_{00}(t) = \lambda[P_{10}(t) - P_{00}(t)], \qquad\qquad (6.10)$$

$$P'_{10}(t) = \mu P_{00}(t) - \mu P_{10}(t) \qquad\qquad (6.11)$$

Multiplying Equation (6.10) by μ and Equation (6.11) by λ and then adding the two equations yields

$$\mu P'_{00}(t) + \lambda P'_{10}(t) = 0$$

By integrating, we obtain

$$\mu P_{00}(t) + \lambda P_{10}(t) = c$$

However, since $P_{00}(0) = 1$ and $P_{10}(0) = 0$, we obtain $c = \mu$ and hence

$$\mu P_{00}(t) + \lambda P_{10}(t) = \mu \qquad\qquad (6.12)$$

or equivalently

$$\lambda P_{10}(t) = \mu[1 - P_{00}(t)]$$

By substituting this result in Equation (6.10), we obtain

$$P'_{00}(t) = \mu[1 - P_{00}(t)] - \lambda P_{00}(t)$$
$$= \mu - (\mu + \lambda)P_{00}(t)$$

Letting

$$h(t) = P_{00}(t) - \frac{\mu}{\mu + \lambda}$$

we have

$$h'(t) = \mu - (\mu + \lambda)\left[h(t) + \frac{\mu}{\mu + \lambda}\right]$$
$$= -(\mu + \lambda)h(t)$$

or

$$\frac{h'(t)}{h(t)} = -(\mu + \lambda)$$

By integrating both sides, we obtain

$$\log h(t) = -(\mu + \lambda)t + c$$

or

$$h(t) = Ke^{-(\mu + \lambda)t}$$

and thus

$$P_{00}(t) = Ke^{-(\mu + \lambda)t} + \frac{\mu}{\mu + \lambda}$$

which finally yields, by setting $t = 0$ and using the fact that $P_{00}(0) = 1$,

$$P_{00}(t) = \frac{\lambda}{\mu + \lambda}e^{-(\mu + \lambda)t} + \frac{\mu}{\mu + \lambda}$$

From Equation (6.12), this also implies that

$$P_{10}(t) = \frac{\mu}{\mu + \lambda} - \frac{\mu}{\mu + \lambda}e^{-(\mu + \lambda)t}$$

Hence, our desired probability $P_{00}(10)$ equals

$$P_{00}(10) = \frac{\lambda}{\mu + \lambda}e^{-10(\mu + \lambda)} + \frac{\mu}{\mu + \lambda} \quad \blacklozenge$$

Another set of differential equations, different from the backward equations, may also be derived. This set of equations, known as *Kolmogorov's forward equations* is derived as follows. From the Chapman–Kolmogorov equations (Lemma 6.2), we have

$$
P_{ij}(t + h) - P_{ij}(t) = \sum_{k=0}^{\infty} P_{ik}(t)P_{kj}(h) - P_{ij}(t)
$$

$$
= \sum_{k \neq j} P_{ik}(t)P_{kj}(h) - [1 - P_{jj}(h)]P_{ij}(t)
$$

and thus

$$
\lim_{h \to 0} \frac{P_{ij}(t + h) - P_{ij}(t)}{h} = \lim_{h \to 0} \left\{ \sum_{k \neq j} P_{ik}(t) \frac{P_{kj}(h)}{h} - \left[\frac{1 - P_{jj}(h)}{h} \right] P_{ij}(t) \right\}
$$

and, assuming that we can interchange limit with summation, we obtain from Lemma 6.1

$$
P'_{ij}(t) = \sum_{k \neq j} q_{kj} P_{ik}(t) - v_j P_{ij}(t)
$$

Unfortunately, we cannot always justify the interchange of limit and summation and thus the above is not always valid. However, they do hold in most models, including all birth and death processes and all finite state models. We thus have the following.

Theorem 6.2 (Kolmogorov's Forward Equations): Under suitable regularity conditions,

$$
P'_{ij}(t) = \sum_{k \neq j} q_{kj} P_{ik}(t) - v_j P_{ij}(t) \tag{6.13}
$$

We shall now solve the forward equations for the pure birth process. For this process, Equation (6.13) reduces to

$$
P'_{ij}(t) = \lambda_{j-1} P_{i,j-1}(t) - \lambda_j P_{ij}(t)
$$

However, by noting that $P_{ij}(t) = 0$ whenever $j < i$ (since no deaths can occur), we can rewrite the preceding equation to obtain

$$
P'_{ii}(t) = -\lambda_i P_{ii}(t),
$$
$$
P'_{ij} = \lambda_{j-1} P_{i,j-1}(t) - \lambda_j P_{ij}(t), \qquad j \geq i + 1 \tag{6.14}
$$

Proposition 6.2 For a pure birth process,

$$
P_{ii}(t) = e^{-\lambda_i t}, \qquad\qquad\qquad\qquad i \geq 0
$$

$$
P_{ij}(t) = \lambda_{j-1} e^{-\lambda_j t} \int_0^t e^{\lambda_j s} P_{i,j-1}(s)\, ds, \qquad j \geq i + 1
$$

Proof   The fact that $P_{ii}(t) = e^{-\lambda_i t}$ follows from Equation (6.14) by integrating and using the fact that $P_{ii}(0) = 1$. To prove the corresponding result for $P_{ij}(t)$, we note by Equation (6.14) that

$$e^{\lambda_j t}[P'_{ij}(t) + \lambda_j P_{ij}(t)] = e^{\lambda_j t}\lambda_{j-1}P_{i,j-1}(t)$$

or

$$\frac{d}{dt}[e^{\lambda_j t}P_{ij}(t)] = \lambda_{j-1}e^{\lambda_j t}P_{i,j-1}(t)$$

Hence, since $P_{ij}(0) = 0$, we obtain the desired results.   ◆

Example 6.12   (Forward Equations for Birth and Death Process): The forward equations (Equation 6.13) for the general birth and death process becomes

$$P'_{i0}(t) = \sum_{k \neq 0} q_{k0}P_{ik}(t) - \lambda_0 P_{i0}(t)$$

$$= \mu_1 P_{i1}(t) - \lambda_0 P_{i0}(t) \tag{6.15}$$

$$P'_{ij}(t) = \sum_{k \neq j} q_{kj}P_{ik}(t) - (\lambda_j + \mu_j)P_{ij}(t)$$

$$= \lambda_{j-1}P_{i,j-1}(t) + \mu_{j+1}P_{i,j+1}(t) - (\lambda_j + \mu_j)P_{ij}(t) \quad ◆ \tag{6.16}$$

6.5. Limiting Probabilities

In analogy with a basic result in discrete-time Markov chains, the probability that a continuous-time Markov chain will be in state j at time t often converges to a limiting value which is independent of the initial state. That is, if we call this value P_j, then

$$P_j \equiv \lim_{t \to \infty} P_{ij}(t)$$

where we are assuming that the limit exists and is independent of the initial state i.

To derive a set of equations for the P_j, consider first the set of forward equations

$$P'_{ij}(t) = \sum_{k \neq j} q_{kj}P_{ik}(t) - v_j P_{ij}(t) \tag{6.17}$$

Now, if we let t approach ∞, then assuming that we can interchange limit and summation, we obtain

$$
\begin{aligned}
\lim_{t \to \infty} P'_{ij}(t) &= \lim_{t \to \infty} \left[\sum_{k \neq j} q_{kj} P_{ik}(t) - v_j P_{ij}(t) \right] \\
&= \sum_{k \neq j} q_{kj} P_k - v_j P_j
\end{aligned}
$$

However, as $P_{ij}(t)$ is a bounded function (being a probability it is always between 0 and 1), it follows that if $P'_{ij}(t)$ converges, then it must converge to 0 (why is this?). Hence, we must have that

$$
0 = \sum_{k \neq j} q_{kj} P_k - v_j P_j
$$

or

$$
v_j P_j = \sum_{k \neq j} q_{kj} P_k, \qquad \text{all states } j \tag{6.18}
$$

The preceding set of equations, along with this equation

$$
\sum_j P_j = 1 \tag{6.19}
$$

can be used to solve for the limiting probabilities.

Remarks   (i) We have assumed that the limiting probabilities P_j exist. A sufficient condition for this is that

(a) all states of the Markov chain communicate in the sense that starting in state i there is a positive probability of ever being in state j, for all i, j and

(b) the Markov chain is positive recurrent in the sense that, starting in any state, the mean time to return to that state is finite.

If conditions (a) and (b) hold, then the limiting probabilities will exist and satisfy Equations (6.18) and (6.19). In addition, P_j also will have the interpretation of being the long-run proportion of time that the process is in state j.

(ii) Equations (6.18) and (6.19) have a nice interpretation: In any interval $(0, t)$ the number of transitions into state j must equal to within 1 the number of transitions out of state j (why?). Hence, in the long run, the rate at which transitions into state j occur must equal the rate at which transitions out of state j occur. Now when the process is in state j, it leaves at rate v_j, and, as P_j is the proportion of time it is in state j, it thus follows that

$$
v_j P_j = \text{rate at which the process leaves state } j
$$

Similarly, when the process is in state k, it enters j at a rate q_{kj}. Hence, as P_k is the proportion of time in state k, we see that the rate at which transitions from k to j occur is just $q_{kj}P_k$; thus

$$\sum_{k \neq j} q_{kj}P_k = \text{rate at which the process enters state } j$$

So, Equation (6.18) is just a statement of the equality of the rates at which the process enters and leaves state j. Because it balances (that is, equates) these rates, the equations (6.18) are sometimes referred to as "balance equations."

(iii) When the limiting probabilities P_j exist, we say that the chain is ergodic. The P_j are sometimes called stationary probabilities since it can be shown that (as in the discrete-time case) if the initial state is chosen according to the distribution $\{P_j\}$, then the probability of being in state j at time t is P_j, for all t.

Let us now determine the limiting probabilities for a birth and death process. From Equation (6.18) or equivalently, by equating the rate at which the process leaves a state with the rate at which it enters that state, we obtain

State	*Rate at which leave = rate at which enter*
0	$\lambda_0 P_0 = \mu_1 P_1$
1	$(\lambda_1 + \mu_1)P_1 = \mu_2 P_2 + \lambda_0 P_0$
2	$(\lambda_2 + \mu_2)P_2 = \mu_3 P_3 + \lambda_1 P_1$
$n, n \geqslant 1$	$(\lambda_n + \mu_n)P_n = \mu_{n+1}P_{n+1} + \lambda_{n-1}P_{n-1}$

By adding to each equation the equation preceding it, we obtain

$$\lambda_0 P_0 = \mu_1 P_1,$$
$$\lambda_1 P_1 = \mu_2 P_2,$$
$$\lambda_2 P_2 = \mu_3 P_3,$$
$$\vdots$$
$$\lambda_n P_n = \mu_{n+1}P_{n+1}, \qquad n \geqslant 0$$

Solving in terms of P_0 yields

$$P_1 = \frac{\lambda_0}{\mu_1}P_0,$$

$$P_2 = \frac{\lambda_1}{\mu_2}P_1 = \frac{\lambda_1 \lambda_0}{\mu_2 \mu_1}P_0,$$

$$P_3 = \frac{\lambda_2}{\mu_3}P_2 = \frac{\lambda_2 \lambda_1 \lambda_0}{\mu_3 \mu_2 \mu_1}P_0,$$

$$\vdots$$

$$P_n = \frac{\lambda_{n-1}}{\mu_n}P_{n-1} = \frac{\lambda_{n-1}\lambda_{n-2} \cdots \lambda_1 \lambda_0}{\mu_n \mu_{n-1} \cdots \mu_2 \mu_1}P_0$$

And by using the fact that $\sum_{n=0}^{\infty} P_n = 1$, we obtain

$$1 = P_0 + P_0 \sum_{n=1}^{\infty} \frac{\lambda_{n-1} \cdots \lambda_1 \lambda_0}{\mu_n \cdots \mu_2 \mu_1}$$

or

$$P_0 = \frac{1}{1 + \sum_{n=1}^{\infty} \dfrac{\lambda_0 \lambda_1 \cdots \lambda_{n-1}}{\mu_1 \mu_2 \cdots \mu_n}}$$

and so

$$P_n = \frac{\lambda_0 \lambda_1 \cdots \lambda_{n-1}}{\mu_1 \mu_2 \cdots \mu_n \left(1 + \sum_{n=1}^{\infty} \dfrac{\lambda_0 \lambda_1 \cdots \lambda_{n-1}}{\mu_1 \mu_2 \cdots \mu_n}\right)}, \qquad n \geq 1 \qquad (6.20)$$

The foregoing equations also show us what condition is necessary for these limiting probabilities to exist. Namely, it is necessary that

$$\sum_{n=1}^{\infty} \frac{\lambda_0 \lambda_1 \cdots \lambda_{n-1}}{\mu_1 \mu_2 \cdots \mu_n} < \infty \qquad (6.21)$$

This condition also may be shown to be sufficient.

In the multiserver exponential queueing system (Example 6.6), Condition (6.21) reduces to

$$\sum_{n=s+1}^{\infty} \frac{\lambda^n}{(s\mu)^n} < \infty$$

which is equivalent to $\lambda/s\mu < 1$.

For the linear growth model with immigration (Example 6.4), Condition (6.21) reduces to

$$\sum_{n=1}^{\infty} \frac{\theta(\theta + \lambda) \cdots (\theta + (n-1)\lambda)}{n! \mu^n} < \infty$$

Using the ratio test, the preceding will converge when

$$\lim_{n \to \infty} \frac{\theta(\theta + \lambda) \cdots (\theta + n\lambda)}{(n+1)! \mu^{n+1}} \frac{n! \mu^n}{\theta(\theta + \lambda) \cdots (\theta + (n-1)\lambda)} \equiv \lim_{n \to \infty} \frac{\theta + n\lambda}{(n+1)\mu}$$

$$\equiv \frac{\lambda}{\mu} < 1$$

That is, the condition is satisfied when $\lambda < \mu$. When $\lambda \geq \mu$ it is easy to show that condition (6.21) is not satisfied.

Example 6.13 (A Machine Repair Model): Consider a job shop that consists of M machines and one serviceman. Suppose that the amount of time each machine runs before breaking down is exponentially distributed with mean $1/\lambda$, and suppose that the amount of time that it takes for the serviceman to fix a machine is exponentially distributed with mean $1/\mu$. We shall attempt to answer these questions: (a) What is the average number of machines not in use? (b) What proportion of time is each machine in use?

Solution: If we say that the system is in state n whenever n machines are not in use, then the above is a birth and death process having parameters

$$\mu_n = \mu \qquad\qquad\qquad n \geqslant 1$$

$$\lambda_n = \begin{cases} (M - n)\lambda, & n \leqslant M \\ 0, & n > M \end{cases}$$

This is so in the sense that a failing machine is regarded as an arrival and a fixed machine as a departure. If any machines are broken down, then since the serviceman's rate is μ, $\mu_n = \mu$. On the other hand, if n machines are not in use, then since the $M - n$ machines in use each fail at a rate λ, it follows that $\lambda_n = (M - n)\lambda$. From Equation (6.20) we have that P_n, the probability that n machines will not be in use, is given by

$$P_0 = \frac{1}{1 + \sum_{n=1}^{M} [M\lambda(M - 1)\lambda \cdots (M - n + 1)\lambda/\mu^n]}$$

$$= \frac{1}{1 + \sum_{n=1}^{M} (\lambda/\mu)^n M!/(M - n)!}$$

$$P_n = \frac{(\lambda/\mu)^n M!/(M - n)!}{1 + \sum_{n=1}^{M} (\lambda/\mu)^n M!/(M - n)!}, \qquad n = 0, 1, \ldots, M$$

Hence, the average number of machines not in use is given by

$$\sum_{n=0}^{M} nP_n = \frac{\sum_{n=0}^{M} n(M!/(M - n)!)(\lambda/\mu)^n}{1 + \sum_{n=1}^{M} (\lambda/\mu)^n M!/(M - n)!} \qquad (6.22)$$

To obtain the long-run proportion of time that a given machine is working we will compute the equivalent limiting probability of its working. To do so, we condition on the number of machines that are not working to obtain

$$P\{\text{machine is working}\} = \sum_{n=0}^{M} P\{\text{machine is working} \,|\, n \text{ not working}\} P_n$$

$$= \sum_{n=0}^{M} \frac{M - n}{M} P_n \qquad \begin{array}{l} \text{(since if } n \text{ are not working,} \\ \text{then } M - n \text{ are working!)} \end{array}$$

$$= 1 - \sum_{0}^{M} \frac{nP_n}{M}$$

where $\sum_{0}^{M} nP_n$ is given by Equation (6.22). ✦

Example 6.14   (The $M/M/1$ Queue): In the $M/M/1$ queue $\lambda_n = \lambda$, $\mu_n = \mu$ and thus, from Equation (6.20),

$$P_n = \frac{(\lambda/\mu)^n}{1 + \sum_{n=1}^{\infty} (\lambda/\mu)^n}$$

$$= (\lambda/\mu)^n \, (1 - \lambda/\mu), \qquad n \geqslant 0$$

provided that $\lambda/\mu < 1$. It is intuitive that λ must be less than μ for limiting probabilities to exist. Customers arrive at rate λ and are served at rate μ, and thus if $\lambda > \mu$, then they arrive at a faster rate than they can be served and the queue size will go to infinity. The case $\lambda = \mu$ behaves much like the symmetric random walk of Section 4.3, which is null recurrent and thus has no limiting probabilities. ✦

Example 6.15   Let us reconsider the shoeshine shop of Example 6.1, and determine the proportion of time the process is in each of the states $0, 1, 2$. Because this is not a birth and death process (since the process can go directly from state 2 to state 0), we start with the balance equations for the limiting probabilities.

State	Rate that the process leaves = rate that the process enters
0	$\lambda P_0 = \mu_2 P_2$
1	$\mu_1 P_1 = \lambda P_0$
2	$\mu_2 P_2 = \mu_1 P_1$

Solving in terms of P_0 yields

$$P_2 = \frac{\lambda}{\mu_2} P_0, \qquad P_1 = \frac{\lambda}{\mu_1} P_0$$

which implies, since $P_0 + P_1 + P_2 = 1$, that

$$P_0 \left[1 + \frac{\lambda}{\mu_2} + \frac{\lambda}{\mu_1} \right] = 1$$

or

$$P_0 = \frac{\mu_1 \mu_2}{\mu_1 \mu_2 + \lambda(\mu_1 + \mu_2)}$$

and

$$P_1 = \frac{\lambda\mu_2}{\mu_1\mu_2 + \lambda(\mu_1 + \mu_2)},$$

$$P_2 = \frac{\lambda\mu_1}{\mu_1\mu_2 + \lambda(\mu_1 + \mu_2)} \quad \blacklozenge$$

Example 6.16 Consider a set of n components along with a single repairman. Suppose that component i functions for an exponentially distributed time with rate λ_i and then fails. The time it then takes to repair component i is exponential with rate μ_i, $i = 1, \ldots, n$. Suppose that when there is more than one failed component the repairman always works on the most recent failure. For instance, if there are at present two failed components — say, components 1 and 2 of which one has failed most recently — then the repairman will be working on component 1. However, if component 3 should fail before 1's repair is completed, then the repairman would stop working on component 1 and switch to component 3 (that is, a newly failed component preempts service).

To analyze the preceding as a continuous-time Markov chain, the state must represent the set of failed components in the order of failure. That is, the state will be $i_1, i_2, \ldots, i_k$ if $i_1, i_2, \ldots, i_k$ are the k failed components (all the other $n - k$ being functional) with i_1 having been the most recent failure (and is thus presently being repaired), i_2 the second most recent, and so on. Because there are $k!$ possible orderings for a fixed set of k failed components and $\binom{n}{k}$ choices of that set, it follows that there are

$$\sum_{k=0}^{n} \binom{n}{k} k! = \sum_{k=0}^{n} \frac{n!}{(n-k)!} = n! \sum_{i=0}^{n} \frac{1}{i!}$$

possible states.

The balance equations for the limiting probabilities are as follows:

$$\left(\mu_{i_1} + \sum_{\substack{i \neq i_j \\ j=1,\ldots,k}} \lambda_i\right) P(i_1, \ldots, i_k) = \sum_{\substack{i \neq i_j \\ j=1,\ldots,k}} P(i, i_1, \ldots, i_k)\mu_i + P(i_2, \ldots, i_k)\lambda_{i_1},$$

$$\sum_{i=1}^{n} \lambda_i P(\phi) = \sum_{i=1}^{n} P(i)\mu_i \tag{6.23}$$

where ϕ is the state when all components are working. The preceding equations follow because state $i_1, \ldots, i_k$ can be left either by a failure of any of the additional components or by a repair completion of component i_1. Also that state can be entered either by a repair completion of component

i when the state is $i, i_1, \ldots, i_k$ or by a failure of component i_1 when the state is $i_2, \ldots, i_k$.

However, if we take

$$P(i_1, \ldots, i_k) = \frac{\lambda_{i_1} \lambda_{i_2} \cdots \lambda_{i_k}}{\mu_{i_1} \mu_{i_2} \cdots \mu_{i_k}} P(\phi) \tag{6.24}$$

then it is easily seen that Equations (6.23) are satisfied. Hence, by uniqueness these must be the limiting probabilities with $P(\phi)$ determined to make their sum equal 1. That is,

$$P(\phi) = \left[1 + \sum_{i_1, \ldots, i_k} \frac{\lambda_{i_1} \cdots \lambda_{i_k}}{\mu_{i_1} \cdots \mu_{i_k}} \right]^{-1}$$

As an illustration, suppose $n = 2$ and so there are 5 states $\phi, 1, 2, 12, 21$. Then from the preceding we would have

$$P(\phi) = \left[1 + \frac{\lambda_1}{\mu_1} + \frac{\lambda_2}{\mu_2} + \frac{2\lambda_1 \lambda_2}{\mu_1 \mu_2} \right]^{-1},$$

$$P(1) = \frac{\lambda_1}{\mu_1} P(\phi),$$

$$P(2) = \frac{\lambda_2}{\mu_2} P(\phi),$$

$$P(1, 2) = P(2, 1) = \frac{\lambda_1 \lambda_2}{\mu_1 \mu_2} P(\phi)$$

It is interesting to note, using Equation (6.24), that given the set of failed components, each of the possible orderings of these components is equally likely. ✦

6.6. Time Reversibility

Consider a continuous-time Markov chain that is ergodic and let us consider the limiting probabilities P_i from a different point of view than previously. If we consider the sequence of states visited, ignoring the amount of time spent in each state during a visit, then this sequence constitutes a discrete-time Markov chain with transition probabilities P_{ij}. Let us assume that this discrete-time Markov chain, called the embedded chain, is ergodic and denote by π_i its limiting probabilities. That is, the π_i are the unique solution of

$$\pi_i = \sum_j \pi_j P_{ji}, \qquad \text{all } i$$

$$\sum_i \pi_i = 1$$

Now since π_i represents the proportion of transitions that take the process into state i, and because $1/v_i$ is the mean time spent in state i during a visit, it seems intuitive that P_i, the proportion of time in state i, should be a weighted average of the π_i where π_i is weighted proportionately to $1/v_i$. That is, it is intuitive that

$$P_i = \frac{\pi_i/v_i}{\Sigma_j \pi_j/v_j} \qquad (6.25)$$

To check the preceding, recall that the limiting probabilities P_i must satisfy

$$v_i P_i = \sum_{j \neq i} P_j q_{ji}, \qquad \text{all } i$$

or equivalently, since $P_{ii} = 0$

$$v_i P_i = \sum_j P_j v_j P_{ji}, \qquad \text{all } i$$

Hence, for the P_is to be given by Equation (6.25), the following would be necessary:

$$\pi_i = \sum_j \pi_j P_{ji}, \qquad \text{all } i$$

But this, of course, follows since it is in fact the defining equation for the π_is.

Suppose now that the continuous-time Markov chain has been in operation for a long time, and suppose that starting at some (large) time T we trace the process going backward in time. To determine the probability structure of this reversed process, we first note that given we are in state i at some time — say, t — the probability that we have been in this state for an amount of time greater than s is just $e^{-v_i s}$. This is so, since

$$P\{\text{process is in state } i \text{ throughout } [t-s,t] \mid X(t) = i\}$$

$$= \frac{P\{\text{process is in state } i \text{ throughout } [t-s,t]\}}{P\{X(t) = i\}}$$

$$= \frac{P\{X(t-s) = i\}e^{-v_i s}}{P\{X(t) = i\}}$$

$$= e^{-v_i s}$$

since for t large $P\{X(t-s) = i\} = P\{X(t) = i\} = P_i$.

In other words, going backward in time, the amount of time the process spends in state i is also exponentially distributed with rate v_i. In addition, as was shown in Section 4.8, the sequence of states visited by the reversed process constitutes a discrete-time Markov chain with transition probabilities Q_{ij} given by

$$Q_{ij} = \frac{\pi_j P_{ji}}{\pi_i}$$

Hence, we see from the preceding that the reversed process is a continuous-time Markov chain with the same transition rates as the forward-time process and with one-stage transition probabilities Q_{ij}. Therefore, the continuous-time Markov chain will be *time reversible,* in the sense that the process reversed in time has the same probabilistic structure as the original process, if the embedded chain is time reversible. That is, if

$$\pi_i P_{ij} = \pi_j P_{ji}, \qquad \text{for all } i, j$$

Now using the fact that $P_i = (\pi_i/v_i)/(\Sigma_j \pi_j/v_j)$, we see that the preceding condition is equivalent to

$$P_i q_{ij} = P_j q_{ji}, \qquad \text{for all } i, j \qquad (6.26)$$

Since P_i is the proportion of time in state i and q_{ij} is the rate when in state i that the process goes to j, the condition of time reversibility is that *the rate at which the process goes directly from state i to state j is equal to the rate at which it goes directly from j to i.* It should be noted that this is exactly the same condition needed for an ergodic discrete-time Markov chain to be time reversible (see Section 4.8).

An application of the preceding condition for time reversibility yields the following proposition concerning birth and death processes.

Proposition 6.3   An ergodic birth and death process is time reversible.

Proof   We must show that the rate at which a birth and death process goes from state i to state $i + 1$ is equal to the rate at which it goes from $i + 1$ to i. Now in any length of time t the number of transitions from i to $i + 1$ must equal to within 1 the number from $i + 1$ to i (since between each transition from i to $i + 1$ the process must return to i, and this can only occur through $i + 1$, and vice versa). Hence, as the number of such transitions goes to infinity as $t \to \infty$, it follows that the rate of transitions from i to $i + 1$ equals the rate from $i + 1$ to i.   ◆

Proposition 6.3 can be used to prove the important result that the output process of an $M/M/s$ queue is a Poisson process. We state this as a corollary.

Corollary 6.4   Consider an $M/M/s$ queue in which customers arrive in accordance with a Poisson process having rate λ and are served by any one of s servers—each having an exponentially distributed service time with

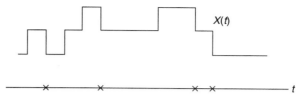

x = times at which going backward in time, $X(t)$ increases
 = times at which going forward in time, $X(t)$ decreases

Figure 6.1.

rate μ. If $\lambda < s\mu$, then the output process of customers departing is, after the process has been in operation for a long time, a Poisson process with rate λ.

Proof   Let $X(t)$ denote the number of customers in the system at time t. Since the $M/M/s$ process is a birth and death process, it follows from Proposition 6.3 that $\{X(t), t \geq 0\}$ is time reversible. Now going forward in time, the time points at which $X(t)$ increases by 1 constitute a Poisson process since these are just the arrival times of customers. Hence, by time reversibility the time points at which the $X(t)$ increases by 1 when we go backward in time also constitute a Poisson process. But these latter points are exactly the points of time when customers depart. (See Figure 6.1.) Hence, the departure times constitute a Poisson process with rate λ. ✦

We have shown that a process is time reversible if and only if

$$P_i q_{ij} = P_j q_{ji} \qquad \text{for all } i \neq j$$

Analogous to the result for discrete-time Markov chains, if one can find a probability vector **P** that satisfies the preceding then the Markov chain is time reversible and the P_is are the long-run probabilities. That is, we have the following proposition.

Proposition 6.5   If for some set $\{P_i\}$

$$\sum_i P_i = 1, \qquad P_i \geq 0$$

and

$$P_i q_{ij} = P_j q_{ji} \qquad \text{for all } i \neq j \qquad (6.27)$$

then the continuous-time Markov chain is time reversible and P_i represents the limiting probability of being in state i.

Proof   For fixed i we obtain upon summing Equation (6.27) over all $j : j \neq i$

$$\sum_{j \neq i} P_i q_{ij} = \sum_{j \neq i} P_j q_{ji}$$

or, since $\sum_{j \neq i} q_{ij} = v_i$,

$$v_i P_i = \sum_{j \neq i} P_j q_{ji}$$

Hence, the P_is satisfy the balance equations and thus represent the limiting probabilities. Because Equation (6.27) holds, the chain is time reversible. ◆

Example 6.17   Consider a set of n machines and a single repair facility to service them. Suppose that when machine $i, i = 1, \ldots, n$, goes down it requires an exponentially distributed amount of work with rate μ_i to get it back up. The repair facility divides its efforts equally among all down components in the sense that whenever there are k down machines $1 \leqslant k \leqslant n$ each receives work at a rate of $1/k$ per unit time. Finally, suppose that each time machine i goes back up it remains up for an exponentially distributed time with rate λ_i.

The preceding can be analyzed as a continuous-time Markov chain having 2^n states where the state at any time corresponds to the set of machines that are down at that time. Thus, for instance, the state will be $(i_1, i_2, \ldots, i_k)$ when machines $i_1, \ldots, i_k$ are down and all the others are up. The instantaneous transition rates are as follows:

$$q_{(i_1, \ldots, i_{k-1}), (i_1, \ldots, i_k)} = \lambda_{i_k},$$

$$q_{(i_1, \ldots, i_k), (i_1, \ldots, i_{k-1})} = \mu_{i_k}/k$$

where $i_1, \ldots, i_k$ are all distinct. This follows since the failure rate of machine i_k is always λ_{i_k} and the repair rate of machine i_k when there are k failed machines is μ_{i_k}/k.

Hence the time reversible equations (6.27) are

$$P(i_1, \ldots, i_k) \mu_{i_k}/k = P(i_1, \ldots, i_{k-1}) \lambda_{i_k}$$

or

$$
\begin{aligned}
P(i_1, \ldots, i_k) &= \frac{k \lambda_{i_k}}{\mu_{i_k}} P(i_1, \ldots, i_{k-1}) \\
&= \frac{k \lambda_{i_k}}{\mu_{i_k}} \frac{(k-1) \lambda_{i_{k-1}}}{\mu_{i_{k-1}}} P(i_1, \ldots, i_{k-2}) \qquad \text{upon iterating} \\
&\vdots \\
&= k! \prod_{j=1}^{k} (\lambda_{i_j}/\mu_{i_j}) P(\phi)
\end{aligned}
$$

where ϕ is the state in which all components are working. Because

$$P(\phi) + \Sigma P(i_1, \ldots, i_k) = 1$$

we see that

$$P(\phi) = \left[1 + \sum_{i_1, \ldots, i_k} k! \prod_{j=1}^{k} (\lambda_{i_j}/\mu_{i_j}) \right]^{-1} \qquad (6.28)$$

where the above sum is over all the $2^n - 1$ nonempty subsets $\{i_1, \ldots, i_k\}$ of $\{1, 2, \ldots, n\}$. Hence as the time reversible equations are satisfied for this choice of probability vector it follows from Proposition 6.5 that the chain is time reversible and

$$P(i_1, \ldots, i_k) = k! \prod_{j=1}^{k} (\lambda_{i_j}/\mu_{i_j})P(\phi)$$

with $P(\phi)$ being given by (6.28).

For instance, suppose there are two machines. Then, from the preceding we would have

$$P(\phi) = \frac{1}{1 + \lambda_1/\mu_1 + \lambda_2/\mu_2 + 2\lambda_1\lambda_2/\mu_1\mu_2},$$

$$P(1) = \frac{\lambda_1/\mu_1}{1 + \lambda_1/\mu_1 + \lambda_2/\mu_2 + 2\lambda_1\lambda_2/\mu_1\mu_2},$$

$$P(2) = \frac{\lambda_2/\mu_2}{1 + \lambda_1/\mu_1 + \lambda_2/\mu_2 + 2\lambda_1\lambda_2/\mu_1\mu_2},$$

$$P(1, 2) = \frac{2\lambda_1\lambda_2}{\mu_1\mu_2[1 + \lambda_1/\mu_1 + \lambda_2/\mu_2 + 2\lambda_1\lambda_2/\mu_1\mu_2]} \quad \blacklozenge$$

Consider a continuous-time Markov chain whose state space is S. We say that the Markov chain is truncated to the set $A \subset S$ if q_{ij} is changed to 0 for all $i \in A$, $j \notin A$. That is, transitions out of the class A are no longer allowed, whereas ones in A continue at the same rates as before. A useful result is that if the chain is time reversible, then so is the truncated one.

Proposition 6.6 A time reversible chain with limiting probabilities P_j, $j \in S$, that is truncated to the set $A \subset S$ and remains irreducible is also time reversible and has limiting probabilities P_j^A given by

$$P_j^A = \frac{P_j}{\sum_{i \in A} P_i}, \quad j \in A$$

Proof By Proposition 6.5 we need to show that, with P_j^A as given,

$$P_i^A q_{ij} = P_j^A q_{ji} \quad \text{for } i \in A, j \in A$$

or, equivalently,

$$P_i q_{ij} = P_j q_{ji} \quad \text{for } i \in A, j \in A$$

But this follows since the original chain is, by assumption, time reversible. $\blacklozenge$

Example 6.18   Consider an $M/M/1$ queue in which arrivals finding N in the system do not enter. This finite capacity system can be regarded as a truncation of the $M/M/1$ queue to the set of states $A = \{0, 1, \ldots, N\}$. Since the number in the system in the $M/M/1$ queue is time reversible and has limiting probabilities $P_j = (\lambda/\mu)^j (1 - \lambda/\mu)$ it follows from Proposition 6.6 that the finite capacity model is also time reversible and has limiting probabilities given by

$$P_j = \frac{(\lambda/\mu)^j}{\sum_{i=0}^{N} (\lambda/\mu)^i}, \qquad j = 0, 1, \ldots, N \quad \blacklozenge$$

Another useful result is given by the following proposition, whose proof is left as an exercise.

Proposition 6.7   If $\{X_i(t), t \geq 0\}$ are, for $i = 1, \ldots, n$, independent time reversible continuous-time Markov chains, then the vector process $\{(X_1(t), \ldots, X_n(t)), t \geq 0\}$ is also a time reversible continuous-time Markov chain.

Example 6.19   Consider an n-component system where component $i, i = 1, \ldots, n$, functions for an exponential time with rate λ_i and then fails; upon failure, repair begins on component i, with the repair taking an exponentially distributed time with rate μ_i. Once repaired, a component is as good as new. The components act independently except that when there is only one working component the system is temporarily shut down until a repair has been completed; it then starts up again with two working components.

(a) What proportion of time is the system shut down?
(b) What is the (limiting) averaging number of components that are being repaired?

Solution: Consider first the system without the restriction that it is shut down when a single component is working. Letting $X_i(t)$, $i = 1, \ldots, n$, equal 1 if component i is working at time t and 0 if it is failed, then $\{X_i(t), t \geq 0\}$, $i = 1, \ldots, n$, are independent birth and death processes. Because a birth and death process is time reversible, it follows from Proposition 6.7 that the process $\{(X_1(t), \ldots, X_n(t)), t \geq 0\}$ is also time reversible. Now, with

$$P_i(j) = \lim_{t \to \infty} P\{X_i(t) = j\}, \qquad j = 0, 1$$

we have

$$P_i(1) = \frac{\mu_i}{\mu_i + \lambda_i}, \qquad P_i(0) = \frac{\lambda_i}{\mu_i + \lambda_i}$$

Also, with

$$P(j_1, \ldots, j_n) = \lim_{t \to \infty} P\{X_i(t) = j_i, i = 1, \ldots, n\}$$

it follows, by independence, that

$$P(j_1, \ldots, j_n) = \prod_{i=1}^{n} P_i(j_i), \qquad j_i = 0, 1, \qquad i = 1, \ldots, n$$

Now, note that shutting down the system when only one component is working is equivalent to truncating the preceding unconstrained system to the set consisting of all states except the one having all components down. Therefore, with P_T denoting a probability for the truncated system, we have from Proposition 6.6 that

$$P_T(j_1, \ldots, j_n) = \frac{P(j_1, \ldots, j_n)}{1 - C}, \qquad \sum_{i=1}^{n} j_i > 0$$

where

$$C = P(0, \ldots, 0) = \prod_{j=1}^{n} \lambda_j/(\mu_j + \lambda_j)$$

Hence, letting $(\mathbf{0}, 1_i) = (0, \ldots, 0, 1, 0, \ldots, 0)$ be the n vector of zeroes and ones whose single 1 is in the ith place, we have

$$P_T(\text{system is shut down}) = \sum_{i=1}^{n} P_T(\mathbf{0}, 1_i)$$

$$= \frac{1}{1 - C} \sum_{i=1}^{n} \left(\frac{\mu_i}{\mu_i + \lambda_i} \right) \Pi_{j \neq i} \left(\frac{\lambda_j}{\mu_j + \lambda_j} \right)$$

$$= \frac{C \sum_{i=1}^{n} \mu_i/\lambda_i}{1 - C}$$

Let R denote the number of components being repaired. Then with I_i equal to 1 if component i is being repaired and 0 otherwise, we have for the unconstrained (nontruncated) system that

$$E[R] = E\left[\sum_{i=1}^{n} I_i \right] = \sum_{i=1}^{n} P_i(0) = \sum_{i=1}^{n} \lambda_i/(\mu_i + \lambda_i)$$

But, in addition,

$E[R] = E[R \mid$ all components are in repair$]C + E[R \mid$ not all components
are in repair$](1 - C)$

$= nC + E_T[R](1 - C)$

implying that

$$E_T[R] = \frac{\Sigma_{i=1}^n \lambda_i/(\mu_i + \lambda_i) - nC}{1 - C} \quad \blacklozenge$$

6.7. Uniformization

Consider a continuous-time Markov chain in which the mean time spent in
a state is the same for all states. That is, suppose that $v_i = v$, for all states i.
In this case since the amount of time spent in each state during a visit is
exponentially distributed with rate v, it follows that if we let $N(t)$ denote the
number of state transitions by time t, then $\{N(t), t \geqslant 0\}$ will be a Poisson
process with rate v.

To compute the transition probabilities $P_{ij}(t)$, we can condition on $N(t)$:

$$P_{ij}(t) = P\{X(t) = j \mid X(0) = i\}$$

$$= \sum_{n=0}^{\infty} P\{X(t) = j \mid X(0) = i, N(t) = n\}P\{N(t) = n \mid X(0) = i\}$$

$$= \sum_{n=0}^{\infty} P\{X(t) = j \mid X(0) = i, N(t) = n\}e^{-vt}\frac{(vt)^n}{n!}$$

Now the fact that there have been n transitions by time t tells us something
about the amounts of time spent in each of the first n states visited, but since
the distribution of time spent in each state is the same for all states, it follows
that knowing that $N(t) = n$ gives us no information about which states were
visited. Hence,

$$P\{X(t) = j \mid X(0) = i, N(t) = n\} = P_{ij}^n$$

where P_{ij}^n is just the n-stage transition probability associated with the
discrete-time Markov chain with transition probabilities P_{ij}; and so when
$v_i \equiv v$

$$P_{ij}(t) = \sum_{n=0}^{\infty} P_{ij}^n e^{-vt}\frac{(vt)^n}{n!} \tag{6.29}$$

Equation (6.29) is often useful from a computational point of view since
it enables us to approximate $P_{ij}(t)$ by taking a partial sum and then

computing (by matrix multiplication of the transition probability matrix) the relevant n stage probabilities P_{ij}^n.

Whereas the applicability of Equation (6.29) would appear to be quite limited since it supposes that $v_i \equiv v$, it turns out that most Markov chains can be put in that form by the trick of allowing fictitious transitions from a state to itself. To see how this works, consider any Markov chain for which the v_i are bounded, and let v be any number such that

$$v_i \leqslant v, \quad \text{for all } i \tag{6.30}$$

Now when in state i, the process actually leaves at rate v_i; but this is equivalent to supposing that transitions occur at rate v, but only the fraction v_i/v of transitions are real ones (and thus real transitions occur at rate v_i) and the remaining fraction $1 - v_i/v$ are fictitious transitions which leave the process in state i. In other words, any Markov chain satisfying condition (6.30) can be thought of as being a process that spends an exponential amount of time with rate v in state i and then makes a transition to j with probability P_{ij}^*, where

$$P_{ij}^* = \begin{cases} 1 - \dfrac{v_i}{v}, & j = i \\ \dfrac{v_i}{v} P_{ij}, & j \neq i \end{cases} \tag{6.31}$$

Hence, from Equation (6.29) we have that the transition probabilities can be computed by

$$P_{ij}(t) = \sum_{n=0}^{\infty} P_{ij}^{*n} e^{-vt} \frac{(vt)^n}{n!}$$

where P_{ij}^* are the n-stage transition probabilities corresponding to Equation (6.31). This technique of uniformizing the rate in which a transition occurs from each state by introducing transitions from a state to itself is known as *uniformization*.

Example 6.20 Let us reconsider Example 6.11, which models the workings of a machine—either on or off—as a two-state continuous-time Markov chain with

$$P_{01} = P_{10} = 1,$$

$$v_0 = \lambda, \qquad v_1 = \mu$$

Letting $v = \lambda + \mu$, the uniformized version of the preceding is to consider it a continuous-time Markov chain with

$$P_{00} = \frac{\mu}{\lambda + \mu} = 1 - P_{01},$$

$$P_{10} = \frac{\mu}{\lambda + \mu} = 1 - P_{11},$$

$$v_i = \lambda + \mu, \qquad i = 1, 2$$

As $P_{00} = P_{10}$, it follows that the probability of a transition into state 0 is equal to $\mu/(\lambda + \mu)$ no matter what the present state. Because a similar result is true for state 1, it follows that the n-stage transition probabilities are given by

$$P_{i0}^n = \frac{\mu}{\lambda + \mu}, \qquad n \geq 1, i = 0, 1$$

$$P_{i1}^n = \frac{\lambda}{\lambda + \mu}, \qquad n \geq 1, i = 0, 1$$

Hence,

$$P_{00}(t) = \sum_{n=0}^{\infty} P_{00}^n e^{-(\lambda+\mu)t} \frac{[(\lambda + \mu)t]^n}{n!}$$

$$= e^{-(\lambda+\mu)t} + \sum_{n=1}^{\infty} \left(\frac{\mu}{\lambda + \mu} \right) e^{-(\lambda+\mu)t} \frac{[(\lambda + \mu)t]^n}{n!}$$

$$= e^{-(\lambda+\mu)t} + [1 - e^{-(\lambda+\mu)t}] \frac{\mu}{\lambda + \mu}$$

$$= \frac{\mu}{\lambda + \mu} + \frac{\lambda}{\lambda + \mu} e^{-(\lambda+\mu)t}$$

Similarly,

$$P_{11}(t) = \sum_{n=0}^{\infty} P_{11}^n e^{-(\lambda+\mu)t} \frac{[(\lambda + \mu)t]^n}{n!}$$

$$= e^{-(\lambda+\mu)t} + [1 - e^{-(\lambda+\mu)t}] \frac{\lambda}{\lambda + \mu}$$

$$= \frac{\lambda}{\lambda + \mu} + \frac{\mu}{\lambda + \mu} e^{-(\lambda+\mu)t}$$

The remaining probabilities are

$$P_{01}(t) = 1 - P_{00}(t) = \frac{\lambda}{\lambda + \mu} [1 - e^{-(\lambda+\mu)t}]$$

$$P_{10}(t) = 1 - P_{11}(t) = \frac{\mu}{\lambda + \mu} [1 - e^{-(\lambda+\mu)t}] \quad \blacklozenge$$

Example 6.21 Consider the two-state chain of Example 6.20 and suppose that the initial state is state 0. Let $O(t)$ denote the total amount of time that the process is in state 0 during the interval $(0, t)$. The random variable $O(t)$ is often called the occupation time. We will now compute its mean.

If we let

$$I(s) = \begin{cases} 1, & \text{if } X(s) = 0 \\ 0, & \text{if } X(s) = 1 \end{cases}$$

then we can represent the occupation time by

$$O(t) = \int_0^t I(s) \, ds$$

Taking expectations and using the fact that we can take the expectation inside the integral sign (since an integral is basically a sum), we obtain

$$E[O(t)] = \int_0^t E[I(s)] \, ds$$

$$= \int_0^t P\{X(s) = 0\} \, ds$$

$$= \int_0^t P_{00}(s) \, ds$$

$$= \frac{\mu}{\lambda + \mu} t + \frac{\lambda}{(\lambda + \mu)^2} \{1 - e^{-(\lambda + \mu)t}\}$$

where the final equality follows by integrating

$$P_{00}(s) = \frac{\mu}{\lambda + \mu} + \frac{\lambda}{\lambda + \mu} e^{-(\lambda + \mu)s}$$

(For another derivation of $E[O(t)]$, see Exercise 38.) ◆

6.8. Computing the Transition Probabilities

For any pair of states i and j, let

$$r_{ij} = \begin{cases} q_{ij}, & \text{if } i \neq j \\ -v_i, & \text{if } i = j \end{cases}$$

Using this notation, we can rewrite the Kolmogorov backward equations

$$P'_{ij}(t) = \sum_{k \neq i} q_{ik} P_{kj}(t) - v_i P_{ij}(t)$$

and the forward equations

$$P'_{ij}(t) = \sum_{k \neq j} q_{kj} P_{ik}(t) - v_j P_{ij}(t)$$

as follows:

$$P'_{ij}(t) = \sum_{k} r_{ik} P_{kj}(t) \qquad \text{(backward)}$$

$$P'_{ij}(t) = \sum_{k} r_{kj} P_{ik}(t) \qquad \text{(forward)}$$

This representation is especially revealing when we introduce matrix nota-tion. Define the matrices $\mathbf{R}$, $\mathbf{P}(t)$, and $\mathbf{P}'(t)$ by letting the element in row i, column j of these matrices be, respectively, r_{ij}, $P_{ij}(t)$, and $P'_{ij}(t)$. Since the backward equations say that the element in row i, column j of the matrix $\mathbf{P}'(t)$ can be obtained by multiplying the ith row of the matrix $\mathbf{R}$ by the jth column of the matrix $\mathbf{P}(t)$, it is equivalent to the matrix equation

$$\mathbf{P}'(t) = \mathbf{R}\mathbf{P}(t) \tag{6.32}$$

Similarly, the forward equations can be written as

$$\mathbf{P}'(t) = \mathbf{P}(t)\mathbf{R} \tag{6.33}$$

Now, just as the solution of the scalar differential equation

$$f'(t) = cf(t)$$

[or, equivalent, $f'(t) = f(t)c$] is

$$f(t) = f(0)e^{ct}$$

it can be shown that the solution of the matrix differential Equations (6.32) and (6.33) is given by

$$\mathbf{P}(t) = \mathbf{P}(0)e^{\mathbf{R}t}$$

Since $\mathbf{P}(0) = \mathbf{I}$ (the identity matrix), this yields that

$$\mathbf{P}(t) = e^{\mathbf{R}t} \tag{6.34}$$

where the matrix $e^{\mathbf{R}t}$ is defined by

$$e^{\mathbf{R}t} = \sum_{n=0}^{\infty} \mathbf{R}^n \frac{t^n}{n!} \tag{6.35}$$

with $\mathbf{R}^n$ being the (matrix) multiplication of $\mathbf{R}$ by itself n times.

The direct use of Equation (6.35) to compute $\mathbf{P}(t)$ turns out to be very inefficient for two reasons. First, since the matrix $\mathbf{R}$ contains both positive and negative elements (remember the off-diagonal elements are the q_{ij} while the ith diagonal element is $-v_i$), there is the problem of computer round-off error when we compute the powers of $\mathbf{R}$. Second, we often have to compute many of the terms in the infinite sum (6.35) to arrive at a good approximation. However, there are certain indirect ways that we can utilize the relation (6.34) to efficiently approximate the matrix $\mathbf{P}(t)$. We now present two of these methods.

Approximation Method 1   Rather than using (6.35) to compute $e^{\mathbf{R}t}$, we can use the matrix equivalent of the identity

$$e^x = \lim_{n \to \infty} \left(1 + \frac{x}{n}\right)^n$$

which states that

$$e^{\mathbf{R}t} = \lim_{n \to \infty} \left(\mathbf{I} + \mathbf{R}\frac{t}{n}\right)^n$$

Thus, if we let n be a power of 2, say, $n = 2^k$, then we can approximate $\mathbf{P}(t)$ by raising the matrix $\mathbf{M} = \mathbf{I} + \mathbf{R}t/n$ to the nth power, which can be accomplished by k matrix multiplications (by first multiplying $\mathbf{M}$ by itself to obtain $\mathbf{M}^2$ and then multiplying that by itself to obtain $\mathbf{M}^4$ and so on). In addition, since only the diagonal elements of $\mathbf{R}$ are negative (and the diagonal elements of the identity matrix $\mathbf{I}$ are equal to 1) by choosing n large enough, we can guarantee that the matrix $\mathbf{I} + \mathbf{R}t/n$ has all nonnegative elements.

Approximation Method 2   A second approach to approximating $e^{\mathbf{R}t}$ uses the identity

$$e^{-\mathbf{R}t} = \lim_{n \to \infty} \left(\mathbf{I} - \mathbf{R}\frac{t}{n}\right)^n$$

$$\approx \left(\mathbf{I} - \mathbf{R}\frac{t}{n}\right)^n \quad \text{for } n \text{ large}$$

and thus

$$\mathbf{P}(t) = e^{\mathbf{R}t} \approx \left(\mathbf{I} - \mathbf{R}\frac{t}{n}\right)^{-n}$$

$$= \left[\left(\mathbf{I} - \mathbf{R}\frac{t}{n}\right)^{-1}\right]^n$$

Hence, if we again choose n to be a large power of 2, say, $n = 2^k$, we can approximate $\mathbf{P}(t)$ by first computing the inverse of the matrix $\mathbf{I} - \mathbf{R}t/n$ and then raising that matrix to the nth power (by utilizing k matrix multiplications). It can be shown that the matrix $(\mathbf{I} - \mathbf{R}t/n)^{-1}$ will have only nonnegative elements.

Remark   Both of the above computational approaches for approximating $P(t)$ have probabilistic interpretations (see Exercises 41 and 42).

Exercises

1.   A population of organisms consists of both male and female members. In a small colony any particular male is likely to mate with any particular female in any time interval of length h, with probability $\lambda h + o(h)$. Each mating immediately produces one offspring, equally likely to be male or female. Let $N_1(t)$ and $N_2(t)$ denote the number of males and females in the population at t. Derive the parameters of the continuous-time Markov chain $\{N_1(t), N_2(t)\}$, i.e., the v_i, P_{ij} of Section 6.2.

*2.   Suppose that a one-celled organism can be in one of two states — either A or B. An individual in state A will change to state B at an exponential rate α; an individual in state B divides into two new individuals of type A at an exponential rate β. Define an appropriate continuous-time Markov chain for a population of such organisms and determine the appropriate parameters for this model.

3.   Consider two machines that are maintained by a single repairman. Machine i functions for an exponential time with rate μ_i before breaking down, $i = 1, 2$. The repair times (for either machine) are exponential with rate μ. Can we analyze this as a birth and death process? If so, what are the parameters? If not, how can we analyze it?

*4.   Potential customers arrive at a single-server station in accordance with a Poisson process with rate λ. However, if the arrival finds n customers already in the station, then he will enter the system with probability α_n. Assuming an exponential service rate μ, set this up as a birth and death process and determine the birth and death rates.

5.   There are N individuals in a population, some of whom have a certain infection that spreads as follows. Contacts between two members of this population occur in accordance with a Poisson process having rate λ. When a contact occurs, it is equally likely to involve any of the $\binom{N}{2}$ pairs of individuals in the population. If a contact involves an infected and a noninfected individual, then with probability p the noninfected individual becomes infected. Once infected, an individual remains infected throughout.

Let $X(t)$ denote the number of infected members of the population at time t.

(a) Is $\{X(t), t \geqslant 0\}$ a continuous-time Markov chain?
(b) Specify the type of stochastic process.
(c) Starting with a single infected individual, what is the expected time until all members are infected?

6. Consider a birth and death process with birth rates $\lambda_i = (i + 1)\lambda$, $i \geqslant 0$, and death rates $\mu_i = i\mu$, $i \geqslant 0$.

(a) Determine the expected time to go from state 0 to state 4.
(b) Determine the expected time to go from state 2 to state 5.
(c) Determine the variances in parts (a) and (b).

***7.** Individuals join a club in accordance with a Poisson process with rate λ. Each new member must pass through k consecutive stages to become a full member of the club. The time it takes to pass through each stage is exponentially distributed with rate μ. Let $N_i(t)$ denote the number of club members at time t that have passed through exactly i stages, $i = 1, \ldots, k - 1$. Also, let $\mathbf{N}(t) = (N_1(t), N_2(t), \ldots, N_{k-1}(t))$.

(a) Is $\{\mathbf{N}(t), t \geqslant 0\}$ a continuous-time Markov chain?
(b) If so, give the infinitesimal transition rates. That is, for any state $\mathbf{n} = (n_1, \ldots, n_{k-1})$ give the possible next states along with their infinitesimal rates.

8. Consider two machines, both of which have an exponential lifetime with mean $1/\lambda$. There is a single repairman that can service machines at an exponential rate μ. Set up the Kolmogorov backward equations; you need not solve them.

9. The birth and death process with parameters $\lambda_n = 0$ and $\mu_n = \mu, n > 0$ is called a pure death process. Find $P_{ij}(t)$.

10. Consider two machines. Machine i operates for an exponential time with rate λ_i and then fails; its repair time is exponential with rate μ_i, $i = 1, 2$. The machines act independently of each other. Define a four-state continuous-time Markov chain which jointly describes the condition of the two machines. Use the assumed independence to compute the transition probabilities for this chain and then verify that these transition probabilities satisfy the forward and backward equations.

***11.** Consider a Yule process starting with a single individual—that is, suppose $X(0) = 1$. Let T_i denote the time it takes the process to go from a population of size i to one of size $i + 1$.

(a) Argue that T_i, $i = 1, \ldots, j$, are independent exponentials with respective rates $i\lambda$.

(b) Let $X_1, \ldots, X_j$ denote independent exponential random variables each having rate λ, and interpret X_i as the lifetime of component i. Argue that $\max(X_1, \ldots, X_j)$ can be expressed as

$$\max(X_1, \ldots, X_j) = \varepsilon_1 + \varepsilon_2 + \cdots + \varepsilon_j$$

where $\varepsilon_1, \varepsilon_2, \ldots, \varepsilon_j$ are independent exponentials with respective rates $j\lambda$, $(j-1)\lambda, \ldots, \lambda$.

Hint: Interpret ε_i as the time between the $i - 1$ and the ith failure.

(c) Using (a) and (b) argue that

$$P\{T_1 + \cdots + T_j \leqslant t\} = (1 - e^{-\lambda t})^j$$

(d) Use (c) to obtain that

$$P_{1j}(t) = (1 - e^{-\lambda t})^{j-1} - (1 - e^{-\lambda t})^j = e^{-\lambda t}(1 - e^{-\lambda t})^{j-1}$$

and hence, given $X(0) = 1$, $X(t)$ has a geometric distribution with parameter $p = e^{-\lambda t}$.

(e) Now conclude that

$$P_{ij}(t) = \binom{j-1}{i-1} e^{-\lambda t i}(1 - e^{-\lambda t})^{j-i}$$

12.   Each individual in a biological population is assumed to give birth at an exponential rate λ, and to die at an exponential rate μ. In addition, there is an exponential rate of increase θ due to immigration. However, immigration is not allowed when the population size is N or larger.

(a) Set this up as a birth and death model.
(b) If $N = 3$, $1 = \theta = \lambda$, $\mu = 2$, determine the proportion of time that immigration is restricted.

13.   A small barbershop, operated by a single barber, has room for at most two customers. Potential customers arrive at a Poisson rate of three per hour, and the successive service times are independent exponential random variables with mean $\frac{1}{4}$ hour. What is

(a) the average number of customers in the shop?
(b) the proportion of potential customers that enter the shop?
(c) If the barber could work twice as fast, how much more business would he do?

14.   Potential customers arrive at a full-service, one-pump gas station at a Poisson rate of 20 cars per hour. However, customers will only enter the station for gas if there are no more than two cars (including the one currently being attended to) at the pump. Suppose the amount of time required to service a car is exponentially distributed with a mean of five minutes.

(a) What fraction of the attendant's time will be spent servicing cars?
(b) What fraction of potential customers are lost?

15. A service center consists of two servers, each working at an exponential rate of two services per hour. If customers arrive at a Poisson rate of three per hour, then, assuming a system capacity of at most three customers,

(a) what fraction of potential customers enter the system?
(b) what would the value of part (a) be if there was only a single server, and his rate was twice as fast (that is, $\mu = 4$)?

*16. The following problem arises in molecular biology. The surface of a bacterium is supposed to consist of several sites at which foreign molecules — some acceptable and some not — become attached. We consider a particular site and assume that molecules arrive at the site according to a Poisson process with parameter λ. Among these molecules a proportion α is acceptable. Unacceptable molecules stay at the site for a length of time which is exponentially distributed with parameter μ_1, whereas an acceptable molecule remains at the site for an exponential time with rate μ_2. An arriving molecule will become attached only if the site is free of other molecules. What percentage of time is the site occupied with an acceptable (unacceptable) molecule?

17. Each time a machine is repaired it remains up for an exponentially distributed time with rate λ. It then fails, and its failure is either of two types. If it is a type 1 failure, then the time to repair the machine is exponential with rate μ_1; if it is a type 2 failure, then the repair time is exponential with rate μ_2. Each failure is, independently of the time it took the machine to fail, a type 1 failure with probability p and a type 2 failure with probability $1 - p$. What proportion of time is the machine down due to a type 1 failure? What proportion of time is it down due to a type 2 failure? What proportion of time is it up?

18. After being repaired, a machine functions for an exponential time with rate λ and then fails. Upon failure, a repair process begins. The repair process proceeds sequentially through k distinct phases. First a phase 1 repair must be performed, then a phase 2, and so on. The times to complete these phases are independent, with phase i taking an exponential time with rate μ_i, $i = 1, \ldots, k$.

(a) What proportion of time is the machine undergoing a phase i repair?
(b) What proportion of time is the machine working?

*19. A single repairperson looks after both machines 1 and 2. Each time it is repaired, machine i stays up for an exponential time with rate λ_i, $i = 1, 2$. When machine i fails, it requires an exponentially distributed amount of work with rate μ_i to complete its repair. The repairperson will always service machine 1 when it is down. For instance, if machine 1 fails while 2 is being repaired, then the repairperson will immediately stop work on machine 2 and start on 1. What proportion of time is machine 2 down?

20. There are two machines, one of which is used as a spare. A working machine will function for an exponential time with rate λ and will then fail. Upon failure, it is immediately replaced by the other machine if that one is in working order, and it goes to the repair facility. The repair facility consists of a single person who takes an exponential time with rate μ to repair a failed machine. At the repair facility, the newly failed machine enters service if the repairperson is free. If the repairperson is busy, it waits until the other machine is fixed; at that time, the newly repaired machine is put in service and repair begins on the other one. Starting with both machines in working condition, find

(a) the expected value and
(b) the variance

of the time until both are in the repair facility.

(c) In the long run, what proportion of time is there a working machine?

21. Suppose that when both machines are down in Exercise 20 a second repairperson is called in to work on the newly failed one. Suppose all repair times remain exponential with rate μ. Now find the proportion of time at least one machine is working, and compare your answer with the one obtained in Exercise 20.

22. Customers arrive at a single-server queue in accordance with a Poisson process having rate λ. However, an arrival that finds n customers already in the system will only join the system with probability $1/(n + 1)$. That is, with probability $n/(n + 1)$ such an arrival will not join the system. Show that the limiting distribution of the number of customers in the system is Poisson with mean λ/μ.

23. A job shop consists of three machines and two repairmen. The amount of time a machine works before breaking down is exponentially distributed with mean 10. If the amount of time it takes a single repairman to fix a machine is exponentially distributed with mean 8, then

(a) what is the average number of machines not in use?
(b) what proportion of time are both repairmen busy?

*24. Consider a taxi station where taxis and customers arrive in accordance with Poisson processes with respective rates of one and two per minute. A taxi will wait no matter how many other taxis are present. However, an arriving customer does not find a taxi waiting leaves. Find

(a) the average number of taxis waiting, and
(b) the proportion of arriving customers that get taxis.

25. Customers arrive at a service station, manned by a single server who serves at an exponential rate μ_1, at a Poisson rate λ. After completion of service the customer then joins a second system where the server serves at an exponential rate μ_2. Such a system is called a *tandem* or *sequential* queueing system. Assuming that $\lambda < \mu_i$, $i = 1, 2$, determine the limiting probabilities.

Hint: Try a solution of the form $P_{n,m} = C\alpha^n\beta^m$, and determine C, α, β.

26. Consider an ergodic $M/M/s$ queue in steady state (that is, after a long time) and argue that the number presently in the system is independent of the sequence of past departure times. That is, for instance, knowing that there have been departures 2, 3, 5, and 10 time units ago does not affect the distribution of the number presently in the system.

27. In the $M/M/s$ queue if you allow the service rate to depend on the number in the system (but in such a way so that it is ergodic), what can you say about the output process? What can you say when the service rate μ remains unchanged but $\lambda > s\mu$?

*28. If $\{X(t)\}$ and $\{Y(t)\}$ are independent continuous-time Markov chains, both of which are time reversible, show that the process $\{X(t), Y(t)\}$ is also a time reversible Markov chain.

29. Consider a set of n machines and a single repair facility to service these machines. Suppose that when machine $i, i = 1, \ldots, n$, fails it requires an exponentially distributed amount of work with rate μ_i to repair it. The repair facility divides its efforts equally among all failed machines in the sense that whenever there are k failed machines each one receives work at a rate of $1/k$ per unit time. If there are a total of r working machines, including machine i, then i fails at an instantaneous rate λ_i/r.

(a) Define an appropriate state space so as to be able to analyze the above system as a continuous-time Markov chain.
(b) Give the instantaneous transition rates (that is, give the q_{ij}).
(c) Write the time reversibility equations.
(d) Find the limiting probabilities and show that the process is time reversible.

30. Consider a graph with nodes $1, 2, \ldots, n$ and the $\binom{n}{2}$ arcs (i, j), $i \neq j$, i,

$j, = 1, \ldots, n$. (See Section 3.6.2 for appropriate definitions.) Suppose that a particle moves along this graph as follows: Events occur along the arcs (i, j) according to independent Poisson processes with rates λ_{ij}. An event along arc (i, j) causes that arc to become excited. If the particle is at node i at the moment that (i, j) becomes excited, it instantaneously moves to node j; $i, j = 1, \ldots, n$. Let P_j denote the proportion of time that the particle is at node j. Show that

$$P_j = \frac{1}{n}$$

Hint: Use time reversibility.

31. A total of N customers move about among r servers in the following manner. When a customer is served by server i, he then goes over to server $j, j \neq i$, with probability $1/(r - 1)$. If the server he goes to is free, then the customer enters service; otherwise he joins the queue. The service times are all independent, with the service times at server i being exponential with rate μ_i, $i = 1, \ldots, r$. Let the state at any time be the vector $(n_1, \ldots, n_r)$, where n_i is the number of customers presently at server i, $i = 1, \ldots, r$, $\Sigma_i n_i = N$.

(a) Argue that if $X(t)$ is the state at time t, then $\{X(t), t \geq 0\}$ is a continuous-time Markov chain.
(b) Give the infinitesimal rates of this chain.
(c) Show that this chain is time reversible, and find the limiting probabilities.

32. Customers arrive at a two-server station in accordance with a Poisson process having rate λ. Upon arriving, they join a single queue. Whenever a server completes a service, the person first in line enters service. The service times of server i are exponential with rate μ_i, $i = 1, 2$, where $\mu_1 + \mu_2 > \lambda$. An arrival finding both servers free is equally likely to go to either one. Define an appropriate continuous-time Markov chain for this model, show it is time reversible, and find the limiting probabilities.

*33. Consider two $M/M/1$ queues with respective parameters λ_i, μ_i, $i = 1, 2$. Suppose they share a common waiting room that can hold at most 3 customers. That is, whenever an arrival finds her server busy and 3 customers in the waiting room, she goes away. Find the limiting probability that there will be n queue 1 customers and m queue 2 customers in the system.

Hint: Use the results of Exercise 28 together with the concept of truncation.

34. Four workers share an office that contains four telephones. At any time, each worker is either "working" or "on the phone." Each "working" period of worker i lasts for an exponentially distributed time with rate λ_i, and each "on the phone" period lasts for an exponentially distributed time with rate $\mu_i, i = 1, 2, 3, 4$.

(a) What proportion of time are all workers "working"?

Let $X_i(t)$ equal 1 if worker i is working at time t, and let it be 0 otherwise. Let $\mathbf{X}(t) = (X_1(t), X_2(t), X_3(t), X_4(t))$.

(b) Argue that $\{\mathbf{X}(t), t \geq 0\}$ is a continuous-time Markov chain and give its infinitesimal rates.
(c) Is $\{\mathbf{X}(t)\}$ time reversible? Why or why not?

Suppose now that one of the phones has broken down. Suppose that a worker who is about to use a phone but finds them all being used begins a new "working" period.

(d) What proportion of time are all workers "working"?

35. Consider a time reversible continuous-time Markov chain having infinitesimal transition rates q_{ij} and limiting probabilities $\{P_i\}$. Let A denote a set of states for this chain, and consider a new continuous-time Markov chain with transition rates q_{ij}^* given by

$$q_{ij}^* = \begin{cases} cq_{ij}, & \text{if } i \in A, j \notin A \\ q_{ij}, & \text{otherwise} \end{cases}$$

where c is an arbitrary positive number. Show that this chain remains time reversible, and find its limiting probabilities.

36. Consider a system of n components such that the working times of component $i, i = 1, \ldots, n$, are exponentially distributed with rate λ_i. When failed, however, the repair rate of component i depends on how many other components are down. Specifically, suppose that the instantaneous repair rate of component i, $i = 1, \ldots, n$, when there are a total of k failed components, is $\alpha^k \mu_i$.

(a) Explain how we can analyze the preceding as a continuous-time Markov chain. Define the states and give the parameters of the chain.
(b) Show that, in steady state, the chain is time reversible and compute the limiting probabilities.

37. For the continuous-time Markov chain of Exercise 3 present a uniformized version.

38.   In Example 6.20, we computed $m(t) = E[O(t)]$, the expected occupation time in state 0 by time t for the two-state continuous-time Markov chain starting in state 0. Another way of obtaining this quantity is by deriving a differential equation for it.

(a) Show that

$$m(t + h) = m(t) + P_{00}(t)h + o(h)$$

(b) Show that

$$m'(t) = \frac{\mu}{\lambda + \mu} + \frac{\lambda}{\lambda + \mu} e^{-(\lambda + \mu)t}$$

(c) Solve for $m(t)$.

39.   Let $O(t)$ be the occupation time for state 0 in the two-state continuous-time Markov chain. Find $E[O(t) \,|\, X(0) = 1]$.

40.   Consider the two-state continuous-time Markov chain. Starting in state 0, find $\text{Cov}[X(s), X(t)]$.

41.   Let Y denote an exponential random variable with rate λ that is independent of the continuous-time Markov chain $\{X(t)\}$ and let

$$\bar{P}_{ij} = P\{X(Y) = j \,|\, X(0) = i\}$$

(a) Show that

$$\bar{P}_{ij} = \frac{1}{v_i + \lambda} \sum_k q_{ik} \bar{P}_{kj} + \frac{\lambda}{v_i + \lambda} \delta_{ij}$$

where δ_{ij} is 1 when $i = j$ and 0 when $i \neq j$.
(b) Show that the solution of the preceding set of equations is given by

$$\bar{P} = (I - R/\lambda)^{-1}$$

where $\bar{P}$ is the matrix of elements $\bar{P}_{ij}$, I is the identity matrix, and R the matrix specified in Section 6.8.
(c) Suppose now that $Y_1, \ldots, Y_n$ are independent exponentials with rate λ that are independent of $\{X(t)\}$. Show that

$$P\{X(Y_1 + \cdots + Y_n) = j \,|\, X(0) = i\}$$

is equal to the element in row i, column j of the matrix $\bar{P}^n$.
(d) Explain the relationship of the preceding to Approximation 2 of Section 6.8.

*42.   (a) Show that Approximation 1 of Section 6.8 is equivalent to uniformizing the continuous-time Markov chain with a value v such that $vt = n$ and then approximating $P_{ij}(t)$ by P_{ij}^{*n}.

(b) Explain why the preceding should make a good approximation.

Hint: What is the standard deviation of a Poisson random variable with mean n?

References

1. D. R. Cox and H. D. Miller, "The Theory of Stochastic Processes," Methuen, London, 1965.
2. A. W. Drake, "Fundamentals of Applied Probability Theory," McGraw-Hill, New York, 1967.
3. S. Karlin and H. Taylor, "A First Course in Stochastic Processes," Second Edition, Academic Press, New York, 1975.
4. E. Parzen, "Stochastic Processes," Holden-Day, San Francisco, California, 1962.
5. S. Ross, "Stochastic Processes," Second Edition, John Wiley, New York, 1996.

Renewal Theory and Its Applications

7

◆◆

7.1. Introduction

We have seen that a Poisson process is a counting process for which the times between successive events are independent and identically distributed exponential random variables. One possible generalization is to consider a counting process for which the times between successive events are independent and identically distributed with an arbitrary distribution. Such a counting process is called a *renewal process*.

Let $\{N(t), t \geq 0\}$ be a counting process and let X_n denote the time between the $(n-1)$st and the nth event of this process, $n \geq 1$.

Definition 7.1 If the sequence of nonnegative random variables $\{X_1 X_2, \ldots\}$ is independent and identically distributed, then the counting process $\{N(t), t \geq 0\}$ is said to be a *renewal process*.

Thus, a renewal process is a counting process such that the time until the first event occurs has some distribution F, the time between the first and second event has, independently of the time of the first event, the same distribution F, and so on. When an event occurs, we say that a renewal has taken place.

For an example of a renewal process, suppose that we have an infinite supply of lightbulbs whose lifetimes are independent and identically distributed. Suppose also that we use a single lightbulb at a time, and when it fails we immediately replace it with a new one. Under these conditions, $\{N(t), t \geq 0\}$ is a renewal process when $N(t)$ represents the number of lightbulbs that have failed by time t.

For a renewal process having interarrival times $X_1, X_2, \ldots$, let

$$S_0 = 0, \qquad S_n = \sum_{i=1}^{n} X_i, \qquad n \geq 1$$

That is, $S_1 = X_1$ is the time of the first renewal; $S_2 = X_1 + X_2$ is the time until the first renewal plus the time between the first and second renewal, that is, S_2 is the time of the second renewal. In general, S_n denotes the time of the nth renewal (see Figure 7.1).

We shall let F denote the interarrival distribution and to avoid trivialities, we assume that $F(0) = P\{X_n = 0\} < 1$. Furthermore, we let

$$\mu = E[X_n], \qquad n \geqslant 1$$

be the mean time between successive renewals. It follows from the non-negativity of X_n and the fact that X_n is not identically 0 that $\mu > 0$.

The first question we shall attempt to answer is whether an infinite number of renewals can occur in a finite amount of time. That is, can $N(t)$ be infinite for some (finite) value of t? To show that this cannot occur, we first note that, as S_n is the time of the nth renewal, $N(t)$ may be written as

$$N(t) = \max\{n : S_n \leqslant t\} \tag{7.1}$$

To understand why Equation (7.1) is valid, suppose, for instance, that $S_4 \leqslant t$ but $S_5 > t$. Hence, the fourth renewal had occurred by time t but the fifth renewal occurred after time t; or in other words, $N(t)$, the number of renewals that occurred by time t, must equal 4. Now by the strong law of large numbers it follows that, with probability 1,

$$\frac{S_n}{n} \to \mu \qquad \text{as } n \to \infty$$

But since $\mu > 0$ this means that S_n must be going to infinity as n goes to infinity. Thus, S_n can be less than or equal to t for at most a finite number of values of n, and hence by Equation (7.1), $N(t)$ must be finite.

However, though $N(t) < \infty$ for each t, it is true that, with probability 1,

$$N(\infty) \equiv \lim_{t \to \infty} N(t) = \infty$$

This follows since the only way in which $N(\infty)$, the total number of renewals that occur, can be finite is for one of the interarrival times to be infinite. Therefore,

$$P\{N(\infty) < \infty\} = P\{X_n = \infty \text{ for some } n\}$$

$$= P\left\{\bigcup_{n=1}^{\infty} \{X_n = \infty\}\right\}$$

$$\leqslant \sum_{n=1}^{\infty} P\{X_n = \infty\}$$

$$= 0$$

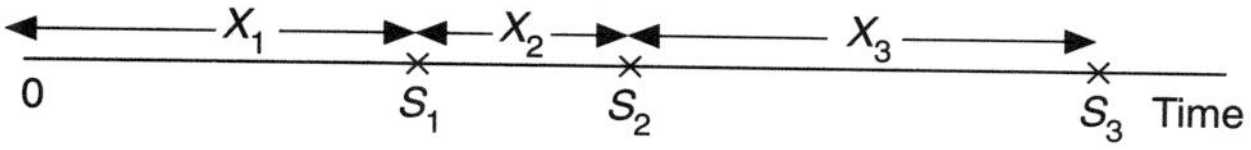

Figure 7.1.

7.2. Distribution of $N(t)$

The distribution of $N(t)$ can be obtained, at least in theory, by first noting the important relationship that *the number of renewals by time t is greater than or equal to n if and only if the nth renewal occurs before or at time t.* That is,

$$N(t) \geq n \Leftrightarrow S_n \leq t \tag{7.2}$$

From Equation (7.2) we obtain

$$P\{N(t) = n\} = P\{N(t) \geq n\} - P\{N(t) \geq n + 1\}$$

$$= P\{S_n \leq t\} - P\{S_{n+1} \leq t\} \tag{7.3}$$

Now since the random variables X_i, $i \geq 1$, are independent and have a common distribution F, it follows that $S_n = \sum_{i=1}^{n} X_i$ is distributed as F_n, the n-fold convolution of F with itself (Section 2.5). Therefore, from Equation (7.3) we obtain

$$P\{N(t) = n\} = F_n(t) - F_{n+1}(t)$$

Example 7.1   Suppose that $P\{X_n = i\} = p(1-p)^{i-1}$, $i \geq 1$. That is, suppose that the interarrival distribution is geometric. Now $S_1 = X_1$ may be interpreted as the number of trials necessary to get a single success when each trial is independent and has a probability p of being a success. Similarly, S_n may be interpreted as the number of trials necessary to attain n successes, and hence has the negative binomial distribution

$$P\{S_n = k\} = \begin{cases} \binom{k-1}{n-1} p^n (1-p)^{k-n}, & k \geq n \\ 0, & k < n \end{cases}$$

Thus, from Equation (7.3) we have that

$$P\{N(t) = n\} = \sum_{k=n}^{[t]} \binom{k-1}{n-1} p^n (1-p)^{k-n}$$

$$- \sum_{k=n+1}^{[t]} \binom{k-1}{n} p^{n+1} (1-p)^{k-n-1}$$

Equivalently, since an event independently occurs with probability p at each of the times $1, 2, \ldots$

$$P\{N(t) = n\} = \binom{[t]}{n} p^n (1 - p)^{[t]-n} \quad \blacklozenge$$

By using Equation (7.2) we can calculate $m(t)$, the mean value of $N(t)$, as

$$m(t) = E[N(t)]$$

$$= \sum_{n=1}^{\infty} P\{N(t) \geq n\}$$

$$= \sum_{n=1}^{\infty} P\{S_n \leq t\}$$

$$= \sum_{n=1}^{\infty} F_n(t)$$

where we have used the fact that if X is nonnegative and integer valued, then

$$E[X] = \sum_{k=1}^{\infty} kP\{X = k\} = \sum_{k=1}^{\infty} \sum_{n=1}^{k} P\{X = k\}$$

$$= \sum_{n=1}^{\infty} \sum_{k=n}^{\infty} P\{X = k\} = \sum_{n=1}^{\infty} P\{X \geq n\}$$

The function $m(t)$ is known as the *mean-value* or the *renewal function*.

It can be shown that the mean-value function $m(t)$ uniquely determines the renewal process. Specifically, there is a one-to-one correspondence between the interarrival distributions F and the mean-value functions $m(t)$.

Example 7.2 Suppose we have a renewal process whose mean-value function is given by

$$m(t) = 2t, \qquad t \geq 0$$

What is the distribution of the number of renewals occurring by time 10?

Solution: Since $m(t) = 2t$ is the mean-value function of a Poisson process with rate 2, it follows, by the one-to-one correspondence of interarrival distributions F and mean-value functions $m(t)$, that F must be exponential with mean $\frac{1}{2}$. Thus, the renewal process is a Poisson process with rate 2 and hence

$$P\{N(10) = n\} = e^{-20} \frac{(20)^n}{n!}, \qquad n \geq 0 \quad \blacklozenge$$

Another interesting result that we state without proof is that

$$m(t) < \infty \qquad \text{for all } t < \infty$$

Remarks (i) Since $m(t)$ uniquely determines the interarrival distribution, it follows that the Poisson process is the only renewal process having a linear mean-value function.

(ii) Some readers might think that the finiteness of $m(t)$ should follow directly from the fact that, with probability 1, $N(t)$ is finite. However, such reasoning is not valid; consider the following: Let Y be a random variable having the following probability distribution

$$Y = 2^n \text{ with probability } (\tfrac{1}{2})^n, \qquad n \geqslant 1$$

Now,

$$P\{Y < \infty\} = \sum_{n=1}^{\infty} P\{Y = 2^n\} = \sum_{n=1}^{\infty} (\tfrac{1}{2})^n = 1$$

But

$$E[Y] = \sum_{n=1}^{\infty} 2^n P\{Y = 2^n\} = \sum_{n=1}^{\infty} 2^n (\tfrac{1}{2})^n = \infty$$

Hence, even when Y is finite, it can still be true that $E[Y] = \infty$.

An integral equation satisfied by the renewal function can be obtained by conditioning on the time of the first renewal. Assuming that the interarrival distribution F is continuous with density function f this yields

$$m(t) = E[N(t)] = \int_0^\infty E[N(t) \,|\, X_1 = x] f(x) \, dx \qquad (7.4)$$

Now suppose that the first renewal occurs at a time x that is less than t. Then, using the fact that a renewal process probabilistically starts over when a renewal occurs, it follows that the number of renewals by time t would have the same distribution as 1 plus the number of renewals in the first $t - x$ time units. Therefore,

$$E[N(t) \,|\, X_1 = x] = 1 + E[N(t - x)] \qquad \text{if } x < t$$

Since, clearly

$$E[N(t) \,|\, X_1 = x] = 0 \qquad \text{when } x > t$$

we obtain from Equation (7.4) that

$$m(t) = \int_0^t [1 + m(t - x)] f(x) \, dx$$

$$= F(t) + \int_0^t m(t - x) f(x) \, dx \qquad (7.5)$$

Equation (7.5) is called the *renewal equation* and can sometimes be solved to obtain the renewal function.

Example 7.3   One instance in which the renewal equation can be solved is when the interarrival distribution is uniform — say, uniform on $(0, 1)$. We will now present a solution in this case when $t < 1$. For such values of t, the renewal function becomes

$$m(t) = t + \int_0^t m(t - x) \, dx$$

$$= t + \int_0^t m(y) \, dy \qquad \text{by the substitution } y = t - x$$

Differentiating the preceding equation yields

$$m'(t) = 1 + m(t)$$

Letting $h(t) = 1 + m(t)$, we obtain

$$h'(t) = h(t)$$

or

$$\log h(t) = t + C$$

or

$$h(t) = Ke^t$$

or

$$m(t) = Ke^t - 1$$

Since $m(0) = 0$, we see that $K = 1$, and so we obtain

$$m(t) = e^t - 1, \qquad 0 \leqslant t \leqslant 1 \quad \blacklozenge$$

7.3. Limit Theorems and Their Applications

We have shown previously that, with probability 1, $N(t)$ goes to infinity as t goes to infinity. However, it would be nice to know the rate at which $N(t)$ goes to infinity. That is, we would like to be able to say something about $\lim_{t \to \infty} N(t)/t$.

As a prelude to determining the rate at which $N(t)$ grows, let us first consider the random variable $S_{N(t)}$. In words, just what does this random variable represent? Proceeding inductively suppose, for instance, that $N(t) = 3$. Then $S_{N(t)} = S_3$ represents the time of the third event. Since there

Figure 7.2.

are only three events that have occurred by time t, S_3 also represents the time of the last event prior to (or at) time t. This is, in fact, what $S_{N(t)}$ represents — namely, the time of the last renewal *prior to or at* time t. Similar reasoning leads to the conclusion that $S_{N(t)+1}$ represents the time of the first renewal *after* time t (see Figure 7.2). We now are ready to prove the following.

Proposition 7.1 With probability 1,

$$\frac{N(t)}{t} \to \frac{1}{\mu} \qquad \text{as } t \to \infty$$

Proof Since $S_{N(t)}$ is the time of the last renewal prior to or at time t, and $S_{N(t)+1}$ is the time of the first renewal after time t, we have

$$S_{N(t)} \leqslant t < S_{N(t)+1}$$

or

$$\frac{S_{N(t)}}{N(t)} \leqslant \frac{t}{N(t)} < \frac{S_{N(t)+1}}{N(t)} \tag{7.6}$$

However, since $S_{N(t)}/N(t) = \sum_{i=1}^{N(t)} X_i/N(t)$ is the average of $N(t)$ independent and identically distributed random variables, it follows by the strong law of large numbers that $S_{N(t)}/N(t) \to \mu$ as $N(t) \to \infty$. But since $N(t) \to \infty$ when $t \to \infty$, we obtain

$$\frac{S_{N(t)}}{N(t)} \to \mu \qquad \text{as } t \to \infty$$

Furthermore, writing

$$\frac{S_{N(t)+1}}{N(t)} = \left(\frac{S_{N(t)+1}}{N(t)+1} \right) \left(\frac{N(t)+1}{N(t)} \right)$$

we have that $S_{N(t)+1}/(N(t)+1) \to \mu$ by the same reasoning as before and

$$\frac{N(t)+1}{N(t)} \to 1 \qquad \text{as } t \to \infty$$

Hence,

$$\frac{S_{N(t)+1}}{N(t)} \to \mu \qquad \text{as } t \to \infty$$

The result now follows by Equation (7.6) since $t/N(t)$ is between two random variables, each of which converges to μ as $t \to \infty$. ◆

Remarks (i) The preceding propositions are true even when μ, the mean time between renewals, is infinite. In this case, we interpret $1/\mu$ to be 0.

(ii) The number $1/\mu$ is called the *rate* of the renewal process.

Proposition 7.1 says that the average renewal rate up to time t will, with probability 1, converge to $1/\mu$ as $t \to \infty$. What about the expected average renewal rate? Is it true that $m(t)/t$ also converges to $1/\mu$? This result, known as the *elementary renewal theorem*, will be stated without proof.

Elementary Renewal Theorem

$$\frac{m(t)}{t} \to \frac{1}{\mu} \qquad \text{as } t \to \infty$$

As before, $1/\mu$ is interpreted as 0 when $\mu = \infty$.

Remark At first glance it might seem that the elementary renewal theorem should be a simple consequence of Proposition 7.1. That is, since the average renewal rate will, with probability 1, converge to $1/\mu$, should this not imply that the expected average renewal rate also converges to $1/\mu$? We must, however, be careful; consider the next example.

Example 7.4 Let U be a random variable which is uniformly distributed on $(0, 1)$; and define the random variables Y_n, $n \geqslant 1$, by

$$Y_n = \begin{cases} 0, & \text{if } U > 1/n \\ n, & \text{if } U \leqslant 1/n \end{cases}$$

Now, since, with probability 1, U will be greater than 0, it follows that Y_n will equal 0 for all sufficiently large n. That is, Y_n will equal 0 for all n large enough so that $1/n < U$. Hence, with probability 1,

$$Y_n \to 0 \qquad \text{as } n \to \infty$$

However,

$$E[Y_n] = nP\left\{U \leqslant \frac{1}{n}\right\} = n\frac{1}{n} = 1$$

Therefore, even though the sequence of random variables Y_n converges to 0, the expected values of the Y_n are all identically 1. ◆

Example 7.5 Beverly has a radio that works on a single battery. As soon as the battery in use fails, Beverly immediately replaces it with a new battery. If the lifetime of a battery (in hours) is distributed uniformly over the interval $(30, 60)$, then at what rate does Beverly have to change batteries?

Solution: If we let $N(t)$ denote the number of batteries that have failed by time t, we have by Proposition 7.1 that the rate at which Beverly replaces batteries is given by

$$\lim_{t \to \infty} \frac{N(t)}{t} = \frac{1}{\mu} = \frac{1}{45}$$

That is, in the long run, Beverly will have to replace one battery every 45 hours. ◆

Example 7.6 Suppose in Example 7.5 that Beverly does not keep any surplus batteries on hand, and so each time a failure occurs she must go and buy a new battery. If the amount of time it takes for her to get a new battery is uniformly distributed over $(0, 1)$, then what is the average rate that Beverly changes batteries?

Solution: In this case the mean time between renewals is given by

$$\mu = EU_1 + EU_2$$

where U_1 is uniform over $(30, 60)$ and U_2 is uniform over $(0, 1)$. Hence,

$$\mu = 45 + \tfrac{1}{2} = 45\tfrac{1}{2}$$

and so in the long run, Beverly will be putting in a new battery at the rate of $\frac{2}{91}$. That is, she will put in two new batteries every 91 hours. ◆

Example 7.7 Suppose that potential customers arrive at a single-server bank in accordance with a Poisson process having rate λ. However, suppose that the potential customer will only enter the bank if the server is free when he arrives. That is, if there is already a customer in the bank, then our arrivee, rather than entering the bank, will go home. If we assume that the amount of time spent in the bank by an entering customer is a random variable having distribution G, then

(a) what is the rate at which customers enter the bank?
(b) what proportion of potential customers actually enter the bank?

Solution: In answering these questions, let us suppose that at time 0 a customer has just entered the bank. (That is, we define the process to start when the first customer enters the bank.) If we let μ_G denote the mean service time, then, by the memoryless property of the Poisson process, it follows that the mean time between entering customers is

$$\mu = \mu_G + \frac{1}{\lambda}$$

Hence, the rate at which customers enter the bank will be given by

$$\frac{1}{\mu} = \frac{\lambda}{1 + \lambda \mu_G}$$

On the other hand, since potential customers will be arriving at a rate λ, it follows that the proportion of them entering the bank will be given by

$$\frac{\lambda/(1 + \lambda \mu_G)}{\lambda} = \frac{1}{1 + \lambda \mu_G}$$

In particular if $\lambda = 2$ (in hours) and $\mu_G = 2$, then only one customer out of five will actually enter the system. ◆

A somewhat unusual application of Proposition 7.1 is provided by our next example.

Example 7.8 A sequence of independent trials, each of which results in outcome number i with probability P_i, $i = 1, \ldots, n$, $\Sigma_1^n P_i = 1$, is observed until the same outcome occurs k times in a row; this outcome then is declared to be the winner of the game. For instance, if $k = 2$ and the sequence of outcomes is 1, 2, 4, 3, 5, 2, 1, 3, 3, then we stop after 9 trials and declare outcome number 3 the winner. What is the probability that i wins, $i = 1, \ldots, n$, and what is the expected number of trials?

Solution: We begin by computing the expected number of coin tosses, call it $E[T]$, until a run of k successive heads occurs when the tosses are independent and each lands on heads with probability p. By conditioning on the time of the first nonhead, we obtain

$$E[T] = \sum_{j=1}^{k} (1 - p)p^{j-1}(j + E[T]) + kp^k$$

Solving this for $E[T]$ yields

$$E[T] = k + \frac{(1 - p)}{p^k} \sum_{j=1}^{k} jp^{j-1}$$

Upon simplifying, we obtain

$$E[T] = \frac{1 + p + \cdots + p^{k-1}}{p^k}$$

$$= \frac{(1 - p^k)}{p^k(1 - p)} \tag{7.7}$$

Now, let us return to our example, and let us suppose that as soon as the winner of a game has been determined we immediately begin playing another game. For each i let us determine the rate at which outcome i wins. Now, every time i wins, everything starts over again and thus wins by i constitute renewals. Hence, from Proposition 7.1, the

$$\text{rate at which } i \text{ wins} = \frac{1}{E[N_i]}$$

where N_i denotes the number of trials played between successive wins of outcome i. Hence, from Equation (7.7) we see that

$$\text{rate at which } i \text{ wins} = \frac{P_i^k(1 - P_i)}{(1 - P_i^k)} \tag{7.8}$$

Hence, the long-run proportion of games which are won by number i is given by

$$\text{proportion of games } i \text{ wins} = \frac{\text{rate at which } i \text{ wins}}{\sum_{j=1}^n \text{rate at which } j \text{ wins}}$$

$$= \frac{P_i^k(1 - P_i)/(1 - P_i^k)}{\sum_{j=1}^n (P_j^k(1 - P_j)/(1 - P_j^k))}$$

However, it follows from the strong law of large numbers that the long-run proportion of games that i wins will, with probability 1, be equal to the probability that i wins any given game. Hence,

$$P\{i \text{ wins}\} = \frac{P_i^k(1 - P_i)/(1 - P_i^k)}{\sum_{j=1}^n (P_j^k(1 - P_j)/(1 - P_j^k))}$$

To compute the expected time of a game, we first note that the

$$\text{rate at which games end} = \sum_{i=1}^n \text{rate at which } i \text{ wins}$$

$$= \sum_{i=1}^n \frac{P_i^k(1 - P_i)}{(1 - P_i^k)} \quad \text{[from Equation (7.8)]}$$

Now, as everything starts over when a game ends, it follows by Proposition 7.1 that the rate at which games end is equal to the reciprocal of the mean time of a game. Hence,

$$E[\text{time of a game}\} = \frac{1}{\text{rate at which games end}}$$

$$= \frac{1}{\sum_{i=1}^{n} (P_i^k(1 - P_i)/(1 - P_i^k))} \qquad \blacklozenge$$

A key element in the proof of the elementary renewal theorem, which is also of independent interest, is the establishment of a relationship between $m(t)$, the mean number of renewals by time t, and $E[S_{N(t)+1}]$, the expected time of the first renewal after t. Letting

$$g(t) = E[S_{N(t)+1}]$$

we will derive an integral equation, similar to the renewal equation, for $g(t)$ by conditioning on the time of the first renewal. This yields

$$g(t) = \int_0^\infty E[S_{N(t)+1} | X_1 = x] f(x) \, dx$$

where we have supposed that the interarrival times are continuous with density f. Now if the first renewal occurs at time x and $x > t$, then clearly the time of the first renewal after t is x. On the other hand, if the first renewal occurs at a time $x < t$, then by regarding x as the new origin, it follows that the expected time, from this origin, of the first renewal occurring after a time $t - x$ from this origin is $g(t - x)$. That is, we see that

$$E[S_{N(t)+1} | X_1 = x] = \begin{cases} g(t - x) + x, & \text{if } x < t \\ x, & \text{if } x > t \end{cases}$$

Substituting this into the preceding equation gives

$$g(t) = \int_0^t (g(t - x) + x) f(x) \, dx + \int_t^\infty x f(x) \, dx$$

$$= \int_0^t g(t - x) f(x) \, dx + \int_0^\infty x f(x) \, dx$$

or

$$g(t) = \mu + \int_0^t g(t - x) f(x) \, dx$$

which is quite similar to the renewal equation

$$m(t) = F(t) + \int_0^t m(t - x) f(x) \, ds$$

Indeed, if we let

$$g_1(t) = \frac{g(t)}{\mu} - 1$$

we see that

$$g_1(t) + 1 = 1 + \int_0^t [g_1(t - x) + 1] f(x) \, dx$$

or

$$g_1(t) = F(t) + \int_0^\infty g_1(t - x) f(x) \, dx$$

That is, $g_1(t) = E[S_{N(t)+1}]/\mu - 1$ satisfies the renewal equation and thus, by uniqueness, must be equal to $m(t)$. We have thus proven the following.

Proposition 7.2

$$E[S_{N(t)+1}] = \mu[m(t) + 1]$$

A second derivation of Proposition 7.2 is given in Exercises 12 and 13. To see how Proposition 7.2 can be used to establish the elementary renewal theorem, let $Y(t)$ denote the time from t until the next renewal. $Y(t)$ is called the excess, or residual life, at t. As the first renewal after t will occur at time $t + Y(t)$, we see that

$$S_{N(t)+1} = t + Y(t)$$

Taking expectations and utilizing Proposition 7.2 yields

$$\mu[m(t) + 1] = t + E[Y(t)] \tag{7.9}$$

which implies that

$$\frac{m(t)}{t} = \frac{1}{\mu} - \frac{1}{t} + \frac{E[Y(t)]}{t\mu}$$

The elementary renewal theorem can now be proven by showing that

$$\lim_{t \to \infty} \frac{E[Y(t)]}{t} = 0$$

(see Exercise 13).

The relation (7.9) shows that if one can determine $E[Y(t)]$, the mean excess at t, then one can compute $m(t)$ and vice versa.

Example 7.9   Consider the renewal process whose interarrival distribution is the convolution of two exponentials; that is,

$$F = F_1 * F_2, \qquad \text{where } F_i(t) = 1 - e^{-\mu_i t}, i = 1, 2$$

We will determine the renewal function by first determining $E[Y(t)]$. To obtain the mean excess at t, imagine that each renewal corresponds to a new machine being put in use, and suppose that each machine has two components — initially component 1 is employed and this lasts an exponential time with rate μ_1, and then component 2, which functions for an exponential time with rate μ_2, is employed. When component 2 fails, a new machine is put in use (that is, a renewal occurs). Now consider the process $\{X(t), t \geqslant 0\}$ where $X(t)$ is i if a type i component is in use at time t. It is easy to see that $\{X(t), t \geqslant 0\}$ is a two-state continuous-time Markov chain, and so, using the results of Example 6.11, its transition probabilities are

$$P_{11}(t) = \frac{\mu_1}{\mu_1 + \mu_2} e^{-(\mu_1 + \mu_2)t} + \frac{\mu_2}{\mu_1 + \mu_2}$$

To compute the remaining life of the machine in use at time t, we condition on whether it is using its first or second component: for if it is still using its first component, then its remaining life is $1/\mu_1 + 1/\mu_2$, whereas if it is already using its second component, then its remaining life is $1/\mu_2$. Hence, letting $p(t)$ denote the probability that the machine in use at time t is using its first component, we have that

$$E[Y(t)] = \left(\frac{1}{\mu_1} + \frac{1}{\mu_2} \right) p(t) + \frac{1 - p(t)}{\mu_2}$$

$$= \frac{1}{\mu_2} + \frac{p(t)}{\mu_1}$$

But, since at time 0 the first machine is utilizing its first component, it follows that $p(t) = P_{11}(t)$, and so, upon using the preceding expression of $P_{11}(t)$, we obtain

$$E[Y(t)] = \frac{1}{\mu_2} + \frac{1}{\mu_1 + \mu_2} e^{-(\mu_1 + \mu_2)t} + \frac{\mu_2}{\mu_1(\mu_1 + \mu_2)} \qquad (7.10)$$

Now it follows from Equation (7.9) that

$$m(t) + 1 = \frac{t}{\mu} + \frac{E[Y(t)]}{\mu} \qquad (7.11)$$

where μ, the mean interarrival time, is given in this case by

$$\mu = \frac{1}{\mu_1} + \frac{1}{\mu_2} = \frac{\mu_1 + \mu_2}{\mu_1 \mu_2}$$

Substituting Equation (7.10) and the preceding equation into (7.11) yields, after simplifying,

$$m(t) = \frac{\mu_1 \mu_2}{\mu_1 + \mu_2} t - \frac{\mu_1 \mu_2}{(\mu_1 + \mu_2)^2} [1 - e^{-(\mu_1 + \mu_2)t}] \quad \blacklozenge$$

Remark Using the relationship of Equation (7.11) and results from the two-state continuous-time Markov chain, the renewal function can also be obtained in the same manner as in Example 7.9 for the interarrival distributions

$$F(t) = pF_1(t) + (1 - p)F_2(t)$$

and

$$F(t) = pF_1(t) + (1 - p)(F_1 * F_2)(t)$$

when $F_i(t) = 1 - e^{-\mu_i t}$, $t > 0$, $i = 1, 2$.

An important limit theorem is the central limit theorem for renewal processes. This states that, for large t, $N(t)$ is approximately normally distributed with mean t/μ and variance $t\sigma^2/\mu^3$, where μ and σ^2 are, respectively, the mean and variance of the interarrival distribution. That is, we have the following theorem which we state without proof.

Central Limit Theorem for Renewal Processes

$$\lim_{t \to \infty} P\left\{ \frac{N(t) - t/\mu}{\sqrt{t\sigma^2/\mu^3}} < x \right\} = \frac{1}{\sqrt{2\pi}} \int_{-\infty}^{x} e^{-x^2/2} \, dx$$

In addition, as might be expected from the central limit theorem for renewal processes, it can be shown that $\mathrm{Var}(N(t))/t$ converges to σ^2/μ^3. That is, it can be shown that

$$\lim_{t \to \infty} \frac{\mathrm{Var}(N(t))}{t} = \sigma^2/\mu^3 \tag{7.12}$$

7.4. Renewal Reward Processes

A large number of probability models are special cases of the following model. Consider a renewal process $\{N(t), t \geq 0\}$ having interarrival times $X_n, n \geq 1$, and suppose that each time a renewal occurs we receive a reward. We denote by R_n, the reward earned at the time of the nth renewal. We shall assume that the R_n, $n \geqslant 1$, are independent and identically distributed. However, we do allow for the possibility that R_n may (and usually will) depend on X_n, the length of the nth renewal interval. If we let

$$R(t) = \sum_{n=1}^{N(t)} R_n$$

then $R(t)$ represents the total reward earned by time t. Let

$$E[R] = E[R_n], \qquad E[X] = E[X_n]$$

Proposition 7.3   If $E[R] < \infty$ and $E[X] < \infty$, then

(a) with probability 1,
$$\lim_{t \to \infty} \frac{R(t)}{t} = \frac{E[R]}{E[X]}$$

(b)
$$\lim_{t \to \infty} \frac{E[R(t)]}{t} = \frac{E[R]}{E[X]}$$

Proof   We give the proof for (a) only. To prove this, write

$$\frac{R(t)}{t} = \frac{\sum_{n=1}^{N(t)} R_n}{t} = \left(\frac{\sum_{n=1}^{N(t)} R_n}{N(t)} \right) \left(\frac{N(t)}{t} \right)$$

By the strong law of large numbers we obtain

$$\frac{\sum_{n=1}^{N(t)} R_n}{N(t)} \to E[R] \qquad \text{as } t \to \infty$$

and by Proposition 7.1

$$\frac{N(t)}{t} \to \frac{1}{E[X]} \qquad \text{as } t \to \infty$$

The result thus follows.

Remarks   (i) If we say that a *cycle* is completed every time a renewal occurs, then Proposition 7.3 states that the long-run average reward per unit time is equal to the expected reward earned during a cycle divided by the expected length of a cycle.

(ii) Although we have supposed that the reward is earned at the time of a renewal, the result remains valid when the reward is earned gradually throughout the renewal cycle. ✦

Example 7.10 In Example 7.7 if we suppose that the amounts that the successive customers deposit in the bank are independent random variables having a common distribution H, then the rate at which deposits accumulate — that is, $\lim_{t \to \infty}$ (total deposits by the time $t)/t$ — is given by

$$\frac{E[\text{deposits during a cycle}]}{E[\text{time of cycle}]} = \frac{\mu_H}{\mu_G + 1/\lambda}$$

where $\mu_G + 1/\lambda$ is the mean time of a cycle, and μ_H is the mean of the distribution H. ✦

Example 7.11 (A Car Buying Model): The lifetime of a car is a continuous random variable having a distribution H and probability density h. Mr. Brown has a policy that he buys a new car as soon as his old one either breaks down or reaches the age of T years. Suppose that a new car costs C_1 dollars and also that an additional cost of C_2 dollars is incurred whenever Mr. Brown's car breaks down. Under the assumption that a used car has no resale value, what is Mr. Brown's long-run average cost?

If we say that a cycle is complete every time Mr. Brown gets a new car, then it follows from Proposition 7.3 (with costs replacing rewards) that his long-run average cost equals

$$\frac{E[\text{cost incurred during a cycle}]}{E[\text{length of a cycle}]}$$

Now letting X be the lifetime of Mr. Brown's car during an arbitrary cycle, then the cost incurred during that cycle will be given by

$$\begin{matrix} C_1, & \text{if } X > T \\ C_1 + C_2, & \text{if } X \leqslant T \end{matrix}$$

so the expected cost incurred over a cycle is

$$C_1 P\{X > T\} + (C_1 + C_2)P\{X \leqslant T\} = C_1 + C_2 H(T)$$

Also, the length of the cycle is

$$\begin{matrix} X, & \text{if } X \leqslant T \\ T, & \text{if } X > T \end{matrix}$$

and so the expected length of a cycle is

$$\int_0^T x h(x) \, dx + \int_T^\infty T h(x) \, dx = \int_0^T x h(x) \, dx + T[1 - H(T)]$$

Therefore, Mr. Brown's long-run average cost will be

$$\frac{C_1 + C_2 H(T)}{\int_0^T x h(x)\,dx + T[1 - H(T)]} \tag{7.13}$$

Now, suppose that the lifetime of a car (in years) is uniformly distributed over $(0, 10)$, and suppose that C_1 is 3 (thousand) dollars and C_2 is $\frac{1}{2}$ (thousand) dollars. What value of T minimizes Mr. Brown's long-run average cost?

If Mr. Brown uses the value $T, T \leqslant 10$, then from Equation (7.13) his long-run average cost equals

$$\frac{3 + \frac{1}{2}(T/10)}{\int_0^T (x/10)\,dx + T(1 - T/10)} = \frac{3 + T/20}{T^2/20 + (10T - T^2)/10}$$

$$= \frac{60 + T}{20T - T^2}$$

We can now minimize this by using the calculus. Toward this end, let

$$g(T) = \frac{60 + T}{20T - T^2}$$

then

$$g'(T) = \frac{(20T - T^2) - (60 + T)(20 - 2T)}{(20T - T^2)^2}$$

Equating to 0 yields

$$20T - T^2 = (60 + T)(20 - 2T)$$

or, equivalently,

$$T^2 + 120T - 1200 = 0$$

which yields the solutions

$$T \approx 9.25 \qquad \text{and} \qquad T \approx -129.25$$

Since $T \leqslant 10$, it follows that the optimal policy for Mr. Brown would be to purchase a new car whenever his old car reaches the age of 9.25 years. ◆

Example 7.12 (Dispatching a Train): Suppose that customers arrive at a train depot in accordance with a renewal process having a mean interarrival time μ. Whenever there are N customers waiting in the depot, a train leaves. If the depot incurs a cost at the rate of nc dollars per unit time whenever there are n customers waiting, what is the average cost incurred by the depot?

If we say that a cycle is completed whenever a train leaves, then the preceding is a renewal reward process. The expected length of a cycle is the expected time required for N customers to arrive and, since the mean interarrival time is μ, this equals

$$E[\text{length of cycle}] = N\mu$$

If we let T_n denote the time between the nth and $(n + 1)$st arrival in a cycle, then the expected cost of a cycle may be expressed as

$$E[\text{cost of a cycle}] = E[cT_1 + 2cT_2 + \cdots + (N - 1)cT_{N-1}]$$

which, since $E[T_n] = \mu$, equals

$$c\mu \frac{N}{2}(N - 1)$$

Hence, the average cost incurred by the depot is

$$\frac{c\mu N(N - 1)}{2N\mu} = \frac{c(N - 1)}{2}$$

Suppose now that each time a train leaves the depot incurs a cost of six units. What value of N minimizes the depot's long-run average cost when $c = 2$, $\mu = 1$?

In this case, we have that the average cost per unit time when the depot uses N is

$$\frac{6 + c\mu N(N - 1)/2}{N\mu} = N - 1 + \frac{6}{N}$$

By treating this as a continuous function of N and using the calculus, we obtain that the minimal value of N is

$$N = \sqrt{6} \approx 2.45$$

Hence, the optimal integral value of N is either 2 which yields a value 4, or 3 which also yields the value 4. Hence, either $N = 2$ or $N = 3$ minimizes the depot's average cost. ✦

Example 7.13 Suppose that customers arrive at a single-server system in accordance with a Poisson process with rate λ. Upon arriving a customer must pass through a door that leads to the server. However, each time someone passes through, the door becomes locked for the next t units of time. An arrival finding a locked door is lost, and a cost c is incurred by the system. An arrival finding the door unlocked passes through to the server. If the server is free, the customer enters service; if the server is busy, the customer departs without service and a cost K is incurred. If the service time of a customer is exponential with rate μ, find the average cost per unit time incurred by the system.

Solution: The preceding can be considered to be a renewal reward process, with a new cycle beginning each time a customer arrives to find the door unlocked. This is so because whether or not the arrival finds the server free, the door will become locked for the next t time units and the server will be busy for a time X that is exponentially distributed with rate μ. (If the server is free, X is the service time of the entering customer; if the server is busy, X is the remaining service time of the customer in service.) Since the next cycle will begin at the first arrival epoch after a time t has passed, it follows that

$$E[\text{time of a cycle}] = t + 1/\lambda$$

Let C_1 denote the cost incurred during a cycle due to arrivals finding the door locked. Then, since each arrival in the first t time units of a cycle will result in a cost c, we have

$$E[C_1] = \lambda t c$$

Also, let C_2 denote the cost incurred during a cycle due to an arrival finding the door unlocked but the server busy. Then because a cost K is incurred if the server is still busy a time t after the cycle began and, in addition, the next arrival after that time occurs before the service completion, we see that

$$E[C_2] = Ke^{-\mu t}\frac{\lambda}{\lambda + \mu}$$

Consequently,

$$\text{average cost per unit time} = \frac{\lambda t c + \lambda K e^{-\mu t}/(\lambda + \mu)}{t + 1/\lambda} \quad \blacklozenge$$

Example 7.14 Consider a manufacturing process that sequentially produces items, each of which is either defective or acceptable. The following type of sampling scheme is often employed in an attempt to detect and eliminate most of the defective items. Initially, each item is inspected and this continues until there are k consecutive items that are acceptable. At this point 100% inspection ends and each successive item is independently inspected with probability α. This partial inspection continues until a defective item is encountered, at which time 100% inspection is reinstituted, and the process begins anew. If each item is, independently, defective with probability q,

(a) what proportion of items are inspected?
(b) if defective items are removed when detected, what proportion of the remaining items are defective?

Remark Before starting our analysis, note that the above inspection scheme was devised for situations in which the probability of producing a defective item changed over time. It was hoped that 100% inspection would correlate with times at which the defect probability was large and partial inspection when it was small. However, it is still important to see how such a scheme would work in the extreme case where the defect probability remains constant throughout.

Solution: We begin our analysis by noting that we can treat the above as a renewal reward process with a new cycle starting each time 100% inspection is instituted. We then have

$$\text{proportion of items inspected} = \frac{E[\text{number inspected in a cycle}]}{E[\text{number produced in a cycle}]}$$

Let N_k denote the number of items inspected until there are k consecutive acceptable items. Once partial inspection begins — that is, after N_k items have been produced — since each inspected item will be defective with probability q, it follows that the expected number that will have to be inspected to find a defective item is $1/q$. Hence,

$$E[\text{number inspected in a cycle}] = E[N_k] + \frac{1}{q}$$

In addition, since at partial inspection each item produced will, independently, be inspected and found to be defective with probability αq, it follows that the number of items produced until one is inspected and found to be defective is $1/\alpha q$, and so

$$E[\text{number produced in a cycle}] = E[N_k] + \frac{1}{\alpha q}$$

Also, as $E[N_k]$ is the expected number of trials needed to obtain k acceptable items in a row when each item is acceptable with probability $p = 1 - q$, it follows from Example 3.14 that

$$E[N_k] = \frac{1}{p} + \frac{1}{p^2} + \cdots + \frac{1}{p^k} = \frac{(1/p)^k - 1}{q}$$

Hence we obtain

$$P_I \equiv \text{proportion of items that are inspected} = \frac{(1/p)^k}{(1/p)^k - 1 + 1/\alpha}$$

To answer (b), note first that since each item produced is defective with probability q it follows that the proportion of items that are both inspected and found to be defective is qP_1. Hence, for N large, out of the first N items produced there will be (approximately) NqP_I that are discovered to be defective and thus removed. As the first N items will contain (approximately) Nq defective items, it follows that there will be $Nq - NqP_I$ defective items not discovered. Hence,

$$\text{proportion of the nonremoved items that are defective} \approx \frac{Nq(1 - P_I)}{N(1 - qP_I)}$$

As the approximation becomes exact as $N \to \infty$, we see that

$$\text{proportion of the nonremoved items that are defective} = \frac{q(1 - P_I)}{(1 - qP_1)} \quad \blacklozenge$$

Example 7.15 (The Average Age of a Renewal Process): Consider a renewal process having interarrival distribution F and define $A(t)$ to be the time at t since the last renewal. If renewals represent old items failing and being replaced by new ones, then $A(t)$ represents the age of the item in use at time t. Since $S_{N(t)}$ represents the time of the last event prior to or at time t, we have that

$$A(t) = t - S_{N(t)}$$

We are interested in the average value of the age — that is, in

$$\lim_{s \to \infty} \frac{\int_0^s A(t)\,dt}{s}$$

To determine the above quantity, we use renewal reward theory in the following way: Let us assume that at any time we are being paid money at a rate equal to the age of the renewal process at that time. That is, at time t, we are being paid at rate $A(t)$, and so $\int_0^s A(t)\,dt$ represents our total earnings by time s. As everything starts over again when a renewal occurs, it follows that

$$\frac{\int_0^s A(t)\,dt}{s} \to \frac{E[\text{reward during a renewal cycle}]}{E[\text{time of a renewal cycle}]}$$

Now since the age of the renewal process a time t into a renewal cycle is just t, we have

$$\text{reward during a renewal cycle} = \int_0^X t\, dt$$

$$= \frac{X^2}{2}$$

where X is the time of the renewal cycle. Hence, we have that

$$\text{average value of age} \equiv \lim_{s\to\infty} \frac{\int_0^s A(t)\, dt}{s}$$

$$= \frac{E[X^2]}{2E[X]} \tag{7.14}$$

where X is an interarrival time having distribution function F. ◆

Example 7.16 (The Average Excess of a Renewal Process): Another quantity associated with a renewal process is $Y(t)$, the excess or residual time at time t. $Y(t)$ is defined to equal the time from t until the next renewal and, as such, represents the remaining (or residual) life of the item in use at time t. The average value of the excess, namely,

$$\lim_{s\to\infty} \frac{\int_0^s Y(t)\, dt}{s}$$

also can be easily obtained by renewal reward theory. To do so, suppose that we are paid at time t at a rate equal to $Y(t)$. Then our average reward per unit time will, by renewal reward theory, be given by

$$\text{average value of excess} \equiv \lim_{s\to\infty} \frac{\int_0^s Y(t)\, dt}{s}$$

$$= \frac{E[\text{reward during a cycle}]}{E[\text{length of a cycle}]}$$

Now, letting X denote the length of a renewal cycle, we have that

$$\text{reward during a cycle} = \int_0^X (X - t)\, dt$$

$$= \frac{X^2}{2}$$

and thus the average value of the excess is

$$\text{average value of excess} = \frac{E[X^2]}{2E[X]}$$

which was the same result obtained for the average value of the age of renewal process. ✦

7.5. Regenerative Processes

Consider a stochastic process $\{X(t), t \geq 0\}$ with state space $0, 1, 2, \ldots$, having the property that there exist time points at which the process (probabilistically) restarts itself. That is, suppose that with probability one, there exists a time T_1, such that the continuation of the process beyond T_1 is a probabilistic replica of the whole process starting at 0. Note that this property implies the existence of further times $T_2, T_3, \ldots$, having the same property as T_1. Such a stochastic process is known as a *regenerative process*.

From the above, it follows that $T_1, T_2, \ldots$, constitute the arrival times of a renewal process, and we shall say that a cycle is completed every time a renewal occurs.

Examples   (1) A renewal process is regenerative, and T_1 represents the time of the first renewal.

(2) A recurrent Markov chain is regenerative, and T_1 represents the time of the first transition into the initial state.

We are interested in determining the long-run proportion of time that a regenerative process spends in state j. To obtain this quantity, let us imagine that we earn a reward at a rate 1 per unit time when the process is in state j and at rate 0 otherwise. That is, if $I(s)$ represents the rate at which we earn at time s, then

$$I(s) = \begin{cases} 1, & \text{if } X(s) = j \\ 0, & \text{if } X(s) \neq j \end{cases}$$

and

$$\text{total reward earned by } t = \int_0^t I(s) \, ds$$

As the preceding is clearly a renewal reward process that starts over again at the cycle time T_1, we see from Proposition 7.3 that

$$\text{average reward per unit time} = \frac{E[\text{reward by time } T_1]}{E[T_1]}$$

However, the average reward per unit is just equal to the proportion of time that the process is in state j. That is, we have the following.

Proposition 7.4 For a regenerative process, the long-run

$$\text{proportion of time in state } j = \frac{E[\text{amount of time in } j \text{ during a cycle}]}{E[\text{time of a cycle}]}$$

Remark If the cycle time T_1 is a continuous random variable, then it can be shown by using an advanced theorem called the "key renewal theorem" that the above is equal also to the limiting probability that the system is in state j at time t. That is, if T_1 is continuous, then

$$\lim_{t \to \infty} P\{X(t) = j\} = \frac{E[\text{amount of time in } j \text{ during a cycle}]}{E[\text{time of a cycle}]}$$

Example 7.17 Consider a positive recurrent continuous time Markov chain that is initially in state i. By the Markovian property, each time the process reenters state i it starts over again. Thus returns to state i are renewals and constitute the beginnings of new cycles. By Proposition 7.4, it follows that the long-run

$$\text{proportion of time in state } j = \frac{E[\text{amount of time in } j \text{ during an } i - i \text{ cycle}]}{\mu_{ii}}$$

where μ_{ii} represents the mean time to return to state i. If we take j to equal i, then we obtain

$$\text{proportion of time in state } i = \frac{1/v_i}{\mu_{ii}} \quad \blacklozenge$$

Example 7.18 (A Queueing System with Renewal Arrivals): Consider a waiting time system in which customers arrive in accordance with an arbitrary renewal process and are served one at time by a single server having an arbitrary service distribution. If we suppose that at time 0 the initial customer has just arrived, then $\{X(t), t \geqslant 0\}$ is a regenerative process, where $X(t)$ denotes the number of customers in the system at time t. The process regenerates each time a customer arrives and finds the server free. $\blacklozenge$

Example 7.19 Although a system needs only a single machine to function, it maintains an additional machine as a backup. A machine in use functions for a random time with density function f and then fails. If a machine fails while the other one is in working condition, then the latter is put in use and, simultaneously, repair begins on the one that just failed. If a machine fails while the other machine is in repair, then the newly failed machine waits until the repair is completed; at that time the repaired machine is put in use and, simultaneously, repair begins on the recently failed one. All repair times have density function g. Find P_0, P_1, P_2, where P_i is the long-run proportion of time that exactly i of the machines are in working condition.

Solution: Let us say that the system is in state i whenever i machines are in working condition $i = 0, 1, 2$. It is then easy to see that every time the system enters state 1 it probabilistically starts over. That is, the system restarts every time that a machine is put in use while, simultaneously, repair begins on the other one. Say that a cycle begins each time the system enters state 1. If we let X denote the working time of the machine put in use at the beginning of a cycle, and let R be the repair time of the other machine, then the length of the cycle, call it T_C, can be expressed as

$$T_C = \max(X, R)$$

The preceding follows when $X \leqslant R$, because, in this case, the machine in use fails before the other one has been repaired, and so a new cycle begins when that repair is completed. Similarly, it follows when $R < X$, because then the repair occurs first, and so a new cycle begins when the machine in use fails. Also, let T_i, $i = 0, 1, 2$, be the amount of time that the system is in state i during a cycle. Then, because the amount of time during a cycle that neither machine is working is $R - X$ provided that this quantity is positive or 0 otherwise, we have

$$T_0 = (R - X)^+$$

Similarly, because the amount of time during the cycle that a single machine is working is $\min(X, R)$, we have

$$T_1 = \min(X, R)$$

Finally, because the amount of time during the cycle that both machines are working is $X - R$ if this quantity is positive or 0 otherwise, we have

$$T_2 = (X - R)^+$$

Hence, we obtain that

$$P_0 = \frac{E[(R - X)^+]}{E[\max(X, R)]}$$

$$P_1 = \frac{E[\min(X, R)]}{E[\max(X, R)]}$$

$$P_2 = \frac{E[(X - R)^+]}{E[\max(X, R)]}$$

That $P_0 + P_1 + P_2 = 1$ follows from the easily checked identity

$$\max(x, r) = \min(x, r) + (x - r)^+ + (r - x)^+$$

The preceding expectations can be computed as follows:

$$E[\max(X, R)] = \int_0^\infty \int_0^\infty \max(x, r) f(x) g(r) \, dx \, dr$$

$$= \int_0^\infty \int_0^r rf(x) g(r) \, dx \, dr + \int_0^\infty \int_r^\infty xf(x) g(r) \, dx \, dr$$

$$E[(R - X)^+] = \int_0^\infty \int_0^\infty (r - x)^+ f(x) g(r) \, dx \, dr$$

$$= \int_0^\infty \int_0^r (r - x) f(x) g(r) \, dx \, dr$$

$$E[\min(X, R)] = \int_0^\infty \int_0^\infty \min(x, r) f(x) g(r) \, dx \, dr$$

$$= \int_0^\infty \int_0^r xf(x) g(r) \, dx \, dr + \int_0^\infty \int_r^\infty rf(x) g(r) \, dx \, dr$$

$$E[(X - R)^+] = \int_0^\infty \int_0^x (x - r) f(x) g(r) \, dr \, dx \quad \blacklozenge$$

7.5.1. Alternating Renewal Processes

Another example of a regenerative process is provided by what is known as an alternating renewal process, which considers a system that can be in one of two states: on or off. Initially it is on, and it remains on for a time Z_1; it then goes off and remains off for a time Y_1. It then goes on for a time Z_2; then off for a time Y_2; then on, and so on.

We suppose that the random vectors (Z_n, Y_n), $n \geq 1$ are independent and identically distributed. That is, both the sequence of random variables $\{Z_n\}$ and the sequence $\{Y_n\}$ are independent and identically distributed; but we allow Z_n and Y_n to be dependent. In other words, each time the process goes on, everything starts over again, but when it then goes off, we allow the length of the off time to depend on the previous on time.

Let $E[Z] = E[Z_n]$ and $E[Y] = E[Y_n]$ denote respectively the mean lengths of an on and off period.

We are concerned with P_{on}, the long-run proportion of time that the system is on. If we let

$$X_n = Y_n + Z_n, \qquad n \geq 1$$

then at time X_1 the process starts over again. That is, the process starts over again after a complete cycle consisting of an on and an off interval. In other words, a renewal occurs whenever a cycle is completed. Therefore, we obtain from Proposition 7.4 that

$$
P_{\text{on}} = \frac{E[Z]}{E[Y] + E[Z]}
$$

$$
= \frac{E[\text{on}]}{E[\text{on}] + E[\text{off}]} \tag{7.15}
$$

Also, if we let P_{off} denote the long-run proportion of time that the system is off, then

$$
P_{\text{off}} = 1 - P_{\text{on}}
$$

$$
= \frac{E[\text{off}]}{E[\text{on}] + E[\text{off}]} \tag{7.16}
$$

Example 7.20 (A Production Process): One example of an alternating renewal process is a production process (or a machine) which works for a time Z_1, then breaks down and has to be repaired (which takes a time Y_1), then works for a time Z_2, then is down for a time Y_2, and so on. If we suppose that the process is as good as new after each repair, then this constitutes an alternating renewal process. It is worthwhile to note that in this example it makes sense to suppose that the repair time will depend on the amount of time the process had been working before breaking down. ◆

Example 7.21 The rate a certain insurance company charges its policy-holders alternates between r_1 and r_0. A new policyholder is initially charged at a rate of r_1 per unit time. When a policyholder paying at rate r_1 has made no claims for the most recent s time units, then the rate charged becomes r_0 per unit time. The rate charged remains at r_0 until a claim is made, at which time it reverts to r_1. Suppose that a given policyholder lives forever and makes claims at times chosen according to a Poisson process with rate λ, and find

(a) P_i, the proportion of time that the policyholder pays at rate r_i, $i = 0, 1$;
(b) the long-run average amount paid per unit time.

Solution: If we say that the system is "on" when the policyholder pays at rate r_1 and "off" when she pays at rate r_0, then this on–off system is an alternating renewal process with a new cycle starting each time a claim is made. If X is the time between successive claims, then the on time in the cycle is the smaller of s and X. (Note that if $X < s$, then the off time in the cycle is 0.) Since X is exponential with rate λ, the preceding yields that

$$E[\text{on time in cycle}] = E[\min(X, s)]$$

$$= \int_0^s x\lambda e^{-\lambda x}\,dx + se^{-\lambda s}$$

$$= \frac{1}{\lambda}(1 - e^{-\lambda s})$$

Since $E[X] = 1/\lambda$, we see that

$$P_1 = \frac{E[\text{on time in cycle}]}{E[X]} = 1 - e^{-\lambda s}$$

and

$$P_0 = 1 - P_1 = e^{-\lambda s}$$

The long-run average amount paid per unit time is

$$r_0 P_0 + r_1 P_1 = r_1 - (r_1 - r_0)e^{-\lambda s} \quad \blacklozenge$$

Example 7.22 (The Age of a Renewal Process): Suppose we are interested in determining the proportion of time that the age of a renewal process is less than some constant c. To do so, let a cycle correspond to a renewal, and say that the system is "on" at time t if the age at t is less than or equal to c, and say it is "off" if the age at t is greater than c. In other words, the system is "on" the first c time units of a renewal interval, and "off" the remaining time. Hence, letting X denote a renewal interval, we have, from Equation (7.15),

$$\text{proportion of time age is less than } c = \frac{E[\min(X, c)]}{E[X]}$$

$$= \frac{\int_0^\infty P\{\min(X, c) > x\}\,dx}{E[X]}$$

$$= \frac{\int_0^c P\{X > x\}\,dx}{E[X]}$$

$$= \frac{\int_0^c (1 - F(x))\,dx}{E[X]} \qquad (7.17)$$

where F is the distribution function of X and where we have used the identity that for a nonnegative random variable Y

$$E[Y] = \int_0^\infty P\{Y > x\} dx \quad \blacklozenge$$

Example 7.23   (The Excess of a Renewal Process): Let us now consider the long-run proportion of time that the excess of a renewal process is less than c. To determine this quantity, let a cycle correspond to a renewal interval and say that the system is on whenever the excess of the renewal process is greater than or equal to c and that it is off otherwise. In other words, whenever a renewal occurs the process goes on and stays on until the last c time units of the renewal interval when it goes off. Clearly this is an alternating renewal process, and so we obtain from Equation (7.16) that

$$\text{long-run proportion of time the excess is less than } c = \frac{E[\text{off time in cycle}]}{E[\text{cycle time}]}$$

Now, if X is the length of a renewal interval, then since the system is off the last c time units of this interval, it follows that the off time in the cycle will equal $\min(X, c)$. Thus,

$$\text{long-run proportion of time the excess is less than } c = \frac{E[\min(X, c)]}{E[X]}$$

$$= \frac{\int_0^c (1 - F(x)) \, dx}{E[X]}$$

where the final equality follows from Equation (7.17). Thus, we see from the result of Example 7.19 that the long-run proportion of time that the excess is less than c and the long-run proportion of time that the age is less than c are equal. One way to understand this equivalence is to consider a renewal process that has been in operation for a long time and then observe it going backwards in time. In doing so, we observe a counting process where the times between successive events are independent random variables having distribution F. That is, when we observe a renewal process going backwards in time we again observe a renewal process having the same probability structure as the original. Since the excess (age) at any time for the backwards process corresponds to the age (excess) at that time for the original renewal process (see Figure 7.3), it follows that all long-run properties of the age and the excess must be equal. $\blacklozenge$

If μ is the mean interarrival time, then the distribution function F_e, defined by

$$F_e(x) = \int_0^x \frac{1 - F(y)}{\mu} \, dy$$

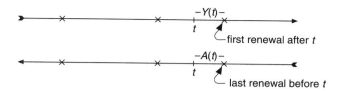

Figure 7.3. Arrowheads indicate direction of time.

is called the *equilibrium distribution* of F. From the preceding, it follows that $F_e(x)$ represents the long-run proportion of time that the age, and the excess, of the renewal process is less than or equal to x.

Example 7.24 (An Inventory Example): Suppose that customers arrive at a specified store in accordance with a renewal process having interarrival distribution F. Suppose that the store stocks a single type of item and that each arriving customer desires a random amount of this commodity, with the amounts desired by the different customers being independent random variables having the common distribution G. The store uses the following (s, S) ordering policy: If its inventory level falls below s then it orders enough to bring its inventory up to S. That is, if the inventory after serving a customer is x, then the amount ordered is

$$S - x, \quad \text{if } x < s$$
$$0, \quad \text{if } x \geqslant s$$

The order is assumed to be instantaneously filled.

For a fixed value y, $s \leqslant y \leqslant S$, suppose that we are interested in determining the long-run proportion of time that the inventory on hand is at least as large as y. To determine this quantity, let us say that the system is "on" whenever the inventory level is at least y and is "off" otherwise. With these definitions, the system will go on each time that a customer's demand causes the store to place an order that results in its inventory level returning to S. Since whenever this occurs a customer must have just arrived it follows that the times until succeeding customers arrive will constitute a renewal process with interarrival distribution F; that is, the process will start over each time the system goes back on. Thus, the on and off periods so defined constitute an alternating renewal process, and from Equation (7.15) we have that

$$\text{long-run proportion of time inventory} \geqslant y = \frac{E[\text{on time in a cycle}]}{E[\text{cycle time}]} \quad (7.18)$$

Now, if we let $D_1, D_2, \ldots$ denote the successive customer demands, and let

$$N_x = \min(n : D_1 + \cdots + D_n > S - x) \quad (7.19)$$

then it is the N_y customer in the cycle that causes the inventory level to fall below y, and it is the N_s customer that ends the cycle. As a result, if we let $X_i, i \geqslant 1$, denote the interarrival times of customers, then

$$\text{on time in a cycle} = \sum_{i=1}^{N_y} X_i \qquad (7.20)$$

$$\text{cycle time} = \sum_{i=1}^{N_s} X_i \qquad (7.21)$$

Assuming that the interarrival times are independent of the successive demands, we have that

$$E\left[\sum_{i=1}^{N_y} X_i\right] = E\left[E\left[\sum_{i=1}^{N_y} X_i \mid N_y\right]\right]$$

$$= E[N_y E[X]]$$

$$= E[X] E[N_y]$$

Similarly,

$$E\left[\sum_{i=1}^{N_s} X_i\right] = E[X] E[N_s]$$

Therefore, from Equations (7.18), (7.20), and (7.21) we see that

$$\text{long-run proportion of time inventory} \geqslant y = \frac{E[N_y]}{E[N_s]} \qquad (7.22)$$

However, as the D_i, $i \geqslant 1$, are independent and identically distributed nonnegative random variables with distribution G, it follows from Equation (7.19) that N_x has the same distribution as the index of the first event to occur after time $S - x$ of a renewal process having interarrival distribution G. That is, $N_x - 1$ would be the number of renewals by time $S - x$ of this process. Hence, we see that

$$E[N_y] = m(S - y) + 1$$

$$E[N_s] = m(S - s) + 1$$

where

$$m(t) = \sum_{n=1}^{\infty} G_n(t)$$

From Equation (7.22), we arrive at

$$\text{long-run proportion of time inventory} \geqslant y = \frac{m(S - y) + 1}{m(S - s) + 1}, \qquad s \leqslant y \leqslant S$$

For instance, if the customer demands are exponentially distributed with mean $1/\mu$, then

$$\text{long-run proportion of time inventory} \geqslant y = \frac{\mu(S-y)+1}{\mu(S-s)+1}, \qquad s \leqslant y \leqslant S \quad \blacklozenge$$

7.6. Semi-Markov Processes

Consider a process that can be in state 1 or state 2 or state 3. It is initially in state 1 where it remains for a random amount of time having mean μ_1, then it goes to state 2 where it remains for a random amount of time having mean μ_2, then it goes to state 3 where it remains for a mean time μ_3, then back to state 1, and so on. What proportion of time is the process in state $i, i = 1, 2, 3$?

If we say that a cycle is completed each time the process returns to state 1, and if we let the reward be the amount of time we spend in state i during that cycle, then the above is a renewal reward process. Hence, from Proposition 7.3 we obtain that P_i, the proportion of time that the process is in state i, is given by

$$P_i = \frac{\mu_i}{\mu_1 + \mu_2 + \mu_3}, \qquad i = 1, 2, 3$$

Similarly, if we had a process which could be in any of N states $1, 2, \ldots, N$ and which moved from state $1 \to 2 \to 3 \to \cdots \to N-1 \to N \to 1$, then the long-run proportion of time that the process spends in state i is

$$P_i = \frac{\mu_i}{\mu_1 + \mu_2 + \cdots + \mu_N}, \qquad i = 1, 2, \ldots, N$$

where μ_i is the expected amount of time the process spends in state i during each visit.

Let us now generalize the above to the following situation. Suppose that a process can be in any one of N states $1, 2, \ldots, N$, and that each time it enters state i it remains there for a random amount of time having mean μ_i and then makes a transition into state j with probability P_{ij}. Such a process is called a *semi-Markov process*. Note that if the amount of time that the process spends in each state before making a transition is identically 1, then the semi-Markov process is just a Markov chain.

Let us calculate P_i for a semi-Markov process. To do so, we first consider π_i the proportion of transitions that take the process into state i. Now if we let X_n denote the state of the process after the nth transition, then $\{X_n, n \geqslant 0\}$ is a Markov chain with transition probabilities $P_{ij}, i, j = 1, 2, \ldots, N$.

Hence, π_i will just be the limiting (or stationary) probabilities for this Markov chain (Section 4.4). That is, π_i will be the unique nonnegative solution of

$$\sum_{i=1}^{N} \pi_i = 1,$$

$$\pi_i = \sum_{j=1}^{N} \pi_j P_{ji}, \qquad i = 1, 2, \dots, N^* \tag{7.23}$$

Now since the process spends an expected time μ_i in state i whenever it visits that state, it seems intuitive that P_i should be a weighted average of the π_i where π_i is weighted proportionately to μ_i. That is,

$$P_i = \frac{\pi_i \mu_i}{\sum_{j=1}^{N} \pi_j \mu_j}, \qquad i = 1, 2, \dots, N \tag{7.24}$$

where the π_i are given as the solution to Equation (7.23).

Example 7.25 Consider a machine that can be in one of three states: *good condition*, *fair condition*, or *broken down*. Suppose that a machine in good condition will remain this way for a mean time μ_1 and then will go to either the fair condition or the broken condition with respective probabilities $\frac{3}{4}$ and $\frac{1}{4}$. A machine in fair condition will remain that way for a mean time μ_2 and then will break down. A broken machine will be repaired, which takes a mean time μ_3, and when repaired will be in good condition with probability $\frac{2}{3}$ and fair condition with probability $\frac{1}{3}$. What proportion of time is the machine in each state?

Solution: Letting the states be 1, 2, 3, we have by Equation (7.23) that the π_i satisfy

$$\pi_1 + \pi_2 + \pi_3 = 1,$$

$$\pi_1 = \tfrac{2}{3}\pi_3,$$

$$\pi_2 = \tfrac{3}{4}\pi_1 + \tfrac{1}{3}\pi_3,$$

$$\pi_3 = \tfrac{1}{4}\pi_1 - \pi_2$$

The solution is

$$\pi_1 = \tfrac{4}{15}, \qquad \pi_2 = \tfrac{1}{3}, \qquad \pi_3 = \tfrac{2}{5}$$

Hence, from Equation (7.24) we obtain that P_i, the proportion of time the

---

* We shall assume that there exists a solution of Equation (7.23). That is, we assume that all of the states in the Markov chain communicate.

machine is in state i, is given by

$$P_1 = \frac{4\mu_1}{4\mu_1 + 5\mu_2 + 6\mu_3},$$

$$P_2 = \frac{5\mu_2}{4\mu_1 + 5\mu_2 + 6\mu_3},$$

$$P_3 = \frac{6\mu_3}{4\mu_1 + 5\mu_2 + 6\mu_3}$$

For instance, if $\mu_1 = 5$, $\mu_2 = 2$, $\mu_3 = 1$, then the machine will be in good condition $\frac{5}{9}$ of the time, in fair condition $\frac{5}{18}$ of the time, in broken condition $\frac{1}{6}$ of the time. ✦

Remark   When the distributions of the amount of time spent in each state during a visit are continuous, then P_i also represents the limiting (as $t \to \infty$) probability that the process will be in state i at time t.

Example 7.26   Consider a renewal process in which the interarrival distribution is discrete and is such that

$$P\{X = i\} = p_i, \qquad i \geqslant 1$$

where X represents an interarrival random variable. Let $L(t)$ denote the length of the renewal interval that contains the point t [that is, if $N(t)$ is the number of renewals by time t and X_n the nth interarrival time, then $L(t) = X_{N(t)+1}$]. If we think of each renewal as corresponding to the failure of a lightbulb (which is then replaced at the beginning of the next period by a new bulb), then $L(t)$ will equal i if the bulb in use at time t dies in its ith period of use.

It is easy to see that $L(t)$ is a semi-Markov process. To determine the proportion of time that $L(t) = j$, note that each time a transition occurs — that is, each time a renewal occurs — the next state will be j with probability p_j. That is, the transition probabilities of the embedded Markov chain are $P_{ij} = p_j$. Hence, the limiting probabilities of this embedded chain are given by

$$\pi_j = p_j$$

and, since the mean time the semi-Markov process spends in state j before a transition occurs is j, it follows that the long-run proportion of time the state is j is

$$P_j = \frac{jp_j}{\Sigma_i \, ip_i} \quad ✦$$

7.7. The Inspection Paradox

Suppose that a piece of equipment, say, a battery, is installed and serves until it breaks down. Upon failure it is instantly replaced by a like battery, and this process continues without interruption. Letting $N(t)$ denote the number of batteries that have failed by time t, we have that $\{N(t), t \geqslant 0\}$ is a renewal process.

Suppose further that the distribution F of the lifetime of a battery is not known and is to be estimated by the following sampling inspection scheme. We fix some time t and observe the total lifetime of the battery that is in use at time t. Since F is the distribution of the lifetime for all batteries, it seems reasonable that it should be the distribution for this battery. However, this is the *inspection paradox* for it turns out that the *battery in use at time t tends to have a larger lifetime than an ordinary battery*.

To understand the preceding so-called paradox, we reason as follows. In renewal theoretic terms what we are interested in is the length of the renewal interval containing the point t. That is, we are interested in $X_{N(t)+1} = S_{N(t)+1} - S_{N(t)}$ (see Figure 7.2). To calculate the distribution of $X_{N(t)+1}$ we condition on the time of the last renewal prior to (or at) time t. That is,

$$P\{X_{N(t)+1} > x\} = E[P\{X_{N(t)+1} > x \mid S_{N(t)} = t - s\}]$$

where we recall (Figure 7.2) that $S_{N(t)}$ is the time of the last renewal prior to (or at) t. Since there are no renewals between $t - s$ and t, it follows that $X_{N(t)+1}$ must be larger than x if $s > x$. That is,

$$P\{X_{N(t)+1} > x \mid S_{N(t)} = t - s\} = 1 \qquad \text{if } s > x \qquad (7.25)$$

On the other hand, suppose that $s \leqslant x$. As before, we know that a renewal occurred at time $t - s$ and no additional renewals occurred between $t - s$ and t, and we ask for the probability that no renewals occur for an additional time $x - s$. That is, we are asking for the probability that an interarrival time will be greater than x given that it is greater than s. Therefore, for $s \leqslant x$,

$$P\{X_{N(t)+1} > x \mid S_{N(t)} = t - s\}$$

$$= P\{\text{interarrival time} > x \mid \text{interarrival time} > s\}$$

$$= P\{\text{interarrival time} > x\}/P[\text{interarrival time} > s]$$

$$= \frac{1 - F(x)}{1 - F(s)}$$

$$\geqslant 1 - F(x) \qquad (7.26)$$

Figure 7.4.

Hence, from Equations (7.25) and (7.26) we see that, for all s,

$$P\{X_{N(t)+1} > x \mid S_{N(t)} = t - s\} \geqslant 1 - F(x)$$

Taking expectations on both sides yields that

$$P\{X_{N(t)+1} > x\} \geqslant 1 - F(x) \tag{7.27}$$

However, $1 - F(x)$ is the probability that an ordinary renewal interval is larger than x, that is, $1 - F(x) = P\{X_n > x\}$, and thus Equation (7.27) is a statement of the inspection paradox that the length of the renewal interval containing the point t tends to be larger than an ordinary renewal interval.

Remark   To obtain an intuitive feel for the so-called inspection paradox, reason as follows. We think of the whole line being covered by renewal intervals, one of which covers the point t. Is it not more likely that a larger interval, as opposed to a shorter interval, covers the point t?

We can explicitly calculate the distribution of $X_{N(t)+1}$ when the renewal process is a Poisson process. [Note that, in the general case, we did not need to calculate explicitly $P\{X_{N(t)+1} > x\}$ to show that it was at least as large as $1 - F(x)$.] To do so we write

$$X_{N(t)+1} = A(t) + Y(t)$$

where $A(t)$ denotes the time from t since the last renewal, and $Y(t)$ denotes the time from t until the next renewal (see Figure 7.4). $A(t)$ is the *age* of the process at time t (in our example it would be the age at time t of the battery in use at time t), and $Y(t)$ is the *excess life* of the process at time t (it is the additional time from t until the battery fails). Of course, it is true that $A(t) = t - S_{N(t)}$, and $Y(t) = S_{N(t)+1} - t$.

To calculate the distribution of $X_{N(t)+1}$ we first note the important fact that, for a Poisson process, $A(t)$ and $Y(t)$ are independent. This follows since by the memoryless property of the Poisson process, the time from t until the next renewal will be exponentially distributed and will be independent of all that has previously occurred [including, in particular, $A(t)$]. In fact, this shows that if $\{N(t), t \geqslant 0\}$ is a Poisson process with rate λ, then

$$P\{Y(t) \leqslant x\} = 1 - e^{-\lambda x} \tag{7.28}$$

The distribution of $A(t)$ may be obtained as follows

$$P\{A(t) > x\} = \begin{cases} P\{0 \text{ renewals in } [t-x, t]\}, & \text{if } x \leqslant t \\ 0, & \text{if } x > t \end{cases}$$

$$= \begin{cases} e^{-\lambda x}, & \text{if } x \leqslant t \\ 0, & \text{if } x > t \end{cases}$$

or, equivalently,

$$P\{A(t) \leqslant x\} = \begin{cases} 1 - e^{-\lambda x}, & x \leqslant t \\ 1, & x > t \end{cases} \tag{7.29}$$

Hence, by the independence of $Y(t)$ and $A(t)$ the distribution of $X_{N(t)+1}$ is just the convolution of the exponential distribution equation (7.28) and the distribution of equation (7.29). It is interesting to note that for t large, $A(t)$ approximately has an exponential distribution. Thus, for t large, $X_{N(t)+1}$ has the distribution of the convolution of two identically distributed exponential random variables, which by Section 5.2.3, is the gamma distribution with parameters $(2, \lambda)$. In particular, for t large, the expected length of the renewal interval containing the point t is approximately *twice* the expected length of an ordinary renewal interval.

7.8. Computing the Renewal Function

The difficulty with attempting to use the identity

$$m(t) = \sum_{n=1}^{\infty} F_n(t)$$

to compute the renewal function is that the determination of $F_n(t) = P\{X_1 + \cdots + X_n \leqslant t\}$ requires the computation of an n-dimensional integral. We present below an effective algorithm which requires as inputs only one-dimensional integrals.

Let Y be an exponential random variable having rate λ, and suppose that Y is independent of the renewal process $\{N(t), t \geqslant 0\}$. We start by determining $E[N(Y)]$, the expected number of renewals by the random time Y. To do so, we first condition on X_1, the time of the first renewal. This yields

$$E[N(Y)] = \int_0^{\infty} E[N(Y) | X_1 = x] f(x) \, dx \tag{7.30}$$

where f is the interarrival density. To determine $E[N(Y) | X_1 = x]$, we now condition on whether or not Y exceeds x. Now, if $Y < x$, then as the first renewal occurs at time x, it follows that the number of renewals by time Y

is equal to 0. On the other hand, if we are given that $x < Y$, then the number of renewals by time Y will equal 1 (the one at x) plus the number of additional renewals between x and Y. But by the memoryless property of exponential random variables, it follows that, given that $Y > x$, the amount by which it exceeds x is also exponential with rate λ, and so given that $Y > x$ the number of renewals between x and Y will have the same distribution as $N(Y)$. Hence,

$$E[N(Y) \mid X_1 = x, Y < x] = 0,$$

$$E[N(Y) \mid X_1 = x, Y > x] = 1 + E[N(Y)]$$

and so,

$$
\begin{aligned}
E[N(Y) \mid X_1 = x] &= E[N(Y) \mid X_1 = x, Y < x]P\{Y < x \mid X_1 = x\} \\
&\quad + E[N(Y) \mid X_1 = x, Y > x]P\{Y > x \mid X_1 = x\} \\
&= E[N(Y) \mid X_1 = x, Y > x]P\{Y > x\}
\end{aligned}
$$

since Y and X_1 are independent

$$= (1 + E[N(Y)])e^{-\lambda x}$$

Substituting this into Equation (7.30) gives

$$E[N(Y)] = (1 + E[N(Y)]) \int_0^\infty e^{-\lambda x} f(x)\, dx$$

or

$$E[N(Y)] = \frac{E[e^{-\lambda X}]}{1 - E[e^{-\lambda X}]} \tag{7.31}$$

where X has the renewal interarrival distribution.

If we let $\lambda = 1/t$, then Equation (7.31) presents an expression for the expected number of renewals (not by time t, but) by a random exponentially distributed time with mean t. However, as such a random variable need not be close to its mean (its variance is t^2), Equation (7.31) need not be particularly close to $m(t)$. To obtain an accurate approximation suppose that $Y_1, Y_2, \ldots, Y_n$ are independent exponentials with rate λ and suppose they are also independent of the renewal process. Let, for $r = 1, \ldots, n$,

$$m_r = E[N(Y_1 + \cdots + Y_r)]$$

To compute an expression for m_r, we again start by conditioning on X_1, the time of the first renewal.

$$m_r = \int_0^\infty E[N(Y_1 + \cdots + Y_r) \mid X_1 = x] f(x)\, dx \tag{7.32}$$

To determine the foregoing conditional expectation, we now condition on the number of partial sums $\Sigma_{i=1}^{j} Y_i$, $j = 1, \ldots, r$, that are less than x. Now, if all r partial sums are less than x — that is, if $\Sigma_{i=1}^{r} Y_i < x$ — then clearly the number of renewals by time $\Sigma_{i=1}^{r} Y_i$ is 0. On the other hand, given that $k, k < r$, of these partial sums are less than x, it follows from the lack of memory property of the exponential that the number of renewals by time $\Sigma_{i=1}^{r} Y_i$ will have the same distribution as 1 plus $N(Y_{k+1} + \cdots + Y_r)$. Hence,

$$E\left[N(Y_1 + \cdots + Y_r) \,\middle|\, X_1 = x, k \text{ of the sums } \sum_{i=1}^{j} Y_i \text{ are less than } x\right]$$

$$= \begin{cases} 0, & \text{if } k = r \\ 1 + m_{r-k}, & \text{if } k < r \end{cases} \tag{7.33}$$

To determine the distribution of the number of the partial sums that are less than x, note that the successive values of these partial sums $\Sigma_{i=1}^{j} Y_i$, $j = 1, \ldots, r$, have the same distribution as the first r event times of a Poisson process with rate λ (since each successive partial sum is the previous sum plus an independent exponential with rate λ). Hence, it follows that, for $k < r$,

$$P\left\{k \text{ of the partial sums } \sum_{i=1}^{j} Y_i \text{ are less than } x \,\middle|\, X_1 = x\right\}$$

$$= \frac{e^{-\lambda x}(\lambda x)^k}{k!} \tag{7.34}$$

Upon substitution of Equations (7.33) and (7.34) into Equation (7.32), we obtain

$$m_r = \int_0^{\infty} \sum_{k=0}^{r-1} (1 + m_{r-k}) \frac{e^{-\lambda x}(\lambda x)^k}{k!} f(x)\, dx$$

or, equivalently,

$$m_r = \frac{\Sigma_{k=1}^{r-1}(1 + m_{r-k})E[X^k e^{-\lambda X}](\lambda^k/k!) + E[e^{-\lambda X}]}{1 - E[e^{-\lambda X}]} \tag{7.35}$$

If we set $\lambda = n/t$, then starting with m_1 given by Equation (7.31), we can use Equation (7.35) to recursively compute $m_2, \ldots, m_n$. The approximation of $m(t) = E[N(t)]$ is given by $m_n = E[N(Y_1 + \cdots + Y_n)]$. Since $Y_1 + \cdots + Y_n$ is the sum of n independent exponential random variables each with mean t/n, it follows that it is (gamma) distributed with mean t and variance $nt^2/n^2 = t^2/n$. Hence, by choosing n large, $\Sigma_{i=1}^{n} Y_i$ will be a random variable having most of its probability concentrated about t, and so $E[N(\Sigma_{i=1}^{n} Y_i)]$ should be quite close to $E[N(t)]$. [Indeed, if $m(t)$ is continuous at t, it can be shown that these approximations converge to $m(t)$ as n goes to infinity.]

Table 7.1

F_i		Exact	Approximation				
i	t	$m(t)$	$n = 1$	$n = 3$	$n = 10$	$n = 25$	$n = 50$
1	1	0.2838	0.3333	0.3040	0.2903	0.2865	0.2852
1	2	0.7546	0.8000	0.7697	0.7586	0.7561	0.7553
1	5	2.250	2.273	2.253	2.250	2.250	2.250
1	10	4.75	4.762	4.751	4.750	4.750	4.750
2	0.1	0.1733	0.1681	0.1687	0.1689	0.1690	—
2	0.3	0.5111	0.4964	0.4997	0.5010	0.5014	—
2	0.5	0.8404	0.8182	0.8245	0.8273	0.8281	0.8283
2	1	1.6400	1.6087	1.6205	1.6261	1.6277	1.6283
2	3	4.7389	4.7143	4.7294	4.7350	4.7363	4.7367
2	10	15.5089	15.5000	15.5081	15.5089	15.5089	15.5089
3	0.1	0.2819	0.2692	0.2772	0.2804	0.2813	—
3	0.3	0.7638	0.7105	0.7421	0.7567	0.7609	—
3	1	2.0890	2.0000	2.0556	2.0789	2.0850	2.0870
3	3	5.4444	5.4000	5.4375	5.4437	5.4442	5.4443

Example 7.27 Table 7.1 compares the approximation with the exact value for the distributions F_i with densities f_i, $i = 1, 2, 3$, which are given by

$$f_1(x) = xe^{-x},$$
$$1 - F_2(x) = 0.3e^{-x} + 0.7e^{-2x},$$
$$1 - F_3(x) = 0.5e^{-x} + 0.5e^{-5x} \quad \blacklozenge$$

7.9. Applications to Patterns

A counting process with independent interarrival times $X_1, X_2, \ldots$ is said to be a *delayed* or *general* renewal process if X_1 has a different distribution from the identically distributed random variables $X_2, X_3, \ldots$. That is, a delayed renewal process is a renewal process in which the first interarrival time has a different distribution than the others. Delayed renewal processes often arise in practice and it is important to note that all of the limiting theorems about $N(t)$, the number of events by time t, remain valid. For instance, it remains true that

$$\frac{E[N(t)]}{t} \to \frac{1}{\mu} \quad \text{and} \quad \frac{\text{Var}(N(t))}{t} \to \sigma^2/\mu^3 \quad \text{as } t \to \infty$$

where μ and σ^2 are the expected value and variance of the interarrivals X_i, $i > 1$.

7.9.1. Patterns of Discrete Random Variables

Let $X_1, X_2, \ldots$ be independent with $P\{X_i = j\} = p(j)$, $j \geq 0$, and let T denote the first time the pattern $x_1, \ldots, x_r$ occurs. If we say that a renewal occurs at time n, $n \geq r$, if $(X_{n-r+1}, \ldots, X_n) = (x_1, \ldots, x_r)$, then $N(n)$, $n \geq 1$, is a delayed renewal process, where $N(n)$ denotes the number of renewals by time n. It follows that

$$\frac{E[N(n)]}{n} \to \frac{1}{\mu} \qquad \text{as } n \to \infty \qquad (7.36)$$

$$\frac{\text{Var}(N(n))}{n} \to \frac{\sigma^2}{\mu^3} \qquad \text{as } n \to \infty \qquad (7.37)$$

where μ and σ are, respectively, the mean and standard deviation of the time between successive renewals. Whereas, in Example 4.20, we showed how to use Markov chain theory to compute the expected value of T, we will now show how to use renewal theory results to compute both its mean and its variance.

To begin, let $I(i)$ equal 1 if there is a renewal at time i and let it be 0 otherwise, $i \geq r$. Also, let $p = \Pi_{i=1}^r p(x_i)$. Since,

$$P\{I(i) = 1\} = P\{X_{i-r+1} = i_1, \ldots, X_i = i_r\} = p$$

it follows that $I(i)$, $i \geq r$, are Bernoulli random variables with parameter p. Now,

$$N(n) = \sum_{i=r}^n I(i)$$

so

$$E[N(n)] = \sum_{i=r}^n E[I(i)] = (n - r + 1)p$$

Dividing by n and then letting $n \to \infty$ gives, from Equation (7.36),

$$\mu = 1/p \qquad (7.38)$$

That is, the mean time between successive occurrences of the pattern is equal to $1/p$. Also,

$$\frac{\text{Var}(N(n))}{n} = \frac{1}{n}\sum_{i=r}^n \text{Var}(I(i)) + \frac{2}{n}\sum_{i=r}^{n-1}\sum_{n \geq j > i} \text{Cov}(I(i), I(j))$$

$$= \frac{n-r+1}{n}p(1-p) + \frac{2}{n}\sum_{i=r}^{n-1}\sum_{i < j \leq \min(i+r-1, n)} \text{Cov}(I(i), I(j))$$

where the final equality used the fact that $I(i)$ and $I(j)$ are independent, and thus have zero covariance, when $|i - j| \geq r$. Letting $n \to \infty$, and using the fact that $\text{Cov}(I(i), I(j))$ depends on i and j only through $|j - i|$, gives

$$\frac{\text{Var}(N(n))}{n} \to p(1-p) + 2\sum_{j=1}^{r-1} \text{Cov}(I(r), I(r+j))$$

Therefore, using Equations (7.37) and (7.38), we see that

$$\sigma^2 = p^{-2}(1 - p) + 2p^{-3} \sum_{j=1}^{r-1} \text{Cov}(I(r), I(r + j)) \tag{7.39}$$

Let us now consider the amount of "overlap" in the pattern. The overlap, equal to the number of values at the end of one pattern that can be used as the beginning part of the next pattern, is said to be of size $k, k > 0$, if

$$k = \max\{j < r : (i_{r-j+1}, \ldots, i_r) = (i_1, \ldots, i_j)\}$$

and is of size 0 if for all $k = 1, \ldots, r - 1$, $(i_{r-k+1}, \ldots, i_r) \neq (i_1, \ldots, i_k)$. Thus, for instance, the pattern $0, 0, 1, 1$ has overlap 0, whereas $0, 0, 1, 0, 0$ has overlap 2. We consider two cases.

Case 1.   The Pattern Has Overlap 0

In this case, $N(n), n \geq 1$, is an ordinary renewal process and T is distributed as an interarrival time with mean μ and variance σ^2. Hence, we have the following from Equation (7.38):

$$E[T] = \mu = \frac{1}{p} \tag{7.40}$$

Also, since two patterns cannot occur within a distance less than r of each other, it follows that $I(r)I(r + j) = 0$ when $1 \leq j \leq r - 1$. Hence,

$$\text{Cov}(I(r), I(r + j)) = -E[I(r)]E[I(r + j)] = -p^2, \qquad \text{if } 1 \leq j \leq r - 1$$

Hence, from Equation (7.39), we obtain

$$\text{Var}(T) = \sigma^2 = p^{-2}(1 - p) - 2p^{-3}(r - 1)p^2 = p^{-2} - (2r - 1)p^{-1} \tag{7.41}$$

Remark   In cases of "rare" patterns, if the pattern hasn't yet occurred by some time n, then it would seem that we would have no reason to believe that the remaining time would be much less than if we were just beginning from scratch. That is, it would seem that the distribution is approximately memoryless and would thus be approximately exponentially distributed. Thus, since the variance of an exponential is equal to its mean squared, we would expect when μ is large that $\text{Var}(T) \approx E^2[T]$, and this is borne out by the preceding, which states that $\text{Var}(T) = E^2[T] - (2r - 1)E[T]$.

Example 7.28   Suppose we are interested in the number of times that a fair coin needs to be flipped before the pattern h, h, t, h, t occurs. For this pattern, $r = 5$, $p = \frac{1}{32}$, and the overlap is 0. Hence, from Equations (7.40) and (7.41)

$$E[T] = 32, \qquad \text{Var}(T) = (32)^2 - 9 \times 32 = 736,$$

and

$$\text{Var}(T)/E^2[T] = 0.71875$$

On the other hand, if $p(i) = i/10$, $i = 1, 2, 3, 4$ and the pattern is $1, 2, 1, 4,$

1, 3, 2 then $r = 7$, $p = 3/625,000$, and the overlap is 0. Thus, again from Equations (7.40) and (7.41), we see that in this case

$$E[T] = 208,333.33, \qquad \mathrm{Var}(T) = 4.34 \times 10^{10},$$

$$\mathrm{Var}(T)/E^2[T] = 0.99994 \quad \blacklozenge$$

Case 2.   The Overlap Is of Size k

In this case,

$$T = T_{i_1,\ldots,i_k} + T^*$$

where $T_{i_1,\ldots,i_k}$ is the time until the pattern $i_1,\ldots,i_k$ appears and T^*, distributed as an interarrival time of the renewal process, is the additional time that it takes, starting with $i_1,\ldots,i_k$, to obtain the pattern $i_1,\ldots,i_r$. Because these random variables are independent, we have

$$E[T] = E[T_{i_1,\ldots,i_k}] + E[T^*] \tag{7.42}$$

$$\mathrm{Var}(T) = \mathrm{Var}(T_{i_1,\ldots,i_k}) + \mathrm{Var}(T^*) \tag{7.43}$$

Now, from Equation (7.38)

$$E[T^*] = \mu = p^{-1} \tag{7.44}$$

Also, since no two renewals can occur within a distance $r - k - 1$ of each other, it follows that $I(r)I(r + j) = 0$ if $1 \leqslant j \leqslant r - k - 1$. Therefore, from Equation (7.39), we see that

$$\mathrm{Var}(T^*) = \sigma^2 = p^{-2}(1 - p) + 2p^{-3}\left(\sum_{j=r-k}^{r-1} E[I(r)I(r + j)] - (r - 1)p^2\right)$$

$$= p^{-2} - (2r - 1)p^{-1} + 2p^{-3}\sum_{j=r-k}^{r-1} E[I(r)I(r + j)] \tag{7.45}$$

The quantities $E[I(r)I(r + j)]$ in Equation (7.45) can be calculated by considering the particular pattern. To complete the calculation of the first two moments of T, we then compute the mean and variance of $T_{i_1,\ldots,i_k}$ by repeating the same method.

Example 7.29   Suppose that we want to determine the number of flips of a fair coin until the pattern h, h, t, h, h occurs. For this pattern, $r = 5$, $p = \frac{1}{32}$, and the overlap parameter is $k = 2$. Because

$$E[I(5)I(8)] = P\{h, h, t, h, h, t, h, h\} = \tfrac{1}{256}$$

$$E[I(5)I(9)] = P\{h, h, t, h, h, h, t, h, h\} = \tfrac{1}{512}$$

we see from Equations (7.44) and (7.45) that

$$E[T^*] = 32$$

$$\mathrm{Var}(T^*) = (32)^2 - 9(32) + 2(32)^3(\tfrac{1}{256} + \tfrac{1}{512}) = 1120$$

Hence, from Equations (7.42) and (7.43) we obtain

$$E[T] = E[T_{h,h}] + 32, \qquad \text{Var}(T) = \text{Var}(T_{h,h}) + 1120$$

Now, consider the pattern h, h. It has $r = 2, p = \frac{1}{4}$, and overlap parameter 1. Since, for this pattern, $E[I(2)I(3)] = \frac{1}{8}$, we obtain, as in the preceding, that

$$E[T_{h,h}] = E[T_h] + 4,$$

$$\text{Var}(T_{h,h}) = \text{Var}(T_h) + 16 - 3(4) + 2(\tfrac{64}{8}) = \text{Var}(T_h) + 20$$

Finally, for the pattern h, which has $r = 1, p = \frac{1}{2}$, we see from Equations (7.40) and (7.41) that

$$E[T_h] = 2, \qquad \text{Var}(T_h) = 2$$

Putting it all together gives

$$E[T] = 38, \qquad \text{Var}(T) = 1142, \qquad \text{Var}(T)/E^2[T] = 0.79086 \quad \blacklozenge$$

Example 7.30 Suppose that $P\{X_n = i\} = p_i$, and consider the pattern $0, 1, 2, 0, 1, 3, 0, 1$. Then $p = p_0^3 p_1^3 p_2 p_3, r = 8$, and the overlap parameter is $k = 2$. Since

$$E[I(8)I(14)] = p_0^5 p_1^5 p_2^2 p_3^2$$

$$E[I(8)I(15)] = 0$$

we see from Equations (7.42) and (7.44) that

$$E[T] = E[T_{0,1}] + p^{-1}$$

and from Equations (7.43) and (7.45) that

$$\text{Var}(T) = \text{Var}(T_{0,1}) + p^{-2} - 15p^{-1} + 2p^{-1}(p_0 p_1)^{-1}$$

Now, the r and p values of the pattern $0, 1$ are $r(0, 1) = 2, p(0, 1) = p_0 p_1$, and this pattern has overlap 0. Hence, from Equations (7.40) and (7.41),

$$E[T_{0,1}] = (p_0 p_1)^{-1}, \qquad \text{Var}(T_{0,1}) = (p_0 p_1)^{-2} - 3(p_0 p_1)^{-1}$$

For instance, if $p_i = 0.2, i = 0, 1, 2, 3$ then

$$E[T] = 25 + 5^8 = 390,650$$

$$\text{Var}(T) = 625 - 75 + 5^{16} + 35 \times 5^8 = 1.526 \times 10^{11}$$

$$\text{Var}(T)/E^2[T] = 0.99996 \quad \blacklozenge$$

Remark It can be shown that T is a type of discrete random variable called *new better than used* (NBU), which loosely means that if the pattern has not yet occurred by some time n then the additional time until it occurs tends to be less than the time it would take the pattern to occur if one started all over at that point. Such a random variable is known to satisfy (see Proposition 9.6.1 of Ref. [4])

$$\text{Var}(T) \leqslant E^2[T] - E[T] \leqslant E^2[T]$$

Now, suppose that there are s patterns, $A(1), \ldots, A(s)$ and that we are interested in the mean time until one of these patterns occurs, as well as the probability mass function of the one that occurs first. Let us assume, without any loss of generality, that none of the patterns is contained in any of the others. [That is, we rule out such trivial cases as $A(1) = h, h$ and $A(2) = h, h, t$.] To determine the quantities of interest, let $T(i)$ denote the time until pattern $A(i)$ occurs, $i = 1, \ldots, s$, and let $T(i, j)$ denote the additional time, starting with the occurrence of pattern $A(i)$, until pattern $A(j)$ occurs, $i \neq j$. Start by computing the expected values of these random variables. We have already shown how to compute $E[T(i)]$, $i = 1, \ldots, s$. To compute $E[T(i, j)]$, use the same approach, taking into account any "overlap" between the latter part of $A(i)$ and the beginning part of $A(j)$. For instance, suppose $A(1) = 0, 0, 1, 2, 0, 3$, and $A(2) = 2, 0, 3, 2, 0$. Then

$$T(2) = T_{2,0,3} + T(1, 2)$$

where $T_{2,0,3}$ is the time to obtain the pattern 2, 0, 3. Hence,

$$E[T(1, 2)] = E[T(2)] - E[T_{2,0,3}]$$
$$= (p_2^2 p_0^2 p_3)^{-1} + (p_0 p_2)^{-1} - (p_2 p_0 p_3)^{-1}$$

So, suppose now that all of the quantities $E[T(i)]$ and $E[T(i, j)]$ have been computed. Let

$$M = \min_i T(i)$$

and let

$$P(i) = P\{M = T(i)\}, \qquad i = 1, \ldots, s$$

That is, $P(i)$ is the probability that pattern $A(i)$ is the first pattern to occur. Now, for each j we will derive an equation that $E[T(j)]$ satisfies as follows:

$$E[T(j)] = E[M] + E[T(j) - M]$$
$$= E[M] + \sum_{i:i \neq j} E[T(ij)]P(i), \qquad j = 1, \ldots, s \qquad (7.46)$$

where the final equality is obtained by conditioning on which pattern occurs first. But the Equations (7.46) along with the equation

$$\sum_{i=1}^{s} P(i) = 1$$

constitute a set of $s + 1$ equations in the $s + 1$ unknowns $E[M]$, $P(i)$, $i = 1, \ldots, s$. Solving them yields the desired quantities.

Example 7.31 Suppose that we continually flip a fair coin. With $A(1) = h, t, t, h, h$ and $A(2) = h, h, t, h, t$, we have

$$E[T(1)] = 32 + E[T_h] = 34$$

$$E[T(2)] = 32$$

$$E[T(1, 2)] = E[T(2)] - E[T_{h,h}] = 32 - (4 + E[T_h]) = 26$$

$$E[T(2, 1)] = E[T(1)] - E[T_{h,t}] = 34 - 4 = 30$$

Hence, we need solve the equations

$$34 = E[M] + 30P(2)$$

$$32 = E[M] + 26P(1)$$

$$1 = P(1) + P(2)$$

These equations are easily solved, and yield the values

$$P(1) = P(2) = \tfrac{1}{2}, \qquad E[M] = 19$$

Note that although the mean time for pattern $A(2)$ is less than that for $A(1)$, each has the same chance of occurring first. ✦

Equations (7.46) are easy to solve when there are no overlaps in any of the patterns. In this case, for all $i \neq j$

$$E[T(i, j)] = E[T(j)]$$

so Equations (7.46) reduce to

$$E[T(j)] = E[M] + (1 - P(j))E[T(j)]$$

or

$$P(j) = E[M]/E[T(j)]$$

Summing the preceding over all j yields

$$E[M] = \frac{1}{\Sigma_{j=1}^{s} 1/E[T(j)]} \tag{7.47}$$

$$P(j) = \frac{1/E[T(j)]}{\Sigma_{j=1}^{s} 1/E[T(j)]} \tag{7.48}$$

In our next example we use the preceding to reanalyze the model of Example 7.8.

Example 7.32 Suppose that each play of a game is, independently of the outcomes of previous plays, won by player i with probability p_i, $i = 1, \ldots, s$. Suppose further that there are specified numbers $n(1), \ldots, n(s)$ such that the first player i to win $n(i)$ consecutive plays is declared the winner of the match. Find the expected number of plays until there is a winner, and also the probability that the winner is i, $i = 1, \ldots, s$.

Solution: Letting $A(i)$, for $i = 1, \ldots, s$, denote the pattern of n_i consecutive values of i, this problem asks for $P(i)$, the probability that pattern $A(i)$ occurs first, and for $E[M]$. Because

$$E[T(i)] = (1/p_i)^{n(i)} + (1/p_i)^{n(i)-1} + \cdots + 1/p_i = \frac{1 - p_i^{n(i)}}{p_i^{n(i)}(1 - p_i)}$$

we obtain, from Equations (7.47) and (7.48), that

$$E[M] = \frac{1}{\sum_{j=1}^{s} \left[p_j^{n(j)}(1 - p_j)/(1 - p_j^{n(j)}) \right]}$$

$$P(i) = \frac{p_i^{n(i)}(1 - p_i)/(1 - p_i^{n(i)})}{\sum_{j=1}^{s} \left[p_j^{n(j)}(1 - p_j)/(1 - p_j^{n(j)}) \right]} \qquad \blacklozenge$$

7.9.2. The Expected Time to a Maximal Run of Distinct Values

Let X_i, $i \geqslant 1$, be independent and identically distributed random variables that are equally likely to take on any of the values $1, 2, \ldots, m$. Suppose that these random variables are observed sequentially, and let T denote the first time that a run of m consecutive values includes all the values $1, \ldots, m$. That is,

$$T = \min\{n : X_{n-m+1}, \ldots, X_n \text{ are all distinct}\}$$

To compute $E[T]$, define a renewal process by letting the first renewal occur at time T. At this point start over and, without using any of the data values up to T, let the next renewal occur the next time a run of m consecutive values are all distinct, and so on. For instance, if $m = 3$ and the data are

$$1, 3, 3, 2, 1, 2, 3, 2, 1, 3, \ldots, \tag{7.49}$$

then there are two renewals by time 10, with the renewals occurring at times 5 and 9. We call the sequence of m distinct values that constitutes a renewal a *renewal run*.

Let us now transform the renewal process into a delayed renewal reward process by supposing that a reward of 1 is earned at time n, $n \geqslant m$, if the values $X_{n-m+1}, \ldots, X_n$ are all distinct. That is, a reward is earned each time the previous m data values are all distinct. For instance, if $m = 3$ and the data values are as in (7.49) then unit rewards are earned at times 5, 7, 9, and 10. If we let R_i denote the reward earned at time i, then by Proposition 7.3,

$$\lim_n \frac{E[\sum_{i=1}^n R_i]}{n} = \frac{E[R]}{E[T]} \tag{7.50}$$

where R is the reward earned between renewal epochs. Now, with A_i equal to the set of the first i data values of a renewal run, and B_i to the set of the first i values following this renewal run, we have the following:

$$E[R] = 1 + \sum_{i=1}^{m-1} E[\text{reward earned a time } i \text{ after a renewal}]$$

$$= 1 + \sum_{i=1}^{m-1} P\{A_i = B_i\}$$

$$= 1 + \sum_{i=1}^{m-1} \frac{i!}{m^i}$$

$$= \sum_{i=0}^{m-1} \frac{i!}{m^i} \tag{7.51}$$

Hence, since for $i \geqslant m$

$$E[R_i] = P\{X_{i-m+1}, \ldots, X_i \text{ are all distinct}\} = \frac{m!}{m^m}$$

it follows from Equation (7.50) that

$$\frac{m!}{m^m} = \frac{E[R]}{E[T]}$$

Thus from Equation (7.51) we obtain

$$E[T] = \frac{m^m}{m!} \sum_{i=0}^{m-1} i!/m^i$$

The preceding delayed renewal reward process approach also gives us another way of computing the expected time until a specified pattern appears. We illustrate by the following example.

Example 7.33 Compute $E[T]$, the expected time until the pattern h, h, h, t, h, h, h appears, when a coin that comes up heads with probability p and tails with probability $q = 1 - p$ is continually flipped.

Solution: Define a renewal process by letting the first renewal occur when the pattern first appears, and then start over. Also, say that a reward of 1 is earned whenever the pattern appears. If R is the reward earned between renewal epochs, we have

$$E[R] = 1 + \sum_{i=1}^{6} E[\text{reward earned } i \text{ units after a renewal}]$$

$$= 1 + 0 + 0 + 0 + p^3 q + pqp^3 + p^2 qp^3$$

Hence, since the expected reward earned at time i is $E[R_i] = p^6 q$, we obtain the following from the renewal reward theorem:

$$\frac{1 + qp^3 + qp^4 + qp^5}{E[T]} = qp^6$$

or

$$E[T] = q^{-1}p^{-6} + p^{-3} + p^{-2} + p^{-1} \quad \blacklozenge$$

7.9.3. Increasing Runs of Continuous Random Variables

Let $X_1, X_2, \ldots$ be a sequence of independent and identically distributed continuous random variables, and let T denote the first time that there is a string of r consecutive increasing values. That is,

$$T = \min\{n \geqslant r : X_{n-r+1} < X_{n-r+2} < \cdots < X_n\}$$

To compute $E[T]$, define a renewal process as follows. Let the first renewal occur at T. Then, using only the data values after T, say that the next renewal occurs when there is again a string of r consecutive increasing values, and continue in this fashion. For instance, if $r = 3$ and the first 15 data values are

$$12, 20, 22, 28, 43, 18, 24, 33, 60, 4, 16, 8, 12, 15, 18$$

then 3 renewals would have occurred by time 15, namely, at times 3, 8, and 14. If we let $N(n)$ denote the number of renewals by time n, then by the elementary renewal theorem

$$\frac{E[N(n)]}{n} \to \frac{1}{E[T]}$$

To compute $E[N(n)]$, define a stochastic process whose state at time k, call it S_k, is equal to the number of consecutive increasing values at time k. That is, for $1 \leqslant j \leqslant k$

$$S_k = j \quad \text{if } X_{k-j} > X_{k-j+1} < \cdots < X_{k-1} < X_k$$

where $X_0 = \infty$. Note that a renewal will occur at time k if and only if $S_k = ir$ for some $i \geqslant 1$. For instance, if $r = 3$ and

$$X_5 > X_6 < X_7 < X_8 < X_9 < X_{10} < X_{11}$$

then

$$S_6 = 1, \quad S_7 = 2, \quad S_8 = 3, \quad S_9 = 4, \quad S_{10} = 5, \quad S_{11} = 6$$

and renewals occur at times 8 and 11. Now, for $k > j$

$$P\{S_k = j\} = P\{X_{k-j} > X_{k-j+1} < \cdots < X_{k-1} < X_k\}$$
$$= P\{X_{k-j+1} < \cdots < X_{k-1} < X_k\}$$
$$\quad - P\{X_{k-j} < X_{k-j+1} < \cdots < X_{k-1} < X_k\}$$
$$= \frac{1}{j!} - \frac{1}{(j+1)!}$$
$$= \frac{j}{(j+1)!}$$

where the next to last equality follows since all possible orderings of the random variables are equally likely.

From the preceding, we see that

$$\lim_{k \to \infty} P\{\text{a renewal occurs at time } k\} = \lim_{k \to \infty} \sum_{i=1}^{\infty} P\{S_k = ir\} = \sum_{i=1}^{\infty} \frac{ir}{(ir+1)!}$$

However,

$$E[N(n)] = \sum_{k=1}^{n} P\{\text{a renewal occurs at time } k\}$$

Because we can show that for any numbers $a_k, k \geq 1$, for which $\lim_{k \to \infty} a_k$ exists

$$\lim_{n \to \infty} \frac{\sum_{k=1}^{n} a_k}{n} = \lim_{k \to \infty} a_k$$

we obtain from the preceding, upon using the elementary renewal theorem,

$$E[T] = \frac{1}{\sum_{i=1}^{\infty} ir/(ir+1)!}$$

Exercises

1. Is is true that

(a) $N(t) < n$ if and only if $S_n > t$?
(b) $N(t) \leq n$ if and only if $S_n \geq t$?
(c) $N(t) > n$ if and only if $S_n < t$?

2. Suppose that the interarrival distribution for a renewal process is Poisson distributed with mean μ. That is, suppose

$$P\{X_n = k\} = e^{-\mu}\frac{\mu^k}{k!}, \qquad k = 0, 1, \ldots$$

(a) Find the distribution of S_n.
(b) Calculate $P\{N(t) = n\}$.

*3. If the mean-value function of the renewal process $\{N(t), t \geqslant 0\}$ is given by $m(t) = t/2$, $t \geqslant 0$, then what is $P\{N(5) = 0\}$?

4. Let $\{N_1(t), t \geqslant 0\}$ and $\{N_2(t), t \geqslant 0\}$ be independent renewal processes. Let $N(t) = N_1(t) + N_2(t)$.

(a) Are the interarrival times of $\{N(t), t \geqslant 0\}$ independent?
(b) Are they identically distributed?
(c) Is $\{N(t), t \geqslant 0\}$ a renewal process?

5. Let $U_1, U_2, \ldots$ be independent uniform $(0, 1)$ random variables, and define N by

$$N = \min\{n : U_1 + U_2 + \cdots + U_n > 1\}$$

What is $E[N]$?

*6. Consider a renewal process $\{N(t), t \geqslant 0\}$ having a gamma (r, λ) interarrival distribution. That is, the interarrival density is

$$f(x) = \frac{\lambda e^{-\lambda x}(\lambda x)^{r-1}}{(r-1)!}, \qquad x > 0$$

(a) Show that

$$P\{N(t) \geqslant n\} = \sum_{i=nr}^{\infty} \frac{e^{-\lambda t}(\lambda t)^i}{i!}$$

(b) Show that

$$m(t) = \sum_{i=r}^{\infty} \left[\frac{i}{r}\right]\frac{e^{-\lambda t}(\lambda t)^i}{i!}$$

where $[i/r]$ is the largest integer less than or equal to i/r.

Hint: Use the relationship between the gamma (r, λ) distribution and the sum of r independent exponentials with rate λ, to define $N(t)$ in terms of a Poisson process with rate λ.

7. Mr. Smith works on a temporary basis. The mean length of each job he gets is three months. If the amount of time he spends between jobs is exponentially distributed with mean 2, then at what rate does Mr. Smith get new jobs?

*8. A machine in use is replaced by a new machine either when it fails or when it reaches the age of T years. If the lifetimes of successive machines are independent with a common distribution F having density f, show that

(a) the long-run rate at which machines are replaced equals

$$\left[\int_0^T xf(x)\, dx + T(1 - F(T))\right]^{-1}$$

(b) the long-run rate at which machines in use fail equals

$$\frac{F(T)}{\int_0^T xf(x)\, dx + T[1 - F(T)]}$$

9. A worker sequentially works on jobs. Each time a job is completed, a new one is begun. Each job, independently, takes a random amount of time having distribution F to complete. However, independently of this, shocks occur according to a Poisson process with rate λ. Whenever a shock occurs, the worker discontinues working on the present job and starts a new one. In the long run, at what rate are jobs completed?

10. Consider a renewal process with mean interarrival time μ. Suppose that each event of this process is independently "counted" with probability p. Let $N_C(t)$ denote the number of counted events by time $t, t > 0$.

(a) Is $N_C(t), t \geq 0$ a renewal process?
(b) What is $\lim_{t \to \infty} N_C(t)/t$?

11. A renewal process for which the time until the initial renewal has a different distribution than the remaining interarrival times is called a delayed (or a general) renewal process. Prove that Proposition 7.1 remains valid for a delayed renewal process. (In general, it can be shown that all of the limit theorems for a renewal process remain valid for a delayed renewal process provided that the time until the first renewal has a finite mean.)

12. Let $X_1, X_2, \ldots$ be a sequence of independent random variables. The nonnegative integer valued random variable N is said to be a *stopping time* for the sequence if the event $\{N = n\}$ is independent of $X_{n+1}, X_{n+2}, \ldots$. The idea being that the X_i are observed one at a time — first X_1, then X_2, and so on — and N represents the number observed when we stop. Hence, the event $\{N = n\}$ corresponds to stopping after having observed $X_1, \ldots, X_n$ and thus must be independent of the values of random variables yet to come, namely, $X_{n+1}, X_{n+2}, \ldots$.

(a) Let $X_1, X_2, \ldots$ be independent with

$$P\{X_i = 1\} = p = 1 - P\{X_i = 0\}, \qquad i \geqslant 1$$

Define

$$N_1 = \min\{n : X_1 + \cdots + X_n = 5\}$$

$$N_2 = \begin{cases} 3, & \text{if } X_1 = 0 \\ 5, & \text{if } X_1 = 1 \end{cases}$$

$$N_3 = \begin{cases} 3, & \text{if } X_4 = 0 \\ 2, & \text{if } X_4 = 1 \end{cases}$$

Which of the N_i are stopping times for the sequence $X_1, \ldots$? An important result, known as *Wald's equation* states that if $X_1, X_2, \ldots$ are independent and identically distributed and have a finite mean $E(X)$, and if N is a stopping time for this sequence having a finite mean, then

$$E\left[\sum_{i=1}^{N} X_i\right] = E[N]E[X]$$

To prove Wald's equation, let us define the indicator variables I_i, $i \geqslant 1$ by

$$I_i = \begin{cases} 1, & \text{if } i \leqslant N \\ 0, & \text{if } i > N \end{cases}$$

(b) Show that

$$\sum_{i=1}^{N} X_i = \sum_{i=1}^{\infty} X_i I_i$$

From part (b) we see that

$$E\left[\sum_{i=1}^{N} X_i\right] = E\left[\sum_{i=1}^{\infty} X_i I_i\right]$$

$$= \sum_{i=1}^{\infty} E[X_i I_i]$$

where the last equality assumes that the expectation can be brought inside the summation (as indeed can be rigorously proven in this case).

(c) Argue that X_i and I_i are independent.

Hint: I_i equals 0 or 1 depending on whether or not we have yet stopped after observing which random variables?

(d) From part (c) we have

$$E\left[\sum_{i=1}^{N} X_i\right] = \sum_{i=1}^{\infty} E[X]E[I_i]$$

Complete the proof of Wald's equation.

(e) What does Wald's equation tell us about the stopping times in part (a)?

13. Wald's equation can be used as the basis of a proof of the elementary renewal theorem. Let $X_1, X_2, \ldots$ denote the interarrival times of a renewal process and let $N(t)$ be the number of renewals by time t.

(a) Show that whereas $N(t)$ is *not* a stopping time, $N(t) + 1$ is.

Hint: Note that

$$N(t) = n \Leftrightarrow X_1 + \cdots + X_n \leqslant t \qquad \text{and} \qquad X_1 + \cdots + X_{n+1} > t$$

(b) Argue that

$$E\left[\sum_{i=1}^{N(t)+1} X_i \right] = \mu[m(t) + 1]$$

(c) Suppose that the X_i are bounded random variables. That is, suppose there is a constant M such that $P\{X_i < M\} = 1$. Argue that

$$t < \sum_{i=1}^{N(t)+1} X_i < t + M$$

(d) Use the previous parts to prove the elementary renewal theorem when the interarrival times are bounded.

14. Consider a miner trapped in a room which contains three doors. Door 1 leads him to freedom after two days of travel; door 2 returns him to his room after a four-day journey; and door 3 returns him to his room after a six-day journey. Suppose at all times he is equally likely to choose any of the three doors, and let T denote the time it takes the miner to become free.

(a) Define a sequence of independent and identically distributed random variables $X_1, X_2 \ldots$ and a stopping time N such that

$$T = \sum_{i=1}^{N} X_i$$

Note: You may have to imagine that the miner continues to randomly choose doors even after he reaches safety.

(b) Use Wald's equation to find $E[T]$.
(c) Compute $E[\sum_{i=1}^{N} X_i | N = n]$ and note that it is not equal to $E[\sum_{i=1}^{n} X_i]$.
(d) Use part (c) for a second derivation of $E[T]$.

15. A deck of 52 playing cards is shuffled and the cards are then turned face up one at a time. Let X_i equal 1 if the ith card turned over is an ace, and let it be 0 otherwise, $i = 1, \ldots, 52$. Also, let N denote the number of cards that need be turned over until all 4 aces appear. That is, the final ace appears on the Nth card to be turned over. Is the equation

$$E\left[\sum_{i=1}^{N} X_i\right] = E[N]E[X_i]$$

valid? If not, why is Wald's equation not applicable?

16. In Example 7.7, suppose that potential customers arrive in accordance with a renewal process having interarrival distribution F. Would the number of events by time t constitute a (possibly delayed) renewal process if an event corresponds to a customer

(a) entering the bank?
(b) leaving the bank?

What if F were exponential?

*17. Compute the renewal function when the interarrival distribution F is such that

$$1 - F(t) = pe^{-\mu_1 t} + (1 - p)e^{-\mu_2 t}$$

18. For the renewal process whose interarrival times are uniformly distributed over $(0, 1)$, determine the expected time from $t = 1$ until the next renewal.

19. For a renewal reward process consider

$$W_n = \frac{R_1 + R_2 + \cdots + R_n}{X_1 + X_2 + \cdots + X_n}$$

where W_n represents the average reward earned during the first n cycles. Show that $W_n \to E[R]/E[X]$ as $n \to \infty$.

20. Consider a single-server bank for which customers arrive in accordance with a Poisson process with rate λ. If a customer will only enter the bank if the server is free when he arrives, and if the service time of a customer has the distribution G, then what proportion of time is the server busy?

*21. The lifetime of a car has a distribution H and probability density h. Ms. Jones buys a new car as soon as her old car either breaks down or reaches the age of T years. A new car costs C_1 dollars and an additional cost of C_2 dollars is incurred whenever a car breaks down. Assuming that a T-year-old car in working order has an expected resale value $R(T)$, what is Ms. Jones' long-run average cost?

22. If H is the uniform distribution over $(2, 8)$ and if $C_1 = 4$, $C_2 = 1$, and

$R(T) = 4 - (T/2)$, then what value of T minimizes Ms. Jones' long-run average cost in Exercise 21?

23. In Exercise 21 suppose that H is exponentially distributed with mean 5, $C_1 = 3$, $C_2 = \frac{1}{2}$, $R(T) = 0$. What value of T minimizes Ms. Jones' long-run average cost?

24. Suppose in Example 7.12 that the arrival process is a Poisson process and suppose that the policy employed is to dispatch the train every t time units.

(a) Determine the average cost per unit time.
(b) Show that the minimal average cost per unit time for such a policy is approximately $c/2$ plus the average cost per unit time for the best policy of the type considered in that example.

25. Consider a train station to which customers arrive in accordance with a Poisson process having rate λ. A train is summoned whenever there are N customers waiting in the station, but it takes K units of time for the train to arrive at the station. When it arrives, it picks up all waiting customers. Assuming that the train station incurs a cost at a rate of nc per unit time whenever there are n customers present, find the long-run average cost.

26. In example 7.14, what proportion of the defective items produced is discovered?

27. Consider a single-server queueing system in which customers arrive in accordance with a renewal process. Each customer brings in a random amount of work, chosen independently according to the distribution G. The server serves one customer at a time. However, the server processes work at rate i per unit time whenever there are i customers in the system. For instance, if a customer with workload 8 enters service when there are 3 other customers waiting in line, then if no one else arrives that customer will spend 2 units of time in service. If another customer arrives after 1 unit of time, then our customer will spend a total of 1.8 units of time in service provided no one else arrives.

Let W_i denote the amount of time customer i spends in the system. Also, define $E[W]$ by

$$E[W] = \lim_{n \to \infty} (W_1 + \cdots + W_n)/n$$

and so $E[W]$ is the average amount of time a customer spends in the system. Let N denote the number of customers that arrive in a busy period.

(a) Argue that

$$E[W] = E[W_1 + \cdots + W_N]/E[N]$$

Let L_i denote the amount of work customer i brings into the system; and so the L_i, $i \geq 1$, are independent random variables having distribution G.

(b) Argue that at any time t, the sum of the times spent in the system by all arrivals prior to t is equal to the total amount of work processed by time t.

Hint: Consider the rate at which the server processes work.

(c) Argue that

$$\sum_{i=1}^{N} W_i = \sum_{i=1}^{N} L_i$$

(d) Use Wald's equation (see Exercise 12) to conclude that

$$E[W] = \mu$$

where μ is the mean of the distribution G. That is, the average time that customers spend in the system is equal to the average work they bring to the system.

*28.   For a renewal process, let $A(t)$ be the age at time t. Prove that if $\mu < \infty$, then with probability 1

$$\frac{A(t)}{t} \to 0 \qquad \text{as } t \to \infty$$

29.   If $A(t)$ and $Y(t)$ are respectively the age and the excess at time t of a renewal process having an interarrival distribution F, calculate

$$P\{Y(t) > x \mid A(t) = s\}$$

30.   Determine the long-run proportion of time that $X_{N(t)+1} < c$.

*31.   Satellites are launched according to a Poisson process with rate λ. Each satellite will, independently, orbit the earth for a random time having distribution F. Let $X(t)$ denote the number of satellites orbiting at time t.

(a) Determine $P\{X(t) = k\}$.

Hint: Relate this to the $M/G/\infty$ queue.

(b) If at least one satellite is orbiting, then messages can be transmitted and we say that the system is functional. If the first satellite is orbited at time $t = 0$, determine the expected time that the system remains functional.

Hint: Make use of part (a) when $k = 0$.

32.   Each of n skiers continually, and independently, climbs up and then skis down a particular slope. The time it takes skier i to climb up has distribution F_i, and it is independent of her time to ski down, which has distribution H_i, $i = 1, \ldots, n$. Let $N(t)$ denote the total number of times members of this group have skied down the slope by time t. Also, let $U(t)$ denote the number of skiers climbing up the hill at time t.

(a) What is $\lim_{t \to \infty} N(t)/t$?
(b) Find $\lim_{t \to \infty} E[U(t)]$.
(c) If all F_i are exponential with rate λ and all G_i are exponential with rate μ, what is $P\{U(t) = k\}$?

33. There are 3 machines, all of which are needed for a system to work. Machine i functions for an exponential time with rate λ_i before it fails, $i = 1, 2, 3$. When a machine fails, the system is shut down and repair begins on the failed machine. The time to fix machine 1 is exponential with rate 5; the time to fix machine 2 is uniform on $(0, 4)$; and the time to fix machine 3 is a gamma random variable with parameters $n = 3$ and $\lambda = 2$. Once a failed machine is repaired, it is as good as new and all machines are restarted.

(a) What proportion of time is the system working?
(b) What proportion of time is machine 1 being repaired?
(c) What proportion of time is machine 2 in a state of suspended animation (that is, neither working nor being repaired)?

34. A truck driver regularly drives round trips from A to B and then back to A. Each time he drives from A to B, he drives at a fixed speed that (in miles per hour) is uniformly distributed between 40 and 60; each time he drives from B to A, he drives at a fixed speed that is equally likely to be either 40 or 60.

(a) In the long run, what proportion of his driving time is spent going to B?
(b) In the long run, for what proportion of his driving time is he driving at a speed of 40 miles per hour?

35. A system consists of two independent machines that each functions for an exponential time with rate λ. There is a single repairperson. If the repairperson is idle when a machine fails, then repair immediately begins on that machine; if the repairperson is busy when a machine fails, then that machine must wait until the other machine has been repaired. All repair times are independent with distribution function G and, once repaired, a machine is as good as new. What proportion of time is the repairperson idle?

36. Three marksmen take turns shooting at a target. Marksman 1 shoots until he misses, then marksman 2 begins shooting until he misses, then marksman 3 until he misses, and then back to marksman 1, and so on. Each time marksman 1 fires he hits the target, independently of the past, with probability P_i, $i = 1, 2, 3$. Determine the proportion of time, in the long run, that each marksman shoots.

37. Each time a certain machine breaks down it is replaced by a new one of the same type. In the long run, what percentage of time is the machine in use less than one year old if the life distribution of a machine is

(a) uniformly distributed over $(0, 2)$?
(b) exponentially distributed with mean 1?

*38. For an interarrival distribution F having mean μ, we defined the equilibrium distribution of F, denoted F_e, by

$$F_e(x) = \frac{1}{\mu} \int_0^x [1 - F(y)] \, dy$$

(a) Show that if F is an exponential distribution, then $F = F_e$.
(b) If for some constant c,

$$F(x) = \begin{cases} 0, & x < c \\ 1, & x \geqslant c \end{cases}$$

show that F_e is the uniform distribution on $(0, c)$. That is, if interarrival times are identically equal to c, then the equilibrium distribution is the uniform distribution on the interval $(0, c)$.
(c) The city of Berkeley, California, allows for two hours parking at all nonmetered locations within one mile of the University of California. Parking officials regularly tour around, passing the same point every two hours. When an official encounters a car he or she marks it with chalk. If the same car is there on the official's return two hours later, then a parking ticket is written. If you park your car in Berkeley and return after 3 hours, what is the probability you will have received a ticket?

39. Consider a renewal process having interarrival distribution F such that

$$\bar{F}(x) = \tfrac{1}{2} e^{-x} + \tfrac{1}{2} e^{-x/2}, \qquad x > 0$$

That is, interarrivals are equally likely to be exponential with mean 1 or exponential with mean 2.

(a) Without any calculations, guess the equilibrium distribution F_e.
(b) Verify your guess in part (a).

40. An airport shuttle bus picks up all passengers waiting at a bus stop and drops them off at the airport terminal; it then returns to the stop and repeats the process. The times between returns to the stop are independent random variables with distribution F, mean μ, and variance σ^2. Passengers arrive at the bus stop in accordance with a Poisson process with rate λ. Suppose the bus has just left the stop, and let X denote the number of passengers it picks up when it returns.

(a) Find $E[X]$.
(b) Find $\text{Var}(X)$.
(c) At what rate does the shuttle bus arrive at the terminal without any passengers?

Suppose that each passenger that has to wait at the bus stop more than c time units writes an angry letter to the shuttle bus manager.

(d) What proportion of passengers write angry letters?
(e) How does your answer in part (d) relate to $F_e(x)$?

41.   Consider a system that can be in either state 1 or 2 or 3. Each time the system enters state i it remains there for a random amount of time having mean μ_i and then makes a transition into state j with probability P_{ij}. Suppose

$$P_{12} = 1, \qquad P_{21} = P_{23} = \tfrac{1}{2}, \qquad P_{31} = 1$$

(a) What proportion of transitions takes the system into state 1?
(b) If $\mu_1 = 1$, $\mu_2 = 2$, $\mu_3 = 3$, then what proportion of time does the system spend in each state?

42.   Consider a semi-Markov process in which the amount of time that the process spends in each state before making a transition into a different state is exponentially distributed. What kind of process is this?

43.   In a semi-Markov process, let t_{ij} denote the conditional expected time that the process spends in state i given that the next state is j.

(a) Present an equation relating μ_i to the t_{ij}.
(b) Show that the proportion of time the process is in i and will next enter j is equal to $P_i P_{ij} t_{ij} / \mu_i$.

Hint: Say that a cycle begins each time state i is entered. Imagine that you receive a reward at a rate of 1 per unit time whenever the process is in i and heading for j. What is the average reward per unit time?

44.   A taxi alternates between three different locations. Whenever it reaches location i, it stops and spends a random time having mean t_i before obtaining another passenger, $i = 1, 2, 3$. A passenger entering the cab at location i will want to go to location j with probability P_{ij}. The time to travel from i to j is a random variable with mean m_{ij}. Suppose that $t_1 = 1$,

$t_2 = 2$, $t_3 = 4$, $P_{12} = 1$, $P_{23} = 1$, $P_{31} = \frac{2}{3} = 1 - P_{32}$, $m_{12} = 10$, $m_{23} = 20$, $m_{31} = 15$, $m_{32} = 25$. Define an appropriate semi-Markov process and determine

(a) the proportion of time the taxi is waiting at location i, and
(b) the proportion of time the taxi is on the road from i to j, $i, j = 1, 2, 3$.

*45. Consider a renewal process having the gamma (n, λ) interarrival distribution, and let $Y(t)$ denote the time from t until the next renewal. Use the theory of semi-Markov processes to show that

$$\lim_{t \to \infty} P\{Y(t) < x\} = \frac{1}{n} \sum_{i=1}^{n} G_{i,\lambda}(x)$$

where $G_{i,\lambda}(x)$ is the gamma (i, λ) distribution function.

46. To prove Equation (7.24), define the following notation:

$X_i^j \equiv$ time spent in state i on the jth visit to this state;

$N_i(m) \equiv$ number of visits to state i in the first m transitions

In terms of this notation, write expressions for

(a) the amount of time during the first m transitions that the process is in state i;
(b) the proportion of time during the first m transitions that the process is in state i.

Argue that, with probability 1,

(c) $$\sum_{j=1}^{N_i(m)} \frac{X_i^j}{N_i(m)} \to \mu_i \qquad \text{as } m \to \infty$$

(d) $N_i(m)/m \to \pi_i$ as $m \to \infty$.
(e) Combine parts (a), (b), (c), and (d) to prove Equation (7.24).

47. In 1984 the country of Morocco in an attempt to determine the average amount of time that tourists spend in that country on a visit tried two different sampling procedures. In one, they questioned randomly chosen tourists as they were leaving the country; in the other, they questioned randomly chosen guests at hotels. (Each tourist stayed at a hotel.) The average visiting time of the 3,000 tourists chosen from hotels was 17.8, whereas the average visiting time of the 12,321 tourists questioned at departure was 9.0. Can you explain this discrepancy? Does it necessarily imply a mistake?

48. Let X_i, $i = 1, 2, \ldots$, be the interarrival times of the renewal process $\{N(t)\}$, and let Y, independent of the X_i, be exponential with rate λ.

(a) Use the lack of memory property of the exponential to argue that

$$P\{X_1 + \cdots + X_n < Y\} = (P\{X < Y\})^n$$

(b) Use part (a) to show that

$$E[N(Y)] = \frac{E[e^{-\lambda X}]}{1 - E[e^{-\lambda X}]}$$

where X has the interarrival distribution.

49. Write a program to approximate $m(t)$ for the interarrival distribution $F * G$, where F is exponential with mean 1 and G is exponential with mean 3.

50. Let X_i, $i \geqslant 1$, be independent random variables with $p_j = P\{X = j\}$, $j \geqslant 1$. If $p_j = j/10$, $j = 1, 2, 3, 4$, find the expected time and the variance of the number of variables that need be observed until the pattern 1, 2, 3, 1, 2 occurs.

51. A coin that comes up heads with probability 0.6 is continually flipped. Find the expected number of flips until either the sequence $thht$ or the sequence ttt occurs, and find the probability that ttt occurs first.

52. Random digits, each of which is equally likely to be any of the digits 0 through 9, are observed in sequence.

(a) Find the expected time until a run of 10 distinct values occurs.
(b) Find the expected time until a run of 5 distinct values occurs.

References

The results in Section 7.9.1 concerning the computation of the variance of the time until a specified pattern appears are new, as are the results of Section 7.9.2. The results of Section 7.9.3 are from Ref [3].

1. D. R. Cox, "Renewal Theory," Methuen, London, 1962.
2. W. Feller, "An Introduction to Probability Theory and Its Applications," Vol. II, John Wiley, New York, 1966.
3. F. Hwang and D. Trietsch, "A Simple Relation Between the Pattern Probability and the Rate of False Signals in Control Charts," *Probability in the Engineering and Informational Sciences*, **10**, 315–323, 1996.
4. S. Ross, "Stochastic Processes," Second Edition, John Wiley, New York, 1996.
5. H. C. Tijms, "Stochastic Models, An Algorithmic Approach," John Wiley, New York, 1994.

Queueing Theory

✦✦

8.1. Introduction

In this chapter we will study a class of models in which customers arrive in some random manner at a service facility. Upon arrival they are made to wait in queue until it is their turn to be served. Once served they are generally assumed to leave the system. For such models we will be interested in determining, among other things, such quantities as the average number of customers in the system (or in the queue) and the average time a customer spends in the system (or spends waiting in the queue).

In Section 8.2 we derive a series of basic queueing identities which are of great use in analyzing queueing models. We also introduce three different sets of limiting probabilities which correspond to what an arrival sees, what a departure sees, and what an outside observer would see.

In Section 8.3 we deal with queueing systems in which all of the defining probability distributions are assumed to be exponential. For instance, the simplest such model is to assume that customers arrive in accordance with a Poisson process (and thus the interarrival times are exponentially distributed) and are served one at a time by a single server who takes an exponentially distributed length of time for each service. These exponential queueing models are special examples of continuous-time Markov chains and so can be analyzed as in Chapter 6. However, at the cost of a (very) slight amount of repetition we shall not assume the reader to be familiar with the material of Chapter 6, but rather we shall redevelop any needed material. Specifically we shall derive anew (by a heuristic argument) the formula for the limiting probabilities.

In Section 8.4 we consider models in which customers move randomly among a network of servers. The model of Section 8.4.1 is an open system in which customers are allowed to enter and depart the system, whereas the one studied in Section 8.4.2 is closed in the sense that the set of customers in the system is constant over time.

In Section 8.5 we study the model $M/G/1$, which while assuming Poisson arrivals, allows the service distribution to be arbitrary. To analyze this model we first introduce in Section 8.5.1 the concept of work, and then use this concept in Section 8.5.2 to help analyze this system. In Section 8.5.3 we derive the average amount of time that a server remains busy between idle periods.

In Section 8.6 we consider some variations of the model $M/G/1$. In particular in Section 8.6.1 we suppose that bus loads of customers arrive according to a Poisson process and that each bus contains a random number of customers. In Section 8.6.2 we suppose that there are two different classes of customers — with type 1 customers receiving service priority over type 2.

In Section 8.6.3 we present an $M/G/1$ optimization example. We suppose that the server goes on break whenever she becomes idle, and then determine, under certain cost assumptions, the optimal time for her to return to service.

In Section 8.7 we consider a model with exponential service times but where the interarrival times between customers is allowed to have an arbitrary distribution. We analyze this model by use of an appropriately defined Markov chain. We also derive the mean length of a busy period and of an idle period for this model.

In Section 8.8 we consider a single server system whose arrival process results from return visits of a finite number of possible sources. Assuming a general service distribution, we show how a Markov chain can be used to analyze this system.

In the final section of the chapter we talk about multiservers systems. We start with loss systems, in which arrivals, finding all servers busy, are assumed to depart and as such are lost to the system. This leads to the famous result known as Erlang's loss formula, which presents a simple formula for the number of busy servers in such a model when the arrival process in Poisson and the service distribution is general. We then discuss multiserver systems in which queues are allowed. However, except in the case where exponential service times are assumed, there are very few explicit formulas for these models. We end by presenting an approximation for the average time a customer waits in queue in a k-server model which assumes Poisson arrivals but allows for a general service distribution.

8.2. Preliminaries

In this section we will derive certain identities which are valid in the great majority of queueing models.

8.2.1. Cost Equations

Some fundamental quantities of interest for queueing models are

L, the average number of customers in the system;
L_Q, the average number of customers waiting in queue;
W, the average amount of time a customer spends in the system;
W_Q, the average amount of time a customer spends waiting in queue.

A large number of interesting and useful relationships between the preceding and other quantities of interest can be obtained by making use of the following idea: Imagine that entering customers are forced to pay money (according to some rule) to the system. We would then have the following basic cost identity:

average rate at which the system earns

$$= \lambda_a \times \text{average amount an entering customer pays} \qquad (8.1)$$

where λ_a is defined to be average arrival rate of entering customers. That is, if $N(t)$ denotes the number of customer arrivals by time t, then

$$\lambda_a = \lim_{t \to \infty} \frac{N(t)}{t}$$

We now present an heuristic proof of Equation (8.1).

Heuristic Proof of Equation (8.1)   Let T be a fixed large number. In two different ways, we will compute the average amount of money the system has earned by time T. On one hand, this quantity approximately can be obtained by multiplying the average rate at which the system earns by the length of time T. On the other hand, we can approximately compute it by multiplying the average amount paid by an entering customer by the average number of customers entering by time T (and this latter factor is approximtely $\lambda_a T$). Hence, both sides of Equation (8.1) when multiplied by T are appoximately equal to the average amount earned by T. The result then follows by letting $T \to \infty$.*

By choosing appropriate cost rules, many useful formulas can be obtained as special cases of Equation (8.1). For instance, by supposing that each customer pays $1 per unit time while in the system, Equation (8.1) yields the so-called Little's formula,

$$L = \lambda_a W \qquad (8.2)$$

---

*This can be made into a rigorous proof provided we assume that the queueing process is regenerative in the sense of Section 7.5. Most models, including all the ones in this chapter, satisfy this condition.

This follows since, under this cost rule, the rate at which the system earns is just the number in the system, and the amount a customer pays is just equal to its time in the system.

Similarly if we suppose that each customer pays $1 per unit time while in queue, then Equation (8.1) yields

$$L_Q = \lambda_a W_Q \tag{8.3}$$

By supposing the cost rule that each customer pays $1 per unit time while in service we obtain from Equation (8.1) that the

$$\text{average number of customers in service} = \lambda_a E[S] \tag{8.4}$$

where $E[S]$ is defined as the average amount of time a customer spends in service.

It should be emphasized that Equations (8.1) through (8.4) are valid for almost all queueing models regardless of the arrival process, the number of servers, or queue discipline.

8.2.2. Steady-State Probabilities

Let $X(t)$ denote the number of customers in the system at time t and define P_n, $n \geqslant 0$, by

$$P_n = \lim_{t \to \infty} P\{X(t) = n\}$$

where we assume the above limit exists. In other words, P_n is the limiting or long-run probability that there will be exactly n customers in the system. It is sometimes referred to as the *steady-state probability* of exactly n customers in the system. It also usually turns out that P_n equals the (long-run) proportion of time that the system contains exactly n customers. For example, if $P_0 = 0.3$, then in the long run, the system will be empty of customers for 30 percent of the time. Similarly, $P_1 = 0.2$ would imply that for 20 percent of the time the system would contain exactly one customer.*

Two other sets of limiting probabilities are $\{a_n, n \geqslant 0\}$ and $\{d_n, n \geqslant 0\}$, where

$a_n = $ proportion of customers that find n
    in the system when they arrive, and

$d_n = $ proportion of customers leaving behind n
    in the system when they depart

That is, P_n is the proportion of time during which there are n in the system;

---

*A sufficient condition for the validity of the dual interpretation of P_n is that the queueing process be regenerative.

a_n is the proportion of arrivals that find n; and d_n is the proportion of departures that leave behind n. That these quantities need not always be equal is illustrated by the following example.

Example 8.1 Consider a queueing model in which all customers have service times equal to 1, and where the times between successive customers are always greater than 1 [for instance, the interarrival times could be uniformly distributed over $(1, 2)$]. Hence, as every arrival finds the system empty and every departure leaves it empty, we have

$$a_0 = d_0 = 1$$

However,

$$P_0 \neq 1$$

as the system is not always empty of customers.   ✦

It was, however, no accident that a_n equaled d_n in the previous example. That arrivals and departures always see the same number of customers is always true as is shown in the next proposition.

Proposition 8.1 In any system in which customers arrive one at a time and are served one at a time

$$a_n = d_n, \qquad n \geq 0$$

Proof An arrival will see n in the system whenever the number in the system goes from n to $n + 1$; similarly, a departure will leave behind n whenever the number in the system goes from $n + 1$ to n. Now in any interval of time T the number of transitions from n to $n + 1$ must equal to within 1 the number from $n + 1$ to n. [For instance, if transitions from 2 to 3 occur 10 times, then 10 times there must have been a transition back to 2 from a higher state (namely, 3).] Hence, the rate of transitions from n to $n + 1$ equals the rate from $n + 1$ to n; or, equivalently, the rate at which arrivals find n equals the rate at which departures leave n. The result now follows since the overall arrival rate must equal the overall departure rate (what goes in eventually goes out.)   ✦

Hence, on the average, arrivals and departures always see the same number of customers. However, as Example 8.1 illustrates, they do not, in general, see the time averages. One important exception where they do is in the case of Poisson arrivals.

Proposition 8.2   Poisson arrivals always see time averages. In particular, for Poisson arrivals,

$$P_n = a_n$$

To understand why Poisson arrivals always see time averages, consider an arbitrary Poisson arrival. If we knew that it arrived at time t, then the conditional distribution of what it sees upon arrival is the same as the unconditional distribution of the system state at time t. For knowing that an arrival occurs at time t gives us no information about what occurred prior to t. (Since the Poisson process has independent increments, knowing that an event occurred at time t does not affect the distribution of what occurred prior to t.) Hence, an arrival would just see the system according to the limiting probabilities.

Contrast the foregoing with the situation of Example 8.1 where knowing that an arrival occurred at time t tells us a great deal about the past; in particular it tells us that there have been no arrivals in $(t-1, t)$. Thus, in this case, we cannot conclude that the distribution of what an arrival at time t observes is the same as the distribution of the system state at time t.

For a second argument as to why Poisson arrivals see time averages, note that the total time the system is in state n by time T is (roughly) $P_n T$. Hence, as Poisson arrivals always arrive at rate λ no matter what the system state, it follows that the number of arrivals in $[0, T]$ that find the system in state n is (roughly) $\lambda P_n T$. In the long run, therefore, the rate at which arrivals find the system in state n is λP_n and, as λ is the overall arrival rate, it follows that $\lambda P_n / \lambda = P_n$ is the proportion of arrivals that find the system in state n.

8.3. Exponential Models

8.3.1. A Single-Server Exponential Queueing System

Suppose that customers arrive at a single-server service station in accordance with a Poisson process having rate λ. That is, the time between successive arrivals are independent exponential random variables having mean $1/\lambda$. Each customer, upon arrival, goes directly into service if the server is free and, if not, the customer joins the queue. When the server finishes serving a customer, the customer leaves the system, and the next customer in line, if there is any, enters service. The successive service times are assumed to be independent exponential random variables having mean $1/\mu$.

The above is called the $M/M/1$ queue. The two Ms refer to the fact that both the interarrival and the service distributions are exponential (and thus memoryless, or Markovian), and the 1 to the fact that there is a single server. To analyze it, we shall begin by determining the limiting probabilities P_n, for $n = 0, 1, \ldots$. To do so, think along the following lines. Suppose that we have an infinite number of rooms numbered $0, 1, 2, \ldots$, and suppose that we instruct an individual to enter room n whenever there are n customers in the system. That is, he would be in room 2 whenever there are two customers in the system; and if another were to arrive, then he would leave room 2 and enter room 3. Similarly, if a service would take place he would leave room 2 and enter room 1 (as there would now be only one customer in the system).

Now suppose that in the long run our individual is seen to have entered room 1 at the rate of ten times an hour. Then at what rate must he have left room 1? Clearly, at this same rate of ten times an hour. For the total number of times that he enters room 1 must be equal to (or one greater than) the total number of times he leaves room 1. This sort of argument thus yields the general principle which will enable us to determine the state probabilities. Namely, for each $n \geqslant 0$, *the rate at which the process enters state n equals the rate at which it leaves state n.* Let us now determine these rates. Consider first state 0. When in state 0 the process can leave only by an arrival as clearly there cannot be a departure when the system is empty. Since the arrival rate is λ and the proportion of time that the process is in state 0 is P_0, it follows that the rate at which the process leaves state 0 is λP_0. On the other hand, state 0 can only be reached from state 1 via a departure. That is, if there is a single customer in the system and he completes service, then the system becomes empty. Since the service rate is μ and the proportion of time that the system has exactly one customer is P_1, it follows that the rate at which the process enters state 0 is μP_1.

Hence, from our rate-equality principle we get our first equation,

$$\lambda P_0 = \mu P_1$$

Now consider state 1. The process can leave this state either by an arrival (which occurs at rate λ) or a departure (which occurs at rate μ). Hence, when in state 1, the process will leave this state at a rate of $\lambda + \mu$.* Since the proportion of time the process is in state 1 is P_1, the rate at which the process leaves state 1 is $(\lambda + \mu)P_1$. On the other hand, state 1 can be entered either from state 0 via an arrival or from state 2 via a departure. Hence, the rate at which the process enters state 1 is $\lambda P_0 + \mu P_2$. Because the reasoning for other states is similar, we obtain the following set of equations:

State	*Rate at which the process leaves = rate at which it enters*	
0	$\lambda P_0 = \mu P_1$	
$n, n \geqslant 1$	$(\lambda + \mu)P_n = \lambda P_{n-1} + \mu P_{n+1}$	(8.5)

---

* If one event occurs at rate λ and another occurs at rate μ, then the total rate at which either event occurs is $\lambda + \mu$. Suppose one man earns \$2 per hour and another earns \$3 per hour; then together they clearly earn \$5 per hour.

The set of Equations (8.5) which balances the rate at which the process enters each state with the rate at which it leaves that state is known as *balance equations*.

In order to solve Equations (8.5), we rewrite them to obtain

$$P_1 = \frac{\lambda}{\mu} P_0,$$

$$P_{n+1} = \frac{\lambda}{\mu} P_n + \left(P_n - \frac{\lambda}{\mu} P_{n-1} \right), \qquad n \geqslant 1$$

Solving in terms of P_0 yields

$$P_0 = P_0,$$

$$P_1 = \frac{\lambda}{\mu} P_0,$$

$$P_2 = \frac{\lambda}{\mu} P_1 + \left(P_1 - \frac{\lambda}{\mu} P_0 \right) = \frac{\lambda}{\mu} P_1 = \left(\frac{\lambda}{\mu} \right)^2 P_0,$$

$$P_3 = \frac{\lambda}{\mu} P_2 + \left(P_2 - \frac{\lambda}{\mu} P_1 \right) = \frac{\lambda}{\mu} P_2 = \left(\frac{\lambda}{\mu} \right)^3 P_0,$$

$$P_4 = \frac{\lambda}{\mu} P_3 + \left(P_3 - \frac{\lambda}{\mu} P_2 \right) = \frac{\lambda}{\mu} P_3 = \left(\frac{\lambda}{\mu} \right)^4 P_0,$$

$$P_{n+1} = \frac{\lambda}{\mu} P_n + \left(P_n - \frac{\lambda}{\mu} P_{n-1} \right) = \frac{\lambda}{\mu} P_n = \left(\frac{\lambda}{\mu} \right)^{n+1} P_0$$

To determine P_0 we use the fact that the P_n must sum to 1, and thus

$$1 = \sum_{n=0}^{\infty} P_n = \sum_{n=0}^{\infty} \left(\frac{\lambda}{\mu} \right)^n P_0 = \frac{P_0}{1 - \lambda/\mu}$$

or

$$P_0 = 1 - \frac{\lambda}{\mu},$$

$$P_n = \left(\frac{\lambda}{\mu} \right)^n \left(1 - \frac{\lambda}{\mu} \right), \qquad n \geqslant 1 \tag{8.6}$$

Notice that for the preceding equations to make sense, it is necessary for λ/μ to be less than 1. For otherwise $\sum_{n=0}^{\infty} (\lambda/\mu)^n$ would be infinite and all the P_n would be 0. Hence, we shall assume that $\lambda/\mu < 1$. Note that it is quite intuitive that there would be no limiting probabilities if $\lambda > \mu$. For suppose that $\lambda > \mu$. Since customers arrive at a Poisson rate λ, it follows that the expected total number of arrivals by time t is λt. On the other hand, what is the expected number of customers served by time t? If there were always customers present, then the number of customers served would be a Poisson process having rate μ since the time between successive services would be independent exponentials having mean $1/\mu$. Hence, the expected number of customers served by time t is no greater than μt; and, therefore, the expected number in the system at time t is at least

$$\lambda t - \mu t = (\lambda - \mu)t$$

Now if $\lambda > \mu$, then the above number goes to infinity at t becomes large. That is, $\lambda/\mu > 1$, the queue size increases without limit and there will be no limiting probabilities. Note also that the condition $\lambda/\mu < 1$ is equivalent to the condition that the mean service time be less than the mean time between successive arrivals. This is the general condition that must be satisfied for limited probabilities to exist in most single-server queueing systems.

Now let us attempt to express the quantities L, L_Q, W, and W_Q in terms of the limiting probabilities P_n. Since P_n is the long-run probability that the system contains exactly n customers, the average number of customers in the system clearly is given by

$$\begin{aligned} L &= \sum_{n=0}^{\infty} nP_n \\ &= \sum_{n=0}^{\infty} n \left(\frac{\lambda}{\mu}\right)^n \left(1 - \frac{\lambda}{\mu}\right) \\ &= \frac{\lambda}{\mu - \lambda} \end{aligned} \tag{8.7}$$

where the last equation followed upon application of the algebraic identity

$$\sum_{n=0}^{\infty} nx^n = \frac{x}{(1-x)^2}$$

The quantities W, W_Q, and L_Q now can be obtained with the help of Equations (8.2) and (8.3). That is, since $\lambda_a = \lambda$, we have from Equation (8.7) that

$$W = \frac{L}{\lambda}$$

$$= \frac{1}{\mu - \lambda},$$

$$W_Q = W - E[S]$$

$$= W - \frac{1}{\mu}$$

$$= \frac{\lambda}{\mu(\mu - \lambda)},$$

$$L_Q = \lambda W_Q$$

$$= \frac{\lambda^2}{\mu(\mu - \lambda)} \tag{8.8}$$

Example 8.2   Suppose that customers arrive at a Poisson rate of one per every 12 minutes, and that the service time is exponential at a rate of one service per 8 minutes. What are L and W?

Solution: Since $\lambda = \frac{1}{12}$, $\mu = \frac{1}{8}$, we have

$$L = 2, \qquad W = 24$$

Hence, the average number of customers in the system is two, and the average time a customer spends in the system is 24 minutes.

Now suppose that the arrival rate increases 20 percent to $\lambda = \frac{1}{10}$. What is the corresponding change in L and W? Again using Equations (8.7), we get

$$L = 4, \qquad W = 40$$

Hence, an increase of 20 percent in the arrival rate *doubled* the average number of customers in the system.

To understand this better, write Equations (8.7) as

$$L = \frac{\lambda/\mu}{1 - \lambda/\mu},$$

$$W = \frac{1/\mu}{1 - \lambda/\mu}$$

From these equations we can see that when λ/μ is near 1, a slight increase in λ/μ will lead to a large increase in L and W. ✦

A Technical Remark   We have used the fact that if one event occurs at an exponential rate λ, and another independent event at an exponential rate μ, then together they occur at an exponential rate $\lambda + \mu$. To check this formally, let T_1 be the time at which the first event occurs, and T_2 the time at which the second event occurs. Then

$$P\{T_1 \leqslant t\} = 1 - e^{-\lambda t},$$

$$P\{T_2 \leqslant t\} = 1 - e^{-\mu t}$$

Now if we are interested in the time until either T_1 or T_2 occurs, then we are interested in $T = \min(T_1, T_2)$. Now

$$P\{T \leqslant t\} = 1 - P\{T > t\}$$

$$= 1 - P\{\min(T_1, T_2) > t\}$$

However, $\min(T_1, T_2) > t$ if and only if both T_1 and T_2 are greater than t;

hence,

$$P\{T \leqslant t\} = 1 - P\{T_1 > t, T_2 > t\}$$
$$= 1 - P\{T_1 > t\}P\{T_2 > t\}$$
$$= 1 - e^{-\lambda t}e^{-\mu t}$$
$$= 1 - e^{-(\lambda + \mu)t}$$

Thus, T has an exponential distribution with rate $\lambda + \mu$, and we are justified in adding the rates. ✦

Let W^* denote the amount of time an arbitrary customer spends in the system. To obtain the distribution of W^*, we condition on the number in the system when the customer arrives. This yields

$$P\{W^* \leqslant a\} = \sum_{n=0}^{\infty} P\{W^* \leqslant a \,|\, n \text{ in the system when he arrives}\}$$

$$\times P\{n \text{ in the system when he arrives}\} \qquad (8.9)$$

Now consider the amount of time that our customer must spend in the system if there are already n customers present when he arrives. If $n = 0$, then his time in the system will just be his service time. When $n \geqslant 1$, there will be one customer in service and $n - 1$ waiting in line ahead of our arrival. The customer in service might have been in service for some time, but due to the lack of memory of the exponential distribution (see Section 5.2), it follows that our arrival would have to wait an exponential amount of time with rate μ for this customer to complete service. As he also would have to wait an exponential amount of time for each of the other $n - 1$ customers in line, it follows, upon adding his own service time, that the amount of time that a customer must spend in the system if there are already n customers present when he arrives is the sum of $n + 1$ independent and identically distributed exponential random variables with rate μ. But it is known (see Section 5.2.3) that such a random variable has a gamma distribution with parameters $(n + 1, \mu)$. That is,

$$P\{W^* \leqslant a \,|\, n \text{ in the system when he arrives}\}$$

$$= \int_0^a \mu e^{-\mu t} \frac{(\mu t)^n}{n!}\, dt$$

Because

$$P\{n \text{ in the system when he arrives}\} = P_n \qquad \text{(since Poisson arrivals)}$$

$$= \left(\frac{\lambda}{\mu}\right)^n \left(1 - \frac{\lambda}{\mu}\right)$$

we have from Equation (8.9) and the preceding that

$$P\{W^* \leqslant a\} = \sum_{n=0}^{\infty} \int_0^a \mu e^{-\mu t} \frac{(\mu t)^n}{n!} \, dt \left(\frac{\lambda}{\mu}\right)^n \left(1 - \frac{\lambda}{\mu}\right)$$

$$= \int_0^a (\mu - \lambda) e^{-\mu t} \sum_{n=0}^{\infty} \frac{(\lambda t)^n}{n!} \, dt \qquad \text{(by interchanging)}$$

$$= \int_0^a (\mu - \lambda) e^{-\mu t} e^{\lambda t} \, dt$$

$$= \int_0^a (\mu - \lambda) e^{-(\mu - \lambda)t} \, dt$$

$$= 1 - e^{-(\mu - \lambda)a}$$

In other words, W^*, the amount of time a customer spends in the system, is an exponential random variable with rate $\mu - \lambda$. (As a check, we note that $E[W^*] = 1/(\mu - \lambda)$ which checks with Equation (8.8) since $W = E[W^*]$.)

Remark Another argument as to why W^* is exponential with rate $\mu - \lambda$ is as follows. If we let N denote the number of customers in the system as seen by an arrival, then this arrival will spend $N + 1$ service times in the system before departing. Now,

$$P\{N + 1 = j\} = P\{N = j - 1\} = (\lambda/\mu)^{j-1}(1 - \lambda/\mu), \qquad j \geqslant 1$$

In words, the number of services that have to be completed before the arrival departs is a geometric random variable with parameter $1 - \lambda/\mu$. Therefore, after each service completion our customer will be the one departing with probability $1 - \lambda/\mu$. Thus, no matter how long the customer has already spent in the system, the probability he will depart in the next h time units is $\mu h + o(h)$, the probability that a service ends in that time, multiplied by $1 - \lambda/\mu$. That is, the customer will depart in the next h time units with probability $(\mu - \lambda)h + o(h)$, which says that the hazard rate function of W^* is the constant $\mu - \lambda$. But only the exponential has a constant hazard rate, and so we can conclude that W^* is exponential with rate $\mu - \lambda$.

8.3.2. A Single-Server Exponential Queueing System Having Finite Capacity

In the previous model, we assumed that there was no limit on the number of customers that could be in the system at the same time. However, in reality there is always a finite system capacity N, in the sense that there can be no more than N customers in the system at any time. By this, we mean that if an arriving customer finds that there are already N customers present, then he does not enter the system.

As before, we let P_n, $0 \leqslant n \leqslant N$, denote the limiting probability that there are n customers in the system. The rate-equality principle yields the following set of balance equations:

State	Rate at which the process leaves = rate at which it enters
0	$\lambda P_0 = \mu P_1$
$1 \leqslant n \leqslant N-1$	$(\lambda + \mu)P_n = \lambda P_{n-1} + \mu P_{n+1}$
N	$\mu P_N = \lambda P_{N-1}$

The argument for state 0 is exactly as before. Namely, when in state 0, the process will leave only via an arrival (which occurs at rate λ) and hence the rate at which the process leaves state 0 is λP_0. On the other hand, the process can enter state 0 only from state 1 via a departure; hence, the rate at which the process enters state 0 is μP_1. The equation for states n, where $1 \leqslant n < N$, is the same as before. The equation for state N is different because now state N can only be left via a departure since an arriving customer will not enter the system when it is in state N; also, state N can now only be entered from state $N - 1$ (as there is no longer a state $N + 1$) via an arrival.

To solve, we again rewrite the preceding system of equations:

$$P_1 = \frac{\lambda}{\mu} P_0,$$

$$P_{n+1} = \frac{\lambda}{\mu} P_n + \left(P_n - \frac{\lambda}{\mu} P_{n-1} \right), \qquad 1 \leqslant n \leqslant N - 1$$

$$P_N = \frac{\lambda}{\mu} P_{N-1}$$

which, solving in terms of P_0, yields

$$P_1 = \frac{\lambda}{\mu} P_0,$$

$$P_2 = \frac{\lambda}{\mu} P_1 + \left(P_1 - \frac{\lambda}{\mu} P_0 \right) = \frac{\lambda}{\mu} P_1 = \left(\frac{\lambda}{\mu} \right)^2 P_0,$$

$$P_3 = \frac{\lambda}{\mu} P_2 + \left(P_2 - \frac{\lambda}{\mu} P_1 \right) = \frac{\lambda}{\mu} P_2 = \left(\frac{\lambda}{\mu} \right)^3 P_0,$$

$$\vdots$$

$$P_{N-1} = \frac{\lambda}{\mu} P_{N-2} + \left(P_{N-2} - \frac{\lambda}{\mu} P_{N-3} \right) = \left(\frac{\lambda}{\mu} \right)^{N-1} P_0,$$

$$P_N = \frac{\lambda}{\mu} P_{N-1} = \left(\frac{\lambda}{\mu} \right)^N P_0 \qquad (8.10)$$

By using the fact that $\sum_{n=0}^{N} P_n = 1$ we obtain

$$
1 = P_0 \sum_{n=0}^{N} \left(\frac{\lambda}{\mu}\right)^n
$$

$$
= P_0 \left[\frac{1 - (\lambda/\mu)^{N+1}}{1 - \lambda/\mu}\right]
$$

or

$$
P_0 = \frac{(1 - \lambda/\mu)}{1 - (\lambda/\mu)^{N+1}}
$$

and hence from Equation (8.10) we obtain

$$
P_n = \frac{(\lambda/\mu)^n (1 - \lambda/\mu)}{1 - (\lambda/\mu)^{N+1}}, \qquad n = 0, 1, \ldots, N \tag{8.11}
$$

Note that in this case, there is no need to impose the condition that $\lambda/\mu < 1$. The queue size is, by definition, bounded so there is no possibility of its increasing indefinitely.

As before, L may be expressed in terms of P_n to yield

$$
L = \sum_{n=0}^{N} nP_n
$$

$$
= \frac{(1 - \lambda/\mu)}{1 - (\lambda/\mu)^{N+1}} \sum_{n=0}^{N} n\left(\frac{\lambda}{\mu}\right)^n
$$

which after some algebra yields

$$
L = \frac{\lambda[1 + N(\lambda/\mu)^{N+1} - (N+1)(\lambda/\mu)^N]}{(\mu - \lambda)(1 - (\lambda/\mu)^{N+1})} \tag{8.12}
$$

In deriving W, the expected amount of time a customer spends in the system, we must be a little careful about what we mean by a customer. Specifically, are we including those "customers" who arrive to find the system full and thus do not spend any time in the system? Or, do we just want the expected time spent in the system by a customer who actually entered the system? The two questions lead, of course, to different answers. In the first case, we have $\lambda_a = \lambda$; whereas in the second case, since the fraction of arrivals that actually enter the system is $1 - P_N$, it follows that $\lambda_a = \lambda(1 - P_N)$. Once it is clear what we mean by a customer, W can be obtained from

$$
W = \frac{L}{\lambda_a}
$$

Example 8.3   Suppose that it costs $c\mu$ dollars per hour to provide service at a rate μ. Suppose also that we incur a gross profit of A dollars for each customer served. If the system has a capacity N, what service rate μ maximizes our total profit?

Solution: To solve this, suppose that we use rate μ. Let us determine the amount of money coming in per hour and subtract from this the amount going out each hour. This will give us our profit per hour, and we can choose μ so as to maximize this.

Now, potential customers arrive at a rate λ. However, a certain proportion of them do not join the system; namely, those who arrive when there are N customers already in the system. Hence, since P_N is the proportion of time that the system is full, it follows that entering customers arrive at a rate of $\lambda(1 - P_N)$. Since each customer pays $\$A$, it follows that money comes in at an hourly rate of $\lambda(1 - P_N)A$ and since it goes out at an hourly rate of $c\mu$, it follows that our total profit per hour is given by

$$\text{profit per hour} = \lambda(1 - P_N)A - c\mu$$

$$= \lambda A\left[1 - \frac{(\lambda/\mu)^N(1 - \lambda/\mu)}{1 - (\lambda/\mu)^{N+1}}\right] - c\mu$$

$$= \frac{\lambda A[1 - (\lambda/\mu)^N]}{1 - (\lambda/\mu)^{N+1}} - c\mu$$

For instance if $N = 2$, $\lambda = 1$, $A = 10$, $c = 1$, then

$$\text{profit per hour} = \frac{10[1 - (1/\mu)^2]}{1 - (1/\mu)^3} - \mu$$

$$= \frac{10(\mu^3 - \mu)}{\mu^3 - 1} - \mu$$

in order to maximize profit we differentiate to obtain

$$\frac{d}{d\mu}[\text{profit per hour}] = 10\frac{(2\mu^3 - 3\mu^2 + 1)}{(\mu^3 - 1)^2} - 1$$

The value of μ that maximizes our profit now can be obtained by equating to zero and solving numerically. ✦

In the previous two models, it has been quite easy to define the state of the system. Namely, it was defined as the number of people in the system. Now we shall consider some examples where a more detailed state space is necessary.

8.3.3. A Shoeshine Shop

Consider a shoeshine shop consisting of two chairs. Suppose that an entering customer first will go to chair 1. When his work is completed in chair 1, he will go either to chair 2 if that chair is empty or else wait in chair 1 until chair 2 becomes empty. Suppose that a potential customer will enter this shop as long as chair 1 is empty. (Thus, for instance, a potential customer might enter even if there is a customer in chair 2.)

If we suppose that potential customers arrive in accordance with a Poisson process at rate λ, and that the service times for the two chairs are independent and have respective exponential rates of μ_1 and μ_2, then

(a) what proportion of potential customers enters the system?
(b) what is the mean number of customers in the system?
(c) what is the average amount of time that an entering customer spends in the system?

To begin we must first decide upon an appropriate state space. It is clear that the state of the system must include more information than merely the number of customers in the system. For instance, it would not be enough to specify that there is one customer in the system as we would also have to know which chair he was in. Further, if we only know that there are two customers in the system, then we would not know if the man in chair 1 is still being served or if he is just waiting for the person in chair 2 to finish. To account for these points, the following state space, consisting of the five states, $(0, 0)$, $(1, 0)$, $(0, 1)$, $(1, 1)$, and $(b, 1)$, will be used. The states have the following interpretation:

State	Interpretation
$(0, 0)$	There are no customers in the system.
$(1, 0)$	There is one customer in the system, and he is in chair 1.
$(0, 1)$	There is one customer in the system, and he is in chair 2.
$(1, 1)$	There are two customers in the system, and both are presently being served.
$(b, 1)$	There are two customers in the system, but the customer in the first chair has completed his work in that chair and is waiting for the second chair to become free.

It should be noted that when the system is in state $(b, 1)$, the person in chair 1, though not being served, is nevertheless "blocking" potential arrivals from entering the system.

As a prelude to writing down the balance equations, it is usually worthwhile to make a transition diagram. This is done by first drawing a circle for each state and then drawing an arrow labeled by the rate at which the process goes from one state to another. The transition diagram for this

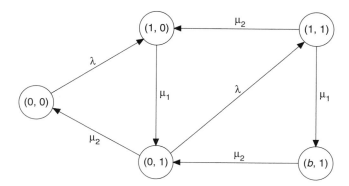

Figure 8.1. A transition diagram.

model is shown in Figure 8.1. The explanation for the diagram is as follows: The arrow from state $(0, 0)$ to state $(1, 0)$ which is labeled λ means that when the process is in state $(0, 0)$, that is, when the system is empty, then it goes to state $(1, 0)$ at a rate λ, that is via an arrival. The arrow from $(0, 1)$ to $(1, 1)$ is similarly explained.

When the process is in state $(1, 0)$, it will go to state $(0, 1)$ when the customer in chair 1 is finished and this occurs at a rate μ_1; hence the arrow from $(1, 0)$ to $(0, 1)$ labeled μ_1. The arrow from $(1, 1)$ to $(b, 1)$ is similarly explained.

When in state $(b, 1)$ the process will go to state $(0, 1)$ when the customer in chair 2 completes his service (which occurs at rate μ_2); hence the arrow from $(b, 1)$ to $(0, 1)$ labeled μ_2. Also when in state $(1, 1)$ the process will go to state $(1, 0)$ when the man in chair 2 finishes and hence the arrow from $(1, 1)$ to $(1, 0)$ labeled μ_2. Finally, if the process is in state $(0, 1)$, then it will go to state $(0, 0)$ when the man in chair 2 completes his service, hence the arrow from $(0, 1)$ to $(0, 0)$ labeled μ_2.

Because there are no other possible transitions, this completes the transition diagram.

To write the balance equations we equate the sum of the arrows (multiplied by the probability of the states where they originate) coming into a state with the sum of the arrows (multiplied by the probability of the state) going out of that state. This gives

State	*Rate that the process leaves = rate that it enters*
$(0, 0)$	$\lambda P_{00} = \mu_2 P_{01}$
$(1, 0)$	$\mu_1 P_{10} = \lambda P_{00} + \mu_2 P_{11}$
$(0, 1)$	$(\lambda + \mu_2)P_{01} = \mu_1 P_{10} + \mu_2 P_{b1}$
$(1, 1)$	$(\mu_1 + \mu_2)P_{11} = \lambda P_{01}$
$(b, 1)$	$\mu_2 P_{b1} = \mu_1 P_{11}$

These along with the equation

$$P_{00} + P_{10} + P_{01} + P_{11} + P_{b1} = 1$$

may be solved to determine the limiting probabilities. Though it is easy to solve the preceding equations, the resulting solutions are quite involved and hence will not be explicitly presented. However, it is easy to answer our questions in terms of these limiting probabilities. First, since a potential customer will enter the system when the state is either $(0, 0)$ or $(0, 1)$, it follows that the proportion of customers entering the system is $P_{00} + P_{01}$. Secondly, since there is one customer in the system whenever the state is $(0, 1)$ or $(1, 0)$ and two customers in the system whenever the state is $(1, 1)$ or $(b, 1)$, it follows that L, the average number in the system, is given by

$$L = P_{01} + P_{10} + 2(P_{11} + P_{b1})$$

To derive the average amount of time that an entering customer spends in the system, we use the relationship $W = L/\lambda_a$. Since a potential customer will enter the system when in state $(0, 0)$ or $(0, 1)$, it follows that $\lambda_a = \lambda(P_{00} + P_{01})$ and hence

$$W = \frac{P_{01} + P_{10} + 2(P_{11} + P_{b1})}{\lambda(P_{00} + P_{01})}$$

Example 8.4   (a) If $\lambda = 1$, $\mu_1 = 1$, $\mu_2 = 2$, then calculate the preceding quantities of interest.
  (b) If $\lambda = 1$, $\mu_1 = 2$, $\mu_2 = 1$, then calculate the preceding.

Solution: (a) Solving the balance equations yields

$$P_{00} = \tfrac{12}{37}, \qquad P_{10} = \tfrac{16}{37}, \qquad P_{11} = \tfrac{2}{37}, \qquad P_{01} = \tfrac{6}{37}, \qquad P_{b1} = \tfrac{1}{37}$$

Hence,

$$L = \tfrac{28}{37}, \qquad W = \tfrac{28}{18}$$

  (b) Solving the balance equations yields

$$P_{00} = \tfrac{3}{11}, \qquad P_{10} = \tfrac{2}{11}, \qquad P_{11} = \tfrac{1}{11}, \qquad P_{b1} = \tfrac{2}{11}, \qquad P_{01} = \tfrac{3}{11}$$

Hence,

$$L = 1, \qquad W = \tfrac{11}{6} \quad \blacklozenge$$

8.3.4.   A Queueing System with Bulk Service

In this model, we consider a single-server exponential queueing system in which the server is able to serve two customers at the same time. Whenever the server completes a service, he then serves the next two customers at the same time. However, if there is only one customer in line, then he serves that customer by himself. We shall assume that his service time is exponential at rate μ whether he is serving one or two customers. As usual, we suppose that customers arrive at an exponential rate λ. One example of such a system might be an elevator or a cable car which can take at most two passengers at any time.

It would seem that the state of the system would have to tell us not only how many customers there are in the system, but also whether one or two are presently being served. However, it turns out that we can more easily solve the problem not by concentrating on the number of customers in the system, but rather on the number in *queue*. So let us define the state as the number of customers waiting in queue, with two states when there is no one in queue. That is, let us have as a state space $0', 0, 1, 2, \ldots$, with the interpretation

State	Interpretation
$0'$	No one in service
0	Server busy; no one waiting
$n, n > 0$	n customers waiting

The transition diagram is shown in Figure 8.2 and the balance equations are

State	Rate at which the process leaves = rate at which it enters
$0'$	$\lambda P_{0'} = \mu P_0$
0	$(\lambda + \mu)P_0 = \lambda P_{0'} + \mu P_1 + \mu P_2$
$n, n \geqslant 1$	$(\lambda + \mu)P_n = \lambda P_{n-1} + \mu P_{n+2}$

Now the set of equations

$$(\lambda + \mu)P_n = \lambda P_{n-1} + \mu P_{n+2} \qquad n = 1, 2, \ldots \qquad (8.13)$$

has a solution of the form

$$P_n = \alpha^n P_0$$

To see this, substitute the preceding in Equation (8.13) to obtain

$$(\lambda + \mu)\alpha^n P_0 = \lambda \alpha^{n-1} P_0 + \mu \alpha^{n+2} P_0$$

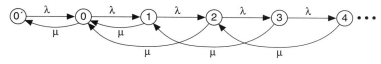

Figure 8.2.

or

$$(\lambda + \mu)\alpha = \lambda + \mu\alpha^3$$

Solving this for α yields the three roots:

$$\alpha = 1, \qquad \alpha = \frac{-1 - \sqrt{1 + 4\lambda/\mu}}{2}, \qquad \text{and} \qquad \alpha = \frac{-1 + \sqrt{1 + 4\lambda/\mu}}{2}$$

As the first two are clearly not possible, it follows that

$$\alpha = \frac{\sqrt{1 + 4\lambda/\mu} - 1}{2}$$

Hence,

$$P_n = \alpha^n P_0,$$

$$P_{0'} = \frac{\mu}{\lambda} P_0$$

where the bottom equation follows from the first balance equation. (We can ignore the second balance equation as one of these equations is always redundant.) To obtain P_0, we use

$$P_0 + P_{0'} + \sum_{n=1}^{\infty} P_n = 1$$

or

$$P_0 \left[1 + \frac{\mu}{\lambda} + \sum_{n=1}^{\infty} \alpha^n \right] = 1$$

or

$$P_0 \left[\frac{1}{1 - \alpha} + \frac{\mu}{\lambda} \right] = 1$$

or

$$P_0 = \frac{\lambda(1 - \alpha)}{\lambda + \mu(1 - \alpha)}$$

and thus

$$P_n = \frac{\alpha^n \lambda(1 - \alpha)}{\lambda + \mu(1 - \alpha)}, \qquad n \geqslant 0$$

$$P_{0'} = \frac{\mu(1 - \alpha)}{\lambda + \mu(1 - \alpha)}$$

(8.14)

where

$$\alpha = \frac{\sqrt{1 + 4\lambda/\mu} - 1}{2}$$

Note that for the preceding to be valid we need $\alpha < 1$, or equivalently $\lambda/\mu < 2$, which is intuitive since the maximum service rate is 2μ, which must be larger than the arrival rate λ to avoid overloading the system.

All the relevant quantities of interest now can be determined. For instance, to determine the proportion of customers that are served alone, we first note that the rate at which customers are served alone is $\lambda P_{0'} + \mu P_1$, since when the system is empty a customer will be served alone upon the next arrival and when there is one customer in queue he will be served alone upon a departure. As the rate at which customers are served is λ, it follows that

$$\text{proportion of customers that are served alone} = \frac{\lambda P_{0'} + \mu P_1}{\lambda}$$

$$= P_{0'} + \frac{\mu}{\lambda} P_1$$

Also,

$$L_Q = \sum_{n=1}^{\infty} n P_n$$

$$= \frac{\lambda(1 - \alpha)}{\lambda + \mu(1 - \alpha)} \sum_{n=1}^{\infty} n\alpha^n \qquad \text{from Equation (8.14)}$$

$$= \frac{\lambda\alpha}{(1 - \alpha)[\lambda + \mu(1 - \alpha)]} \qquad \text{by algebraic identity } \sum_{1}^{\infty} n\alpha^n = \frac{\alpha}{(1 - \alpha)^2}$$

and

$$W_Q = \frac{L_Q}{\lambda},$$

$$W = W_Q + \frac{1}{\mu},$$

$$L = \lambda W$$

8.4. Network of Queues

8.4.1. Open Systems

Consider a two-server system in which customers arrive at a Poisson rate λ at server 1. After being served by server 1 they then join the queue in front of server 2. We suppose there is infinite waiting space at both servers. Each server serves one customer at a time with server i taking an exponential time with rate μ_i for a service, $i = 1, 2$. Such a system is called a *tandem* or *sequential* system (see Figure 8.3).

Figure 8.3. A tandem queue.

To analyze this system we need to keep track of the number of customers at server 1 and the number at server 2. So let us define the state by the pair (n, m)—meaning that there are n customers at server 1 and m at server 2. The balance equations are

State	Rate that the process leaves = rate that it enters
$0, 0$	$\lambda P_{0,0} = \mu_2 P_{0,1}$
$n, 0; \ n > 0$	$(\lambda + \mu_1)P_{n,0} = \mu_2 P_{n,1} + \lambda P_{n-1,0}$
$0, m; \ m > 0$	$(\lambda + \mu_2)P_{0,m} = \mu_2 P_{0,m+1} + \mu_1 P_{1,m-1}$
$n, m; \ nm > 0$	$(\lambda + \mu_1 + \mu_2)P_{n,m} = \mu_2 P_{n,m+1} + \mu_1 P_{n+1,m-1}$
	$\qquad\qquad\qquad\qquad + \lambda P_{n-1,m}$ (8.15)

Rather than directly attempting to solve these (along with the equation $\sum_{n,m} P_{n,m} = 1$) we shall guess at a solution and then verify that it indeed satisfies the preceding. We first note that the situation at server 1 is just as in an $M/M/1$ model. Similarly, as it was shown in Section 6.6 that the departure process of an $M/M/1$ queue is a Poisson process with rate λ, it follows that what server 2 faces is also an $M/M/1$ queue. Hence, the probability that there are n customers at server 1 is

$$P\{n \text{ at server } 1\} = \left(\frac{\lambda}{\mu_1}\right)^n \left(1 - \frac{\lambda}{\mu_1}\right)$$

and, similarly,

$$P\{m \text{ at server } 2\} = \left(\frac{\lambda}{\mu_2}\right)^m \left(1 - \frac{\lambda}{\mu_2}\right)$$

Now if the numbers of customers at servers 1 and 2 were independent random variables, then it would follow that

$$P_{n,m} = \left(\frac{\lambda}{\mu_1}\right)^n \left(1 - \frac{\lambda}{\mu_1}\right)\left(\frac{\lambda}{\mu_2}\right)^m \left(1 - \frac{\lambda}{\mu_2}\right) \qquad (8.16)$$

To verify that $P_{n,m}$ is indeed equal to the preceding (and thus that the number of customers at server 1 is independent of the number at server 2), all we need do is verify that the preceding satisfies the set of Equations (8.15)—this suffices since we know that the $P_{n,m}$ are the unique solution of

Equations (8.15). Now, for instance, if we consider the first equation of (8.15), we need to show that

$$\lambda\left(1 - \frac{\lambda}{\mu_1}\right)\left(1 - \frac{\lambda}{\mu_2}\right) = \mu_2\left(1 - \frac{\lambda}{\mu_1}\right)\left(\frac{\lambda}{\mu_2}\right)\left(1 - \frac{\lambda}{\mu_2}\right)$$

which is easily verified. We leave it as an exercise to show that the $P_{n,m}$, as given by Equation (8.16), satisfy all of the Equations (8.15), and are thus the limiting probabilities.

From the preceding we see that L, the average number of customers in the system, is given by

$$L = \sum_{n,m} (n + m)P_{n,m}$$

$$= \sum_n n\left(\frac{\lambda}{\mu_1}\right)^n\left(1 - \frac{\lambda}{\mu_1}\right) + \sum_m m\left(\frac{\lambda}{\mu_2}\right)^m\left(1 - \frac{\lambda}{\mu_2}\right)$$

$$= \frac{\lambda}{\mu_1 - \lambda} + \frac{\lambda}{\mu_2 - \lambda}$$

and from this we see that the average time a customer spends in the system is

$$W = \frac{L}{\lambda} = \frac{1}{\mu_1 - \lambda} + \frac{1}{\mu_2 - \lambda}$$

Remarks (i) The result (Equation 8.15) could have been obtained as a direct consequence of the time reversibility of an $M/M/1$ (see Section 6.6). For not only does time reversibility imply that the output from server 1 is a Poisson process, but it also implies (Exercise 26 of Chapter 6) that the number of customers at server 1 is independent of the past departure times from server 1. As these past departure times constitute the arrival process to server 2, the independence of the numbers of customers in the two systems follows.

(ii) Since a Poisson arrival sees time averages, it follows that in a tandem queue the numbers of customers an arrival (to server 1) sees at the two servers are independent random variables. However, it should be noted that this does not imply that the waiting times of a given customer at the two servers are independent. For a counter example suppose that λ is very small with respect to $\mu_1 = \mu_2$; and thus almost all customers have zero wait in queue at both servers. However, given that the wait in queue of a customer at server 1 is positive, his wait in queue at server 2 also will be positive with probability at least as large as $\frac{1}{2}$ (why?). Hence, the waiting times in queue are not independent. Remarkably enough, however, it turns out that the total times (that is, service time plus wait in queue) that an arrival spends at the two servers are indeed independent random variables.

The preceding result can be substantially generalized. To do so, consider a system of k servers. Customers arrive from outside the system to server i, $i = 1, \ldots, k$, in accordance with independent Poisson processes at rate r_i; they then join the queue at i until their turn at service comes. Once a customer is served by server i, he then joins the queue in front of server $j, j = 1, \ldots, k$, with probability P_{ij}. Hence, $\Sigma_{j=1}^k P_{ij} \leqslant 1$, and $1 - \Sigma_{j=1}^k P_{ij}$ represents the probability that a customer departs the system after being served by server i.

If we let λ_j denote the total arrival rate of customers to server j, then the λ_j can be obtained as the solution of

$$\lambda_j = r_j + \sum_{i=1}^k \lambda_i P_{ij}, \qquad i = 1, \ldots, k \qquad (8.17)$$

Equation (8.17) follows since r_j is the arrival rate of customers to j coming from outside the system and, as λ_i is the rate at which customers depart server i (rate in must equal rate out), $\lambda_i P_{ij}$ is the arrival rate to j of those coming from server i.

It turns out that the number of customers at each of the servers is independent and of the form

$$P\{n \text{ customers at server } j\} = \left(\frac{\lambda_j}{\mu_j}\right)^n \left(1 - \frac{\lambda_j}{\mu_j}\right), \qquad n \geqslant 1$$

where μ_j is the exponential service rate at server j and the λ_j are the solution to Equation (8.17). Of course, it is necessary that $\lambda_j/\mu_j < 1$ for all j. To prove this, we first note that it is equivalent to asserting that the limiting probabilities $P(n_1, n_2, \ldots, n_k) = P\{n_j \text{ at server } j, j = 1, \ldots, k\}$ are given by

$$P(n_1 \, n_2, \ldots, n_k) = \prod_{j=1}^k \left(\frac{\lambda_j}{\mu_j}\right)^{n_j} \left(1 - \frac{\lambda_j}{\mu_j}\right) \qquad (8.18)$$

which can be verified by showing that it satisfies the balance equations for this model.

The average number of customers in the system is

$$L = \sum_{j=1}^k \text{average number at server } j$$

$$= \sum_{j=1}^k \frac{\lambda_j}{\mu_j - \lambda_j}$$

The average time a customer spends in the system can be obtained from $L = \lambda W$ with $\lambda = \Sigma_{j=1}^{k} r_j$. (Why not $\lambda = \Sigma_{j=1}^{k} \lambda_j$?) This yields

$$W = \frac{\Sigma_{j=1}^{k} \lambda_j / (\mu_j - \lambda_j)}{\Sigma_{j=1}^{k} r_j}$$

Remarks   The result embodied in Equation (8.18) is rather remarkable in that it says that the distribution of the number of customers at server i is the same as in an $M/M/1$ system with rates λ_i and μ_i. What is remarkable is that in the network model the arrival process at node i need *not* be a Poisson process. For if there is a possibility that a customer may visit a server more than once (a situation called *feedback*), then the arrival process will not be Poisson. An easy example illustrating this is to suppose that there is a single server whose service rate is very large with respect to the arrival rate from outside. Suppose also that with probability $p = 0.9$ a customer upon completion of service is fed back into the system. Hence, at an arrival time epoch there is a large probability of another arrival in a short time (namely, the feedback arrival); whereas at an arbitrary time point there will be only a very slight chance of an arrival occurring shortly (since λ is so very small). Hence, the arrival process does not possess independent increments and so cannot be Poisson. In fact even though it is straightforward to verify Equation (8.18) there does not appear to be, at present, any simple explanation as to why it is, in fact, true.

Thus, we see that when feedback is allowed the steady-state probabilities of the number of customers at any given station have the same distribution as in an $M/M/1$ model even though the model is not $M/M/1$. (Presumably such quantities as the joint distribution of the number at the station at two different time points will not be the same as for an $M/M/1$.)

Example 8.5   Consider a system of two servers where customers from outside the system arrive at server 1 at a Poisson rate 4 and at server 2 at a Poisson rate 5. The service rates of 1 and 2 are respectively 8 and 10. A customer upon completion of service at server 1 is equally likely to go to server 2 or to leave the system (i.e., $P_{11} = 0, P_{12} = \frac{1}{2}$); whereas a departure from server 2 will go 25 percent of the time to server 1 and will depart the system otherwise (i.e., $P_{21} = \frac{1}{4}, P_{22} = 0$). Determine the limiting probabilities, L, and W.

Solution: The total arrival rates to servers 1 and 2 — call them λ_1 and λ_2 — can be obtained from Equation (8.17). That is, we have

$$\lambda_1 = 4 + \tfrac{1}{4}\lambda_2,$$

$$\lambda_2 = 5 + \tfrac{1}{2}\lambda_1$$

implying that

$$\lambda_1 = 6, \qquad \lambda_2 = 8$$

Hence,

$$P\{n \text{ at server } 1, m \text{ at server } 2\} = (\tfrac{3}{4})^n\tfrac{1}{4}(\tfrac{4}{5})^m\tfrac{1}{5}$$

$$= \tfrac{1}{20}(\tfrac{3}{4})^n(\tfrac{4}{5})^m$$

and

$$L = \frac{6}{8-6} + \frac{8}{10-8} = 7,$$

$$W = \frac{L}{9} = \frac{7}{9} \quad \blacklozenge$$

8.4.2. Closed Systems

The queueing systems described in Section 8.4.1 are called *open systems* since customers are able to enter and depart the system. A system in which new customers never enter and existing ones never depart is called a *closed system*.

Let us suppose that we have m customers moving among a system of k servers, where the service times at server i are exponential with rate $\mu_i, i = 1, \ldots, k$. When a customer completes service at server i, she then joins the queue in front of server j, $j = 1, \ldots, k$, with probability P_{ij}, where we now suppose that $\Sigma_{j=1}^k P_{ij} = 1$ for all $i = 1, \ldots, k$. That is, $\mathbf{P} = [P_{ij}]$ is a Markov transition probability matrix, which we shall assume is irreducible. Let $\pi = (\pi_1, \ldots, \pi_k)$ denote the stationary probabilities for this Markov chain; that is, π is the unique positive solution of

$$\pi_j = \sum_{i=1}^{k} \pi_i P_{ij},$$

$$\sum_{j=1}^{k} \pi_j = 1$$

(8.19)

If we denote the average arrival rate (or equivalently the average service completion rate) at server j by $\lambda_m(j)$, $j = 1, \ldots, k$ then, analogous to Equation (8.17), the $\lambda_m(j)$ satisfy

$$\lambda_m(j) = \sum_{i=1}^{k} \lambda_m(i) P_{ij}$$

Hence, from (8.19) we can conclude that

$$\lambda_m(j) = \lambda_m \pi_j, \qquad j = 1, 2, \ldots, k$$

(8.20)

where

$$\lambda_m = \sum_{j=1}^{k} \lambda_m(j) \tag{8.21}$$

From Equation (8.21), we see that λ_m is the average service completion rate of the entire system, that is, it is the system *throughput* rate.*

If we let $P_m(n_1, n_2, \ldots, n_k)$ denote the limiting probabilities

$$P_m(n_1, n_2, \ldots, n_k) = P\{n_j \text{ customers at server } j, j = 1, \ldots, k\}$$

then, by verifying that they satisfy the balance equation, it can be shown that

$$P_m(n_1, n_2, \ldots, n_k) = \begin{cases} K_m \prod_{j=1}^{k} (\lambda_m(j)/\mu_j)^{n_j}, & \text{if } \sum_{j=1}^{k} n_j = m \\ 0, & \text{otherwise} \end{cases}$$

But from Equation (8.20) we thus obtain that

$$P_m(n_1, n_2, \ldots, n_k) = \begin{cases} C_m \prod_{j=1}^{k} (\pi_j/\mu_j)^{n_j}, & \text{if } \sum_{j=1}^{k} n_j = m \\ 0, & \text{otherwise} \end{cases} \tag{8.22}$$

where

$$C_m = \left[\sum_{\substack{n_1, \ldots, n_k: \\ \Sigma n_j = m}} \prod_{j=1}^{k} (\pi_j/\mu_j)^{n_j} \right]^{-1} \tag{8.23}$$

Equation (8.22) is not as useful as one might suppose, for in order to utilize it we must determine the normalizing constant C_m given by Equation (8.23) which requires summing the products $\prod_{j=1}^{k} (\pi_j/\mu_j)^{n_j}$ over all the feasible vectors $(n_1, \ldots, n_k)$: $\sum_{j=1}^{k} n_j = m$. Hence, since there are $\binom{m+k-1}{m}$ vectors this is only computationally feasible for relatively small values of m and k.

We will now present an approach that will enable us to determine recursively many of the quantities of interest in this model without first computing the normalizing constants. To begin, consider a customer who has just left server i and is headed to server j, and let us determine the probability of the system as seen by this customer. In particular, let us determine the probability that this customer observes, at that moment, n_l

---

* We are using the notation of $\lambda_m(j)$ and λ_m to indicate the dependence on the number of customers in the closed system. This will be used in recursive relations we will develop.

customers at server l, $l = 1, \ldots, k$, $\Sigma_{l=1}^{k} n_l = m - 1$. This is done as follows:

$P\{$customer observes n_l at server l, $l = 1, \ldots, k\,|\,$customer goes from i to $j\}$

$$= \frac{P\{\text{state is } (n_1, \ldots, n_i + 1, \ldots, n_j, \ldots, n_k), \text{ customer goes from } i \text{ to } j\}}{P\{\text{customer goes from } i \text{ to } j\}}$$

$$= \frac{P_m(n_1, \ldots, n_i + 1, \ldots, n_j, \ldots, n_k)\mu_i P_{ij}}{\Sigma_{\mathbf{n}: \Sigma n_j = m-1} P_m(n_1, \ldots, n_i + 1, \ldots, n_k)\mu_i P_{ij}}$$

$$= \frac{(\pi_i/\mu_i)\, \Pi_{j=1}^{k}\, (\pi_j/\mu_j)^{n_j}}{K} \qquad \text{from (8.22)}$$

$$= C \prod_{j=1}^{k} (\pi_j/\mu_j)^{n_j}$$

where C does not depend on $n_1, \ldots, n_k$. But because the above is a probability density on the set of vectors $(n_1, \ldots, n_k)$, $\Sigma_{j=1}^{k} n_j = m - 1$, it follows from (8.22) that it must equal $P_{m-1}(n_1, \ldots, n_k)$. Hence,

$P\{$customer observes n_l at server l, $l = 1, \ldots, k\,|\,$customer goes from i to $j\}$

$$= P_{m-1}(n_1, \ldots, n_k), \qquad \sum_{i=1}^{k} n_i = m - 1$$

$$(8.24)$$

As (8.24) is true for all i, we thus have proven the following proposition, known as the arrival theorem.

Proposition 8.3   (The Arrival Theorem): In the closed network system with m customers, the system as seen by arrivals to server j is distributed as the stationary distribution in the same network system when there are only $m - 1$ customers.

Denote by $L_m(j)$ and $W_m(j)$ the average number of customers and the average time a customer spends at server j when there are m customers in the network. Upon conditioning on the number of customers found at server j by an arrival to that server, it follows that

$$W_m(j) = \frac{1 + E_m[\text{number at server } j \text{ as seen by an arrival}]}{\mu_j}$$

$$(8.25)$$

$$= \frac{1 + L_{m-1}(j)}{\mu_j}$$

where the last equality follows from the arrival theorem. Now when there are $m-1$ customers in the system, then, from Equation (8.20), $\lambda_{m-1}(j)$, the average arrival rate to server j_i, satisfies

$$\lambda_{m-1}(j) = \lambda_{m-1}\pi_j$$

Now, applying the basic cost identity Equation (8.1) with the cost rule being that each customer in the network system of $m-1$ customers pays one per unit time while at server j, we obtain

$$L_{m-1}(j) = \lambda_{m-1}\pi_j W_{m-1}(j) \tag{8.26}$$

Using Equation (8.25), this yields

$$W_m(j) = \frac{1 + \lambda_{m-1}\pi_j W_{m-1}(j)}{\mu_j} \tag{8.27}$$

Also using the fact that $\sum_{j=1}^{k} L_{m-1}(j) = m-1$ (why?) we obtain, from Equation (8.26):

$$m - 1 = \lambda_{m-1} \sum_{j=1}^{k} \pi_j W_{m-1}(j)$$

or

$$\lambda_{m-1} = \frac{m-1}{\sum_{i=1}^{k} \pi_i W_{m-1}(i)} \tag{8.28}$$

Hence, from Equation (8.27), we obtain the recursion

$$W_m(j) = \frac{1}{\mu_j} + \frac{(m-1)\pi_j W_{m-1}(j)}{\mu_j \sum_{i=1}^{k} \pi_i W_{m-1}(i)} \tag{8.29}$$

Starting with the stationary probabilities π_j, $j = 1, \ldots, k$, and $W_1(j) = 1/\mu_j$ we can now use Equation (8.29) to determine recursively $W_2(j)$, $W_3(j), \ldots, W_m(j)$. We can then determine the throughput rate λ_m by using Equation (8.28), and this will determine $L_m(j)$ by Equation (8.26). This recursive approach is called *mean value analysis*.

Example 8.6   Consider a k-server network in which the customers move in a cyclic permutation. That is,

$$P_{i,i+1} = 1, \qquad i = 1, 2, \ldots, k-1, \qquad P_{k,1} = 1$$

Let us determine the average number of customers at server j when there are two customers in the system. Now, for this network

$$\pi_i = 1/k, \qquad i = 1, \ldots, k$$

and as

$$W_1(j) = \frac{1}{\mu_j}$$

we obtain from Equation (8.29) that

$$W_2(j) = \frac{1}{\mu_j} + \frac{(1/k)(1/\mu_j)}{\mu_j \sum_{i=1}^k (1/k)(1/\mu_i)}$$

$$= \frac{1}{\mu_j} + \frac{1}{\mu_j^2 \sum_{i=1}^k 1/\mu_i}$$

Hence, from Equation (8.28),

$$\lambda_2 = \frac{2}{\sum_{l=1}^k \frac{1}{k} W_2(l)} = \frac{2k}{\sum_{l=1}^k \left(\frac{1}{\mu_l} + \frac{1}{\mu_l^2 \sum_{i=1}^k 1/\mu_i} \right)}$$

and finally, using Equation (8.26),

$$L_2(j) = \lambda_2 \frac{1}{k} W_2(j)$$

$$= \frac{2 \left(\frac{1}{\mu_j} + \frac{1}{\mu_j^2 \sum_{i=1}^k 1/\mu_i} \right)}{\sum_{l=1}^k \left(\frac{1}{\mu_l} + \frac{1}{\mu_l^2 \sum_{i=1}^k 1/\mu_i} \right)} \qquad \blacklozenge$$

Another approach to learning about the stationary probabilities specified by Equation (8.22), which finesses the computational difficulties of computing the constant C_m, is to use the Gibbs sampler of Section 4.9 to generate a Markov chain having these stationary probabilities. To begin, note that since there are always a total of m customers in the system, Equation (8.22) may equivalently be written as a joint mass function of the numbers of customers at each of the servers $1, \ldots, k-1$, as follows:

$$P_m(n_1, \ldots, n_{k-1}) = C_m (\pi_k/\mu_k)^{m - \Sigma n_j} \prod_{j=1}^{k-1} (\pi_j/\mu_j)^{n_j}$$

$$= K \prod_{j=1}^{k-1} (a_j)^{n_j}, \qquad \sum_{j=1}^{k-1} n_j \leqslant m$$

where $a_j = (\pi_j \mu_k)/(\pi_k \mu_j)$, $j = 1, \ldots, k-1$. Now, if $\mathbf{N} = (N_1, \ldots, N_{k-1})$ has the preceding joint mass function then,

$$P\{N_i = n \mid N_1 = n_1, \ldots, N_{i-1} = n_{i-1}, N_{i+1} = n_{i+1}, \ldots, N_{k-1} = n_{k-1}\}$$

$$= \frac{P_m(n_1, \ldots, n_{i-1}, n, n_{i+1}, \ldots, n_{k-1})}{\sum_r P_m(n_1, \ldots, n_{i-1}, r, n_{i+1}, \ldots, n_{k-1})}$$

$$= C a_i^n, \qquad n \leqslant m - \sum_{j \neq i} n_j$$

It follows from the preceding that we may use the Gibbs sampler to generate the values of a Markov chain whose limiting probability mass function is $P_m(n_1, \ldots, n_{k-1})$ as follows:

1. Let $(n_1, \ldots, n_{k-1})$ be arbitrary nonnegative integers satisfying $\sum_{j=1}^{k-1} n_j \leqslant m$.
2. Generate a random variable I that is equally likely to be any of $1, \ldots, k-1$.
3. If $I = i$, set $s = m - \sum_{j \neq i} n_j$, and generate the value of a random variable X having probability mass function

$$P\{X = n\} = C a_i^n, \qquad n = 0, \ldots, s$$

4. Let $n_I = X$ and go to step 2.

The successive values of the state vector $(n_1, \ldots, n_{k-1}, m - \sum_{j=1}^{k-1} n_j)$ constitute the sequence of states of a Markov chain with the limiting distribution P_m. All quantities of interest can be estimated from this sequence. For instance, the average of the values of the jth coordinate of these vectors will converge to the mean number of individuals at station j, the proportion of vectors whose jth coordinate is less than r will converge to the limiting probability that the number of individuals at station j is less than r, and so on.

Other quantities of interest can also be obtained from the simulation. For instance, suppose we want to estimate W_j, the average amount of time a customer spends at server j on each visit. Then, as noted in the preceding, L_j, the average number of customers at server j, can be estimated. To estimate W_j, we use the identity

$$L_j = \lambda_j W_j$$

where λ_j is the rate at which customers arrive at server j. Setting λ_j equal to the service completion rate at server j shows that

$$\lambda_j = P\{j \text{ is busy}\}\mu_j$$

Using the Gibbs sampler simulation to estimate $P\{j \text{ is busy}\}$ then leads to an estimator of W_j.

8.5. The System M/G/1

8.5.1   Preliminaries: Work and Another Cost Identity

For an arbitrary queueing system, let us define the work in the system at any time t to be the sum of the remaining service times of all customers in the system at time t. For instance, suppose there are three customers in the system — the one in service having been there for three of his required five units of service time, and both people in queue having service times of six units. Then the work at that time is $2 + 6 + 6 = 14$. Let V denote the (time) average work in the system.

Now recall the fundamental cost equation (8.1), which states that the

> average rate at which the system earns
>
> $= \lambda_a \times$ average amount a customer pays

and consider the following cost rule: *Each customer pays at a rate of y/unit time when his remaining service time is y, whether he is in queue or in service.* Thus, the rate at which the system earns is just the work in the system; so the basic identity yields that

$$V = \lambda_a E[\text{amount paid by a customer}]$$

Now, let S and W_Q^* denote respectively the service time and the time a given customer spends waiting in queue. Then, since the customer pays at a constant rate of S per unit time while he waits in queue and at a rate of $S - x$ after spending an amount of time x in service, we have

$$E[\text{amount paid by a customer}] = E\left[SW_Q^* + \int_0^S (S - x)\, dx \right]$$

and thus

$$V = \lambda_a E[SW_Q^*] + \frac{\lambda_a E[S^2]}{2} \tag{8.30}$$

It should be noted that the preceding is a basic queueing identity [like Equations (8.2)–(8.4)] and as such is valid in almost all models. In addition, if a customer's service time is independent of his wait in queue (as is usually, but not always the case),* then we have from Equation (8.30) that

$$V = \lambda_a E[S] W_Q + \frac{\lambda_a E[S^2]}{2} \tag{8.31}$$

*For an example where it is not true, see Section 8.6.2.

8.5.2. Application of Work to M/G/1

The $M/G/1$ model assumes (i) Poisson arrivals at rate λ; (ii) a general service distribution; and (iii) a single server. In addition, we will suppose that customers are served in the order of their arrival.

Now, for an arbitrary customer in an $M/G/1$ system.

Customer's wait in queue = work in the system when he arrives   (8.32)

this follows since there is only a single server (think about it!). Taking expectations of both sides of Equation (8.32) yields

$$W_Q = \text{average work as seen by an arrival}$$

But, due to Poisson arrivals, the average work as seen by an arrival will equal V, the time average work in the system. Hence, for the model $M/G/1$,

$$W_Q = V$$

The preceding in conjunction with the identity

$$V = \lambda E[S]W_Q + \frac{\lambda E[S^2]}{2}$$

yields the so-called Pollaczek–Khintchine formula,

$$W_Q = \frac{\lambda E[S^2]}{2(1 - \lambda E[S])} \qquad (8.33)$$

where $E[S]$ and $E[S^2]$ are the first two moments of the service distribution. The quantities L, L_Q, and W can be obtained from Equation (8.33) as

$$L_Q = \lambda W_Q = \frac{\lambda^2 E[S^2]}{2(1 - \lambda E[S])},$$

$$W = W_Q + E[S] = \frac{\lambda E[S^2]}{2(1 - \lambda E[S])} + E[S], \qquad (8.34)$$

$$L = \lambda W = \frac{\lambda^2 E[S^2]}{2(1 - \lambda E[S])} + \lambda E[S]$$

Remarks   (i) For the preceding quantities to be finite, we need $\lambda E[S] < 1$. This condition is intuitive since we know from renewal theory that if the server was always busy, then the departure rate would be $1/E[S]$ (see Section 7.3), which must be larger than the arrival rate λ to keep things finite.

(ii) Since $E[S^2] = \text{Var}(S) + (E[S])^2$, we see from Equation (8.33) and (8.34) that, for fixed mean service time, L, L_Q, W, and W_Q all increase as the variance of the service distribution increases.

(iii) Another approach to obtain W_Q is presented in Exercise 38.

8.5.3. Busy Periods

The system alternates between idle periods (when there are no customers in the system, and so the server is idle) and busy periods (when there is at least one customer in the system, and so the server is busy).

Let us denote by I_n and B_n, respectively, the lengths of the nth idle and the nth busy period, $n \geqslant 1$. Hence, in the first $\sum_{j=1}^{n}(I_j + B_j)$ time units the server will be idle for a time $\sum_{j=1}^{n} I_j$, and so the proportion of time that the server will be idle, which of course is just P_0, can be expressed as

$$P_0 = \text{proportion of idle time}$$
$$= \lim_{n \to \infty} \frac{I_1 + \cdots + I_n}{I_1 + \cdots + I_n + B_1 + \cdots + B_n}$$

Now it is easy to see that the $I_1, I_2, \ldots$ are independent and identically distributed as are the $B_1, B_2, \ldots$. Hence, by dividing the numerator and the denominator of the right side of the above by n, and then applying the strong law of large numbers, we obtain

$$P_0 = \lim_{n \to \infty} \frac{(I_1 + \cdots + I_n)/n}{(I_1 + \cdots + I_n)/n + (B_1 + \cdots + B_n)/n}$$
$$= \frac{E[I]}{E[I] + E[B]} \tag{8.35}$$

where I and B represent idle and busy time random variables.

Now I represents the time from when a customer departs and leaves the system empty until the next arrival. Hence, from Poisson arrivals, it follows that I is exponential with rate λ, and so

$$E[I] = \frac{1}{\lambda} \tag{8.36}$$

To compute P_0, we note from Equation (8.4) (obtained from the fundamental cost equation by supposing that a customer pays at a rate of one per unit time while in service) that

$$\text{average number of busy servers} = \lambda E[S]$$

However, as the left-hand side of the above equals $1 - P_0$ (why?), we have

$$P_0 = 1 - \lambda E[S] \tag{8.37}$$

and, from Equations (8.35)–(8.37),

$$1 - \lambda E[S] = \frac{1/\lambda}{1/\lambda + E[B]}$$

or

$$E[B] = \frac{E[S]}{1 - \lambda E[S]}$$

Another quantity of interest is C, the number of customers served in a busy period. The mean of C can be computed by noting that, on the average, for every $E[C]$ arrivals exactly one will find the system empty (namely, the first customer in the busy period). Hence,

$$a_0 = \frac{1}{E[C]}$$

and, as $a_0 = P_0 = 1 - \lambda E[S]$ because of Poisson arrivals, we see that

$$E[C] = \frac{1}{1 - \lambda E[S]}$$

8.6. Variations on the *M/G/*1

8.6.1. The *M/G/*1 with Random-Sized Batch Arrivals

Suppose that, as in the $M/G/1$, arrivals occur in accordance with a Poisson process having rate λ. But now suppose that each arrival consists not of a single customer but of a random number of customers. As before there is a single server whose service times have distribution G.

Let us denote by $\alpha_j, j \geqslant 1$, the probability that an arbitrary batch consists of j customers; and let N denote a random variable representing the size of a batch and so $P\{N = j\} = \alpha_j$. Since $\lambda_a = \lambda E(N)$, the basic formula for work [Equation (8.31)] becomes

$$V = \lambda E[N] \left[E(S)W_Q + \frac{E(S^2)}{2} \right] \tag{8.38}$$

To obtain a second equation relating V to W_Q, consider an average customer. We have that

his wait in queue = work in system when he arrives

+ his waiting time due to those in his batch

Taking expectations and using the fact that Poisson arrivals see time averages yields

$$W_Q = V + E[\text{waiting time due to those in his batch}]$$

$$= V + E[W_B] \tag{8.39}$$

Now, $E(W_B)$ can be computed by conditioning on the number in the batch, but we must be careful because the probability that our average customer comes from a batch of size j is *not* α_j. For α_j is the proportion of batches that are of size j, and if we pick a customer at random, it is more likely that he comes from a larger rather than a smaller batch. (For instance, suppose $\alpha_1 = \alpha_{100} = \frac{1}{2}$, then half the batches are of size 1 but 100/101 of the customers will come from a batch of size 100!)

To determine the probability that our average customer came from a batch of size j we reason as follows: Let M be a large number. Then of the first M batches approximately $M\alpha_j$ will be of size j, $j \geqslant 1$, and thus there would have been approximately $jM\alpha_j$ customers that arrived in a batch of size j. Hence, the proportion of arrivals in the first M batches that were from batches of size j is approximately $jM\alpha_j/\Sigma_j jM\alpha_j$. This proportion becomes exact as $M \to \infty$, and so we see that

$$\text{proportion of customers from batches of size } j = \frac{j\alpha_j}{\Sigma_j j\alpha_j}$$

$$= \frac{j\alpha_j}{E[N]}$$

We are now ready to compute $E(W_B)$, the expected wait in queue due to others in the batch:

$$E[W_B] = \sum_j E[W_B \,|\, \text{batch of size } j] \frac{j\alpha_j}{E[N]} \tag{8.40}$$

Now if there are j customers in his batch, then our customer would have to wait for $i - 1$ of them to be served if he was ith in line among his batch members. As he is equally likely to be either 1st, 2nd, ..., or jth in line we see that

$$E[W_B \,|\, \text{batch is of size } j] = \sum_{i=1}^{j} (i - 1)E(S) \frac{1}{j}$$

$$= \frac{j - 1}{2} E[S]$$

Substituting this in Equation (8.40) yields

$$E[W_B] = \frac{E[S]}{2E[N]} \sum_j (j-1)j\alpha_j$$

$$= \frac{E[S](E[N^2] - E[N])}{2E[N]}$$

and from Equations (8.38) and (8.39) we obtain

$$W_Q = \frac{E[S](E[N^2] - E[N])/2E[N] + \lambda E[N]E[S^2]/2}{1 - \lambda E[N]E[S]}$$

Remarks   (i) Note that the condition for W_Q to be finite is that

$$\lambda E(N) < \frac{1}{E[S]}$$

which again says that the arrival rate must be less than the service rate (when the server is busy).

(ii) For fixed value of $E[N]$, W_Q is increasing in $\text{Var}[N]$, again indicating that "single-server queues do not like variation."

(iii) The other quantities L, L_Q, and W can be obtained by using

$$W = W_Q + E[S],$$

$$L = \lambda_a W = \lambda E[N]W,$$

$$L_Q = \lambda E[N]W_Q$$

8.6.2. Priority Queues

Priority queueing systems are ones in which customers are classified into types and then given service priority according to their type. Consider the situation where there are two types of customers, which arrive according to independent Poisson processes with respective rates λ_1 and λ_2, and have service distributions G_1 and G_2. We suppose that type 1 customers are given service priority, in that service will never begin on a type 2 customer if a type 1 is waiting. However, if a type 2 is being served and a type 1 arrives, we assume that the service of the type 2 is continued until completion. That is, there is no preemption once service has begun.

Let W_Q^i denote the average wait in queue of a type i customer, $i = 1, 2$. Our objective is to compute the W_Q^i.

First, note that the total work in the system at any time would be exactly the same no matter what priority rule was employed (as long as the server is always busy whenever there are customers in the system). This is so since the work will always decrease at a rate of one per unit time when the server is busy (no matter who is in service) and will always jump by the service time of an arrival. Hence, the work in the system is exactly as it would be if there was no priority rule but rather a first-come, first-served (called FIFO) ordering. However, under FIFO the above model is just $M/G/1$ with

$$\lambda = \lambda_1 + \lambda_2$$

$$G(x) = \frac{\lambda_1}{\lambda} G_1(x) + \frac{\lambda_2}{\lambda} G_2(x) \tag{8.41}$$

which follows since the combination of two independent Poisson processes is itself a Poisson process whose rate is the sum of the rates of the component processes. The service distribution G can be obtained by conditioning on which priority class the arrival is from — as is done in Equation (8.41).

Hence, from the results of Section 8.5, it follows that V, the average work in the priority queueing system, is given by

$$V = \frac{\lambda E[S^2]}{2(1 - \lambda E[S])}$$

$$= \frac{\lambda((\lambda_1/\lambda)E[S_1^2] + (\lambda_2/\lambda)E[S_2^2])}{2[1 - \lambda((\lambda_1/\lambda)E[S_1] + (\lambda_2/\lambda)E[S_2])]}$$

$$= \frac{\lambda_1 E[S_1^2] + \lambda_2 E[S_2^2]}{2(1 - \lambda_1 E[S_1] - \lambda_2 E[S_2])} \tag{8.42}$$

where S_i has distribution G_i, $i = 1, 2$.

Continuing in our quest for W_Q^*, let us note that S and W_Q^*, the service and wait in queue of an arbitrary customer, are not independent in the priority model since knowledge about S gives us information as to the type of customer which in turn gives us information about W_Q^*. To get around this we will compute separately the average amount of type 1 and type 2 work in the system. Denoting V^i as the average amount of type i work we have, exactly as in Section 8.5.1,

$$V^i = \lambda_i E[S_i]W_Q^i + \frac{\lambda_i E[S_i^2]}{2}, \qquad i = 1, 2 \tag{8.43}$$

If we define

$$V_Q^i \equiv \lambda_i E[S_i]W_Q^i,$$

$$V_S^i \equiv \frac{\lambda_i E[S_i^2]}{2}$$

then we may interpret V_Q^i as the average amount of type i work in queue, and V_S^i as the average amount of type i work in service (why?).

Now we are ready to compute W_Q^1. To do so, consider an arbitrary type 1 arrival. Then

his delay = amount of type 1 work in the system when he arrives
+ amounts of type 2 work in service when he arrives

Taking expectations and using the fact that Poisson arrivals see time average yields

$$W_Q^1 = V^1 + V_S^2$$

$$= \lambda_1 E[S_1]W_Q^1 + \frac{\lambda_1 E[S_1^2]}{2} + \frac{\lambda_2 E[S_2^2]}{2} \tag{8.44}$$

or

$$W_Q^1 = \frac{\lambda_1 E[S_1^2] + \lambda_2 E[S_2^2]}{2(1 - \lambda_1 E[S_1])} \tag{8.45}$$

To obtain W_Q^2 we first note that since $V = V^1 + V^2$, we have from Equations (8.42) and (8.43) that

$$\frac{\lambda_1 E[S_1^2] + \lambda_2 E[S_2^2]}{2(1 - \lambda_1 E[S_1] - \lambda_2 E[S_2])} = \lambda_1 E[S_1]W_Q^1 + \lambda_2 E[S_2]W_Q^2$$

$$+ \frac{\lambda_1 E[S_1^2]}{2} + \frac{\lambda_2 E[S_2^2]}{2}$$

$$= W_Q^1 + \lambda_2 E[S_2]W_Q^2 \quad \text{[from Equation (8.44)]}$$

Now, using Equation (8.45), we obtain

$$\lambda_2 E[S_2]W_Q^2 = \frac{\lambda_1 E[S_1^2] + \lambda_2 E[S_2^2]}{2}\left[\frac{1}{1 - \lambda_1 E[S_1] - \lambda_2 E[S_2]} - \frac{1}{1 - \lambda_1 E[S_1]}\right]$$

or

$$W_Q^2 = \frac{\lambda_1 E[S_1^2] + \lambda_2 E[S_2^2]}{2(1 - \lambda_1 E[S_1] - \lambda_2 E[S_2])(1 - \lambda_1 E[S_1])} \tag{8.46}$$

Remarks (i) Note that from Equation (8.45), the condition for W_Q^1 to be finite is that $\lambda_1 E[S_1] < 1$, which is independent of the type 2 parameters. (Is this intuitive?) For W_Q^2 to be finite, we need, from Equation (8.46), that

$$\lambda_1 E[S_1] + \lambda_2 E[S_2] < 1$$

Since the arrival rate of all customers is $\lambda = \lambda_1 + \lambda_2$, and the average service time of a customer is $(\lambda_1/\lambda)E[S_1] + (\lambda_2/\lambda)E[S_2]$, the preceding condition is just that the average arrival rate be less than the average service rate.

(ii) If there are n types of customers, we can solve for $V^j, j = 1, \ldots, n$, in a similar fashion. First, note that the total amount of work in the system of customers of types $1, \ldots, j$ is independent of the internal priority rule concerning types $1, \ldots, j$ and only depends on the fact that each of them is given priority over any customers of types $j + 1, \ldots, n$. (Why is this? Reason it out!) Hence, $V^1 + \cdots + V^j$ is the same as it would be if types $1, \ldots, j$ were considered as a single type I priority class and types $j + 1, \ldots, n$ as a single type II priority class. Now, from Equations (8.43) and (8.45),

$$V^{\mathrm{I}} = \frac{\lambda_{\mathrm{I}} E[S_{\mathrm{I}}^2] + \lambda_{\mathrm{I}} \lambda_{\mathrm{II}} E[S_{\mathrm{I}}] E[S_{\mathrm{II}}^2]}{2(1 - \lambda_{\mathrm{I}} E[S_{\mathrm{I}}])}$$

where

$$\lambda_{\mathrm{I}} = \lambda_1 + \cdots + \lambda_j,$$

$$\lambda_{\mathrm{II}} = \lambda_{j+1} + \cdots + \lambda_n,$$

$$E[S_{\mathrm{I}}] = \sum_{i=1}^{j} \frac{\lambda_i}{\lambda_{\mathrm{I}}} E[S_i],$$

$$E[S_{\mathrm{I}}^2] = \sum_{i=1}^{j} \frac{\lambda_i}{\lambda_{\mathrm{I}}} E[S_i^2],$$

$$E[S_{\mathrm{II}}^2] = \sum_{i=j+1}^{n} \frac{\lambda_i}{\lambda_{\mathrm{II}}} E[S_i^2]$$

Hence, as $V^{\mathrm{I}} = V^1 + \cdots + V^j$, we have an expression for $V^1 + \cdots + V^j$, for each $j = 1, \ldots, n$, which then can be solved for the individual $V^1, V^2, \ldots, V^n$. We now can obtain W_Q^i from Equation (8.43). The result of all this (which we leave for an exercise) is that

$$W_Q^i = \frac{\lambda_1 E[S_1^2] + \cdots + \lambda_n E[S_n^2]}{2 \, \Pi_{j=i-1}^{i} (1 - \lambda_1 E[S_1] - \cdots - \lambda_j E[S_j])}, \qquad i = 1, \ldots, n \quad (8.47)$$

8.6.3. An M/G/1 Optimization Example

Consider a single-server system where customers arrive according to a Poisson process with rate λ, and where the service times are independent and have distribution function G. Let $\rho = \lambda E[S]$, where S represents a service time random variable, and suppose that $\rho < 1$. Suppose that the server departs whenever a busy period ends and does not return until there are n customers waiting. At that time the server returns and continues serving until the system is once again empty. If the system facility incurs costs at a rate of c per unit time per customer in the system, as well as a cost K each time the server returns, what value of $n, n \geqslant 1$, minimizes the long-run average cost per unit time incurred by the facility, and what is this minimal cost?

To answer the preceding, let us first determine $A(n)$, the average cost per unit time for the policy that returns the server whenever there are n customers waiting. To do so, say that a new cycle begins each time the server returns. As it is easy to see that everything probabilistically starts over when a cycle begins, it follows from the theory of renewal reward processes that if $C(n)$ is the cost incurred in a cycle and $T(n)$ is the time of a cycle, then

$$A(n) = \frac{E[C(n)]}{E[T(n)]}$$

To determine $E[C(n)]$ and $E[T(n)]$, consider the time interval of length, say, T_i, starting from the first time during a cycle that there are a total of i customers in the system until the first time afterward that there are only $i - 1$. Therefore, $\Sigma_{i=1}^{n} T_i$ is the amount of time that the server is busy during a cycle. Adding the additional mean idle time until n customers are in the system gives that

$$E[T(n)] = \sum_{i=1}^{n} E[T_i] + n/\lambda$$

Now, consider the system at the moment when a service is about to begin and there are $i - 1$ customers waiting in queue. Since service times do not depend on the order in which customers are served, suppose that the order of service is last come first served, implying that service does not begin on the $i - 1$ presently in queue until these $i - 1$ are the only ones in the system. Thus, we see that the time that it takes to go from i customers in the system to $i - 1$ has the same distribution as the time it takes the $M/G/1$ system to go from a single customer (just beginning service) to empty; that is, its distribution is that of B, the length of an $M/G/1$ busy period. (Essentially the same argument was made in Example 5.23.) Hence,

$$E[T_i] = E[B] = \frac{E[S]}{1 - \rho}$$

implying that

$$E[T(n)] = \frac{nE[S]}{1 - \lambda E[S]} + \frac{n}{\lambda} = \frac{n}{\lambda(1 - \rho)} \tag{8.48}$$

To determine $E[C(n)]$, let C_i denote the cost incurred during the interval of length T_i that starts with $i - 1$ in queue and a service just beginning and ends when the $i - 1$ in queue are the only customers in the system. Thus, $K + \Sigma_{i=1}^{n} C_i$ represents the total cost incurred during the busy part of the cycle. In addition, during the idle part of the cycle there will be i customers in the system for an exponential time with rate λ, $i = 1, \ldots, n - 1$, resulting in an expected cost of $c(1 + \cdots + n - 1)/\lambda$. Consequently,

$$E[C(n)] = K + \sum_{i=1}^{n} E[C_i] + \frac{n(n-1)c}{2\lambda} \tag{8.49}$$

To find $E[C_i]$, consider the moment when the interval of length T_i begins, and let W_i be the sum of the initial service time plus the sum of the times spent in the system by all the customers that arrive (and are served) until the moment when the interval ends and there are only $i - 1$ customers in the system. Then,

$$C_i = (i - 1)cT_i + cW_i$$

where the first term refers to the cost incurred due to the $i - 1$ customers in queue during the interval of length T_i. As it is easy to see that W_i has the same distribution as W_b, the sum of the times spent in the system by all arrivals in an $M/G/1$ busy period, we obtain that

$$E[C_i] = (i - 1)c\frac{E[S]}{1 - \rho} + cE[W_b] \tag{8.50}$$

Using Equation (8.49), this yields

$$E[C(n)] = K + \frac{n(n-1)cE[S]}{2(1-\rho)} + ncE[W_b] + \frac{n(n-1)c}{2\lambda}$$

$$= K + ncE[W_b] + \frac{n(n-1)c}{2\lambda}\left(\frac{\rho}{1-\rho} + 1\right)$$

$$= K + ncE[W_b] + \frac{n(n-1)c}{2\lambda(1-\rho)}$$

Utilizing the preceding in conjunction with Equation (8.48) shows that

$$A(n) = \frac{K\lambda(1-\rho)}{n} + \lambda c(1-\rho)E[W_b] + \frac{c(n-1)}{2} \tag{8.51}$$

To determine $E[W_b]$, we use the result that the average amount of time spent in the system by a customer in the $M/G/1$ system is

$$W = W_Q + E[S] = \frac{\lambda E[S^2]}{2(1-\rho)} + E[S]$$

However, if we imagine that on day j, $j \geqslant 1$, we earn an amount equal to the total time spent in the system by the jth arrival at the $M/G/1$ system, then it follows from renewal reward processes (since everything probabilistically restarts at the end of a busy period) that

$$W = \frac{E[W_b]}{E[N]}$$

where N is the number of customers served in an $M/G/1$ busy period. Since $E[N] = 1/(1 - \rho)$ we see that

$$(1 - \rho)E[W_b] = W = \frac{\lambda E[S^2]}{2(1 - \rho)} + E[S]$$

Therefore, using Equation (8.51), we obtain

$$A(n) = \frac{K\lambda(1 - \rho)}{n} + \frac{c\lambda^2 E[S^2]}{2(1 - \rho)} + c\rho + \frac{c(n - 1)}{2}$$

To determine the optimal value of n, treat n as a continuous variable and differentiate the preceding to obtain

$$A'(n) = \frac{-K\lambda(1 - \rho)}{n^2} + \frac{c}{2}$$

Setting this equal to 0 and solving yields that the optimal value of n is

$$n^* = \sqrt{\frac{2K\lambda(1 - \rho)}{c}}$$

and the minimal average cost per unit time is

$$A(n^*) = \sqrt{2\lambda K(1 - \rho)c} + \frac{c\lambda^2 E[S^2]}{2(1 - \rho)} + c\rho - \frac{c}{2}$$

It is interesting to see how close we can come to the minimal average cost when we use a simpler policy of the following form: Whenever the server finds the system empty of customers she departs and then returns after a fixed time t has elapsed. Again let us say that a new cycle begins each time the server departs. Both the expected costs incurred during the idle and the busy parts of a cycle are obtained by conditioning on $N(t)$, the number of arrivals in the time t that the server is gone. With $\bar{C}(t)$ being the cost incurred during a cycle, we obtain

$$E[\bar{C}(t) \mid N(t)] = K + \sum_{i=1}^{N(t)} E[C_i] + cN(t)\frac{t}{2}$$

$$= K + \frac{N(t)(N(t) - 1)cE[S]}{2(1 - \rho)} + N(t)cE[W_b] + cN(t)\frac{t}{2}$$

The final term of the first equality is the conditional expected cost during the idle time in the cycle and is obtained by using that, given the number of arrivals in the time t, the arrival times are independent and uniformly distributed on $(0, t)$; the second equality used Equation (8.50). Since $N(t)$ is Poisson with mean λt, it follows that $E[N(t)(N(t) - 1)] = E[N^2(t)] - E[N(t)] = \lambda^2 t^2$. Thus, taking the expected value of the preceding gives

$$E[\bar{C}(t)] = K + \frac{\lambda^2 t^2 c E[S]}{2(1 - \rho)} + \lambda t c E[W_b] + \frac{c\lambda t^2}{2}$$

$$= K + \frac{c\lambda t^2}{2(1 - \rho)} + \lambda t c E[W_b]$$

Similarly, if $\bar{T}(t)$ is the time of a cycle, then

$$E[\bar{T}(t)] = E[E[\bar{T}(t) \mid N(t)]]$$

$$= E[t + N(t)E[B]]$$

$$= t + \frac{\rho t}{1 - \rho}$$

$$= \frac{t}{1 - \rho}$$

Hence, the average cost per unit time, call it $\bar{A}(t)$, is

$$\bar{A}(t) = \frac{E[\bar{C}(t)]}{E[\bar{T}(t)]}$$

$$= \frac{K(1 - \rho)}{t} + \frac{c\lambda t}{2} + c\lambda(1 - \rho)E[W_b]$$

Thus, from Equation (8.51), we see that

$$\bar{A}(n/\lambda) - A(n) = c/2$$

which shows that allowing the return decision to depend on the number presently in the system can reduce the average cost only by the amount $c/2$. ✦

8.7. The Model *G/M*/1

The model $G/M/1$ assumes that the times between successive arrivals have an arbitrary distribution G. The service times are exponentially distributed with rate μ and there is a single server.

The immediate difficulty in analyzing this model stems from the fact that the number of customers in the system is not informative enough to serve as a state space. For in summarizing what has occurred up to the present we would need to know not only the number in the system, but also the amount of time that has elapsed since the last arrival (since G is not memoryless). (Why need we not be concerned with the amount of time the person being served has already spent in service?) To get around this problem we shall only look at the system when a customer arrives; and so let us define X_n, $n \geq 1$, by

$$X_n \equiv \text{the number in the system as seen by the } n\text{th arrival}$$

It is easy to see that the process $\{X_n, n \geq 1\}$ is a Markov chain. To compute the transition probabilities P_{ij} for this Markov chain let us first note that, as long as there are customers to be served, the number of services in any length of time t is a Poisson random variable with mean μt. This is true since the time between successive services is exponential and, as we know, this implies that the number of services thus constitutes a Poisson process. Hence,

$$P_{i,i+1-j} = \int_0^\infty e^{-\mu t} \frac{(\mu t)^j}{j!} \, dG(t), \qquad j = 0, 1, \dots, i$$

which follows since if an arrival finds i in the system, then the next arrival will find $i + 1$ minus the number served, and the probability that j will be served is easily seen to equal the right side of the above (by conditioning on the time between the successive arrivals).

The formula for P_{i0} is a little different (it is the probability that *at least* $i + 1$ Poisson events occur in a random length of time having distribution G) and can be obtained from

$$P_{i0} = 1 - \sum_{j=0}^i P_{i,i+1-j}$$

The limiting probabilities π_k, $k = 0, 1, \dots$, can be obtained as the unique solution of

$$\pi_k = \sum_i \pi_i P_{ik}, \qquad k \geq 0$$

$$\sum_k \pi_k = 1$$

which, in this case, reduce to

$$\pi_k = \sum_{i=k-1}^\infty \pi_i \int_0^\infty e^{-\mu t} \frac{(\mu t)^{i+1-k}}{(i + 1 - k)!} \, dG(t), \qquad k \geq 1$$

$$\sum_0^\infty \pi_k = 1$$

(8.52)

(We have not included the equation $\pi_0 = \Sigma \pi_i P_{i0}$ since one of the equations is always redundant.)

To solve the above, let us try a solution of the form $\pi_k = c\beta^k$. Substitution into Equation (8.52) leads to

$$c\beta^k = c \sum_{i=k-1}^{\infty} \beta^i \int_0^{\infty} e^{-\mu t} \frac{(\mu t)^{i+1-k}}{(i+1-k)!} \, dG(t)$$

$$= c \int_0^{\infty} e^{-\mu t} \beta^{k-1} \sum_{i=k-1}^{\infty} \frac{(\beta \mu t)^{i+1-k}}{(i+1-k)!} \, dG(t) \qquad (8.53)$$

However,

$$\sum_{i=k-1}^{\infty} \frac{(\beta \mu t)^{i+1-k}}{(i+1-k)!} = \sum_{j=0}^{\infty} \frac{(\beta \mu t)^j}{j!}$$

$$= e^{\beta \mu t}$$

and thus Equation (8.53) reduces to

$$\beta^k = \beta^{k-1} \int_0^{\infty} e^{-\mu t(1-\beta)} \, dG(t)$$

or

$$\beta = \int_0^{\infty} e^{-\mu t(1-\beta)} \, dG(t) \qquad (8.54)$$

The constant c can be obtained from $\Sigma_k \pi_k = 1$, which implies that

$$c \sum_0^{\infty} \beta^k = 1$$

or

$$c = 1 - \beta$$

As the π_k is the *unique* solution to Equation (8.52), and $\pi_k = (1-\beta)\beta^k$ satisfies, it follows that

$$\pi_k = (1-\beta)\beta^k, \qquad k = 0, 1, \ldots$$

where β is the solution of Equation (8.54). [It can be shown that if the mean of G is greater than the mean service time $1/\mu$, then there is a unique value of β satisfying Equation (8.54) which is between 0 and 1.] The exact value of β usually can only be obtained by numerical methods.

As π_k is the limiting probability that an arrival sees k customers, it is just the a_k as defined in Section 8.2. Hence,

$$a_k = (1-\beta)\beta^k, \qquad k \geq 0 \qquad (8.55)$$

We can obtain W by conditioning on the number in the system when a customer arrives. This yields

$$W = \sum_k E[\text{time in system} \mid \text{arrival sees } k](1 - \beta)\beta^k$$

$$= \sum_k \frac{k+1}{\mu}(1 - \beta)\beta^k \quad \begin{array}{l}\text{(Since if an arrival sees } k \text{, then he spends}\\ k + 1 \text{ service periods in the system.}\end{array}$$

$$= \frac{1}{\mu(1 - \beta)} \quad \left(\text{by using } \sum_0^\infty kx^k = \frac{x}{(1 - x)^2}\right)$$

and

$$W_Q = W - \frac{1}{\mu} = \frac{\beta}{\mu(1 - \beta)},$$

$$L = \lambda W = \frac{\lambda}{\mu(1 - \beta)}, \qquad (8.56)$$

$$L_Q = \lambda W_Q = \frac{\lambda\beta}{\mu(1 - \beta)}$$

where λ is the reciprocal of the mean interarrival time. That is,

$$\frac{1}{\lambda} = \int_0^\infty x \, dG(x)$$

In fact, in exactly the same manner as shown for the $M/M/1$ in Section 8.3.1 and Exercise 4 we can show that

W^* is exponential with rate $\mu(1 - \beta)$,

$$W_Q^* = \begin{cases} 0 \text{ with probability } 1 - \beta \\ \text{exponential with rate } \mu(1 - \beta) \text{ with probability } \beta \end{cases}$$

where W^* and W_Q^* are the amounts of time that a customer spends in system and queue, respectively (their means are W and W_Q).

Whereas $a_k = (1 - \beta)\beta^k$ is the probability that an arrival sees k in the system, it is not equal to the proportion of time during which there are k in the system (since the arrival process is not Poisson). To obtain the P_k we first note that the rate at which the number in the system changes from $k - 1$ to k must equal the rate at which it changes from k to $k - 1$ (why?). Now the rate at which it changes from $k - 1$ to k is equal to the arrival rate λ multiplied by the proportion of arrivals finding $k - 1$ in the system. That is,

$$\text{rate number in system goes from } k - 1 \text{ to } k = \lambda a_{k-1}$$

Similarly, the rate at which the number in the system changes from k to $k - 1$ is equal to the proportion of time during which there are k in the system multiplied by the (constant) service rate. That is,

$$\text{rate number in system goes from } k \text{ to } k - 1 = P_k \mu$$

Equating these rates yields

$$P_k = \frac{\lambda}{\mu} a_{k-1}, \qquad k \geqslant 1$$

and so, from Equation (8.55),

$$P_k = \frac{\lambda}{\mu}(1 - \beta)\beta^{k-1}, \qquad k \geqslant 1$$

and, as $P_0 = 1 - \sum_{k=1}^{\infty} P_k$, we obtain

$$P_0 = 1 - \frac{\lambda}{\mu}$$

Remark   In the foregoing analysis we guessed at a solution of the stationary probabilities of the Markov chain of the form $\pi_k = c\beta^k$, then verified such a solution by substituting in the stationary Equation (8.52). However, it could have been argued directly that the stationary probabilities of the Markov chain are of this form. To do so, define β_i to be the expected number of times that state $i + 1$ is visited in the Markov chain between two successive visits to state i, $i \geqslant 0$. Now it is not difficult to see (and we will let the reader argue it out for him or herself) that

$$\beta_0 = \beta_1 = \beta_2 = \cdots = \beta$$

Now it can be shown by using renewal reward processes that

$$\pi_{i+1} = \frac{E[\text{number of visits to state } i+1 \text{ in an } i-i \text{ cycle}]}{E[\text{number of transtions in an } i-i \text{ cycle}]}$$

$$= \frac{\beta_i}{1/\pi_i}$$

and so,

$$\pi_{i+1} = \beta_i \pi_i = \beta \pi_i, \qquad i \geqslant 0$$

implying, since $\sum_0^{\infty} \pi_i = 1$, that

$$\pi_i = \beta^i(1 - \beta), \qquad i \geqslant 0$$

8.7.1. The G/M/1 Busy and Idle Periods

Suppose that an arrival has just found the system empty — and so initiates a busy period — and let N denote the number of customers served in that busy period. Since the Nth arrival (after the initiator of the busy period) will also find the system empty, it follows that N is the number of transitions for the Markov chain (of Section 8.7) to go from state 0 to state 0. Hence, $1/E[N]$ is the proportion of transitions that take the Markov chain into state 0; or equivalently, it is the proportion of arrivals that find the system empty. Therefore,

$$E[N] = \frac{1}{a_0} = \frac{1}{1 - \beta}$$

Also, as the next busy period begins after the Nth interarrival, it follows that the cycle time (that is, the sum of a busy and idle period) is equal to the time until the Nth interarrival. In other words, the sum of a busy and idle period can be expressed as the sum of N interarrival times. Thus, if T_i is the ith interarrival time after the busy period begins, then

$$E[\text{Busy}] + E[\text{Idle}] = E\left[\sum_{i=1}^{N} T_i\right]$$
$$= E[N]E[T] \quad \text{(by Wald's equation)}$$
$$= \frac{1}{\lambda(1 - \beta)} \tag{8.57}$$

For a second relation between $E[\text{Busy}]$ and $E[\text{Idle}]$, we can use the same argument as in Section 8.5.3 to conclude that

$$1 - P_0 = \frac{E[\text{Busy}]}{E[\text{Idle}] + E[\text{Busy}]}$$

and since $P_0 = 1 - \lambda/\mu$, we obtain, upon combining this with (8.57), that

$$E[\text{Busy}] = \frac{1}{\mu(1 - \beta)},$$

$$E[\text{Idle}] = \frac{\mu - \lambda}{\lambda\mu(1 - \beta)}$$

8.8. A Finite Source Model

Consider a system of m machines, whose working times are independent exponential random variables with rate λ. Upon failure, a machine instantly goes to a repair facility that consists of a single repairperson. If the repairperson is free, repair begins on the machine; otherwise, the machine joins the queue of failed machines. When a machine is repaired it becomes a working machine, and repair begins on a new machine from the queue of failed machines (provided the queue is nonempty). The successive repair times are independent random variables having density function g, with mean

$$\mu_R = \int_0^\infty xg(x)\,dx$$

To analyze this system, so as to determine such quantities as the average number of machines that are down and the average time that a machine is down, we will exploit the exponentially distributed working times to obtain a Markov chain. Specifically, let X_n denote the number of failed machines immediately after the nth repair occurs, $n \geqslant 1$. Now, if $X_n = i > 0$, then the situation when the nth repair has just occurred is that repair is about to begin on a machine, there are $i - 1$ other machines waiting for repair, and there are $m - i$ working machines, each of which will (independently) continue to work for an exponential time with rate λ. Similarly, if $X_n = 0$, then all m machines are working and will (independently) continue to do so for exponentially distributed times with rate λ. Consequently, any information about earlier states of the system will not affect the probability distribution of the number of down machines at the moment of the next repair completion; hence, $\{X_n, n \geqslant 1\}$ is a Markov chain. To determine its transition probabilities $P_{i,j}$, suppose first that $i > 0$. Conditioning on R, the length of the next repair time, and making use of the independence of the $m - i$ remaining working times, yields that for $j \leqslant m - i$

$$P_{i,i-1+j} = P\{j \text{ failures during } R\}$$

$$= \int_0^\infty P\{j \text{ failures during } R \mid R = r\}g(r)\,dr$$

$$= \int_0^\infty \binom{m-i}{j}(1 - e^{-\lambda r})^j(e^{-\lambda r})^{m-i-j}g(r)\,dr$$

If $i = 0$, then, because the next repair will not begin until one of the machines fails,

$$P_{0,j} = P_{1,j}, \qquad j \leqslant m - 1$$

Let π_j, $j = 0, \ldots, m - 1$, denote the stationary probabilities of this Markov chain. That is, they are the unique solution of

$$\pi_j = \sum_i \pi_i P_{i,j}$$

$$\sum_{j=0}^{m-1} \pi_j = 1$$

Therefore, after explicitly determining the transition probabilities and solving the preceding equations, we would know the value of π_0, the proportion of repair completions that leaves all machines working. Let us say that the system is "on" when all machines are working and "off" otherwise. (Thus, the system is on when the repairperson is idle and off when he is busy.) As all machines are working when the system goes back on, it follows from the lack of memory property of the exponential that the system probabilistically starts over when it goes on. Hence, this on–off system is an alternating renewal process. Suppose that the system has just become on, thus starting a new cycle, and let R_i, $i \geqslant 1$, be the time of the ith repair from that moment. Also, let N denote the number of repairs in the off (busy) time of the cycle. Then, it follows that B, the length of the off period, can be expressed as

$$B = \sum_{i=1}^{N} R_i$$

Although N is not independent of the sequence $R_1, R_2, \ldots$, it is easy to check that it is a stopping time for this sequence, and thus by Wald's equation (see Exercise 12 of Chapter 7) we have that

$$E[B] = E[N]E[R] = E[N]\mu_R$$

Also, since an on time will last until one of the machines fails, and since the minimum of independent exponential random variables is exponential with a rate equal to the sum of their rates, it follows that $E[I]$, the mean on (idle) time in a cycle, is given by

$$E[I] = 1/(m\lambda)$$

Hence, P_B, the proportion of time that the repairperson is busy, satisfies

$$P_B = \frac{E[N]\mu_R}{E[N]\mu_R + 1/(m\lambda)}$$

However, since, on average, one out of every $E[N]$ repair completions will leave all machines working, it follows that

$$\pi_0 = \frac{1}{E[N]}$$

Consequently,

$$P_B = \frac{\mu_R}{\mu_R + \pi_0/(m\lambda)} \qquad (8.58)$$

Now focus attention on one of the machines, call it machine number 1, and let $P_{1,R}$ denote the proportion of time that machine 1 is being repaired. Since the proportion of time that the repairperson is busy is P_B, and since all machines fail at the same rate and have the same repair distribution, it follows that

$$P_{1,R} = \frac{P_B}{m} = \frac{\mu_R}{m\mu_R + \pi_0/\lambda} \tag{8.59}$$

However, machine 1 alternates between time periods when it is working, when it is waiting in queue, and when it is in repair. Let W_i, Q_i, S_i denote, respectively, the ith working time, the ith queueing time, and the ith repair time of machine 1, $i \geqslant 1$. Then, the proportion of time that machine 1 is being repaired during its first n working–queue–repair cycles is:

proportion of time in the first n cycles that machine 1 is being repaired

$$= \frac{\sum_{i=1}^{n} S_i}{\sum_{i=1}^{n} W_i + \sum_{i=1}^{n} Q_i + \sum_{i=1}^{n} S_i}$$

$$= \frac{\sum_{i=1}^{n} S_i/n}{\sum_{i=1}^{n} W_i/n + \sum_{i=1}^{n} Q_i/n + \sum_{i=1}^{n} S_i/n}$$

Letting $n \to \infty$ and using the strong law of large numbers to conclude that the averages of the W_i and of the S_i converge, respectively, to $1/\lambda$ and μ_R, yields that

$$P_{1,R} = \frac{\mu_R}{1/\lambda + \bar{Q} + \mu_R}$$

where $\bar{Q}$ is the average amount of time that machine 1 spends in queue when it fails. Using Equation (8.59), the preceding gives that

$$\frac{\mu_R}{m\mu_R + \pi_0/\lambda} = \frac{\mu_R}{1/\lambda + \bar{Q} + \mu_R}$$

or, equivalently, that

$$\bar{Q} = (m - 1)\mu_R - (1 - \pi_0)/\lambda$$

Moreover, since all machines are probabilistically equivalent it follows that $\bar{Q}$ is equal to W_Q, the average amount of time that a failed machine spends in queue. To determine the average number of machines in queue, we will make use of the basic queueing identity

$$L_Q = \lambda_a W_Q = \lambda_a \bar{Q}$$

where λ_a is the average rate at which machines fail. To determine λ_a, again focus attention on machine 1 and suppose that we earn one per unit time whenever machine 1 is being repaired. It then follows from the basic cost identity of Equation (8.1) that

$$P_{1,R} = r_1 \mu_R$$

where r_1 is the average rate at which machine 1 fails. Thus, from Equation (8.59), we obtain that

$$r_1 = \frac{1}{m\mu_R + \pi_0/\lambda}$$

Because all m machines fail at the same rate, the preceding implies that

$$\lambda_a = mr_1 = \frac{m}{m\mu_R + \pi_0/\lambda}$$

which gives that the average number of machines in queue is

$$L_Q = \frac{m(m-1)\mu_R - m(1-\pi_0)/\lambda}{m\mu_R + \pi_0/\lambda}$$

Since the average number of machines being repaired is P_B, the preceding, along with Equation (8.58), shows that the average number of down machines is

$$L = L_Q + P_B = \frac{m^2\mu_R - m(1-\pi_0)/\lambda}{m\mu_R + \pi_0/\lambda}$$

8.9. Multiserver Queues

By and large, systems that have more than one server are much more difficult to analyze than those with a single server. In Section 8.9.1 we start first with a Poisson arrival system in which no queue is allowed, and then consider in Section 8.9.2 the infinite capacity $M/M/k$ system. For both of these models we are able to present the limiting probabilities. In Section 8.9.3 we consider the model $G/M/k$. The analysis here is similar to that of the $G/M/1$ (Section 8.7) except that in place of a single quantity β given as the solution of an integral equation, we have k such quantities. We end in Section 8.9.4 with the model $M/G/k$ for which unfortunately our previous technique (used in $M/G/1$) no longer enables us to derive W_Q, and we content ourselves with an approximation.

8.9.1. Erlang's Loss System

A loss system is a queueing system in which arrivals that find all servers busy do not enter but rather are lost to the system. The simplest such system is the $M/M/k$ loss system in which customers arrive according to a Poisson process having rate λ, enter the system if at least one of the k servers is free, and then spend an exponential amount of time with rate μ being served. The balance equations for this system are

State	Rate leave = rate enter
0	$\lambda P_0 = \mu P_1$
1	$(\lambda + \mu)P_1 = 2\mu P_2 + \lambda P_0$
2	$(\lambda + 2\mu)P_2 = 3\mu P_3 + \lambda P_1$
$i, 0 < i < k$	$(\lambda + i\mu)P_i = (i + 1)\mu P_{i+1} + \lambda P_{i-1}$
k	$k\mu P_k = \lambda P_{k-1}$

Rewriting gives

$$\lambda P_0 = \mu P_1,$$
$$\lambda P_1 = 2\mu P_2,$$
$$\lambda P_2 = 3\mu P_3,$$
$$\vdots$$
$$\lambda P_{k-1} = k\mu P_k$$

or

$$P_1 = \frac{\lambda}{\mu} P_0$$

$$P_2 = \frac{\lambda}{2\mu} P_1 = \frac{(\lambda/\mu)^2}{2} P_0,$$

$$P_3 = \frac{\lambda}{3\mu} P_2 = \frac{(\lambda/\mu)^3}{3!} P_0,$$

$$\vdots$$

$$P_k = \frac{\lambda}{k\mu} P_{k-1} = \frac{(\lambda/\mu)^k}{k!} P_0$$

and using $\Sigma_0^k P_i = 1$, we obtain

$$P_i = \frac{(\lambda/\mu)^i/i!}{\Sigma_{j=0}^k (\lambda/\mu)^j/j!}, \qquad i = 0, 1, \ldots, k$$

Since $E[S] = 1/\mu$, where $E[S]$ is the mean service time, the preceding can be written as

$$P_i = \frac{(\lambda E[S])^i/i!}{\Sigma_{j=0}^k (\lambda E[S])^j/j!}, \qquad i = 0, 1, \ldots, k \qquad (8.60)$$

Consider now the same system except that the service distribution is general — that is, consider the $M/G/k$ with no queue allowed. This model is sometimes called the *Erlang loss system*. It can be shown (though the proof is advanced) that Equation (8.60) (which is called Erlang's loss formula) remains valid for this more general system.

8.9.2. The *M/M/k* Queue

The $M/M/k$ infinite capacity queue can be analyzed by the balance equation technique. We leave it for the reader to verify that

$$
P_i = \begin{cases}
\dfrac{\dfrac{(\lambda/\mu)^i}{i!}}{\displaystyle\sum_{i=0}^{k-1} \dfrac{(\lambda/\mu)^i}{i!} + \dfrac{(\lambda/\mu)^k}{k!}\dfrac{k\mu}{k\mu - \lambda}}, & i \leqslant k \\[6ex]
\dfrac{(\lambda/k\mu)^i k^k}{k!} P_0, & i > k
\end{cases}
$$

We see from the preceding that we need to impose the condition $\lambda < k\mu$.

8.9.3. The *G/M/k* Queue

In this model we again suppose that there are k servers, each of whom serves at an exponential rate μ. However, we now allow the time between successive arrivals to have an arbitrary distribution G. To ensure that a steady-state (or limiting) distribution exists, we assume the condition $1/\mu_G < k\mu$ where μ_G is the mean of G.*

The analysis for this model is similar to that presented in Section 8.7 for the case $k = 1$. Namely, to avoid having to keep track of the time since the last arrival, we look at the system only at arrival epochs. Once again, if we define X_n as the number in the system at the moment of the nth arrival, then $\{X_n, n \geqslant 0\}$ is a Markov chain.

To derive the transition probabilities of the Markov chain, it helps to first note the relationship

$$
X_{n+1} = X_n + 1 - Y_n, \qquad n \geqslant 0
$$

where Y_n denotes the number of departures during the interarrival time between the nth and $(n+1)$st arrival. The transition probabilities P_{ij} can now be calculated as follows:

Case (i) $j > i + 1$.

In this case it easily follows that $P_{ij} = 0$.

---

*It follows from renewal theory (Proposition 7.1) that customers arrive at rate $1/\mu_G$, and as the maximum service rate is $k\mu$, we clearly need that $1/\mu_G < k\mu$ for limiting probabilities to exist.

Case (ii) $j \leqslant i + 1 \leqslant k$.

In this case if an arrival finds i in the system, then as $i < k$ the new arrival will also immediately enter service. Hence, the next arrival will find j if of the $i + 1$ services exactly $i + 1 - j$ are completed during the interarrival time. Conditioning on the length of this interarrival time yields

$$P_{ij} = P\{i + 1 - j \text{ of } i + 1 \text{ services are completed in an interarrival time}\}$$

$$= \int_0^\infty P\{i + 1 - j \text{ of } i + 1 \text{ are completed} \,|\, \text{interarrival time is } t\} \, dG(t)$$

$$= \int_0^\infty \binom{i + 1}{j} (1 - e^{-\mu t})^{i + 1 - j} (e^{-\mu t})^j \, dG(t)$$

where the last equality follows since the number of service completions in a time t will have a binomial distribution.

Case (iii) $i + 1 \geqslant j \geqslant k$.

To evaluate P_{ij} in this case we first note that when all servers are busy, the departure process is a Poisson process with rate $k\mu$ (why?). Hence, again conditioning on the interarrival time we have

$$P_{ij} = P\{i + 1 - j \text{ departures}\}$$

$$= \int_0^\infty P\{i + 1 - j \text{ departures in time } t\} \, dG(t)$$

$$= \int_0^\infty e^{-k\mu t} \frac{(k\mu t)^{i + 1 - j}}{(i + 1 - j)!} \, dG(t)$$

Case (iv) $i + 1 \geqslant k > j$.

In this case since when all servers are busy the departure process is a Poisson process, it follows that the length of time until there will only be k in the system will have a gamma distribution with parameters $i + 1 - k$, $k\mu$ (the time until $i + 1 - k$ events of a Poisson process with rate $k\mu$ occur is gamma distributed with parameters $i + 1 - k$, $k\mu$). Conditioning first on the interarrival time and then on the time until there are only k in the system (call this latter random variable T_k) yields

$$P_{ij} = \int_0^\infty P\{i + 1 - j \text{ departures in time } t\} \, dG(t)$$

$$= \int_0^\infty \int_0^t P\{i + 1 - j \text{ departures in } t \,|\, T_k = s\} k\mu e^{-k\mu s} \frac{(k\mu s)^{i - k}}{(i - k)!} \, ds \, dG(t)$$

$$= \int_0^\infty \int_0^t \binom{k}{j} (1 - e^{-\mu(t - s)})^{k - j} (e^{-\mu(t - s)})^j k\mu e^{-k\mu s} \frac{(k\mu s)^{i - k}}{(i - k)!} \, ds \, dG(t)$$

where the last equality follows since of the k people in service at time s the number whose service will end by time t is binomial with parameters k and $1 - e^{-\mu(t-s)}$.

We now can verify either by a direct substitution into the equations $\pi_j = \Sigma_i \pi_i P_{ij}$, or by the same argument as presented in the remark at the end of Section 8.7, that the limiting probabilities of this Markov chain are of the form

$$\pi_{k-1+j} = c\beta^j, \qquad j = 0, 1, \dots.$$

Substitution into any of the equations $\pi_j = \Sigma_i \pi_i P_{ij}$ when $j > k$ yields that β is given as the solution of

$$\beta = \int_0^\infty e^{-k\mu t(1-\beta)} \, dG(t)$$

The values $\pi_0, \pi_1, \dots, \pi_{k-2}$, can be obtained by recursively solving the first $k-1$ of the steady-state equations, and c can then be computed by using $\Sigma_0^\infty \pi_i = 1$.

If we let W_Q^* denote the amount of time that a customer spends in queue, then in exactly the same manner as in $G/M/1$ we can show that

$$W_Q^* = \begin{cases} 0, & \text{with probability } \sum_0^{k-1} \pi_i = 1 - \dfrac{c\beta}{1-\beta} \\ \text{Exp}(k\mu(1-\beta)), & \text{with probability } \sum_k^\infty \pi_i = \dfrac{c\beta}{1-\beta} \end{cases}$$

where $\text{Exp}(k\mu(1-\beta))$ is an exponential random variable with rate $k\mu(1-\beta)$.

8.9.4. The $M/G/k$ Queue

In this section we consider the $M/G/k$ system in which customers arrive at a Poisson rate λ and are served by any of k servers, each of whom has the service distribution G. If we attempt to mimic the analysis presented in Section 8.5 for the $M/G/1$ system, then we would start with the basic identity

$$V = \lambda E[S]W_Q + \lambda E[S^2]/2 \tag{8.61}$$

and then attempt to derive a second equation relating V and W_Q.

Now if we consider an arbitrary arrival, then we have the following identity:

work in system when customer arrives

$$= k \times \text{time customer spends in queue} + R \tag{8.62}$$

where R is the sum of the remaining service times of all other customers in service at the moment when our arrival enters service.

The foregoing follows because while the arrival is waiting in queue, work is being processed at a rate k per unit time (since all servers are busy). Thus, an amount of work $k \times$ time in queue is processed while he waits in queue. Now, all of this work was present when he arrived and in addition the remaining work on those still being served when he enters service was also present when he arrived — so we obtain Equation (8.62). For an illustration, suppose that there are three servers all of whom are busy when the customer arrives. Suppose, in addition, that there are no other customers in the system and also that the remaining service times of the three people in service are 3, 6, and 7. Hence, the work seen by the arrival is $3 + 6 + 7 = 16$. Now the arrival will spend 3 time units in queue, and at the moment he enters service, the remaining times of the other two customers are $6 - 3 = 3$ and $7 - 3 = 4$. Hence, $R = 3 + 4 = 7$ and as a check of Equation (8.62) we see that $16 = 3 \times 3 + 7$.

Taking expectations of Equation (8.61) and using the fact that Poisson arrivals see time averages, we obtain

$$V = kW_Q + E[R]$$

which, along with Equation (8.61), would enable us to solve for W_Q if we could compute $E[R]$. However there is no known method for computing $E[R]$ and in fact, there is no known exact formula for W_Q. The following approximation for W_Q was obtained in Reference 6 by using the foregoing approach and then approximating $E[R]$:

$$W_Q \approx \frac{\lambda^k E[S^2](E[S])^{k-1}}{2(k-1)!(k - \lambda E[S])^2 \left[\sum_{n=0}^{k-1} \frac{(\lambda E[S])^n}{n!} + \frac{(\lambda E[S])^k}{(k-1)!(k - \lambda E[S])} \right]} \qquad (8.63)$$

The preceding approximation has been shown to be quite close to the W_Q when the service distribution is gamma. It is also exact when G is exponential.

Exercises

1. For the $M/M/1$ queue, compute

(a) the expected number of arrivals during a service period and
(b) the probability that no customers arrive during a service period.

Hint: "Condition."

*2. Machines in a factory break down at an exponential rate of six per hour. There is a single repairman who fixes machines at an exponential rate of eight per hour. The cost incurred in lost production when machines are out of service is $10 per hour per machine. What is the average cost rate incurred due to failed machines?

3. The manager of a market can hire either Mary or Alice. Mary, who gives service at an exponential rate of 20 customers per hour, can be hired at a rate of $3 per hour. Alice, who gives service at an exponential rate of 30 customers per hour, can be hired at a rate of $C per hour. The manager estimates that, on the average, each customer's time is worth $1 per hour and should be accounted for in the model. If customers arrive at a Poisson rate of 10 per hour, then

    (a) what is the average cost per hour if Mary is hired? if Alice is hired?
    (b) find C if the average cost per hour is the same for Mary and Alice.

4. For the $M/M/1$ queue, show that the probability that a customer spends an amount of time x or less in queue is given by

$$1 - \frac{\lambda}{\mu}, \qquad \text{if } x = 0$$

$$1 - \frac{\lambda}{\mu} + \frac{\lambda}{\mu}(1 - e^{-(\mu - \lambda)x}), \qquad \text{if } x > 0$$

5. It follows from Exercise 4 that if, in the $M/M/1$ model, W_Q^* is the amount of time that a customer spends waiting in queue, then

$$W_Q^* = \begin{cases} 0, & \text{with probability } 1 - \lambda/\mu \\ \text{Exp}(\mu - \lambda), & \text{with probability } \lambda/\mu \end{cases}$$

where $\text{Exp}(\mu - \lambda)$ is an exponential random variable with rate $\mu - \lambda$. Using this, find $\text{Var}(W_Q^*)$.

6. Two customers move about among three servers. Upon completion of service at server i, the customer leaves that server and enters service at whichever of the other two servers is free. (Therefore, there are always two busy servers.) If the service times at server i are exponential with rate μ_i, $i = 1, 2, 3$, what proportion of time is server i idle?

*7. Show that W is smaller in an $M/M/1$ model having arrivals at rate λ and service at rate 2μ than it is in a two-server $M/M/2$ model with arrivals at rate λ and with each server at rate μ. Can you give an intuitive explanation for this result? Would it also be true for W_Q?

8. A group of n customers moves around among two servers. Upon completion of service, the served customer then joins the queue (or enters service if the server is free) at the other server. All service times are exponential with rate μ. Find the proportion of time that there are j customers at server 1, $j = 0, \ldots, n$.

9. A facility produces items according to a Poisson process with rate λ. However, it has shelf space for only k items and so it shuts down production whenever k items are present. Customers arrive at the facility according to a Poisson process with rate μ. Each customer wants one item and will immediately depart either with the item or empty handed if there is no item available.

(a) Find the proportion of customers that go away empty handed.
(b) Find the average time that an item is on the shelf.
(c) Find the average number of items on the shelf.

Suppose now that when a customer does not find any available items it joins the "customers' queue" as long as there are no more than $n - 1$ other customers waiting at that time. If there are n waiting customers then the new arrival departs without an item.

(d) Set up the balance equations.
(e) In terms of the solution of the balance equations, what is the average number of customers in the system?

10. A group of m customers frequents a single-server station in the following manner. When a customer arrives, he or she either enters service if the server is free or joins the queue otherwise. Upon completing service the customer departs the system, but then returns after an exponential time with rate θ. All service times are exponentially distributed with rate μ.

(a) Define states and set up the balance equations.

In terms of the solution of the balance equations, find

(b) the average rate at which customers enter the station.
(c) the average time that a customer spends in the station per visit.

11. Consider a single-server queue with Poisson arrivals and exponential service times having the following variation: Whenever a service is completed a departure occurs only with probability α. With probability $1 - \alpha$ the customer, instead of leaving, joins the end of the queue. Note that a customer may be serviced more than once.

(a) Set up the balance equations and solve for the steady-state probabilities, stating conditions for it to exist.
(b) Find the expected waiting time of a customer from the time he arrives until he enters service for the first time.
(c) What is the probability that a customer enters service exactly n times, $n = 1, 2, \ldots$?
(d) What is the expected amount of time that a customer spends in service (which does not include the time he spends waiting in line)?

Hint: Use part (c).

(e) What is the distribution of the total length of time a customer spends being served?

Hint: Is it memoryless?

12.   Whenever there are n customers in the system, the probability of an arrival in a small time h is $\lambda_n h + o(h)$ whereas the probability of a departure is $\mu_n h + o(h)$. Let the state be the number of customers in the system.

(a) Write down the balance equations.
(b) Show that the balance equations imply that

$$\lambda_n P_n = \mu_{n+1} P_{n+1}, \qquad n \geqslant 0$$

(c) Solve the preceding equation. [In doing so, you will discover the condition needed for a solution. Assume this condition is satisfied for part (d).]
(d) If $\mu_n = \mu$, $n \geqslant 1$ (as would be the case in a single-server model in which all service times are exponential with rate μ), argue (without doing any algebra) that $\Sigma_n \lambda_n P_n = \mu(1 - P_0)$.

*13.   A supermarket has two exponential checkout counters, each operating at rate μ. Arrivals are Poisson at rate λ. The counters operate in the following way:

(i) One queue feeds both counters.
(ii) One counter is operated by a permanent checker and the other by a stock clerk who instantaneously begins checking whenever there are two or more customers in the system. The clerk returns to stocking whenever he completes a service, and there are fewer than two customers in the system.

(a) Let $P_n =$ proportion of time there are n in the system. Set up equations for P_n and solve.
(b) At what rate does the number in the system go from 0 to 1? from 2 to 1?
(c) What proportion of time is the stock clerk checking?

Hint: Be a little careful when there is one in the system.

14.   Customers arrive at a single-service facility at a Poisson rate of 40 per hour. When two or fewer customers are present, a single attendant operates the facility, and the service time for each customer is exponentially distributed with a mean value of two minutes. However, when there are three or more customers at the facility, the attendant is joined by an assistant and, working together, they reduce the mean service time to one minute.

Assuming a system capacity of four customers,

(a) what proportion of time are both servers free?
(b) each man is to receive a salary proportional to the amount of time he is actually at work servicing customers, the rate being the same for both. If together they earn $100 per day, how should this money be split?

15. Consider a sequential-service system consisting of two servers, A and B. Arriving customers will enter this system only if server A is free. If a customer does enter, then he is immediately served by server A. When his service by A is completed, he then goes to B if B is free, or if B is busy, he leaves the system. Upon completion of service at server B, the customer departs. Assuming that the (Poisson) arrival rate is two customers an hour, and that A and B serve at respective (exponential) rates of four and two customers an hour,

(a) what proportion of customers enter the system?
(b) what proportion of entering customers receive service from B?
(c) what is the average number of customers in the system?
(d) what is the average amount of time that an entering customer spends in the system?

16. Customers arrive at a two-server system according to a Poisson process having rate $\lambda = 5$. An arrival finding server 1 free will begin service with that server. An arrival finding server 1 busy and server 2 free will enter service with server 2. An arrival finding both servers busy goes away. Once a customer is served by either server, he departs the system. The service times at server i are exponential with rates μ_i, where $\mu_1 = 4$, $\mu_2 = 2$.

(a) What is the average time an entering customer spends in the system?
(b) What proportion of time is server 2 busy?

17. Customers arrive at a two-server station in accordance with a Poisson process with a rate of two per hour. Arrivals finding server 1 free begin service with that server. Arrivals finding server 1 busy and server 2 free begin service with server 2. Arrivals finding both servers busy are lost. When a customer is served by server 1, she then either enters service with server 2 if 2 is free or departs the system if 2 is busy. A customer completing service at server 2 departs the system. The service times at server 1 and server 2 are exponential random variables with respective rates of four and six per hour.

(a) What fraction of customers do not enter the system?
(b) What is the average amount of time that an entering customer spends in the system?
(c) What fraction of entering customers receives service from server 1?

18. Customers arrive at a two-server system at a Poisson rate λ. An arrival finding the system empty is equally likely to enter service with either server. An arrival finding one customer in the system will enter service with the idle server. An arrival finding two others in the system will wait in line for the first free server. An arrival finding three in the system will not enter. All service times are exponential with rate μ, and once a customer is served (by either server), he departs the system.

(a) Define the states.
(b) Find the long-run probabilities.
(c) Suppose a customer arrives and finds two others in the system. What is the expected times he spends in the system?
(d) What proportion of customers enter the system?
(e) What is the average time an entering customer spends in the system?

19. The economy alternates between good and bad periods. During good times customers arrive at a certain single-server queueing system in accordance with a Poisson process with rate λ_1, and during bad times they arrive in accordance with a Poisson process with rate λ_2. A good time period lasts for an exponentially distributed time with rate α_1, and a bad time period lasts for an exponential time with rate α_2. An arriving customer will only enter the queueing system if the server is free; an arrival finding the server busy goes away. All service times are exponential with rate μ.

(a) Define states so as to be able to analyze this system.
(b) Give a set of linear equations whose solution will yield the long run proportion of time the system is in each state.

In terms of the solutions of the equations in part (b),

(c) what proportion of time is the system empty?
(d) what is the average rate at which customers enter the system?

20. There are two types of customers. Type i customers arrive in accordance with independent Poisson processes with respective rate λ_1 and λ_2. There are two servers. A type 1 arrival will enter service with server 1 if that server is free; if server 1 is busy and server 2 is free, then the type 1 arrival will enter service with server 2. If both servers are busy, then the type 1 arrival will go away. A type 2 customer can only be served by server 2; if server 2 is free when a type 2 customer arrives, then the customer enters service with that server. If server 2 is busy when a type 2 arrives, then that customer goes away. Once a customer is served by either server, he departs the system. Service times at server i are exponential with rate μ_i, $i = 1, 2$.

Suppose we want to find the average number of customers in the system.

(a) Define states.
(b) Give the balance equations. Do not attempt to solve them.

In terms of the long-run probabilities, what is

(c) the average number of customers in the system?
(d) the average time a customer spends in the system?

*21. Suppose in Exercise 20 we want to find out the proportion of time there is a type 1 customer with server 2. In terms of the long-run probabilities given in Exercise 20, what is

(a) the rate at which a type 1 customer enters service with server 2?
(b) the rate at which a type 2 customer enters service with server 2?
(c) the fraction of server 2's customers that are type 1?
(d) the proportion of time that a type 1 customer is with server 2?

22. Customers arrive at a single-server station in accordance with a Poisson process with rate λ. All arrivals that find the server free immediately enter service. All service times are exponentially distributed with rate μ. An arrival that finds the server busy will leave the system and roam around "in orbit" for an exponential time with rate θ at which time it will then return. If the server is busy when an orbiting customer returns, then that customer returns to orbit for another exponential time with rate θ before returning again. An arrival that finds the server busy and N other customers in orbit will depart and not return. That is, N is the maximum number of customers in orbit.

(a) Define states.
(b) Give the balance equations.

In terms of the solution of the balance equations, find

(c) the proportion of all customers that are eventually served.
(d) the average time that a served customer spends waiting in orbit.

23. Consider the $M/M/1$ system in which customers arrive at rate λ and the server serves at rate μ. However, suppose that in any interval of length h in which the server is busy there is a probability $\alpha h + o(h)$ that the server will experience a breakdown, which causes the system to shut down. All customers that are in the system depart, and no additional arrivals are allowed to enter until the breakdown is fixed. The time to fix a breakdown is exponentially distributed with rate β.

(a) Define appropriate states.
(b) Give the balance equations.

In terms of the long-run probabilities,

(c) what is the average amount of time that an entering customer spends in the system?

(d) what proportion of entering customers complete their service?
(e) what proportion of customers arrive during a breakdown?

*24. Reconsider Exercise 23, but this time suppose that a customer that is in the system when a breakdown occurs remains there while the server is being fixed. In addition, suppose that new arrivals during a breakdown period are allowed to enter the system. What is the average time a customer spends in the system?

25. Poisson (λ) arrivals join a queue in front of two parallel servers A and B, having exponential service rates μ_A and μ_B (see Figure 8.4). When the system is empty, arrivals go into server A with probability α and into B with probability $1 - \alpha$. Otherwise, the head of the queue takes the first free server.

(a) Define states and set up the balance equations. Do not solve,
(b) In terms of the probabilities in part (a), what is the average number in the system? Average number of servers idle?
(c) In terms of the probabilities in part (a), what is the probability that an arbitrary arrival will get serviced in A?

26. In a queue with unlimited waiting space, arrivals are Poisson (parameter λ) and service times are exponentially distributed (parameter μ). However, the server waits until K people are present before beginning service on the first customer; thereafter, he services one at a time until all K units, and all subsequent arrivals, are serviced. The server is then "idle" until K new arrivals have occurred.

(a) Define an appropriate state space, draw the transition diagram, and set up the balance equations.
(b) In terms of the limiting probabilities, what is the average time a customer spends in queue?
(c) What conditions on λ and μ are necessary?

27. Consider a single-server exponential system in which ordinary customers arrive at a rate λ and have service rate μ. In addition, there is a special customer who has a service rate μ_1. Whenever this special customer arrives, she goes directly into service (if anyone else is in service, then this person is bumped back into queue). When the special customer is not being serviced, she spends an exponential amount of time (with mean $1/\theta$) out of the system.

Figure 8.4.

(a) What is the average arrival rate of the special customer?
(b) Define an appropriate state space and set up balance equations.
(c) Find the probability that an ordinary customer is bumped n times.

*28. Let D denote the time between successive departures in a stationary $M/M/1$ queue with $\lambda < \mu$. Show, by conditioning on whether or not a departure has left the system empty, that D is exponential with rate λ.

Hint: By conditioning on whether or not the departure has left the system empty we see that

$$D = \begin{cases} \text{Exponential } (\mu), & \text{with probability } \lambda/\mu \\ \text{Exponential } (\lambda) * \text{Exponential } (\mu), & \text{with probability } 1 - \lambda/\mu \end{cases}$$

where Exponential $(\lambda) *$ Exponential (μ) represents the sum of two independent exponential random variables having rates μ and λ. Now use moment-generating functions to show that D has the required distribution.

Note that the above does not prove that the departure process is Poisson. To prove this we need show not only that the interdeparture times are all exponential with rate λ, but also that they are independent.

29. For the tandem queue model verify that

$$P_{n,m} = (\lambda/\mu_1)^n (1 - \lambda/\mu_1)(\lambda/\mu_2)^m (1 - \lambda/\mu_2)$$

satisfies the balance equation (8.15).

30. Verify Equation (8.18) for a system of two servers by showing that it satisfies the balance equations for this model.

31. Consider a network of three stations. Customers arrive at stations 1, 2, 3 in accordance with Poisson processes having respective rates, 5, 10, 15. The service times at the three stations are exponential with respective rates 10, 50, 100. A customer completing service at station 1 is equally likely to (i) go to station 2, (ii) go to station 3, or (iii) leave the system. A customer departing service at station 2 always goes to station 3. A departure from service at station 3 is equally likely to either go to station 2 or leave the system.

(a) What is the average number of customers in the system (consisting of all three stations)?
(b) What is the average time a customer spends in the system?

32. Consider a closed queueing network consisting of two customers moving among two servers, and suppose that after each service completion the customer is equally likely to go to either server — that is, $P_{1,2} = P_{2,1} = \frac{1}{2}$.

Let μ_i denote the exponential service rate at server i, $i = 1, 2$.

(a) Determine the average number of customers at each server.
(b) Determine the service completion rate for each server.

33. Explain how a Markov chain Monte Carlo simulation using the Gibbs sampler can be utilized to estimate.

(a) the distribution of the amount of time spent at server j on a visit.

Hint: Use the arrival theorem.

(b) the proportion of time a customer is with server j (i.e., either in server j's queue or in service with j).

34. State and prove the equivalent of the arrival theorem for open queueing networks.

35. Customers arrive at a single-server station in accordance with a Poison process having rate λ. Each customer has a value. The successive values of customers are independent and come from a uniform distribution on $(0, 1)$. The service time of a customer having value x is a random variable with mean $3 + 4x$ and variance 5.

(a) What is the average time a customer spends in the system?
(b) What is the average time a customer having value x spends in the system?

*36. Compare the $M/G/1$ system for first-come, first-served queue discipline with one of last-come, first-served (for instance, in which units for service are taken from the top of a stack). Would you think that the queue size, waiting time, and busy-period distribution differ? What about their means? What if the queue discipline was always to choose at random among those waiting? Intuitively which discipline would result in the smallest variance in the waiting time distribution?

37. In an $M/G/1$ queue,

(a) what proportion of departures leave behind 0 work?
(b) what is the average work in the system as seen by a departure?

38. For the $M/G/1$ queue, let X_n denote the number in the system left behind by the nth departure.

(a) If

$$X_{n+1} = \begin{cases} X_n - 1 + Y_n, & \text{if } X_n \geqslant 1 \\ Y_n, & \text{if } X_n = 0 \end{cases}$$

what does Y_n represent?

(b) Rewrite the preceding as

$$X_{n+1} = X_n - 1 + Y_n + \delta_n \tag{8.64}$$

where

$$\delta_n = \begin{cases} 1, & \text{if } X_n = 0 \\ 0, & \text{if } X_n \geqslant 1 \end{cases}$$

Take expectations and let $n \to \infty$ in Equation (8.64) to obtain

$$E[\delta_\infty] = 1 - \lambda E[S]$$

(c) Square both sides of Equation (8.64), take expectations, and then let $n \to \infty$ to obtain

$$E[X_\infty] = \frac{\lambda^2 E[S^2]}{2(1 - \lambda E[S])} + \lambda E[S]$$

(d) Argue that $E[X_\infty]$, the average number as seen by a departure, is equal to L.

*39. Consider an $M/G/1$ system in which the first customer in a busy period has the service distribution G_1 and all others have distribution G_2. Let C denote the number of customers in a busy period, and let S denote the service time of a customer chosen at random.
   Argue that

(a) $a_0 = P_0 = 1 - \lambda E[S]$.
(b) $E[S] = a_0 E[S_1] + (1 - a_0) E[S_2]$ where S_i has distribution G_i.
(c) Use (a) and (b) to show that $E[B]$, the expected length of a busy period, is given by

$$E[B] = \frac{E[S_1]}{1 - \lambda E[S_2]}$$

(d) Find $E[C]$.

40. Consider a $M/G/1$ system with $\lambda E[S] < 1$.

(a) Suppose that service is about to begin at a moment when there are n customers in the system.
   (i) Argue that the additional time until there are only $n - 1$ customers in the system has the same distribution as a busy period.
   (ii) What is the expected additional time until the system is empty?
(b) Suppose that the work in the system at some moment is A. We are interested in the expected additional time until the system is empty — call it $E[T]$. Let N denote the number of arrivals during the first A units of time.

(i) Compute $E[T|N]$.
(ii) Compute $E[T]$.

41. Carloads of customers arrive at a single-server station in accordance with a Poisson process with rate 4 per hour. The service times are exponentially distributed with rate 20 per hour. If each carload contains either 1, 2, or 3 customers with respective probabilities $\frac{1}{4}, \frac{1}{2}, \frac{1}{4}$, compute the average customer delay in queue.

42. In the two-class priority queueing model of Section 8.6.2, what is W_Q? Show that W_Q is less than it would be under FIFO if $E[S_1] < E[S_2]$ and greater than under FIFO if $E[S_1] > E[S_2]$.

43. In a two-class priority queueing model suppose that a cost of C_i per unit time is incurred for each type i customer that waits in queue, $i = 1, 2$. Show that type 1 customers should be given priority over type 2 (as opposed to the reverse) if

$$\frac{E[S_1]}{C_1} < \frac{E[S_2]}{C_2}$$

44. Consider the priority queueing model of Section 8.6.2 but now suppose that if a type 2 customer is being served when a type 1 arrives then the type 2 customer is bumped out of service. This is called the preemptive case. Suppose that when a bumped type 2 customer goes back in service his service begins at the point where it left off when he was bumped.

(a) Argue that the work in the system at any time is the same as in the nonpreemptive case.
(b) Derive W_Q^1.

Hint: How do type 2 customers affect type 1s?

(c) Why is it not true that

$$V_Q^2 = \lambda_2 E[S_2]W_Q^2$$

(d) Argue that the work seen by a type 2 arrival is the same as in the nonpreemptive case, and so

$$W_Q^2 = W_Q^2(\text{nonpreemptive}) + E[\text{extra time}]$$

where the extra time is due to the fact that he may be bumped.
(e) Let N denote the number of times a type 2 customer is bumped. Why is

$$E[\text{extra time}|N] = \frac{NE[S_1]}{1 - \lambda_1 E[S_1]}$$

Hint: When a type 2 is bumped, relate the time until he gets back in service to a "busy period."

(f) Let S_2 denote the service time of a type 2. What is $E[N \mid S_2]$?
(g) Combine the preceding to obtain

$$W_Q^2 = W_Q^2(\text{nonpreemptive}) + \frac{\lambda_1 E[S_1] E[S_2]}{1 - \lambda_1 E[S_1]}$$

*45. Calculate explicitly (not in terms of limiting probabilities) the average time a customer spends in the system in Exercise 24.

46. In the $G/M/1$ model if G is exponential with rate λ show that $\beta = \lambda/\mu$.

47. Verify Erlang's loss formula, Equation (8.60), when $k = 1$.

48. Verify the formula given for the P_i of the $M/M/k$.

49. In the Erlang loss system suppose the Poisson arrival rate is $\lambda = 2$, and suppose there are three servers each of whom has a service distribution that is uniformly distributed over $(0, 2)$. What proportion of potential customers is lost?

50. In the $M/M/k$ system,

(a) what is the probability that a customer will have to wait in queue?
(b) determine L and W.

51. Verify the formula for the distribution of W_Q^* given for the $G/M/k$ model.

*52. Consider a system where the interarrival times have an arbitrary distribution F, and there is a single server whose service distribution is G. Let D_n denote the amount of time the nth customer spends waiting in queue. Interpret S_n, T_n so that

$$D_{n+1} = \begin{cases} D_n + S_n - T_n, & \text{if } D_n + S_n - T_n \geqslant 0 \\ 0, & \text{if } D_n + S_n - T_n < 0 \end{cases}$$

53. Consider a model in which the interarrival times have an arbitrary distribution F, and there are k servers each having service distribution G. What condition on F and G do you think would be necessary for there to exist limiting probabilities?

References

1. J. Cohen, "The Single Server Queue," North-Holland, Amsterdam, 1969.
2. R. B. Cooper, "Introduction to Queueing Theory," Second Edition, Macmillan, New York, 1984.

3. D. R. Cox and W. L. Smith, "Queues," Wiley, New York, 1961.
4. F. Kelly, "Reversibility and Stochastic Networks," Wiley, New York, 1979.
5. L. Kleinrock, "Queueing Systems," Vol. I, Wiley, New York, 1975.
6. S. Nozaki and S. Ross, "Approximations in Finite Capacity Multiserver Queues with Poisson Arrivals," *J. Appl. Prob.* **13**, 826–834 (1978).
7. L. Takacs, "Introduction to the Theory of Queues," Oxford University Press, London and New York, 1962.
8. H. Tijms, "Stochastic Models: An Algorithmic Approach," Wiley, New York, 1994.
9. P. Whittle, "Systems in Stochastic Equilibrium," Wiley, New York, 1986.
10. Wolff, "Stochastic Modeling and the Theory of Queues," Prentice Hall, New Jersey, 1989.

Reliability Theory 9

◆◆◆

9.1. Introduction

Reliability theory is concerned with determining the probability that a system, possibly consisting of many components, will function. We shall suppose that whether or not the system functions is determined solely from a knowledge of which components are functioning. For instance, a *series* system will function if and only if all of its components are functioning, while a *parallel* system will function if and only if at least one of its components is functioning. In Section 9.2, we explore the possible ways in which the functioning of the system may depend upon the functioning of its components. In Section 9.3, we suppose that each component will function with some known probability (independently of each other) and show how to obtain the probability that the system will function. As this probability often is difficult to explicitly compute, we also present useful upper and lower bounds in Section 9.4. In Section 9.5 we look at a system dynamically over time by supposing that each component initially functions and does so for a random length of time at which it fails. We then discuss the relationship between the distribution of the amount of time that a system functions and the distributions of the component lifetimes. In particular, it turns out that if the amount of time that a component functions has an *increasing failure rate on the average* (IFRA) distribution, then so does the distribution of system lifetime. In Section 9.6 we consider the problem of obtaining the mean lifetime of a system. In the final section we analyze the system when failed components are subjected to repair.

9.2. Structure Functions

Consider a system consisting of n components, and suppose that each component is either functioning or has failed. To indicate whether or not the ith component is functioning, we define the indicator variable x_i by

$$x_i = \begin{cases} 1, & \text{if the } i\text{th component is functioning} \\ 0, & \text{if the } i\text{th component has failed} \end{cases}$$

The vector $\mathbf{x} = (x_1, \ldots, x_n)$ is called the state vector. It indicates which of the components are functioning and which have failed.

We further suppose that whether or not the system as a whole is functioning is completely determined by the state vector $\mathbf{x}$. Specifically, it is supposed that there exists a function $\phi(\mathbf{x})$ such that

$$\phi(\mathbf{x}) = \begin{cases} 1, & \text{if the system is functioning when the state vector is } \mathbf{x} \\ 0, & \text{if the system has failed when the state vector is } \mathbf{x} \end{cases}$$

The function $\phi(\mathbf{x})$ is called the *structure function* of the system.

Example 9.1   (The Series Structure): A series system functions if and only if all of its components are functioning. Hence, its structure function is given by

$$\phi(\mathbf{x}) = \min(x_1, \ldots, x_n) = \prod_{i=1}^{n} x_i$$

We shall find it useful to represent the structure of a system in terms of a diagram. The relevant diagram for the series structure is shown in Figure 9.1. The idea is that if a signal is initiated at the left end of the diagram then in order for it to successfully reach the right end, it must pass through all of the components; hence, they must all be functioning. ✦

Figure 9.1.

Example 9.2   (The Parallel Structure): A parallel system functions if and only if at least one of its components is functioning. Hence its structure function is given by

$$\phi(\mathbf{x}) = \max(x_1, \ldots, x_n)$$

A parallel structure may be pictorially illustrated by Figure 9.2. This follows since a signal at the left end can successfully reach the right end as long as at least one component is functioning. ✦

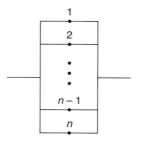

Figure 9.2.

Example 9.3   (The k-Out-of-n Structure): The series and parallel systems are both special cases of a k-out-of-n system. Such a system functions if and only if at least k of the n components are functioning. As $\sum_{i=1}^{n} x_i$ equals the number of functioning components, the structure function of a k-out-of-n system is given by

$$\phi(\mathbf{x}) = \begin{cases} 1, & \text{if } \sum_{i=1}^{n} x_i \geq k \\ 0, & \text{if } \sum_{i=1}^{n} x_i < k \end{cases}$$

Series and parallel systems are respectively n-out-of-n and 1-out-of-n systems. The two-out-of-three system may be diagramed as shown in Figure 9.3. ◆

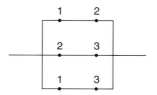

Figure 9.3.

Example 9.4   (A Four-Component Structure): Consider a system consisting of four components, and suppose that the system functions if and only if components 1 and 2 both function and at least one of components 3 and 4 function. Its structure function is given by

$$\phi(\mathbf{x}) = x_1 x_2 \, \text{matrix}(x_3, x_4)$$

Pictorially, the system is as shown in Figure 9.4. A useful identity, easily checked, is that for binary variables,* x_i, $i = 1, \ldots, n$,

$$\max(x_1, \ldots, x_n) = 1 - \prod_{i=1}^{n} (1 - x_i)$$

* A binary variable is one that assumes either the value 0 or 1.

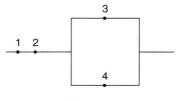

Figure 9.4.

When $n = 2$, this yields

$$\max(x_1, x_2) = 1 - (1 - x_1)(1 - x_2) = x_1 + x_2 - x_1 x_2$$

Hence, the structure function in the example may be written as

$$\phi(\mathbf{x}) = x_1 x_2 (x_3 + x_4 - x_3 x_4) \quad \blacklozenge$$

It is natural to assume that replacing a failed component by a functioning one will never lead to a deterioration of the system. In other words, it is natural to assume that the structure function $\phi(\mathbf{x})$ is an increasing function of $\mathbf{x}$, that is, if $x_i \leqslant y_i$, $i = 1, \ldots, n$, then $\phi(\mathbf{x}) \leqslant \phi(\mathbf{y})$. Such an assumption shall be made in this chapter and the system will be called *monotone*.

9.2.1. Minimal Path and Minimal Cut Sets

In this section we show how any system can be represented both as a series arrangement of parallel structures and as a parallel arrangement of series structures. As a preliminary, we need the following concepts.

A state vector $\mathbf{x}$ is called a *path vector* if $\phi(\mathbf{x}) = 1$. If, in addition, $\phi(\mathbf{y}) = 0$ for all $\mathbf{y} < \mathbf{x}$, then $\mathbf{x}$ is said to be a *minimal path vector*.* If $\mathbf{x}$ is a minimal path vector, then the set $A = \{i: x_i = 1\}$ is called a *minimal path set*. In other words, a minimal path set is a minimal set of components whose functioning ensures the functioning of the system.

Example 9.5 Consider a five-component system whose structure is illustrated by Figure 9.5. Its structure function equals

$$\phi(\mathbf{x}) = \max(x_1, x_2) \max(x_3 x_4, x_5)$$

$$= (x_1 + x_2 - x_1 x_2)(x_3 x_4 + x_5 - x_3 x_4 x_5)$$

There are four minimal path sets, namely, $\{1, 3, 4\}, \{2, 3, 4\}, \{1, 5\}, \{2, 5\}$. $\blacklozenge$

---

* We say that $\mathbf{y} < \mathbf{x}$ if $y_i \leqslant x_i$, $i = 1, \ldots, n$, with $y_i < x_i$ for some i.

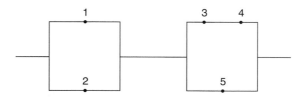

Figure 9.5.

Example 9.6 In a k-out-of-n system, there are $\binom{n}{k}$ minimal path sets, namely, all of the sets consisting of exactly k components. ◆

Let $A_1, \ldots, A_s$ denote the minimal path sets of a given system. We define $\alpha_j(\mathbf{x})$, the indicator function of the jth minimal path set, by

$$\alpha_j(\mathbf{x}) = \begin{cases} 1, & \text{if all the components of } A_j \text{ are functioning} \\ 0, & \text{otherwise} \end{cases}$$

$$= \prod_{i \in A_j} x_i$$

By definition, it follows that the system will function if all the components of at least one minimal path set are functioning; that is, if $\alpha_j(\mathbf{x}) = 1$ for some j. On the other hand, if the system functions, then the set of functioning components must include a minimal path set. Therefore, *a system will function if and only if all the components of at least one minimal path set are functioning*. Hence,

$$\phi(\mathbf{x}) = \begin{cases} 1, & \text{if } \alpha_j(\mathbf{x}) = 1 \text{ for some } j \\ 0, & \text{if } \alpha_j(\mathbf{x}) = 0 \text{ for all } j \end{cases}$$

or equivalently

$$\phi(\mathbf{x}) = \max_j \alpha_j(\mathbf{x})$$

$$= \max_j \prod_{i \in A_j} x_i$$

(9.1)

Since $\alpha_j(\mathbf{x})$ is a series structure function of the components of the jth minimal path set, Equation (9.1) expresses an arbitrary system as a parallel arrangement of series systems.

Example 9.7 Consider the system of Example 9.5. Because its minimal path sets are $A_1 = \{1, 3, 4\}$, $A_2 = \{2, 3, 4\}$, $A_3 = \{1, 5\}$, and $A_4 = \{2, 5\}$, we

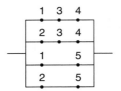

Figure 9.6.

have by Equation (9.1) that

$$\phi(\mathbf{x}) = \max\{x_1x_3x_4, x_2x_3x_4, x_1x_5, x_2x_5\}$$
$$= 1 - (1 - x_1x_3x_4)(1 - x_2x_3x_4)(1 - x_1x_5)(1 - x_2x_5)$$

The reader should verify that this equals the value of $\phi(\mathbf{x})$ given in Example 9.5. (Make use of the fact that, since x_i equals 0 or 1, $x_i^2 = x_i$.) This representation may be pictured as shown in Figure 9.6. ◆

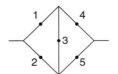

Figure 9.7.

Example 9.8   The system whose structure is as pictured in Figure 9.7 is called the *bridge system*. Its minimal path sets are $\{1, 4\}$, $\{1, 3, 5\}$, $\{2, 5\}$, and $\{2, 3, 4\}$. Hence, by Equation (9.1), its structure function may be expressed as

$$\phi(\mathbf{x}) = \max\{x_1x_4, x_1x_3x_5, x_2x_5, x_2x_3x_4\}$$

$$= 1 - (1 - x_1x_4)(1 - x_1x_3x_5)(1 - x_2x_5)(1 - x_2x_3x_4)$$

This representation $\phi(\mathbf{x})$ is diagramed as shown in Figure 9.8. ◆

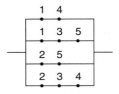

Figure 9.8.

A state vector $\mathbf{x}$ is called a *cut vector* if $\phi(\mathbf{x}) = 0$. If, in addition, $\phi(\mathbf{y}) = 1$ for all $\mathbf{y} > \mathbf{x}$, then $\mathbf{x}$ is said to be a *minimal cut vector*. If $\mathbf{x}$ is a minimal cut vector, then the set $C = \{i : x_i = 0\}$ is called a *minimal cut set*. In other words, a minimal cut set is a minimal set of components whose failure ensures the failure of the system.

Let $C_1, \ldots, C_k$ denote the minimal cut sets of a given system. We define $\beta_j(\mathbf{x})$, the indicator function of the jth minimal cut set, by

$$\beta_j(\mathbf{x}) = \begin{cases} 1, & \text{if at least one component of the } j\text{th minimal} \\ & \text{cut set is functioning} \\ 0, & \text{if all of the components of the } j\text{th minimal} \\ & \text{cut set are not functioning} \end{cases}$$

$$= \max_{i \in C_j} x_i$$

Since a system is not functioning if and only if all the components of at least one minimal cut set are not functioning, it follows that

$$\phi(\mathbf{x}) = \prod_{j=1}^{k} \beta_j(\mathbf{x}) = \prod_{j=1}^{k} \max_{i \in C_j} x_i \tag{9.2}$$

Since $\beta_j(\mathbf{x})$ is a parallel structure function of the components of the jth minimal cut set, Equation (9.2) represents an arbitrary system as a series arrangement of parallel systems.

Example 9.9 The minimal cut sets of the bridge structure shown in Figure 9.9 are $\{1, 2\}$, $\{1, 3, 5\}$, $\{2, 3, 4\}$, and $\{4, 5\}$. Hence, from Equation (9.2), we may express $\phi(\mathbf{x})$ by

$$\phi(\mathbf{x}) = \max(x_1, x_2) \max(x_1, x_3, x_5) \max(x_2, x_3, x_4) \max(x_4, x_5)$$

$$= [1 - (1 - x_1)(1 - x_2)][1 - (1 - x_1)(1 - x_3)(1 - x_5)]$$

$$\times [1 - (1 - x_2)(1 - x_3)(1 - x_4)][1 - (1 - x_4)(1 - x_5)]$$

This representation of $\phi(\mathbf{x})$ is pictorially expressed as Figure 9.10. ✦

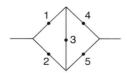

Figure 9.9.

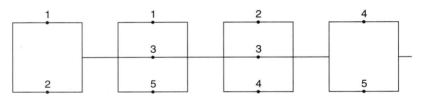

Figure 9.10.

9.3. Reliability of Systems of Independent Components

In this section, we suppose that X_i, the state of the ith component, is a random variable such that

$$P\{X_i = 1\} = p_i = 1 - P\{X_i = 0\}$$

The value p_i, which equals the probability that the ith component is functioning, is called the *reliability* of the ith component. If we define r by

$$r = P\{\phi(\mathbf{X}) = 1\}, \qquad \text{where } \mathbf{X} = (X_1, \ldots, X_n)$$

then r is called the *reliability* of the system. When the components, that is, the random variables X_i, $i = 1, \ldots, n$, are independent, we may express r as a function of the component reliabilities. That is,

$$r = r(\mathbf{p}), \qquad \text{where } \mathbf{p} = (p_1, \ldots, p_n)$$

The function $r(\mathbf{p})$ is called the *reliability function*. We shall assume throughout the remainder of this chapter that the components are independent.

Example 9.10 (The Series System): The reliability function of the series system of n independent components is given by

$$
\begin{aligned}
r(\mathbf{p}) &= P\{\phi(\mathbf{X}) = 1\} \\
&= P\{X_i = 1 \text{ for all } i = 1, \ldots, n\} \\
&= \prod_{i=1}^{n} p_i \quad \blacklozenge
\end{aligned}
$$

Example 9.11 (The Parallel System): The reliability function of the parallel system of n independent components is given by

$$
\begin{aligned}
r(\mathbf{p}) &= P\{\phi(\mathbf{X}) = 1\} \\
&= P\{X_i = 1 \text{ for some } i = 1, \ldots, n\}
\end{aligned}
$$

$$= 1 - P\{X_i = 0 \text{ for all } i = 1, \ldots, n\}$$

$$= 1 - \prod_{i=1}^{n} (1 - p_i) \quad \blacklozenge$$

Example 9.12 (The k-out-of-n System with Equal Probabilities): Consider a k-out-of-n system. If $p_i = p$ for all $i = 1, \ldots, n$, then the reliability function is given by

$$r(p, \ldots, p) = P\{\phi(\mathbf{X}) = 1\}$$

$$= P\left\{ \sum_{i=1}^{n} X_i \geq k \right\}$$

$$= \sum_{i=k}^{n} \binom{n}{i} p^i (1 - p)^{n-i} \quad \blacklozenge$$

Example 9.13 (The Two-out-of-Three System): The reliability function of a two-out-of-three system is given by

$$r(\mathbf{p}) = P\{\phi(\mathbf{X}) = 1\}$$

$$= P\{\mathbf{X} = (1, 1, 1)\} + P\{\mathbf{X} = (1, 1, 0)\} + P\{\mathbf{X} = (1, 0, 1)\} + P\{\mathbf{X} = (0, 1, 1)\}$$

$$= p_1 p_2 p_3 + p_1 p_2 (1 - p_3) + p_1 (1 - p_2) p_3 + (1 - p_1) p_2 p_3$$

$$= p_1 p_2 + p_1 p_3 + p_2 p_3 - 2 p_1 p_2 p_3 \quad \blacklozenge$$

Example 9.14 (The Three-out-of-Four System): The reliability function of a three-out-of-four system is given by

$$r(\mathbf{p}) = P\{\mathbf{X} = (1, 1, 1, 1)\} + P\{\mathbf{X} = (1, 1, 1, 0)\} + P\{\mathbf{X} = (1, 1, 0, 1)\}$$

$$\quad + P\{\mathbf{X} = (1, 0, 1, 1)\} + P\{\mathbf{X} = (0, 1, 1, 1)\}$$

$$= p_1 p_2 p_3 p_4 + p_1 p_2 p_3 (1 - p_4) + p_1 p_2 (1 - p_3) p_4$$

$$\quad + p_1 (1 - p_2) p_3 p_4 + (1 - p_1) p_2 p_3 p_4$$

$$= p_1 p_2 p_3 + p_1 p_2 p_4 + p_1 p_3 p_4 + p_2 p_3 p_4 - 3 p_1 p_2 p_3 p_4 \quad \blacklozenge$$

Example 9.15 (A Five-Component System): Consider a five-component system that functions if and only if component 1, component 2, and at least one of the remaining components function. Its reliability function is given by

$$r(\mathbf{p}) = P\{X_1 = 1, X_2 = 1, \max(X_3, X_4, X_5) = 1\}$$

$$= P\{X_1 = 1\} P\{X_2 = 1\} P\{\max(X_3, X_4, X_5) = 1\}$$

$$= p_1 p_2 [1 - (1 - p_3)(1 - p_4)(1 - p_5)] \quad \blacklozenge$$

Since $\phi(\mathbf{X})$ is a $0-1$ (that is, a Bernoulli) random variable, we may also compute $r(\mathbf{p})$ by taking its expectation. That is,

$$r(\mathbf{p}) = P\{\phi(\mathbf{X}) = 1\}$$
$$= E[\phi(\mathbf{X})]$$

Example 9.16 (A Four-Component System): A four-component system that functions when both components 1, 4, and at least one of the other components function has its stucture function given by

$$\phi(\mathbf{x}) = x_1 x_4 \max(x_2, x_3)$$

Hence,

$$r(\mathbf{p}) = E[\phi(\mathbf{X})]$$
$$= E[X_1 X_4(1 - (1 - X_2)(1 - X_3))]$$
$$= p_1 p_4[1 - (1 - p_2)(1 - p_3)] \quad \blacklozenge$$

An important and intuitive property of the reliability function $r(\mathbf{p})$ is given by the following proposition.

Proposition 9.1 If $r(\mathbf{p})$ is the reliability function of a system of independent components, then $r(\mathbf{p})$ is an increasing function of $\mathbf{p}$.

Proof By conditioning on X_i and using the independence of the components, we obtain

$$r(\mathbf{p}) = E[\phi(\mathbf{X})]$$
$$= p_i E[\phi(\mathbf{X}) \,|\, X_i = 1] + (1 - p_i) E[\phi(\mathbf{X}) \,|\, X_i = 0]$$
$$= p_i E[\phi(1_i, \mathbf{X})] + (1 - p_i) E[\phi(0_i, \mathbf{X})]$$

where

$$(1_i, \mathbf{X}) = (X_1, \ldots, X_{i-1}, 1, X_{i+1}, \ldots, X_n),$$
$$(0_i, \mathbf{X}) = (X_1, \ldots, X_{i-1}, 0, X_{i+1}, \ldots, X_n)$$

Thus,

$$r(\mathbf{p}) = p_i E[\phi(1_i, \mathbf{X}) - \phi(0_i, \mathbf{X})] + E[\phi(0_i, \mathbf{X})]$$

However, since ϕ is an increasing function, it follows that

$$E[\phi(1_i, \mathbf{X}) - \phi(0_i, \mathbf{X})] \geqslant 0$$

and so the preceding is increasing in p_i for all i. Hence the result is proven. $\blacklozenge$

Let us now consider the following situation: A system consisting of n different components is to be built from a stockpile containing exactly two of each type of component. How should we use the stockpile so as to maximize our probability of attaining a functioning system? In particular, should we build two separate systems, in which case the probability of attaining a functioning one would be

$$P\{\text{at least one of the two systems function}\}$$

$$= 1 - P\{\text{neither of the systems function}\}$$

$$= 1 - [(1 - r(\mathbf{p}))(1 - r(\mathbf{p}'))]$$

where $p_i(p_i')$ is the probability that the first (second) number i component functions; or should we build a single system whose ith component functions if at least one of the number i components functions? In this latter case, the probability that the system will function equals

$$r[\mathbf{1} - (\mathbf{1} - \mathbf{p})(\mathbf{1} - \mathbf{p}')]$$

since $1 - (1 - p_i)(1 - p_i')$ equals the probability that the ith component in the single system will function.* We now show that replication at the component level is more effective than replication at the system level.

Theorem 9.1 For any reliability function r and vectors $\mathbf{p}, \mathbf{p}'$,

$$r[\mathbf{1} - (\mathbf{1} - \mathbf{p})(\mathbf{1} - \mathbf{p}')] \geqslant 1 - [1 - r(\mathbf{p})][1 - r(\mathbf{p}')]$$

Proof Let $X_1, \ldots, X_n, X_1', \ldots, X_n'$ be mutually independent $0-1$ random variables with

$$p_i = P\{X_i = 1\}, \qquad p_i' = P\{X_i' = 1\}$$

Since $P\{\max(X_i, X_i') = 1\} = 1 - (1 - p_i)(1 - p_i')$, it follows that

$$r[\mathbf{1} - (\mathbf{1} - \mathbf{p})(\mathbf{1} - \mathbf{p}')] = E(\phi[\max(\mathbf{X}, \mathbf{X}')])$$

However, by the monotonicity of ϕ, we have that $\phi[\max(\mathbf{X}, \mathbf{X}')]$ is greater than or equal to both $\phi(\mathbf{X})$ and $\phi(\mathbf{X}')$ and hence is at least as large as $\max[\phi(\mathbf{X}), \phi(\mathbf{X}')]$. Hence, from the preceding we have that

$$r[\mathbf{1} - (\mathbf{1} - \mathbf{p})(\mathbf{1} - \mathbf{p}')] \geqslant E[\max(\phi(\mathbf{X}), \phi(\mathbf{X}'))]$$

$$= P\{\max[\phi(\mathbf{X}), \phi(\mathbf{X}')] = 1\}$$

$$= 1 - P\{\phi(\mathbf{X}) = 0, \phi(\mathbf{X}') = 0\}$$

$$= 1 - [1 - r(\mathbf{p})][1 - r(\mathbf{p}')]$$

* Notation: If $\mathbf{x} = (x_1, \ldots, x_n)$, $\mathbf{y} = (y_1, \ldots, y_n)$, then $\mathbf{xy} = (x_1 y_1, \ldots, x_n y_n)$. Also, $\max(\mathbf{x}, \mathbf{y}) = (\max(x_1, y_1), \ldots, \max(x_n, y_n))$ and $\min(\mathbf{x}, \mathbf{y}) = (\min(x_1, y_1), \ldots, \min(x_n, y_n))$.

where the first equality follows from the fact that $\max[\phi(\mathbf{X}), \phi(\mathbf{X}')]$ is a $0-1$ random variable and hence its expectation equals the probability that it equals 1. ✦

As an illustration of the preceding theorem, suppose that we want to build a series system of two different types of components from a stockpile consisting of two of each of the kinds of components. Suppose that the reliability of each component is $\frac{1}{2}$. If we use the stockpile to build two separate systems, then the probability of attaining a working system is

$$1 - (\tfrac{3}{4})^2 = \tfrac{7}{16}$$

while if we build a single system, replicating components, then the probability of attaining a working system is

$$(\tfrac{3}{4})^2 = \tfrac{9}{16}$$

Hence, replicating components leads to a higher reliability than replicating systems (as, of course, it must by Theorem 9.1).

9.4. Bounds on the Reliability Function

Consider the bridge system of Example 9.8, which is represented by Figure 9.11. Using the minimal path representation, we have that

$$\phi(\mathbf{x}) = 1 - (1 - x_1 x_4)(1 - x_1 x_3 x_5)(1 - x_2 x_5)(1 - x_2 x_3 x_4)$$

Hence,

$$r(\mathbf{p}) = 1 - E[(1 - X_1 X_4)(1 - X_1 X_3 X_5)(1 - X_2 X_5)(1 - X_2 X_3 X_4)]$$

However, since the minimal path sets overlap (that is, they have components in common), the random variables $(1 - X_1 X_4)$, $(1 - X_1 X_3 X_5)$, $(1 - X_2 X_5)$, and $(1 - X_2 X_3 X_4)$ are not independent, and thus the expected value of their product is not equal to the product of their expected values. Therefore, in order to compute $r(\mathbf{p})$, we must first multiply the four random variables and

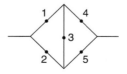

Figure 9.11.

take the expected value. Doing so, we obtain

$$r(\mathbf{p}) = E[X_1X_4 + X_2X_5 + X_1X_3X_5 + X_2X_3X_4 - X_1X_2X_3X_4$$
$$\qquad - X_1X_2X_3X_5 - X_1X_2X_4X_5 - X_1X_3X_4X_5 - X_2X_3X_4X_5$$
$$\qquad + 2X_1X_2X_3X_4X_5]$$
$$= p_1p_4 + p_2p_5 + p_1p_3p_5 + p_2p_3p_4 - p_1p_2p_3p_4 - p_1p_2p_3p_5$$
$$\qquad - p_1p_2p_4p_5 - p_1p_3p_4p_5 - p_2p_3p_4p_5 + 2p_1p_2p_3p_4p_5$$

As can be seen by the above example, it is often suite tedious to evaluate $r(\mathbf{p})$, and thus it would be useful if we had a simple way of obtaining bounds. We now consider two methods for this.

9.4.1. Method of Inclusion and Exclusion

The following is a well-known formula for the probability of the union of the events $E_1, E_2, \ldots, E_n$:

$$P\left(\bigcup_{i=1}^{n} E_i\right) = \sum_{i=1}^{n} P(E_i) - \sum\sum_{i<j} P(E_iE_j) + \sum\sum\sum_{i<j<k} P(E_iE_jE_k)$$
$$\qquad - \cdots + (-1)^{n+1} P(E_1E_2 \cdots E_n) \tag{9.3}$$

A result, not as well known, is the following set of inequalities:

$$P\left(\bigcup_{1}^{n} E_i\right) \leq \sum_{i=1}^{n} P(E_i),$$

$$P\left(\bigcup_{1}^{n} E_i\right) \geq \sum_{i} P(E_i) - \sum_{i<j} P(E_iE_j),$$

$$P\left(\bigcup_{1}^{n} E_i\right) \leq \sum_{i} P(E_i) - \sum\sum_{i<j} P(E_iE_j) + \sum\sum\sum_{i<j<k} P(E_iE_jE_k),$$

$$\geq \cdots$$

$$\leq \cdots \tag{9.4}$$

where the inequality always changes direction as we add an additional term of the expansion of $P(\bigcup_{i=1}^{n} E_i)$.

The equality (9.3) is usually proven by induction on the number of events. However, let us now present another approach that will not only prove Equation (9.3) but also establish the inequalities (9.4).

To begin, define the indicator variables $I_j, j = 1, \ldots, n$, by

$$I_j = \begin{cases} 1, & \text{if } E_j \text{ occurs} \\ 0, & \text{otherwise} \end{cases}$$

Letting

$$N = \sum_{j=1}^{n} I_j$$

then N denotes the number of the E_j, $1 \leqslant j \leqslant n$, that occur. Also, let

$$I = \begin{cases} 1, & \text{if } N > 0 \\ 0, & \text{if } N = 0 \end{cases}$$

Then, as

$$1 - I = (1 - 1)^N$$

we obtain, upon application of the binomial theorem, that

$$1 - I = \sum_{i=0}^{N} \binom{N}{i}(-1)^i$$

or

$$I = N - \binom{N}{2} + \binom{N}{3} - \cdots \pm \binom{N}{N} \tag{9.5}$$

We now make use of the following combinatorial identity (which is easily established by induction on i):

$$\binom{n}{i} - \binom{n}{i+1} + \cdots \pm \binom{n}{n} = \binom{n-1}{i-1} \geqslant 0, \qquad i \leqslant n$$

The preceding thus implies that

$$\binom{N}{i} - \binom{N}{i+1} + \cdots \pm \binom{N}{N} \geqslant 0 \tag{9.6}$$

From Equations (9.5) and (9.6) we obtain

$$\begin{aligned} & I \leqslant N, & & \text{by letting } i = 2 \text{ in (9.6)} \\ & I \geqslant N - \binom{N}{2}, & & \text{by letting } i = 3 \text{ in (9.6)} \\ & I \leqslant N - \binom{N}{2} + \binom{N}{3}, \\ & \quad\vdots \end{aligned} \tag{9.7}$$

and so on. Now, since $N \leqslant n$ and $\binom{m}{i} = 0$ whenever $i > m$, we can rewrite Equation (9.5) as

$$I = \sum_{i=1}^{n} \binom{N}{i}(-1)^{i+1} \tag{9.8}$$

The equality (9.3) and inequalities (9.4) now follow upon taking expectations of (9.7) and (9.8). This is the case since

$$E[I] = P\{N > 0\} = P\{\text{at least one of the } E_j \text{ occurs}\} = P\left(\bigcup_{1}^{n} E_j\right),$$

$$E[N] = E\left[\sum_{j=1}^{n} I_j\right] = \sum_{j=1}^{n} P(E_j)$$

Also,

$$E\left[\binom{N}{2}\right] = E[\text{number of pairs of the } E_j \text{ that occur}]$$

$$= E\left[\sum\sum_{i<j} I_i I_j\right]$$

$$= \sum\sum_{i<j} P(E_i E_j)$$

and, in general

$$E\left[\binom{N}{i}\right] = E[\text{number of sets of size } i \text{ that occur}]$$

$$= E\left[\sum\sum_{j_1 < j_2 < \cdots < j_i} I_{j_1} I_{j_2} \cdots I_{j_i}\right]$$

$$= \sum\sum_{j_1 < j_2 < \cdots < j_i} P(E_{j_1} E_{j_2} \cdots E_{j_i})$$

The bounds expressed in Equation (9.4) are commonly called the *inclusion–exclusion bounds*. To apply them in order to obtain bounds on the reliability function, let $A_1 A_2, \ldots, A_s$ denote the minimal path sets of a given structure ϕ, and define the events $E_1, E_2, \ldots, E_s$ by

$$E_i = \{\text{all components in } A_i \text{ function}\}$$

Now, since the system functions if and only if at least one of the events E_i occur, we have

$$r(\mathbf{p}) = P\left(\bigcup_{1}^{s} E_i\right)$$

Applying (9.4) yields the desired bounds on $r(\mathbf{p})$. The terms in the summation are computed thusly:

$$P(E_i) = \prod_{l \in A_i} p_l,$$

$$P(E_i E_j) = \prod_{l \in A_i \cup A_j} p_l,$$

$$P(E_i E_j E_k) = \prod_{l \in A_i \cup A_j \cup A_k} p_l$$

and so forth for intersections of more than three of the events. (The preceding follows since, for instance, in order for the event $E_i E_j$ to occur, all of the components in A_i and all of them in A_j must function; or, in other words, all components in $A_i \cup A_j$ must function.)

When the p_is are small the probabilities of the intersection of many of the events E_i should be quite small and the convergence should be relatively rapid.

Example 9.17 Consider the bridge structure with identical component probabilities. That is, take p_i to equal p for all i. Letting $A_1 = \{1, 4\}$, $A_2 = \{1, 3, 5\}$, $A_3 = \{2, 5\}$, and $A_4 = \{2, 3, 4\}$ denote the minimal path sets, we have that

$$P(E_1) = P(E_3) = p^2,$$

$$P(E_2) = P(E_4) = p^3$$

Also, because exactly five of the six $= \binom{4}{2}$ unions of A_i and A_j contain four components (the exception being $A_2 \cup A_4$ which contains all five components), we have

$$P(E_1 E_2) = P(E_1 E_3) = P(E_1 E_4) = P(E_2 E_3) = P(E_3 E_4) = p^4,$$

$$P(E_2 E_4) = p^5$$

Hence, the first two inclusion–exclusion bounds yield

$$2(p^2 + p^3) - 5p^4 - p^5 \leqslant r(p) \leqslant 2(p^2 + p^3)$$

where $r(p) = r(p, p, p, p, p)$. For instance, when $p = 0.2$, we have

$$0.08768 \leqslant r(0.2) \leqslant 0.09600$$

and, when $p = 0.01$,

$$0.02149 \leqslant r(0.1) \leqslant 0.02200 \quad \blacklozenge$$

Just as we can define events in terms of the minimal path sets whose union is the event that the system functions, so can we define events in terms of the minimal cut sets whose union is the event that the system fails. Let $C_1, C_2, \ldots, C_r$ denote the minimal cut sets and define the events $F_1, \ldots, F_r$ by

$$F_i = \{\text{all components in } C_i \text{ are failed}\}$$

Now, because the system is failed if and only if all of the components of at least one minimal cut set are failed, we have that

$$1 - r(\mathbf{p}) = P\left(\bigcup_{1}^{r} F_i\right),$$

$$1 - r(\mathbf{p}) \leqslant \sum_{i} P(F_i),$$

$$1 - r(\mathbf{p}) \geqslant \sum_{i} P(F_i) - \sum_{i<j}\sum P(F_i F_j),$$

$$1 - r(\mathbf{p}) \leqslant \sum_{i} P(F_i) - \sum_{i<j}\sum P(F_i F_j) + \sum_{i<j<k}\sum\sum P(F_i F_j F_k),$$

and so on. As

$$P(F_i) = \prod_{l \in C_i} (1 - p_l),$$

$$P(F_i F_j) = \prod_{l \in C_i \cup C_j} (1 - p_l),$$

$$P(F_i F_j F_k) = \prod_{l \in C_i \cup C_j \cup C_k} (1 - p_l)$$

the convergence should be relatively rapid when the p_is are large.

Example 9.18   (A Random Graph): Let us recall from Section 3.6.2 that graph consists of a set N of nodes and a set A of pairs of nodes, called arcs. For any two nodes i and j we say that the sequence of arcs $(i, i_1)(i_1, i_2), \ldots, (i_k, j)$ constitutes an $i - j$ path. If there is an $i - j$ path between all the $\binom{n}{2}$ pairs of nodes i and j, $i \neq j$, then the graph is said to be connected. If we think of the nodes of a graph as representing geographical locations and the arcs as representing direct communication links between the nodes, then the graph will be connected if any two nodes can communicate with each other—if not directly, then at least through the use of intermediary nodes.

A graph can always be subdivided into nonoverlapping connected subgraphs called components. For instance, the graph in Figure 9.12 with nodes

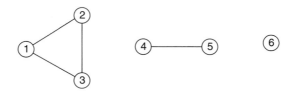

Figure 9.12.

$N = \{1, 2, 3, 4, 5, 6\}$ and arcs $A = \{(1, 2), (1, 3), (2, 3), (4, 5)\}$ consists of three components (a graph consisting of a single node is considered to be connected).

Consider now the random graph having nodes $1, 2, \ldots, n$ which is such that there is an arc from node i to node j with probability P_{ij}. Assume in addition that the occurrences of these arcs constitute independent events. That is, assume that the $\binom{n}{2}$ random variables $X_{ij}, i \neq j$, are independent where

$$X_{ij} = \begin{cases} 1, & \text{if } (i, j) \text{ is an arc} \\ 0, & \text{otherwise} \end{cases}$$

We are interested in the probability that this graph will be connected.

We can think of the preceding as being a reliability system of $\binom{n}{2}$ components—each component corresponding to a potential arc. The component is said to work if the corresponding arc is indeed an arc of the network, and the system is said to work if the corresponding graph is connected. As the addition of an arc to a connected graph cannot disconnect the graph, it follows that the structure so defined is monotone.

Let us start by determining the minimal path and minimal cut sets. It is easy to see that a graph will not be connected if and only if the sets of arcs can be partitioned into two nonempty subsets X and X^c in such a way that there is no arc connecting a node from X with one from X^c. For instance, if there are six nodes and if there are no arcs connecting any of the nodes 1, 2, 3, 4 with either 5 or 6, then clearly the graph will not be connected. Thus, we see that any partition of the nodes into two nonempty subsets X and X^c corresponds to the minimal cut set defined by

$$\{(i, j): i \in X, j \in X^c\}$$

As there are $2^{n-1} - 1$ such partitions (there are $2^n - 2$ ways of choosing a nonempty proper subset X and, as the partition X, X^c is the same as X^c, X, we must divide by 2) there are therefore this number of minimal cut sets.

To determine the minimal path sets, we must characterize a minimal set of arcs which result in a connected graph. Now the graph in Figure 9.13 is conected but it would remain connected if any one of the arcs from the cycle shown in Figure 9.14 were removed. In fact it is not difficult to see that the minimal path sets are exactly those sets of arcs which result in a graph being connected but not having any cycles (a cycle being a path from a node to itself). Such sets of arcs are called *spanning trees* (Figure 9.15). It is easily verified that any spanning tree contains exactly $n - 1$ arcs, and it is a famous result in graph theory (due to Cayley) that there are exactly n^{n-2} of these minimal path sets.

Because of the large number of minimal path and minimal cut sets (n^{n-2}

Figure 9.13.   **Figure 9.14.**

and $2^{n-1} - 1$, respectively), it is difficult to obtain any useful bounds without making further restrictions. So, let us assume that all the P_{ij} equal the common value p. That is, we suppose that each of the possible arcs exists, independently, with the same probability p. We shall start by deriving a recursive formula for the probability that the graph is connected, which is computationally useful when n is not too large, and then we shall present an asymptotic formula for this probability when n is large.

Let us denote by P_n the probability that the random graph having n nodes is connected. To derive a recursive formula for P_n we first concentrate attention on a single node—say, node 1—and try to determine the probability that node 1 will be part of a component of size k in the resultant graph. Now for given set of $k - 1$ other nodes these nodes along with node 1 will form a component if

(i) there are no arcs connecting any of these k nodes with any of the remaining $n - k$ nodes;

(ii) the random graph, restricted to these k nodes [and $\binom{k}{2}$ potential arcs—each independently appearing with probability p] is connected.

The probability that (i) and (ii) both occur is

$$q^{k(n-k)}P_k$$

where $q = 1 - p$. As there are $\binom{n-1}{k-1}$ ways of choosing $k - 1$ other nodes (to form along with node 1 a component of size k) we see that

$$P\{\text{node 1 is part of a component of size } k\}$$

$$= \binom{n-1}{k-1} q^{k(n-k)}P_k, \qquad k = 1, 2, \ldots, n$$

Figure 9.15. Two spanning trees (minimal path sets) when $n = 4$.

Now since the sum of the foregoing probabilities as k ranges from 1 through n clearly must equal 1, and as the graph is connected if and only if node 1 is part of a component of size n, we see that

$$P_n = 1 - \sum_{k=1}^{n-1} \binom{n-1}{k-1} q^{k(n-k)} P_k, \qquad n = 2, 3, \ldots \qquad (9.9)$$

Starting with $P_1 = 1$, $P_2 = p$, Equation (9.9) can be used to determine P_n recursively when n is not too large. It is particularly suited for numerical computation.

To determine an asymptotic formula for P_n when n is large, first note from Equation (9.9) that since $P_k \leqslant 1$, we have

$$1 - P_n \leqslant \sum_{k=1}^{n-1} \binom{n-1}{k-1} q^{k(n-k)}$$

As it can be shown that for $q < 1$ and n sufficiently large

$$\sum_{k=1}^{n-1} \binom{n-1}{k-1} q^{k(n-k)} \leqslant (n+1) q^{n-1}$$

we have that for n large

$$1 - P_n \leqslant (n+1) q^{n-1} \qquad (9.10)$$

To obtain a bound in the other direction, we concentrate our attention on a particular type of minimal cut set — namely, those that separate one node from all others in the graph. Specifically, define the minimal cut set C_i as

$$C_i = \{(i, j) : j \neq i\}$$

and define F_i to be the event that all arcs in C_i are not working (and thus, node i is isolated from the other nodes). Now,

$$1 - P_n = P(\text{graph is not connected}) \geqslant P\left(\bigcup_i F_i\right)$$

since, if any of the events F_i occur, then the graph will be disconnected. By the inclusion–exclusion bounds, we have that

$$P\left(\bigcup_i F_i\right) \geqslant \sum_i P(F_i) - \sum\sum_{i<j} P(F_i F_j)$$

As $P(F_i)$ and $P(F_i F_j)$ are just the respective probabilities that a given set of $n - 1$ arcs and that a given set of $2n - 3$ arcs are not in the graph (why?), it follows that

$$P(F_i) = q^{n-1},$$

$$P(F_i F_j) = q^{2n-3}, \qquad i \neq j$$

and so

$$1 - P_n \geqslant nq^{n-1} - \binom{n}{2}q^{2n-3}$$

Combining this with Equation (9.10) yields that for n sufficiently large

$$nq^{n-1} - \binom{n}{2}q^{2n-3} \leqslant 1 - P_n \leqslant (n+1)q^{n-1}$$

and as

$$\binom{n}{2}\frac{q^{2n-3}}{nq^{n-1}} \to 0$$

as $n \to \infty$, we see that, for large n,

$$1 - P_n \approx nq^{n-1}$$

Thus, for instance, when $n = 20$ and $p = \frac{1}{2}$, the probability that the random graph will be connected is approximately given by

$$P_{20} \approx 1 - 20(\tfrac{1}{2})^{19} = 0.99998 \quad \blacklozenge$$

9.4.2. Second Method for Obtaining Bounds on $r(p)$

Our second approach to obtaining bounds on $r(\mathbf{p})$ is based on expressing the desired probability as the probability of the intersection of events. To do so, let $A_1, A_2, \ldots, A_s$ denote the minimal path sets as before, and define the events, D_i, $i = 1, \ldots$ by

$$D_i = \{\text{at least one component in } A_i \text{ has failed}\}$$

Now since the system will have failed if and only if at least one component in each of the minimal path sets has failed we have that

$$1 - r(\mathbf{p}) = P(D_1 D_2 \cdots D_s)$$

$$= P(D_1)P(D_2 \mid D_1) \cdots P(D_s \mid D_1 D_2 \cdots D_{s-1}) \qquad (9.11)$$

Now it is quite intuitive that the information that at least one component of A_1 is down can only increase the probability that at least one component of A_2 is down (or else leave the probability unchanged if A_1 and A_2 do not overlap). Hence, intuitively

$$P(D_2 \mid D_1) \geqslant P(D_2)$$

To prove this inequality, we write

$$P(D_2) = P(D_2 \mid D_1)P(D_1) + P(D_2 \mid D_1^c)(1 - P(D_1)) \qquad (9.12)$$

and note that

$$P(D_2 \mid D_1^c) = P\{\text{at least one failed in } A_2 \mid \text{all functioning in } A_1\}$$

$$= 1 - \prod_{\substack{j \in A_2 \\ j \notin A_1}} p_j$$

$$\leqslant 1 - \prod_{j \in A_2} p_j$$

$$= P(D_2)$$

Hence, from Equation (9.12) we see that

$$P(D_2) \leqslant P(D_2 \mid D_1)P(D_1) + P(D_2)(1 - P(D_1))$$

or

$$P(D_2 \mid D_1) \geqslant P(D_2)$$

By the same reasoning, it also follows that

$$P(D_i \mid D_1 \cdots D_{i-1}) \geqslant P(D_i)$$

and so from Equation (9.11) we have

$$1 - r(\mathbf{p}) \geqslant \prod_i P(D_i)$$

or, equivalently,

$$r(\mathbf{p}) \leqslant 1 - \prod_i \left(1 - \prod_{j \in A_i} p_j \right)$$

To obtain a bound in the other direction, let $C_1, \ldots, C_r$ denote the minimal cut sets and define the events $U_1, \ldots, U_r$ by

$$U_i = \{\text{at least one component in } C_i \text{ is functioning}\}$$

Then, since the system will function if and only if all of the events U_i occur, we have

$$r(\mathbf{p}) = P(U_1 U_2 \cdots U_r)$$

$$= P(U_1)P(U_2 \mid U_1) \cdots P(U_r \mid U_1 \cdots U_{r-1})$$

$$\geqslant \prod_i P(U_i)$$

where the last inequality is established in exactly the same manner as for the D_i. Hence,

$$r(\mathbf{p}) \geqslant \prod_i \left[1 - \prod_{j \in C_i} (1 - p_j) \right]$$

and we thus have the following bounds for the reliability function

$$\prod_i \left[1 - \prod_{j \in C_i} (1 - p_j) \right] \leqslant r(\mathbf{p}) \leqslant 1 - \prod_i \left(1 - \prod_{j \in A_i} p_j \right) \qquad (9.13)$$

It is to be expected that the upper bound should be close to the actual $r(\mathbf{p})$ if there is not too much overlap in the minimal path sets, and the lower bound to be close if there is not too much overlap in the minimal cut sets.

Example 9.19 For the three-out-of-four system the minimal path sets are $A_1 = \{1, 2, 3\}$, $A_2 = \{1, 2, 4\}$, $A_3 = \{1, 3, 4\}$, and $A_4 = \{2, 3, 4\}$; and the minimal cut sets are $C_1 = \{1, 2\}$, $C_2 = \{1, 3\}$, $C_3 = \{1, 4\}$, $C_4 = \{2, 3\}$, $C_5 = \{2, 4\}$, and $C_6 = \{3, 4\}$. Hence, by Equation (9.13) we have

$$(1 - q_1 q_2)(1 - q_1 q_3)(1 - q_1 q_4)(1 - q_2 q_3)(1 - q_2 q_4)(1 - q_3 q_4)$$

$$\leqslant r(\mathbf{p}) \leqslant 1 - (1 - p_1 p_2 p_3)(1 - p_1 p_2 p_4)(1 - p_1 p_3 p_4)(1 - p_2 p_3 p_4)$$

where $q_i \equiv 1 - p_i$. For instance, if $p_i = \frac{1}{2}$ for all i, then the preceding yields that

$$0.18 \leqslant r(\tfrac{1}{2}, \dots, \tfrac{1}{2}) \leqslant 0.59$$

The exact value for this structure is easily computed to be

$$r(\tfrac{1}{2}, \dots, \tfrac{1}{2}) = \tfrac{5}{16} = 0.31 \quad \blacklozenge$$

9.5. System Life as a Function of Component Lives

For a random variable having distribution function G, we define $\bar{G}(a) \equiv 1 - G(a)$ to be the probability that the random variable is greater than a.

Consider a system in which the ith component functions for a random length of time having distribution F_i and then fails. Once failed it remains in that state forever. Assuming that the individual component lifetimes are independent, how can we express the distribution of system lifetime as a function of the system reliability function $r(\mathbf{p})$ and the individual component lifetime distributions F_i, $i = 1, \dots, n$?

To answer the above we first note that the system will function for a length of time t or greater if and only if it is still functioning at time t. That is, letting F denote the distribution of system lifetime, we have

$$\bar{F}(t) = P\{\text{system life} > t\}$$

$$= P\{\text{system is functioning at time } t\}$$

But, by the definition of $r(\mathbf{p})$ we have that

$$P\{\text{system is functioning at time } t\} = r(P_1(t), \ldots, P_n(t))$$

where

$$P_i(t) = P\{\text{component } i \text{ is functioning at } t\}$$
$$= P\{\text{lifetime of } i > t\}$$
$$= \bar{F}_i(t)$$

Hence we see that

$$\bar{F}(t) = r(\bar{F}_1(t), \ldots, \bar{F}_n(t)) \tag{9.14}$$

Example 9.20 In a series system, $r(\mathbf{p}) = \Pi_1^n p_i$ and so from Equation (9.14)

$$\bar{F}(t) = \prod_1^n \bar{F}_i(t)$$

which is, of course, quite obvious since for a series system the system life is equal to the minimum of the component lives and so will be greater than t if and only if all component lives are greater than t. ✦

Example 9.21 In a parallel system $r(\mathbf{p}) = 1 - \Pi_1^n (1 - p_i)$ and so

$$\bar{F}(t) = 1 - \prod_1^n F_i(t)$$

The preceding is also easily derived by noting that, in the case of a parallel system, the system life is equal to the maximum of the component lives. ✦

For a continuous distribution G, we define $\lambda(t)$, the *failure rate function* of G, by

$$\lambda(t) = \frac{g(t)}{\bar{G}(t)}$$

where $g(t) = d/dt \, G(t)$. In Section 5.2.2, it is shown that if G is the distribution of the lifetime of an item, then $\lambda(t)$ represents the probability intensity that a t-year-old item will fail. We say that G is an *increasing failure rate* (IFR) distribution if $\lambda(t)$ is an increasing function of t. Similarly, we say that G is a *decreasing failure rate* (DFR) distribution if $\lambda(t)$ is a decreasing function of t.

Example 9.22 (The Weibull Distribution): A random variable is said to have the *Weibull* distribution if its distribution is given, for some $\lambda > 0$, $\alpha > 0$, by

$$G(t) = 1 - e^{-(\lambda t)^{\alpha}}, \qquad t \geq 0$$

The failure rate function for a Weibull distribution equals

$$\lambda(t) = \frac{e^{-(\lambda t)^{\alpha}} \alpha (\lambda t)^{\alpha - 1} \lambda}{e^{-(\lambda t)^{\alpha}}} = \alpha \lambda (\lambda t)^{\alpha - 1}$$

Thus, the Weibull distribution is IFR when $\alpha \geq 1$, and DFR when $0 < \alpha \leq 1$; when $\alpha = 1$, $G(t) = 1 - e^{-\lambda t}$, the exponential distribution, which is both IFR and DFR. ✦

Example 9.23 (The Gamma Distribution): A random variable is said to have a *gamma* distribution if its density is given, for some $\lambda > 0$, $\alpha > 0$, by

$$g(t) = \frac{\lambda e^{-\lambda t} (\lambda t)^{\alpha - 1}}{\Gamma(\alpha)} \qquad \text{for } t \geq 0$$

where

$$\Gamma(\alpha) \equiv \int_{0}^{\infty} e^{-t} t^{\alpha - 1} \, dt$$

For the gamma distribution,

$$\frac{1}{\lambda(t)} = \frac{\bar{G}(t)}{g(t)} = \frac{\int_{t}^{\infty} \lambda e^{-\lambda x} (\lambda x)^{\alpha - 1} \, dx}{\lambda e^{-\lambda t} (\lambda t)^{\alpha - 1}}$$

$$= \int_{t}^{\infty} e^{-\lambda(x - t)} \left(\frac{x}{t} \right)^{\alpha - 1} dx$$

With the change of variables $u = x - t$, we obtain

$$\frac{1}{\lambda(t)} = \int_{0}^{\infty} e^{-\lambda u} \left(1 + \frac{u}{t} \right)^{\alpha - 1} du$$

Hence, G is IFR when $\alpha \geq 1$ and is DFR when $0 < \alpha \leq 1$. ✦

Suppose that the lifetime distribution of each component in a monotone system is IFR. Does this imply that the system lifetime is also IFR? To answer this, let us at first suppose that each component has the same lifetime distribution, which we denote by G. That is, $F_i(t) = G(t)$, $i = 1, \ldots, n$. To determine whether the system lifetime is IFR, we must compute $\lambda_F(t)$, the failure rate function of F. Now, by definition,

$$\lambda_F(t) = \frac{(d/dt)F(t)}{\bar{F}(t)}$$

$$= \frac{(d/dt)[1 - r(\bar{G}(t))]}{r(\bar{G}(t))}$$

where

$$r(\bar{G}(t)) \equiv r(\bar{G}(t), \ldots, \bar{G}(t))$$

Hence,

$$\lambda_F(t) = \frac{r'(\bar{G}(t))}{r(\bar{G}(t))} G'(t)$$

$$= \frac{\bar{G}(t)r'(\bar{G}(t))}{r(\bar{G}(t))} \frac{G'(t)}{\bar{G}(t)}$$

$$= \lambda_G(t) \left. \frac{pr'(p)}{r(p)} \right|_{p = \bar{G}(t)} \tag{9.15}$$

Since $\bar{G}(t)$ is a decreasing function of t, it follows from Equation (9.15) that *if each component of a coherent system has the same IFR lifetime distribution, then the distribution of system lifetime will be IFR if $pr'(p)/r(p)$ is a decreasing function of p.*

Example 9.24 (The k-out-of-n System with Identical Components): Consider the k-out-of-n system which will function if and only if k or more components function. When each component has the same probability p of functioning, the number of functioning components will have a binomial distribution with parameters n and p. Hence,

$$r(p) = \sum_{i=k}^{n} \binom{n}{i} p^i(1 - p)^{n-i}$$

which, by continual integration by parts, can be shown to be equal to

$$r(p) = \frac{n!}{(k-1)!(n-k)!} \int_0^p x^{k-1}(1 - x)^{n-k} \, dx$$

Upon differentiation, we obtain

$$r'(p) = \frac{n!}{(k-1)!(n-k)!} p^{k-1}(1 - p)^{n-k}$$

Therefore,

$$\frac{pr'(p)}{r(p)} = \left[\frac{r(p)}{pr'(p)}\right]^{-1}$$

$$= \left[\frac{1}{p}\int_0^p \left(\frac{x}{p}\right)^{k-1}\left(\frac{1-x}{1-p}\right)^{n-k}dx\right]^{-1}$$

Letting $y = x/p$, yields

$$\frac{pr'(p)}{r(p)} = \left[\int_0^1 y^{k-1}\left(\frac{1-yp}{1-p}\right)^{n-k}dy\right]^{-1}$$

Since $(1 - yp)/(1 - p)$ is increasing in p, it follows that $pr'(p)/r(p)$ is decreasing in p. Thus, if a k-out-of-n system is composed of independent, like components having an increasing failure rate, the system itself has an increasing failure rate. ✦

It turns out, however, that for a k-out-of-n system, in which the independent components have different IFR lifetime distributions, the system lifetime need not be IFR. Consider the following example of a two-out-of-two (that is, a parallel) system.

Example 9.25 (A Parallel System That is Not IFR): The life distribution of a parallel system of two independent components, the ith component having an exponential distribution with mean $1/i$, $i = 1, 2$, is given by

$$\bar{F}(t) = 1 - (1 - e^{-t})(1 - e^{-2t})$$

$$= e^{-2t} + e^{-t} - e^{-3t}$$

Therefore,

$$\lambda(t) = \frac{f(t)}{\bar{F}(t)}$$

$$= \frac{2e^{-2t} + e^{-t} - 3e^{-3t}}{e^{-2t} + e^{-t} - e^{-3t}}$$

It easily follows upon differentiation, that the sign of $\lambda'(t)$ is determined by $e^{-5t} - e^{-3t} + 3e^{-4t}$, which is positive for small values and negative for large values of t. Therefore, $\lambda(t)$ is initially strictly increasing, and then strictly decreasing. Hence, F is not IFR. ✦

Remark The result of the preceding example is quite surprising at first glance. To obtain a better feel for it we need the concept of a mixture of distribution functions. The distribution function G is said to be a *mixture of* the distributions G_1 and G_2 if for some $p, 0 < p < 1$,

$$G(x) = pG_1(x) + (1 - p)G_2(x) \qquad (9.16)$$

Mixtures occur when we sample from a population made up of two distinct groups. For example, suppose we have a stockpile of items of which the fraction p are type 1 and the fraction $1 - p$ are type 2. Suppose that the lifetime distribution of type 1 items is G_1 and of type 2 items is G_2. If we choose an item at random from the stockpile, then its life distribution is as given by Equation (9.16).

Consider now a mixture of two exponential distributions having rates λ_1 and λ_2 where $\lambda_1 < \lambda_2$. We are interested in determining whether or not this mixture distribution is IFR. To do so, we note that if the item selected has survived up to time t, then its distribution of remaining life is still a mixture of the two exponential distributions. This is so since its remaining life will still be exponential with rate λ_1 if it is type 1 or with rate λ_2 if it is a type 2 item. However, the probability that it is a type 1 item is no longer the (prior) probability p but is now a conditional probability given that it has survived to time t. In fact, its probability of being a type 1 is

$$P\{\text{type } 1 \,|\, \text{life} > t\} = \frac{P\{\text{type } 1, \text{life} > t\}}{P\{\text{life} > t\}}$$

$$= \frac{pe^{-\lambda_1 t}}{pe^{-\lambda_1 t} + (1 - p)e^{-\lambda_2 t}}$$

As the preceding is increasing in t, it follows that the larger t is, the more likely it is that the item in use is a type 1 (the better one, since $\lambda_1 < \lambda_2$). Hence, the older the item is, the less likely it is to fail, and thus the mixture of exponentials far from being IFR is, in fact, DFR.

Now, let us return to the parallel system of two exponential components having respective rates λ_1 and λ_2. The lifetime of such a system can be expressed as the sum of two independent random variables, namely,

$$\text{system life} = \text{Exp}(\lambda_1 + \lambda_2) + \begin{cases} \text{Exp}(\lambda_1) \text{ with probability } \dfrac{\lambda_2}{\lambda_1 + \lambda_2} \\ \\ \text{Exp}(\lambda_2) \text{ with probability } \dfrac{\lambda_1}{\lambda_1 + \lambda_2} \end{cases}$$

The first random variable whose distribution is exponential with rate $\lambda_1 + \lambda_2$ represents the time until one of the components fails, and the second, which is a mixture of exponentials, is the additional time until the other component fails. (Why are these two random variables independent?)

Now, given that the system has survived a time t, it is very unlikely when t is large that both components are still functioning, but instead it is far more likely that one of the components has failed. Hence, for large t, the distribution of remaining life is basically a mixture of two exponentials — and so as t becomes even larger its failure rate should decrease (as indeed occurs). ✦

Recall that the failure rate function of a distribution $F(t)$ having density $f(t) = F'(t)$ is defined by

$$\lambda(t) = \frac{f(t)}{1 - F(t)}$$

By integrating both sides, we obtain

$$\int_0^t \lambda(s) \, ds = \int_0^t \frac{f(s)}{1 - F(s)} \, ds$$
$$= -\log \bar{F}(t)$$

Hence,

$$\bar{F}(t) = e^{-\Lambda(t)} \tag{9.17}$$

where

$$\Lambda(t) = \int_0^t \lambda(s) \, ds$$

The function $\Lambda(t)$ is called the *hazard function* of the distribution F.

Definition 9.1 A distribution F is said to have *increasing failure on the average* (IFRA) if

$$\frac{\Lambda(t)}{t} = \frac{\int_0^t \lambda(s) \, ds}{t} \tag{9.18}$$

increases in t for $t \geqslant 0$.

In other words, Equation (9.18) states that the average failure rate up to time t increases as t increases. It is not difficult to show that if F is IFR, then F is IFRA; but the reverse need not be true.

Note that F is IFRA if $\Lambda(s)/s \leqslant \Lambda(t)/t$ whenever $0 \leqslant s \leqslant t$, which is equivalent to

$$\frac{\Lambda(\alpha t)}{\alpha t} \leqslant \frac{\Lambda(t)}{t} \qquad \text{for } 0 \leqslant \alpha \leqslant 1, \text{ all } t \geqslant 0$$

But by Equation (9.17) we see that $\Lambda(t) = -\log \bar{F}(t)$, and so the preceding is equivalent to

$$-\log \bar{F}(\alpha t) \leqslant -\alpha \log \bar{F}(t)$$

or equivalently

$$\log \bar{F}(\alpha t) \geqslant \log \bar{F}^\alpha(t)$$

which, since $\log x$ is a monotone function of x, shows that F is *IFRA if and only if*

$$\bar{F}(\alpha t) \geqslant \bar{F}^\alpha(t) \qquad \text{for } 0 \leqslant \alpha \leqslant 1, \text{ all } t \geqslant 0 \tag{9.19}$$

For a vector $\mathbf{p} = (p_1, \ldots, p_n)$ we define $\mathbf{p}^\alpha = (p_1^\alpha, \ldots, p_n^\alpha)$. We shall need the following proposition.

Proposition 9.2   Any reliability function $r(\mathbf{p})$ satisfies

$$r(\mathbf{p}^\alpha) \geqslant [r(\mathbf{p})]^\alpha, \qquad 0 \leqslant \alpha \leqslant 1$$

Proof   We prove this by induction on n, the number of components in the system. If $n = 1$, then either $r(p) \equiv 0$, $r(p) \equiv 1$, or $r(p) \equiv p$. Hence, the proposition follows in this case.

So assume that Proposition 9.2 is valid for all monotone systems of $n - 1$ components and consider a system of n components having structure function ϕ. By conditioning upon whether or not the nth component is functioning, we obtain

$$r(\mathbf{p}^\alpha) = p_n^\alpha r(1_n, \mathbf{p}^\alpha) + (1 - p_n^\alpha) r(0_n, \mathbf{p}^\alpha) \tag{9.20}$$

Now consider a system of components 1 through $n - 1$ having a structure function $\phi_1(\mathbf{x}) = \phi(1_n, \mathbf{x})$. The reliability function for this system is given by $r_1(\mathbf{p}) = r(1_n, \mathbf{p})$; hence, from the induction assumption (valid for all monotone systems of $n - 1$ components), we have

$$r(1_n, \mathbf{p}^\alpha) \geqslant [r(1_n, \mathbf{p})]^\alpha$$

Similarly, by considering the system of components 1 through $n - 1$ and structure function $\phi_0(\mathbf{x}) = \phi(0_n, \mathbf{x})$, we obtain

$$r(0_n, \mathbf{p}^\alpha) \geqslant [r(0_n, \mathbf{p})]^\alpha$$

Thus, from Equation (9.20), we obtain

$$r(\mathbf{p}^\alpha) \geqslant p_n^\alpha [r(1_n, \mathbf{p})]^\alpha + (1 - p_n^\alpha)[r(0_n, \mathbf{p})]^\alpha$$

which, by using the lemma to follow [with $\lambda = p_n$, $x = r(1_n, \mathbf{p})$, $y = r(0_n, \mathbf{p})$], implies that

$$r(\mathbf{p}^\alpha) \geqslant [p_n r(1_n, \mathbf{p}) + (1 - p_n) r(0_n, \mathbf{p})]^\alpha$$

$$= [r(\mathbf{p})]^\alpha$$

which proves the result. ✦

Lemma 9.3 If $0 \leqslant \alpha \leqslant 1$, $0 \leqslant \lambda \leqslant 1$, then

$$h(y) = \lambda^\alpha x^\alpha + (1 - \lambda^\alpha) y^\alpha - (\lambda x + (1 - \lambda) y)^\alpha \geqslant 0$$

for all $0 \leqslant y \leqslant x$.

Proof The proof is left as an exercise. ✦

We are now ready to prove the following important theorem.

Theorem 9.2 For a monotone system of independent components, if each component has an IFRA lifetime distribution, then the distribution of system lifetime is itself IFRA.

Proof The distribution of system lifetime F is given by

$$\bar{F}(\alpha t) = r(\bar{F}_1(\alpha t), \ldots, \bar{F}_n(\alpha t))$$

Hence, since r is a monotone function, and since each of the component distributions $\bar{F}_i$ is IFRA, we obtain from Equation (9.19)

$$\bar{F}(\alpha t) \geqslant r(\bar{F}_1^\alpha(t), \ldots, \bar{F}_n^\alpha(t))$$

$$\geqslant [r(\bar{F}_1(t), \ldots, \bar{F}_n(t))]^\alpha$$

$$= \bar{F}^\alpha(t)$$

which by Equation (9.19) proves the theorem. The last inequality followed, of course, from Proposition 9.2. ✦

9.6. Expected System Lifetime

In this section, we show how the mean lifetime of a system can be determined, at least in theory, from a knowledge of the reliability function $r(\mathbf{p})$ and the component lifetime distributions F_i, $i = 1, \ldots, n$.

Since the system's lifetime will be t or larger if and only if the system is still functioning at time t, we have that

$$P\{\text{system life} > t\} = r(\bar{\mathbf{F}}(t))$$

where $\bar{\mathbf{F}}(t) = (\bar{F}_1(t), \ldots, \bar{F}_n(t))$. Hence, by a well-known formula that states that for any nonnegative random variable X,

$$E[X] = \int_0^\infty P\{X > x\}\, dx,$$

we see that*

$$E[\text{system life}] = \int_0^\infty r(\bar{\mathbf{F}}(t))\, dt \qquad (9.21)$$

Example 9.26 (A Series System of Uniformly Distributed Components): Consider a series system of three independent components each of which functions for an amount of time (in hours) uniformly distributed over $(0, 10)$. Hence, $r(\mathbf{p}) = p_1 p_2 p_3$ and

$$F_i(t) = \begin{cases} t/10, & 0 \leqslant t \leqslant 10 \\ 1, & t > 10 \end{cases} \qquad i = 1, 2, 3$$

Therefore,

$$r(\bar{\mathbf{F}}(t)) = \begin{cases} \left(\dfrac{10 - t}{10}\right)^3, & 0 \leqslant t \leqslant 10 \\ 0, & t > 10 \end{cases}$$

and so from Equation (9.21) we obtain

$$E[\text{system life}] = \int_0^{10} \left(\frac{10 - t}{10}\right)^3 dt$$

$$= 10 \int_0^1 y^3\, dy$$

$$= \tfrac{5}{2} \qquad \blacklozenge$$

Example 9.27 (A Two-Out-of-Three System): Consider a two-out-of-three system of independent components, in which each component's lifetime is (in months) uniformly distributed over $(0, 1)$. As was shown in Example 9.13, the reliability of such a system is given by

$$r(\mathbf{p}) = p_1 p_2 + p_1 p_3 + p_2 p_3 - 2 p_1 p_2 p_3$$

---

* That $E[X] = \int_0^\infty P\{X > x\}\, dx$ can be shown as follows when X has density f:

$$\int_0^\infty P\{X > x\}\, dx = \int_0^\infty \int_x^\infty f(y)\, dy\, dx = \int_0^\infty \int_0^y f(y)\, dx\, dy = \int_0^\infty y f(y)\, dy = E[X]$$

Since

$$F_i(t) = \begin{cases} t, & 0 \leqslant t \leqslant 1 \\ 1, & t > 1 \end{cases}$$

we see from Equation (9.21) that

$$E[\text{system life}] = \int_0^1 [3(1-t)^2 - 2(1-t)^3]\, dt$$

$$= \int_0^1 (3y^2 - 2y^3)\, dy$$

$$= 1 - \tfrac{1}{2}$$

$$= \tfrac{1}{2} \quad \blacklozenge$$

Example 9.28 (A Four-Component System): Consider the four-component system that functions when components 1 and 2 and at least one of components 3 and 4 functions. Its structure function is given by

$$\phi(\mathbf{x}) = x_1 x_2 (x_3 + x_4 - x_3 x_4)$$

and thus its reliability function equals

$$r(\mathbf{p}) = p_1 p_2 (p_3 + p_4 - p_3 p_4)$$

Let us compute the mean system lifetime when the ith component is uniformly distributed over $(0, i)$, $i = 1, 2, 3, 4$. Now,

$$\bar{F}_1(t) = \begin{cases} 1 - t, & 0 \leqslant t \leqslant 1 \\ 0, & t > 1 \end{cases}$$

$$\bar{F}_2(t) = \begin{cases} 1 - t/2, & 0 \leqslant t \leqslant 2 \\ 0, & t > 2 \end{cases}$$

$$\bar{F}_3(t) = \begin{cases} 1 - t/3, & 0 \leqslant t \leqslant 3 \\ 0, & t > 3 \end{cases}$$

$$\bar{F}_4(t) = \begin{cases} 1 - t/4, & 0 \leqslant t \leqslant 4 \\ 0, & t > 4 \end{cases}$$

Hence,

$$r(\bar{\mathbf{F}}(t)) = \begin{cases} (1-t)\left(\dfrac{2-t}{2}\right)\left[\dfrac{3-t}{3} + \dfrac{4-t}{4} - \dfrac{(3-t)(4-t)}{12}\right], & 0 \leqslant t \leqslant 1 \\ 0, & t > 1 \end{cases}$$

Therefore,

$$E[\text{system life}] = \frac{1}{24} \int_0^1 (1 - t)(2 - t)(12 - t^2) \, dt$$

$$= \frac{593}{(24)(60)}$$

$$\approx 0.41 \quad \blacklozenge$$

We end this section by obtaining the mean lifetime of a k-out-of-n system of independent identically distributed exponential components. If θ is the mean lifetime of each component, then

$$\bar{F}_i(t) = e^{-t/\theta}$$

Hence, since for a k-out-of-n system,

$$r(p, p, \ldots, p) = \sum_{i=k}^{n} \binom{n}{i} p^i (1 - p)^{n-i}$$

we obtain from Equation (9.21)

$$E[\text{system life}] = \int_0^\infty \sum_{i=k}^{n} \binom{n}{i} (e^{-t/\theta})^i (1 - e^{-t/\theta})^{n-i} \, dt$$

Making the substitution

$$y = e^{-t/\theta}, \qquad dy = -\frac{1}{\theta} e^{-t/\theta} \, dt = -\frac{y}{\theta} \, dt$$

yields

$$E[\text{system life}] = \theta \sum_{i=k}^{n} \binom{n}{i} \int_0^1 y^{i-1} (1 - y)^{n-i} \, dy$$

Now, it is not difficult to show that*

$$\int_0^1 y^n (1 - y)^m \, dy = \frac{m! \, n!}{(m + n + 1)!} \tag{9.22}$$

---

* Let

$$C(n, m) = \int_0^1 y^n (1 - y)^m \, dy$$

Integration by parts yields that $C(n, m) = [m/(n + 1)]C(n + 1, m - 1)$. Starting with $C(n, 0) = 1/(n + 1)$, Equation (9.22) follows by mathematical induction.

Thus, the foregoing equals

$$E[\text{system life}] = \theta \sum_{i=k}^{n} \frac{n!}{(n-i)!i!} \frac{(i-1)!(n-i)!}{n!}$$

$$= \theta \sum_{i=k}^{n} \frac{1}{i} \qquad (9.23)$$

Remark Equation (9.23) could have been proven directly by making use of special properties of the exponential distribution. First note that the lifetime of a k-out-of-n system can be written as $T_1 + \cdots + T_{n-k+1}$, where T_i represents the time between the $(i-1)$st and ith failure. This is true since $T_1 + \cdots + T_{n-k+1}$ equals the time at which the $(n-k+1)$st component fails, which is also the first time that the number of functioning components is less than k. Now, when all n components are functioning, the rate at which failures occur is n/θ. That is, T_1 is exponentially distributed with mean θ/n. Similarly, since T_i represents the time until the next failure when there are $n - (i-1)$ functioning components, it follows that T_i is exponentially distributed with mean $\theta/(n-i+1)$. Hence, the mean system lifetime equals

$$E[T_1 + \cdots + T_{n-k+1}] = \theta \left[\frac{1}{n} + \cdots + \frac{1}{k} \right]$$

Note also that it follows, from the lack of memory of the exponential, that the T_i, $i = 1, \ldots, n-k+1$, are independent random variables.

9.6.1. An Upper Bound on the Expected Life of a Parallel System

Consider a parallel system of n components, whose lifetimes are not necessarily independent. The system lifetime can be expressed as

$$\text{system life} = \max_i X_i$$

where X_i is the lifetime of component i, $i = 1, \ldots, n$. We can bound the expected system lifetime by making use of the following inequality. Namely, for any constant c

$$\max_i X_i \leq c + \sum_{i=1}^{n} (X_i - c)^+ \qquad (9.24)$$

where x^+, the positive part of x, is equal to x if $x > 0$ and is equal to 0 if $x \leq 0$. The validity of inequality (9.24) is immediate since if $\max X_i < c$ then the left side is equal to $\max X_i$ and the right side is equal to c. On the other hand, if $X_{(n)} = \max X_i > c$ then the right side is at least as large as $c + (X_{(n)} - c) = X_{(n)}$. It follows from inequality (9.24), upon taking expectations, that

$$E\left[\max_i X_i\right] \leqslant c + \sum_{i=1}^{n} E[(X_i - c)^+] \qquad (9.25)$$

Now, $(X_i - c)^+$ is a nonnegative random variable and so

$$E[(X_i - c)^+] = \int_0^{\infty} P\{(X_i - c)^+ > x\}\,dx$$

$$= \int_0^{\infty} P\{X_i - c > x\}\,dx$$

$$= \int_c^{\infty} P\{X_i > y\}\,dy$$

Thus, we obtain

$$E\left[\max_i X_i\right] \leqslant c + \sum_{i=1}^{n} \int_c^{\infty} P\{X_i > y\}\,dy \qquad (9.26)$$

Because the preceding is true for all c, it follows that we obtain the best bound by letting c equal the value that minimizes the right side of the above. To determine that value, differentiate the right side of the preceding and set the result equal to 0, to obtain:

$$1 - \sum_{i=1}^{n} P\{X_i > c\} = 0$$

That is, the minimizing value of c is that value c^* for which

$$\sum_{i=1}^{n} P\{X_i > c^*\} = 1$$

Since $\sum_{i=1}^{n} P\{X_i > c\}$ is a decreasing function of c, the value of c^* can be easily approximated and then utilized in inequality (9.26). Also, it is interesting to note that c^* is such that the expected number of the X_i that exceed c^* is equal to 1 (see Exercise 32). That the optimal value of c has this property is interesting and somewhat intuitive in as much as inequality 9.24 is an equality when exactly one of the X_i exceeds c.

Example 9.29 Suppose the lifetime of component i is exponentially distributed with rate λ_i, $i = 1, \ldots, n$. Then the minimizing value of c is such that

$$1 = \sum_{i=1}^{n} P\{X_i > c^*\} = \sum_{i=1}^{n} e^{-\lambda_i c^*}$$

and the resulting bound of the mean system life is

$$E\left[\max_i X_i\right] \leqslant c^* + \sum_{i=1}^{n} E[(X_i - c^*)^+]$$

$$= c^* + \sum_{i=1}^{n} (E[(X_i - c^*)^+ \mid X_i > c^*]P\{X_i > c^*\}$$

$$+ E[(X_i - c^*)^+ \mid X_i \leqslant c^*]P\{X_i \leqslant c^*\})$$

$$= c^* + \sum_{i=1}^{n} \frac{1}{\lambda_i} e^{-\lambda_i c^*}$$

In the special case where all the rates are equal, say, $\lambda_i = \lambda$, $i = 1, \ldots, n$, then

$$1 = ne^{-\lambda c^*} \quad \text{or} \quad c^* = \frac{1}{\lambda} \log(n)$$

and the bound is

$$E\left[\max_i X_i\right] \leqslant \frac{1}{\lambda}(\log(n) + 1)$$

That is, if $X_1, \ldots, X_n$ are identically distributed exponential random variables with rate λ, then the preceding gives a bound on the expected value of their maximum. In the special case where these random variables are also independent, the following exact expression, given by Equation (9.25), is not much less than the preceding upper bound:

$$E\left[\max_i X_i\right] = \frac{1}{\lambda} \sum_{i=1}^{n} 1/i \approx \frac{1}{\lambda} \int_1^n \frac{1}{x} dx \approx \frac{1}{\lambda} \log(n) \quad \blacklozenge$$

9.7. Systems with Repair

Consider an n component system having reliability function $r(\mathbf{p})$. Suppose that component i functions for an exponentially distributed time with rate λ_i and then fails; once failed it takes an exponential time with rate μ_i to be repaired, $i = 1, \ldots, n$. All components act independently.

Let us suppose that all components are initially working, and let

$$A(t) = P\{\text{system is working at } t\}$$

$A(t)$ is called the *availability* at time t. Since the components act independently, $A(t)$ can be expressed in terms of the reliability function as follows:

$$A(t) = r(A_1(t), \ldots, A_n(t)) \tag{9.27}$$

where
$$A_i(t) = P\{\text{component } i \text{ is functioning at } t\}$$

Now the state of component i—either on or off—changes in accordance with a two-state continuous time Markov chain. Hence, from the results of Example 6.12 we have

$$A_i(t) = P_{00}(t) = \frac{\mu_i}{\mu_i + \lambda_i} + \frac{\lambda_i}{\mu_i + \lambda_i} e^{-(\lambda_i + \mu_i)t}$$

Thus, we obtain

$$A(t) + r\left(\frac{\mu}{\mu + \lambda} + \frac{\lambda}{\mu + \lambda} e^{-(\lambda + \mu)t}\right)$$

If we let t approach ∞, then we obtain the limiting availability—call it A—which is given by

$$A = \lim_{t \to \infty} A(t) = r\left(\frac{\mu}{\lambda + \mu}\right)$$

Remarks   (i) If the on and off distribution for component i are arbitrary continuous distributions with respective means $1/\lambda_i$ and $1/\mu_i$, $i = 1, \ldots, n$, then it follows from the theory of alternating renewal processes (see Section 7.5.1) that

$$A_i(t) \to \frac{1/\lambda_i}{1/\lambda_i + 1/\mu_i} = \frac{\mu_i}{\mu_i + \lambda_i}$$

and thus using the continuity of the reliability function, it follows from (9.27) that the limiting availability is

$$A = \lim_{t \to \infty} A(t) = r\left(\frac{\mu}{\mu + \lambda}\right)$$

Hence, A depends only on the on and off distributions through their means.

(ii) It can be shown (using the theory of regenerative processes as presented in Section 7.5) that A will also equal the long-run proportion of time that the system will be functioning.

Example 9.30   For a series system, $r(\mathbf{p}) = \Pi_{i=1}^n p_i$ and so

$$A(t) = \prod_{i=1}^n \left[\frac{\mu_i}{\mu_i + \lambda_i} + \frac{\lambda_i}{\mu_i + \lambda_i} e^{-(\lambda_i + \mu_i)t}\right]$$

and

$$A = \prod_{i=1}^n \frac{\mu_i}{\mu_i + \lambda_i} \quad \blacklozenge$$

Example 9.31   For a parallel system, $r(\mathbf{p}) = 1 - \Pi_{i=1}^n (1 - p_i)$ and thus

$$A(t) = 1 - \prod_{i=1}^n \left[\frac{\lambda_i}{\mu_i + \lambda_i}(1 - e^{-(\lambda_i + \mu_i)t})\right]$$

and

$$A = 1 - \prod_{i=1}^{n} \frac{\lambda_i}{\mu_i + \lambda_i} \quad \blacklozenge$$

The preceding system will alternate between periods when it is up and periods when it is down. Let us denote by U_i and D_i, $i \geqslant 1$, the lengths of the ith up and down period respectively. For instance in a two-out-of-three system, U_1 will be the time until two components are down; D_1, the additional time until two are up; U_2 the additional time until two are down, and so on. Let

$$\bar{U} = \lim_{n \to \infty} \frac{U_1 + \cdots + U_n}{n},$$

$$\bar{D} = \lim_{n \to \infty} \frac{D_1 + \cdots + D_n}{n}$$

denote the average length of an up and down period respectively.*

To determine $\bar{U}$ and $\bar{D}$, note first that in the first n up–down cycles — that is, in time $\Sigma_{i=1}^{n} (U_i + D_i)$ — the system will be up for a time $\Sigma_{i=1}^{n} U_i$. Hence, the proportion of time the system will be up in the first n up–down cycles is

$$\frac{U_1 + \cdots + U_n}{U_1 + \cdots + U_n + D_1 + \cdots + D_n} = \frac{\Sigma_{i=1}^{n} U_i / n}{\Sigma_{i=1}^{n} U_i / n + \Sigma_{i=1}^{n} D_i / n}$$

As $n \to \infty$, this must converge to A, the long-run proportion of time the system is up. Hence,

$$\frac{\bar{U}}{\bar{U} + \bar{D}} = A = r\left(\frac{\mu}{\lambda + \mu}\right) \tag{9.28}$$

However, to solve for $\bar{U}$ and $\bar{D}$ we need a second equation. To obtain one consider the rate at which the system fails. As there will be n failures in time $\Sigma_{i=1}^{n} (U_i + D_i)$, it follows that the rate at which the system fails is

$$\text{rate at which system fails} = \lim_{n \to \infty} \frac{n}{\Sigma_1^n U_i + \Sigma_1^n D_1}$$

$$= \lim_{n \to \infty} \frac{1}{\Sigma_1^n U_i / n + \Sigma_1^n D_i / n} = \frac{1}{\bar{U} + \bar{D}} \tag{9.29}$$

That is, the foregoing yields the intuitive result that, on average, there is one failure every $\bar{U} + \bar{D}$ time units. To utilize this, let us determine the rate at which a failure of component i causes the system to go from up to down.

---

*It can be shown using the theory of regenerative processes that, with probability 1, the preceding limits will exist and will be constants.

Now, the system will go from up to down when component i fails if the states of the other components $x_1, \ldots, x_{i-1}, x_{i-1}, \ldots, x_n$ are such that $\phi(1_i, \mathbf{x}) = 1$, $\phi(0_i, \mathbf{x}) = 0$. That is the states of the other components must be such that

$$\phi(1_i, \mathbf{x}) - \phi(0_i, \mathbf{x}) = 1 \tag{9.30}$$

Since component i will, on average, have one failure every $1/\lambda_i + 1/\mu_i$ time units, it follows that the rate at which component i fails is equal to $(1/\lambda_i + 1/\mu_i)^{-1} = \lambda_i \mu_i / (\lambda_i + \mu_i)$. In addition, the states of the other components will be such that (9.30) holds with probability

$$
\begin{aligned}
P\{&\phi(1_i, X(\infty)) - \phi(0_i, X(\infty)) = 1\} \\
&= E[\phi(1_i, X(\infty)) - \phi(0_i, X(\infty))] \qquad \text{since } \phi(1_i, X(\infty)) - \phi(0_i, X(\infty)) \\
&\hspace{6.3cm} \text{is a Bernoulli random variable} \\
&= r\left(1_i, \frac{\mu}{\lambda + \mu}\right) - r\left(0_i, \frac{\mu}{\lambda + \mu}\right)
\end{aligned}
$$

Hence, putting the preceding together we see that

$$\begin{matrix} \text{rate at which component} \\ i \text{ causes the system to fail} \end{matrix} = \frac{\lambda_i \mu_i}{\lambda_i + \mu_i}\left[r\left(1_i, \frac{\mu}{\lambda + \mu}\right) - r\left(0_i, \frac{\mu}{\lambda + \mu}\right)\right]$$

Summing this over all components i thus gives

$$\text{rate at which system fails} = \sum_i \frac{\lambda_i \mu_i}{\lambda_i + \mu_i}\left[r\left(1_i, \frac{\mu}{\lambda + \mu}\right) - r\left(0_i, \frac{\mu}{\lambda + \mu}\right)\right]$$

Finally, equating the preceding with (9.29) yields

$$\frac{1}{\bar{U} + \bar{D}} = \sum_i \frac{\lambda_i \mu_i}{\lambda_i + \mu_i}\left[r\left(1_i, \frac{\mu}{\lambda + \mu}\right) - r\left(0_i, \frac{\mu}{\lambda + \mu}\right)\right] \tag{9.31}$$

Solving (9.28) and (9.31), we obtain

$$\bar{U} = \frac{r\left(\dfrac{\mu}{\lambda + \mu}\right)}{\displaystyle\sum_{i=1}^{n} \frac{\lambda_i \mu_i}{\lambda_i + \mu_i}\left[r\left(1_i, \dfrac{\mu}{\lambda + \mu}\right) - r\left(0_i, \dfrac{\mu}{\lambda + \mu}\right)\right]}, \tag{9.32}$$

$$\bar{D} = \frac{\left[1 - r\left(\dfrac{\mu}{\lambda + \mu}\right)\right]\bar{U}}{r\left(\dfrac{\mu}{\lambda + \mu}\right)} \tag{9.33}$$

Also, (9.31) yields the rate at which the system fails.

Remark In establishing the formulas for $\bar{U}$ and $\bar{D}$, we did not make use of the assumption of exponential on and off times and in fact, our derivation is valid and Equations (9.32) and (9.33) hold whenever $\bar{U}$ and $\bar{D}$ are well defined (a sufficient condition is that all on and off distributions are continuous). The quantities λ_i, μ_i, $i = 1, \ldots, n$, will represent, respectively, the reciprocals of the mean lifetimes and mean repair times.

Example 9.32 For a series system,

$$\bar{U} = \frac{\prod_i \dfrac{\mu_i}{\mu_i + \lambda_i}}{\sum_i \dfrac{\lambda_i \mu_i}{\lambda_i + \mu_i} \prod_{j \neq i} \dfrac{\mu_j}{\mu_j + \lambda_j}} = \frac{1}{\sum_i \lambda_i},$$

$$\bar{D} = \frac{1 - \prod_i \dfrac{\mu_i}{\mu_i + \lambda_i}}{\prod_i \dfrac{\mu_i}{\mu_i + \lambda_i}} = \frac{1}{\sum_i \lambda_i}$$

whereas for a parallel system,

$$\bar{U} = \frac{1 - \prod_i \dfrac{\lambda_i}{\mu_i + \lambda_i}}{\sum_i \dfrac{\lambda_i \mu_i}{\lambda_i + \mu_i} \prod_{j \neq i} \dfrac{\lambda_j}{\lambda_j + \mu_j}} = \frac{1 - \prod_i \dfrac{\lambda_i}{\mu_i + \lambda_i}}{\prod_j \dfrac{\lambda_j}{\lambda_j + \mu_j}} = \frac{1}{\sum_i \mu_i}$$

$$\bar{D} = \frac{\left[\prod_i \dfrac{\lambda_i}{\mu_i + \lambda_i} \right]}{1 - \prod_i \dfrac{\lambda_i}{\mu_i + \lambda_i}} \bar{U} = \frac{1}{\sum_i \mu_i}$$

The preceding formulas hold for arbitrary continuous up and down distributions with $1/\lambda_i$ and $1/\mu_i$ denoting respectively the mean up and down times of component i, $i = 1, \ldots, n$. ✦

9.7.1. A Series Model with Suspended Animation

Consider a series consisting of n components, and suppose that whenever a component (and thus the system) goes down, repair begins on that component and each of the other components enters a state of suspended animation. That is, after the down component is repaired, the other components resume operation in exactly the same condition as when the failure occurred. If two or more components go down simultaneously, one of them is arbitrarily chosen as being the failed component and repair on that component begins; the others that went down at the same time are considered to be in a state of suspended animation, and they will instantaneously go down when the repair is completed. We suppose that (not counting any time in suspended animation) the distribution of time that component i functions is F_i with mean u_i, whereas its repair distribution is G_i with mean d_i, $i = 1, \ldots, n$.

To determine the long-run proportion of time this system is working, we reason as follows. To begin, consider the time, call it T, at which the system has been up for a time t. Now, when the system is up, the failure times of component i constitute a renewal process with mean interarrival time u_i. Therefore, it follows that

$$\text{number of failures of } i \text{ in time } T \approx \frac{t}{u_i}$$

As the average repair time of i is d_i, the preceding implies that

$$\text{total repair time of } i \text{ in time } T \approx \frac{t d_i}{u_i}$$

Therefore, in the period of time in which the system has been up for a time t, the total system downtime has approximately been

$$t \sum_{i=1}^{n} d_i/u_i$$

Hence, the proportion of time that the system has been up is approximately

$$\frac{t}{t + t \sum_{i=1}^{n} d_i/u_i}$$

Because this approximation should become exact as we let t become larger, it follows that

$$\text{proportion of time the system is up} = \frac{1}{1 + \Sigma_i d_i/u_i} \tag{9.34}$$

which also shows that

proportion of time the system is down $= 1 - $ proportion of time system is up

$$= \frac{\Sigma_i d_i/u_i}{1 + \Sigma_i d_i/u_i}$$

Moreover, in the time interval from 0 to T, the proportion of the repair time that has been devoted to component i is approximately

$$\frac{t d_i/u_i}{\Sigma_i t d_i/u_i}$$

Thus, in the long run,

$$\text{proportion of down time that is due to component } i = \frac{d_i/u_i}{\Sigma_i d_i/u_i}$$

Multiplying the preceding by the proportion of time the system is down gives

$$\text{proportion of time component } i \text{ is being repaired} = \frac{d_i/u_i}{1 + \Sigma_i d_i/u_i}$$

Also, since component j will be in suspended animation whenever any of the other components is in repair, we see that

$$\text{proportion of time component } j \text{ is in suspended animation} = \frac{\Sigma_{i \neq j} d_i/u_i}{1 + \Sigma_i d_i/u_i}$$

Another quantity of interest is the long-run rate at which the system fails. Since component i fails at rate $1/u_i$ when the system is up, and does not fail when the system is down, it follows that

$$\text{rate at which } i \text{ fails} = \frac{\text{proportion of time system is up}}{u_i}$$

$$= \frac{1/u_i}{1 + \Sigma_i d_i/u_i}$$

Since the system fails when any of its components fail, the preceding yields that

$$\text{rate at which the system fails} = \frac{\Sigma_i 1/u_i}{1 + \Sigma_i d_i/u_i} \tag{9.35}$$

If we partition the time axis into periods when the system is up and those when it is down, we can determine the average length of an up period by noting that if $U(t)$ is the total amount of time that the system is up in the interval $[0, t]$, and if $N(t)$ is the number of failures by time t, then

$$\text{average length of an up period} = \lim_{t \to \infty} \frac{U(t)}{N(t)}$$

$$= \lim_{t \to \infty} \frac{U(t)/t}{N(t)/t}$$

$$= \frac{1}{\Sigma_i 1/u_i}$$

where the final equality used Equations (9.34) and (9.35). Also, in a similar manner it can be shown that

$$\text{average length of a down period} = \frac{\Sigma_i d_i/u_i}{\Sigma_i 1/u_i} \tag{9.36}$$

Exercises

1.   Prove that, for any structure function ϕ,

$$\phi(\mathbf{x}) = x_i \phi(1_i, \mathbf{x}) + (1 - x_i)\phi(0_i, \mathbf{x})$$

where

$$(1_i, \mathbf{x}) = (x_1, \ldots, x_{i-1}, 1, x_{i+1}, \ldots, x_n),$$
$$(0_i, \mathbf{x}) = (x_1, \ldots, x_{i-1}, 0, x_{i+1}, \ldots, x_n)$$

2.   Show that

(a) if $\phi(0, 0, \ldots, 0) = 0$ and $\phi(1, 1, \ldots, 1) = 1$, then

$$\min x_i \leqslant \varphi(\mathbf{x}) \leqslant \max x_i$$

(b) $\phi(\max(\mathbf{x}, \mathbf{y})) \geqslant \max(\phi(\mathbf{x}), \phi(\mathbf{y}))$
(c) $\phi(\min(\mathbf{x}, \mathbf{y})) \leqslant \min(\phi(\mathbf{x}), \phi(\mathbf{y}))$

3.   For any structure function ϕ, we define the dual structure ϕ^D by

$$\phi^D(\mathbf{x}) = 1 - \phi(1 - \mathbf{x})$$

(a) Show that the dual of a parallel (series) system is a series (parallel) system.
(b) Show that the dual of a dual structure is the original structure.
(c) What is the dual of a k-out-of-n structure?
(d) Show that a minimal path (cut) set of the dual system is a minimal cut (path) set of the original structure.

*4.   Write the structure function corresponding to the following:

(a)

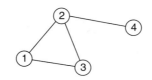

Figure 9.16.

(b)

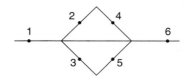

Figure 9.17.

(c)

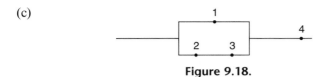

Figure 9.18.

5. Find the minimal path and minimal cut sets for:

(a)

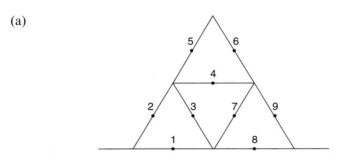

Figure 9.19.

(b)

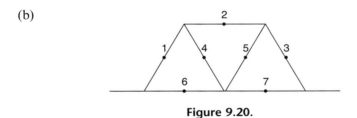

Figure 9.20.

*6. The minimal path sets are $\{1, 2, 4\}$, $\{1, 3, 5\}$, and $\{5, 6\}$. Give the minimal cut sets.

7. The minimal cut sets are $\{1, 2, 3\}$, $\{2, 3, 4\}$, and $\{3, 5\}$. What are the minimal path sets?

8. Give the minimal path sets and the minimal cut sets for the structure given by Figure 9.21 on page 544.

9. Component i is said to be *relevant* to the system if for some state vector $\mathbf{x}$,

$$\phi(1_i, \mathbf{x}) = 1, \qquad \phi(0_i, \mathbf{x}) = 0$$

Otherwise, it is said to be *irrelevant*.

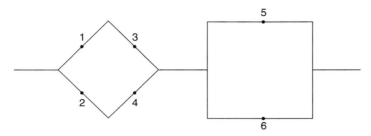

Figure 9.21.

(a) Explain in words what it means for a component to be irrelevant.

(b) Let $A_1, \ldots, A_s$ be the minimal path sets of a system, and let S denote the set of components. Show that $S = \bigcup_{i=1}^{s} A_i$ if and only if all components are relevant.

(c) Let $C_1, \ldots, C_k$ denote the minimal cut sets. Show that $S = \bigcup_{i=1}^{k} C_i$ if and only if all components are relevant.

10. Let t_i denote the time of failure of the ith component; let $\tau_\phi(t)$ denote the time to failure of the system ϕ as a function of the vector $\mathbf{t} = (t_1, \ldots, t_n)$. Show that

$$\max_{1 \leq j \leq s} \min_{i \in A_j} t_i = \tau_\phi(\mathbf{t}) = \min_{1 \leq j \leq k} \max_{i \in C_j} t_i$$

where $C_1, \ldots, C_k$ are the minimal cut sets, and $A_1, \ldots, A_s$ the minimal path sets.

11. Give the reliability function of the structure of Exercise 8.

*12. Give the minimal path sets and the reliability function for the structure in Figure 9.22.

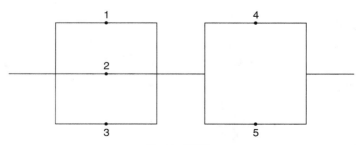

Figure 9.22.

13. Let $r(\mathbf{p})$ be the reliability function. Show that

$$r(\mathbf{p}) = p_i r(1_i, \mathbf{p}) + (1 - p_i)r(0_i, \mathbf{p})$$

14. Compute the reliability function of the bridge system (see Figure 9.11) by conditioning upon whether or not component 3 is working.

15. Compute upper and lower bounds of the reliability function (using Method 2) for the systems given in Exercise 4, and compare them with the exact values when $p_i \equiv \frac{1}{2}$.

16. Compute the upper and lower bounds of $r(\mathbf{p})$ using both methods for the

    (a) two-out-of-three system and
    (b) two-out-of-four system.
    (c) Compare these bounds with the exact reliability when
        (i) $p_i \equiv 0.5$
        (ii) $p_i \equiv 0.8$
        (iii) $p_i \equiv 0.2$

*17. Let N be a nonnegative, integer-valued random variable. Show that

$$P\{N > 0\} \geqslant \frac{(E[N])^2}{E[N^2]}$$

and explain how this inequality can be used to derive additional bounds on a reliability function.

Hint:

$$E[N^2] = E[N^2 \mid N > 0]P\{N > 0\} \qquad \text{(Why?)}$$
$$\geqslant (E[N \mid N > 0])^2 P\{N > 0\} \qquad \text{(Why?)}$$

Now multiply both sides by $P\{N > 0\}$.

18. Consider a structure in which the minimal path sets are $\{1, 2, 3\}$ and $\{3, 4, 5\}$.

    (a) What are the minimal cut sets?
    (b) If the component lifetimes are independent uniform $(0, 1)$ random variables, determine the probability that the system life will be less than $\frac{1}{2}$.

19. Let $X_1, X_2, \ldots, X_n$ denote independent and identically distributed random variables and define the order statistics $X_{(1)}, \ldots, X_{(n)}$ by

$$X_{(i)} \equiv i\text{th smallest of } X_1, \ldots, X_n$$

Show that if the distribution of X_j is IFR, then so is the distribution of $X_{(i)}$.

Hint: Relate this to one of the examples of this chapter.

20. Let F be a continuous distribution function. For some positive α, define the distribution function G by

$$\bar{G}(t) = (\bar{F}(t))^{\alpha}$$

Find the relationship between $\lambda_G(t)$ and $\lambda_F(t)$, the respective failure rate functions of G and F.

21. Consider the following four structures:

(i)

(ii)

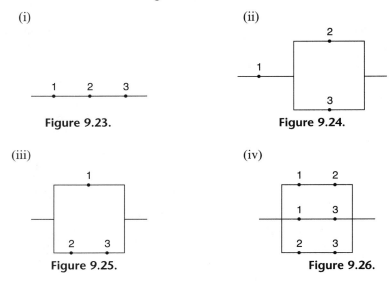

Figure 9.23.

Figure 9.24.

(iii)

(iv)

Figure 9.25.

Figure 9.26.

Let F_1, F_2, and F_3 be the corresponding component failure distributions; each of which is assumed to be IFR (increasing failure rate). Let F be the system failure distribution. All components are independent.

(a) For which structures is F necessarily IFR if $F_1 = F_2 = F_3$? Give reasons.

(b) For which structures is F necessarily IFR if $F_2 = F_3$? Give reasons.

(c) For which structures is F necessarily IFR if $F_1 \neq F_2 \neq F_3$? Give reasons.

*22. Let X denote the lifetime of an item. Suppose the item has reached the age of t. Let X_t denote its remaining life and define

$$\bar{F}_t(a) = P\{X_t > a\}$$

In words, $\bar{F}_t(a)$ is the probability that a t-year old item survives an additional time a. Show that

(a) $\bar{F}_t(a) = \bar{F}(t + a)/\bar{F}(t)$ where F is the distribution function of X.

(b) Another definition of IFR is to say that F is IFR if $\bar{F}_t(a)$ decreases in t, for all a. Show that this definition is equivalent to the one given in the text when F has a density.

23. Show that if each (independent) component of a series system has an IFR distribution, then the system lifetime is itself IFR by

(a) showing that

$$\lambda_F(t) = \sum_i \lambda_i(t)$$

where $\lambda_F(t)$ is the failure rate function of the system; and $\lambda_i(t)$ the failure rate function of the lifetime of component i.
(b) using the definition of IFR given in Exercise 22.

24. Show that if F is IFR, then it is also IFRA, and show by counter-example that the reverse is not true.

*25. We say that ζ is a p-percentile of the distribution F if $F(\zeta) = p$. Show that if ζ is a p-percentile of the IFRA distribution F, then

$$\bar{F}(x) \leqslant e^{-\theta x}, \qquad x \geqslant \zeta$$
$$\bar{F}(x) \geqslant e^{-\theta x}, \qquad x \leqslant \zeta$$

where

$$\theta = \frac{-\log(1 - p)}{\zeta}$$

26. Prove Lemma 9.3.

Hint: Let $x = y + \delta$. Note that $f(t) = t^\alpha$ is a concave function when $0 \leqslant \alpha \leqslant 1$, and use the fact that for a concave function $f(t + h) - f(t)$ is decreasing in t.

27. Let $r(p) = r(p, p, \ldots, p)$. Show that if $r(p_0) = p_0$, then

$$r(p) \geqslant p \qquad \text{for } p \geqslant p_0$$
$$r(p) \leqslant p \qquad \text{for } p \leqslant p_0$$

Hint: Use Proposition 9.2.

28. Find the mean lifetime of a series system of two components when the component lifetimes are respectively uniform on $(0, 1)$ and uniform on $(0, 2)$. Repeat for a parallel system.

29. Show that the mean lifetime of a parallel system of two components is

$$\frac{1}{\mu_1 + \mu_2} + \frac{\mu_1}{(\mu_1 + \mu_2)\mu_2} + \frac{\mu_2}{(\mu_1 + \mu_2)\mu_1}$$

when the first component is exponentially distributed with mean $1/\mu_1$ and the second is exponential with mean $1/\mu_2$.

*30.   Compute the expected system lifetime of a three-out-of-four system when the first two component lifetimes are uniform on $(0, 1)$ and the second two are uniform on $(0, 2)$.

31.   Show that the variance of the lifetime of a k-out-of-n system of components, each of whose lifetimes is exponential with mean θ, is given by

$$\theta^2 \sum_{i=k}^{n} \frac{1}{i^2}$$

32.   In Section 9.6.1 show that the expected number of X_i that exceed c^* is equal to 1.

33.   Let X_i be an exponential random variable with mean $8 + 2i$, for $i = 1, 2, 3$. Use the results of Section 9.6.1 to obtain an upper bound on $E[\max X_i]$, and then compare this with the exact result when the X_i are independent.

34.   For the model of Section 9.7, compute for a k-out-of-n structure (i) the average up time, (ii) the average down time, and (iii) the system failure rate.

35.   Prove the combinatorial identity

$$\binom{n-1}{i-1} = \binom{n}{i} - \binom{n}{i+1} + \cdots \pm \binom{n}{n}, \qquad i \leq n$$

(a) by induction on i
(b) by a backwards induction argument on i — that is, prove it first for $i = n$, then assume it for $i = k$ and show that this implies that it is true for $i = k - 1$.

36.   Verify Equation (9.36).

References

1. R. E. Barlow and F. Proschan, "Statistical Theory of Reliability and Life Testing," Holt, New York, 1975.
2. H. Frank and I. Frisch, "Communication, Transmission, and Transportation Network," Addison-Wesley, Reading, Massachusetts, 1971.
3. I. B. Gertsbakh, "Statistical Reliability Theory," Marcel Dekker, New York and Basel, 1989.

Brownian Motion and Stationary Processes

10

10.1. Brownian Motion

Let us start by considering the symmetric random walk which in each time unit is equally likely to take a unit step either to the left or to the right. That is, it is a Markov chain with $P_{i,i+1} = \frac{1}{2} = P_{i,i-1}$, $i = 0, \pm 1, \ldots$. Now suppose that we speed up this process by taking smaller and smaller steps in smaller and smaller time intervals. If we now go to the limit in the right manner what we obtain is Brownian motion.

More precisely, suppose that each Δt time unit we take a step of size Δx either to the left or the right with equal probabilities. If we let $X(t)$ denote the position at time t then

$$X(t) = \Delta x (X_1 + \cdots + X_{[t/\Delta t]}) \tag{10.1}$$

where

$$X_i = \begin{cases} +1, & \text{if the } i\text{th step of length } \Delta x \text{ is to the right} \\ -1, & \text{if it is to the left} \end{cases}$$

and $[t/\Delta t]$ is the largest integer less than or equal to $t/\Delta t$, and where the X_i are assumed independent with

$$P\{X_i = 1\} = P\{X_i = -1\} = \frac{1}{2}$$

As $E[X_i] = 0$, $\text{Var}(X_i) = E[X_i^2] = 1$, we see from Equation (10.1) that

$$E[X(t)] = 0$$

$$\text{Var}(X(t)) = (\Delta x)^2 \left[\frac{t}{\Delta t} \right] \tag{10.2}$$

We shall now let Δx and Δt go to 0. However, we must do it in a way such that the resulting limiting process is nontrivial (for instance, if we let

$\Delta x = \Delta t$ and let $\Delta t \to 0$, then from the preceding we see that $E[X(t)]$ and $\text{Var}(X(t))$ would both converge to 0 and thus $X(t)$ would equal 0 with probability 1). If we let $\Delta x = \sigma \sqrt{\Delta t}$ for some positive constant σ then from Equation (10.2) we see that as $\Delta t \to 0$

$$E[X(t)] = 0,$$

$$\text{Var}(X(t)) \to \sigma^2 t$$

We now list some intuitive properties of this limiting process obtained by taking $\Delta x = \sigma \sqrt{\Delta t}$ and then letting $\Delta t \to 0$. From Equation (10.1) and the central limit theorem the following seems reasonable:

(i) $X(t)$ is normal with mean 0 and variance $\sigma^2 t$. In addition, because the changes of value of the random walk in nonoverlapping time intervals are independent, we have
(ii) $\{X(t),\ t \geq 0\}$ has independent increments, in that for all $t_1 < t_2 < \cdots < t_n$

$$X(t_n) - X(t_{n-1}), X(t_{n-1}) - X(t_{n-2}), \ldots, X(t_2) - X(t_1), X(t_1)$$

are independent. Finally, because the distribution of the change in position of the random walk over any time interval depends only on the length of that interval, it would appear that
(iii) $\{X(t),\ t \geq 0\}$ has stationary increments, in that the distribution of $X(t + s) - X(t)$ does not depend on t. We are now ready for the following formal definition.

Definition 10.1 A stochastic process $\{X(t),\ t \geq 0\}$ is said to be a *Brownian motion* process if

(i) $X(0) = 0$;
(ii) $\{X(t),\ t \geq 0\}$ has stationary and independent increments;
(iii) for every $t > 0$, $X(t)$ is normally distributed with mean 0 and variance $\sigma^2 t$.

The Brownian motion process, sometimes called the Wiener process, is one of the most useful stochastic processes in applied probability theory. It originated in physics as a description of Brownian motion. This phenomenon, named after the English botanist Robert Brown who discovered it, is the motion exhibited by a small particle which is totally immersed in a liquid or gas. Since then, the process has been used beneficially in such areas as statistical testing of goodness of fit, analyzing the price levels on the stock market, and quantum mechanics.

The first explanation of the phenomenon of Brownian motion was given by Einstein in 1905. He showed that Brownian motion could be explained by assuming that the immersed particle was continually being subjected to bombardment by the molecules of the surrounding medium. However, the preceding concise definition of this stochastic process underlying Brownian motion was given by Wiener in a series of papers originating in 1918.

When $\sigma = 1$, the process is called *standard Brownian motion*. Because any Brownian motion can be converted to the standard process by letting $B(t) = X(t)/\sigma$ we shall, unless otherwise stated, suppose throughout this chapter that $\sigma = 1$.

The interpretation of Brownian motion as the limit of the random walks [Equation (10.1)] suggests that $X(t)$ should be a continuous function of t. This turns out to be the case, and it may be proven that, with probability 1, $X(t)$ is indeed a continuous function of t. This fact is quite deep, and no proof shall be attempted.

As $X(t)$ is normal with mean 0 and variance t, its density function is given by

$$f_t(x) = \frac{1}{\sqrt{2\pi t}} \, e^{-x^2/2t}$$

To obtain the joint density function of $X(t_1)$, $X(t_2), \ldots, X(t_n)$ for $t_1 < \cdots < t_n$, note first that the set of equalities

$$X(t_1) = x_1,$$
$$X(t_2) = x_2,$$
$$\vdots$$
$$X(t_n) = x_n$$

is equivalent to

$$X(t_1) = x_1,$$
$$X(t_2) - X(t_1) = x_2 - x_1,$$
$$\vdots$$
$$X(t_n) - X(t_{n-1}) = x_n - x_{n-1}$$

However, by the independent increment assumption it follows that $X(t_1)$, $X(t_2) - X(t_1), \ldots, X(t_n) - X(t_{n-1})$, are independent and, by the stationary increment assumption, that $X(t_k) - X(t_{k-1})$ is normal with mean 0 and variance $t_k - t_{k-1}$. Hence, the joint density of $X(t_1), \ldots, X(t_n)$ is given by

$$
\begin{aligned}
f(x_1, x_2, \ldots, x_n) &= f_{t_1}(x_1) f_{t_2-t_1}(x_2 - x_1) \cdots f_{t_n-t_{n-1}}(x_n - x_{n-1}) \\[2mm]
&= \frac{\exp\left\{-\dfrac{1}{2}\left[\dfrac{x_1^2}{t_1} + \dfrac{(x_2 - x_1)^2}{t_2 - t_1} + \cdots + \dfrac{(x_n - x_{n-1})^2}{t_n - t_{n-1}}\right]\right\}}{(2\pi)^{n/2}[t_1(t_2 - t_1)\cdots(t_n - t_{n-1})]^{1/2}}
\end{aligned}
\tag{10.3}
$$

From this equation, we can compute in principle any desired probabilities. For instance, suppose we require the conditional distribution of $X(s)$ given that $X(t) = B$ where $s < t$. The conditional density is

$$f_{s|t}(x \mid B) = \frac{f_s(x) f_{t-s}(B - x)}{f_t(B)}$$

$$= K_1 \exp\{-x^2/2s - (B - x)^2/2(t - s)\}$$

$$= K_2 \exp\left\{-x^2 \left(\frac{1}{2s} + \frac{1}{2(t - s)}\right) + \frac{Bx}{t - s}\right\}$$

$$= K_2 \exp\left\{-\frac{t}{2s(t - s)} \left(x^2 - 2\frac{sB}{t} x\right)\right\}$$

$$= K_3 \exp\left\{-\frac{(x - Bs/t)^2}{2s(t - s)/t}\right\}$$

where K_1, K_2, and K_3 do not depend on x. Hence, we see from the above that the conditional distribution of $X(s)$ given that $X(t) = B$ is, for $s < t$, normal with mean and variance given by

$$E[X(s) \mid X(t) = B] = \frac{s}{t} B,$$

$$\text{Var}[X(s) \mid X(t) = B] = \frac{s}{t} (t - s) \tag{10.4}$$

Example 10.1   In a bicycle race between two competitors, let $Y(t)$ denote the amount of time (in seconds) by which the racer that started in the inside position is ahead when $100t$ percent of the race has been completed, and suppose that $\{Y(t),\ 0 \leqslant t \leqslant 1\}$ can be effectively modeled as a Brownian motion process with variance parameter σ^2.

(a) If the inside racer is leading by σ seconds at the midpoint of the race, what is the probability that she is the winner?

(b) If the inside racer wins the race by a margin of σ seconds, what is the probability that she was ahead at the midpoint?

Solution:

(a) $P\{Y(1) > 0 \mid Y(1/2) = \sigma\}$

$\quad = P\{Y(1) - Y(1/2) > -\sigma \mid Y(1/2) = \sigma\}$

$\quad = P\{Y(1) - Y(1/2) > -\sigma\}$ by independent increments

$\quad = P\{Y(1/2) > -\sigma\}$ by stationary increments

$\quad = P\left\{\frac{Y(1/2)}{\sigma/\sqrt{2}} > -\sqrt{2}\right\}$

$\quad = \Phi(\sqrt{2})$

$\quad \approx 0.9213$

where $\Phi(x) = P\{N(0, 1) \leqslant x\}$ is the standard normal distribution function.

(b) Because we must compute $P\{Y(1/2) > 0 \,|\, Y(1) = \sigma\}$, let us first determine the conditional distribution of $Y(s)$ given that $Y(t) = C$, when $s < t$. Now, since $\{X(t), t \geqslant 0\}$ is standard Brownian motion when $X(t) = Y(t)/\sigma$, we obtain from Equation (10.4) that the conditional distribution of $X(s)$, given that $X(t) = C/\sigma$, is normal with mean $sC/t\sigma$ and variance $s(t - s)/t$. Hence, the conditional distribution of $Y(s) = \sigma X(s)$ given that $Y(t) = C$ is normal with mean sC/t and variance $\sigma^2 s(t - s)/t$. Hence,

$$P\{Y(1/2) > 0 \,|\, Y(1) = \sigma\} = P\{N(\sigma/2,\, \sigma^2/4) > 0\}$$

$$= \Phi(1)$$

$$\approx 0.8413 \quad \blacklozenge$$

10.2. Hitting Times, Maximum Variable, and the Gambler's Ruin Problem

Let T_a denote the first time the Brownian motion process hits a. When $a > 0$ we will compute $P\{T_a \leqslant t\}$ by considering $P\{X(t) \geqslant a\}$ and conditioning on whether or not $T_a \leqslant t$. This gives

$$P\{X(t) \geqslant a\} = P\{X(t) \geqslant a \,|\, T_a \leqslant t\} P\{T_a \leqslant t\}$$
$$+ P\{X(t) \geqslant a \,|\, T_a > t\} P\{T_a > t\} \qquad (10.5)$$

Now if $T_a \leqslant t$, then the process hits a at some point in $[0, t]$ and, by symmetry, it is just as likely to be above a or below a at time t. That is

$$P\{X(t) \geqslant a \,|\, T_a \leqslant t\} = \tfrac{1}{2}$$

As the second right-hand term of Equation (10.5) is clearly equal to 0 (since, by continuity, the process value cannot be greater than a without having yet hit a), we see that

$$P\{T_a \leqslant t\} = 2P\{X(t) \geqslant a\}$$

$$= \frac{2}{\sqrt{2\pi t}} \int_a^\infty e^{-x^2/2t} \, dx$$

$$= \frac{2}{\sqrt{2\pi}} \int_{a/\sqrt{t}}^\infty e^{-y^2/2} \, dy, \qquad a > 0 \qquad (10.6)$$

For $a < 0$, the distribution of T_a is, by symmetry, the same as that of T_{-a}.

Hence, from Equation (10.6) we obtain

$$P\{T_a \leqslant t\} = \frac{2}{\sqrt{2\pi}} \int_{|a|/\sqrt{t}}^{\infty} e^{-y^2/2} \, dy \tag{10.7}$$

Another random variable of interest is the maximum value the process attains in $[0, t]$. Its distribution is obtained as follows: For $a > 0$

$$P\left\{\max_{0 \leqslant s \leqslant t} X(s) \geqslant a\right\} = P\{T_a \leqslant t\} \qquad \text{by continuity}$$

$$= 2P\{X(t) \geqslant a\} \qquad \text{from (10.6)}$$

$$= \frac{2}{\sqrt{2\pi}} \int_{a/\sqrt{t}}^{\infty} e^{-y^2/2} \, dy$$

Let us now consider the probability that Brownian motion hits A before $-B$ where $A > 0$, $B > 0$. To compute this we shall make use of the interpretation of Brownian motion as being a limit of the symmetric random walk. To start let us recall from the results of the gambler's ruin problem (see Section 4.5.1) that the probability that the symmetric random walk goes up A before going down B when each step is equally likely to be either up or down a distance Δx is [by Equation (4.12) with $N = (A + B)/\Delta x$, $i = B/\Delta x$] equal to $B \Delta x/(A + B) \Delta x = B/(A + B)$.

Hence, upon letting $\Delta x \to 0$, we see that

$$P\{\text{up } A \text{ before down } B\} = \frac{B}{A + B}$$

10.3.   Variations on Brownian Motion

10.3.1.   Brownian Motion with Drift

We say that $\{X(t), \ t \geqslant 0\}$ is a Brownian motion process with drift coefficient μ and variance parameter σ^2 if

   (i) $X(0) = 0$;
   (ii) $\{X(t), \ t \geqslant 0\}$ has stationary and independent increments;
   (iii) $X(t)$ is normally distributed with mean μt and variance $t\sigma^2$.

An equivalent definition is to let $\{B(t), \ t \geqslant 0\}$ be standard Brownian motion and then define

$$X(t) = \sigma B(t) + \mu t$$

10.3.2. Geometric Brownian Motion

If $\{Y(t),\ t \geq 0\}$ is a Brownian motion process with drift coefficient μ and variance parameter σ^2, then the process $\{X(t),\ t \geq 0\}$ defined by

$$X(t) = e^{Y(t)}$$

is called *geometric Brownian motion.*

For a geometric Brownian motion process $\{X(t)\}$, let us compute the expected value of the process at time t given the history of the process up to time s. That is, for $s < t$, consider $E[X(t) \mid X(u),\ 0 \leq u \leq s]$. Now,

$$
\begin{aligned}
E[X(t) \mid X(u),\ 0 \leq u \leq s] &= E[e^{Y(t)} \mid Y(u),\ 0 \leq u \leq s] \\
&= E[e^{Y(s) + Y(t) - Y(s)} \mid Y(u),\ 0 \leq u \leq s] \\
&= e^{Y(s)} E[e^{Y(t) - Y(s)} \mid Y(u),\ 0 \leq u \leq s] \\
&= X(s) E[e^{Y(t) - Y(s)}]
\end{aligned}
$$

where the next to last equality follows from the fact that $Y(s)$ is given, and the last equality from the independent increment property of Brownian motion. Now, the moment generating function of a normal random variable W is given by

$$E[e^{aW}] = e^{aE[W] + a^2 \operatorname{Var}(W)/2}$$

Hence, since $Y(t) - Y(s)$ is normal with means $\mu(t - s)$ and variance $(t - s)\sigma^2$, it follows by setting $a = 1$ that

$$E[e^{Y(t) - Y(s)}] = e^{\mu(t - s) + (t - s)\sigma^2/2}$$

Thus, we obtain

$$E[X(t) \mid X(u),\ 0 \leq u \leq s] = X(s) e^{(t - s)(\mu + \sigma^2/2)} \tag{10.8}$$

Geometric Brownian motion is useful in the modeling of stock prices over time when one feels that the percentage changes are independent and identically distributed. For instance, suppose that X_n is the price of some stock at time n. Then, it might be reasonable to suppose that X_n/X_{n-1}, $n \geq 1$, are independent and identically distributed. Let

$$Y_n = X_n/X_{n-1}$$

and so

$$X_n = Y_n X_{n-1}$$

Iterating this equality gives

$$
\begin{aligned}
X_n &= Y_n Y_{n-1} X_{n-2} \\
&= Y_n Y_{n-1} Y_{n-2} X_{n-3} \\
&\ \ \vdots \\
&= Y_n Y_{n-1} \cdots Y_1 X_0
\end{aligned}
$$

Thus,

$$\log(X_n) = \sum_{i=1}^{n} \log(Y_i) + \log(X_0)$$

Since $\log(Y_i)$, $i \geq 1$, are independent and identically distributed, $\{\log(X_n)\}$ will, when suitably normalized, approximately be Brownian motion with a drift, and so $\{X_n\}$ will be approximately geometric Brownian motion.

10.4. Pricing Stock Options

10.4.1. An Example in Options Pricing

In situations in which money is to be received or paid out in differing time periods, one must take into account the time value of money. That is, to be given the amount v a time t in the future is not worth as much as being given v immediately. The reason for this is that if one was immediately given v, then it could be loaned out with interest and so be worth more than v at time t. To take this into account, we will suppose that the time 0 value, also called the *present value*, of the amount v to be earned at time t is $ve^{-\alpha t}$. The quantity α is often called the discount factor. In economic terms, the assumption of the discount function $e^{-\alpha t}$ is equivalent to the assumption that one can earn interest at a continuously compounded rate of 100α percent per unit time.

We will now consider a simple model for pricing an option to purchase a stock at a future time at a fixed price.

Suppose the present price of a stock is $100 per unit share, and suppose we know that after one time period it will be, in present value dollars, either $200 or $50 (see Figure 10.1). It should be noted that the prices at time 1 are the present value (or time 0) prices. That is, if the discount factor is α, then the actual possible prices at time 1 are either $200e^{\alpha}$ or $50e^{\alpha}$. To keep the notation simple, we will suppose that all prices given are time 0 prices.

Suppose that for any y, at a cost of cy, you can purchase at time 0 the option to buy y shares of the stock at time 1 at a (time 0) cost of $150 per

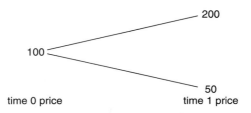

Figure 10.1.

share. Thus, for instance, if you do purchase this option and the stock rises to $200, then you would exercise the option at time 1 and realize a gain of $200 − 150 = 50 for each of the y option units purchased. On the other hand, if the price at time 1 was $50, then the option would be worthless at time 1. In addition, at a cost of $100x$ you can purchase x units of the stock at time 0, and this will be worth either $200x$ or $50x$ at time 1.

We will suppose that both x or y can be either positive or negative (or zero). That is, you can either buy or sell both the stock and the option. For instance, if x were negative then you would be selling $−x$ shares of the stock, yielding you a return of $−100x$, and you would then be responsible for buying $−x$ shares of the stock at time 1 at a cost of either $200 or $50 per share.

We are interested in determining the appropriate value of c, the unit cost of an option. Specifically, we will show that unless $c = 50/3$ there will be a combination of purchases that will always result in a positive gain.

To show this, suppose that at time 0 we

buy x units of stock, and

buy y units of options

where x and y (which can be either positive or negative) are to be determined. The value of our holding at time 1 depends on the price of the stock at that time; and it is given by the following

$$\text{value} = \begin{cases} 200x + 50y, & \text{if price is } 200 \\ 50x, & \text{if price is } 50 \end{cases}$$

The preceding formula follows by noting that if the price is 200 then the x units of the stock are worth $200x$, and the y units of the option to buy the stock at a unit price of 150 are worth $(200 − 150)y$. On the other hand, if the stock price is 50, then the x units are worth $50x$ and the y units of the option are worthless. Now, suppose we choose y to be such that the preceding value is the same no matter what the price at time 1. That is, we choose y so that

$$200x + 50y = 50x$$

or

$$y = -3x$$

(Note that y has the opposite sign of x, and so if x is positive and as a result x units of the stock are purchased at time 0, then $3x$ units of stock options are also *sold* at that time. Similarly, if x is negative, then $−x$ units of stock are sold and $−3x$ units of stock options are purchased at time 0.)

Thus, with $y = -3x$, the value of our holding at time 1 is

$$\text{value} = 50x$$

Since the original cost of purchasing x units of the stock and $-3x$ units of options is

$$\text{original cost} = 100x - 3xc,$$

we see that our gain on the transaction is

$$\text{gain} = 50x - (100x - 3xc) = x(3c - 50)$$

Thus, if $3c - 50$, then the gain is 0; on the other hand if $3c \neq 50$, we can guarantee a positive gain (no matter what the price of the stock at time 1) by letting x be positive when $3c > 50$ and letting it be negative when $3c < 50$.

For instance, if the unit cost per option is $c = 20$, then purchasing 1 unit of the stock $(x = 1)$ and simultaneously selling 3 units of the option $(y = -3)$ initially costs us $100 - 60 = 40$. However, the value of this holding at time 1 is 50 whether the stock goes up to 200 or down to 50. Thus, a guaranteed profit of 10 is attained. Similarly, if the unit cost per option is $c = 15$, then selling 1 unit of the stock $(x = -1)$ and buying 3 units of the option $(y = 3)$ leads to an initial gain of $100 - 45 = 55$. On the other hand, the value of this holding at time 1 is -50. Thus, a guaranteed profit of 5 is attained.

A sure win betting scheme is called an *arbitrage*. Thus, as we have just seen, the only option cost c that does not result in an arbitrage is $c = 50/3$.

10.4.2.   The Arbitrage Theorem

Consider an experiment whose set of possible outcomes is $S = \{1, 2, \ldots, m\}$. Suppose that n wagers are available. If the amount x is bet on wager i, then the return $xr_i(j)$ is earned if the outcome of the experiment is j. In other words, $r_i(\cdot)$ is the return function for a unit bet on wager i. The amount bet on a wager is allowed to be either positive or negative or zero.

A betting scheme is a vector $\mathbf{x} = (x_1, \ldots, x_n)$ with the interpretation that x_1 is bet on wager 1, x_2 on wager 2, ..., and x_n on wager n. If the outcome of the experiment is j, then the return from the betting scheme $\mathbf{x}$ is

$$\text{return from } \mathbf{x} = \sum_{i=1}^{n} x_i r_i(j)$$

The following theorem states that either there exists a probability vector $\mathbf{p} = (p_1, \ldots, p_m)$ on the set of possible outcomes of the experiment under which each of the wagers has expected return 0, or else there is a betting scheme that guarantees a positive win.

Theorem 10.1 (The Arbitrage Theorem): Exactly one of the following is true: Either

(i) There exists a probability vector $\mathbf{p} = (p_1, \ldots, p_m)$ for which

$$\sum_{j=1}^{m} p_j r_i(j) = 0, \qquad \text{for all } i = 1, \ldots, n$$

or

(ii) there exists a betting scheme $\mathbf{x} = (x_1, \ldots, x_n)$ for which

$$\sum_{i=1}^{n} x_i r_i(j) > 0, \qquad \text{for all } j = 1, \ldots, m$$

In other words, if X is the outcome of the experiment, then the arbitrage theorem states that either there is a probability vector $\mathbf{p}$ for X such that

$$E_{\mathbf{p}}[r_i(X)] = 0, \qquad \text{for all } i = 1, \ldots, n$$

or else there is a betting scheme that leads to a sure win.

Remark This theorem is a consequence of the (linear algebra) theorem of the separating hyperplane, which is often used as a mechanism to prove the duality theorem of linear programming.

The theory of linear programming can be used to determine a betting strategy that guarantees the greatest return. Suppose that the absolute value of the amount bet on each wager must be less than or equal to 1. To determine the vector $\mathbf{x}$ that yields the greatest guaranteed win — call this win v — we need to choose $\mathbf{x}$ and v so as to maximize v, subject to the constraints

$$\sum_{i=1}^{n} x_i r_i(j) \geqslant v, \qquad \text{for } j = 1, \ldots, m$$

$$-1 \leqslant x_i \leqslant 1, \qquad i = 1, \ldots, n$$

This optimization problem is a linear program and can be solved by standard techniques (such as by using the simplex algorithm). The arbitrage theorem yields that the optimal v will be positive unless there is a probability vector $\mathbf{p}$ for which $\sum_{j=1}^{m} p_j r_i(j) = 0$ for all $i = 1, \ldots, n$.

Example 10.2 In some situations, the only type of wagers allowed are to choose one of the outcomes i, $i = 1, \ldots, m$, and bet that i is the outcome of the experiment. The return from such a bet is often quoted in terms of "odds." If the odds for outcome i are o_i (often written as "o_i to 1") then a 1 unit bet will return o_i if the outcome of the experiment is i and will return

−1 otherwise. That is,

$$r_i(j) = \begin{cases} o_i, & \text{if } j = i \\ -1 & \text{otherwise} \end{cases}$$

Suppose the odds $o_1, \ldots, o_m$ are posted. In order for there not to be a sure win there must be a probability vector $\mathbf{p} = (p_1, \ldots, p_m)$ such that

$$0 \equiv E_{\mathbf{p}}[r_i(X)] = o_i p_i - (1 - p_i)$$

That is, we must have

$$p_i = \frac{1}{1 + o_i}$$

Since the p_i must sum to 1, this means that the condition for there not to be an arbitrage is that

$$\sum_{i=1}^{m} (1 + o_i)^{-1} = 1$$

Thus, if the posted odds are such that $\sum_i (1 + o_i)^{-1} \neq 1$, then a sure win is possible. For instance, suppose there are three possible outcomes and the odds are as follows:

Outcome	Odds
1	1
2	2
3	3

That is, the odds for outcome 1 are $1-1$, the odds for outcome 2 are $2-1$, and that for outcome 3 are $3-1$. Since

$$\tfrac{1}{2} + \tfrac{1}{3} + \tfrac{1}{4} > 1$$

a sure win is possible. One possibility is to bet -1 on outcome 1 (and so you either win 1 if the outcome is not 1 and lose 1 if the outcome is 1) and bet -0.7 on outcome 2, and -0.5 on outcome 3. If the experiment results in outcome 1, then we win $-1 + 0.7 + 0.5 = 0.2$; if it results in outcome 2, then we win $1 - 1.4 + 0.5 = 0.1$; if it results in outcome 3, then we win $1 + 0.7 - 1.5 = 0.2$. Hence, in all cases we win a positive amount. ✦

Remark   If $\sum_i (1 + o_i)^{-1} \neq 1$, then the betting scheme

$$x_i = \frac{(1 + o_i)^{-1}}{1 - \sum_i (1 - o_i)^{-1}}, \qquad i = 1, \ldots, n$$

will always yield a gain of exactly 1.

Example 10.3 Let us reconsider the option pricing example of the previous section, where the initial price of a stock is 100 and the present value of the price at time 1 is either 200 or 50. At a cost of c per share we can purchase at time 0 the option to buy the stock at time 1 at a present value price of 150 per share. The problem is to set the value of c so that no sure win is possible.

In the context of this section, the outcome of the experiment is the value of the stock at time 1. Thus, there are two possible outcomes. There are also two different wagers: to buy (or sell) the stock, and to buy (or sell) the option. By the arbitrage theorem, there will be no sure win if there is a probability vector $(p, 1 - p)$ that makes the expected return under both wagers equal to 0.

Now, the return from purchasing 1 unit of the stock is

$$\text{return} = \begin{cases} 200 - 100 = 100, & \text{if the price is 200 at time 1} \\ 50 - 100 = -50, & \text{if the price is 50 at time 1} \end{cases}$$

Hence, if p is the probability that the price is 200 at time 1, then

$$E[\text{return}] = 100p - 50(1 - p)$$

Setting this equal to 0 yields that

$$p = \tfrac{1}{3}$$

That is, the only probability vector $(p, 1 - p)$ for which wager 1 yields an expected return 0 is the vector $(\tfrac{1}{3}, \tfrac{2}{3})$.

Now, the return from purchasing one share of the option is

$$\text{return} = \begin{cases} 50 - c, & \text{if price is 200} \\ -c, & \text{if price is 50} \end{cases}$$

Hence, the expected return when $p = \tfrac{1}{3}$ is

$$E[\text{return}] = (50 - c)\tfrac{1}{3} - c\tfrac{2}{3}$$

$$= \tfrac{50}{3} - c$$

Thus, it follows from the arbitrage theorem that the only value of c for which there will not be a sure win is $c = \tfrac{50}{3}$, which verifies the result of section 10.4.1. ✦

10.4.3. The Black-Scholes Option Pricing Formula

Suppose the present price of a stock is $X(0) = x_0$, and let $X(t)$ denote its price at time t. Suppose we are interested in the stock over the time interval 0 to T. Assume that the discount factor is α (equivalently, the interest rate is 100α percent compounded continuously), and so the present value of the stock price at time t is $e^{-\alpha t}X(t)$.

We can regard the evolution of the price of the stock over time as our experiment, and thus the outcome of the experiment is the value of the function $X(t)$, $0 \leqslant t \leqslant T$. The types of wagers available are that for any $s < t$ we can observe the process for a time s and then buy (or sell) shares of the stock at price $X(s)$ and then sell (or buy) these shares at time t for the price $X(t)$. In addition, we will suppose that we may purchase any of N different options at time 0. Option i, costing c_i per share, gives us the option of purchasing shares of the stock at time t_i for the fixed price of K_i per share, $i = 1, \ldots, N$.

Suppose we want to determine values of the c_i for which there is no betting strategy that leads to a sure win. Assuming that the arbitrage theorem can be generalized (to handle the preceding situation, where the outcome of the experiment is a function), it follows that there will be no sure win if and only if there exists a probability measure over the set of outcomes under which all of the wagers have expected return 0. Let $\mathbf{P}$ be a probability measure on the set of outcomes. Consider first the wager of observing the stock for a time s and then purchasing (or selling) one share with the intention of selling (or purchasing) it at time t, $0 \leqslant s < t \leqslant T$. The present value of the amount paid for the stock is $e^{-\alpha s}X(s)$, whereas the present value of the amount received is $e^{-\alpha t}X(t)$. Hence, in order for the expected return of this wager to be 0 when $\mathbf{P}$ is the probability measure on $X(t)$, $0 \leqslant t \leqslant T$, we must have that

$$E_{\mathbf{P}}[e^{-\alpha t}X(t) \mid X(u), \ 0 \leqslant u \leqslant s] = e^{-\alpha s}X(s) \tag{10.9}$$

Consider now the wager of purchasing an option. Suppose the option gives one the right to buy one share of the stock at time t for a price K. At time t, the worth of this option will be as follows:

$$\text{worth of option at } t = \begin{cases} X(t) - K, & \text{if } X(t) \geqslant K \\ 0, & \text{if } X(t) < K \end{cases}$$

That is, the time t worth of the option is $(X(t) - K)^{+}$. Hence, the present value of the worth of the option is $e^{-\alpha t}(X(t) - K)^{+}$. If c is the (time 0) cost of the option, we see that, in order for purchasing the option to have expected (present value) return 0, we must have that

$$E_{\mathbf{P}}[e^{-\alpha t}(X(t) - K)^{+}] = c \tag{10.10}$$

By the arbitrage theorem, if we can find a probability measure $\mathbf{P}$ on the set of outcomes that satisfies Equation (10.9), then if c, the cost of an option to purchase one share at time t at the fixed price K, is as given in Equation (10.10), then no arbitrage is possible. On the other hand, if for given prices

c_i, $i = 1, \ldots, N$, there is no probability measure **P** that satisfies both (10.9) and the equality

$$c_i = E_{\mathbf{P}}[e^{-\alpha t_i}(X(t_i) - K_i)^+], \qquad i = 1, \ldots, N$$

then a sure win is possible.

We will now present a probability measure **P** on the outcome $X(t)$, $0 \leqslant t \leqslant T$, that satisfies Equation (10.9).

Suppose that

$$X(t) = x_0 e^{Y(t)}$$

where $\{Y(t), \ t \geqslant 0\}$ is a Brownian motion process with drift coefficient μ and variance parameter σ^2. That is, $\{X(t), \ t \geqslant 0\}$ is a geometric Brownian motion process (see Section 10.3.2). From Equation (10.8) we have that, for $s < t$,

$$E[X(t) \mid X(u), \ 0 \leqslant u \leqslant s\} = X(s)e^{(t-s)(\mu + \sigma^2/2)}$$

Hence, if we choose μ and σ^2 so that

$$\mu + \sigma^2/2 = \alpha$$

then Equation (10.9) will be satisfied. That is, by letting **P** be the probability measure governing the stochastic process $\{x_0 e^{Y(t)}, 0 \leqslant t \leqslant T\}$, where $\{Y(t)\}$ is Brownian motion with drift parameter μ and variance parameter σ^2, and where $\mu + \sigma^2/2 = \alpha$, Equation (10.9) is satisfied.

It follows from the preceding that if we price an option to purchase a share of the stock at time t for a fixed price K by

$$c = E_{\mathbf{P}}[e^{-\alpha t}(X(t) - K)^+]$$

then no arbitrage is possible. Since $X(t) = x_0 e^{Y(t)}$, where $Y(t)$ is normal with mean μt and variance $t\sigma^2$, we see that

$$ce^{\alpha t} = \int_{-\infty}^{\infty} (x_0 e^y - K)^+ \frac{1}{\sqrt{2\pi t \sigma^2}} e^{-(y-\mu t)^2/2t\sigma^2} \, dy$$

$$= \int_{\log(K/x_0)}^{\infty} (x_0 e^y - K) \frac{1}{\sqrt{2\pi t \sigma^2}} e^{-(y-\mu t)^2/2t\sigma^2} \, dy$$

Making the change of variable $w = (y - \mu t)/(\sigma t^{1/2})$ yields

$$ce^{\alpha t} = x_0 e^{\mu t} \frac{1}{\sqrt{2\pi}} \int_a^\infty e^{\sigma w \sqrt{t}} e^{-w^2/2} \, dw - K \frac{1}{\sqrt{2\pi}} \int_a^\infty e^{-w^2/2} \, dw \qquad (10.11)$$

where

$$a = \frac{\log(K/x_0) - \mu t}{\sigma \sqrt{t}}$$

Now,

$$\frac{1}{\sqrt{2\pi}} \int_a^\infty e^{\sigma w \sqrt{t}} e^{-w^2/2} \, dw = e^{t\sigma^2/2} \frac{1}{\sqrt{2\pi}} \int_a^\infty e^{-(w - \sigma\sqrt{(t)^2/2})} \, dw$$

$$= e^{t\sigma^2/2} P\{N(\sigma\sqrt{t}, 1) \geq a\}$$

$$= e^{t\sigma^2/2} P\{N(0, 1) \geq a - \sigma\sqrt{t}\}$$

$$= e^{t\sigma^2/2} P\{N(0, 1) \leq -(a - \sigma\sqrt{t})\}$$

$$= e^{t\sigma^2/2} \phi(\sigma\sqrt{t} - a)$$

where $N(m, v)$ is a normal random variable with mean m and variance v, and ϕ is the standard normal distribution function.

Thus, we see from Equation (10.11) that

$$ce^{\alpha t} = x_0 e^{\mu t + \sigma^2 t/2} \phi(\sigma\sqrt{t} - a) - K\phi(-a)$$

Using that

$$\mu + \sigma^2/2 = \alpha$$

and letting $b = -a$, we can write this as follows:

$$c = x_0 \phi(\sigma\sqrt{t} + b) - Ke^{-\alpha t} \phi(b) \qquad (10.12)$$

where

$$b = \frac{\alpha t - \sigma^2 t/2 - \log(K/x_0)}{\sigma\sqrt{t}}$$

The option price formula given by Equation (10.12) depends on the initial price of the stock x_0, the option exercise time t, the option exercise price K, the discount (or interest rate) factor α, and the value σ^2. Note that for *any* value of σ^2, if the options are priced according to the formula of Equation (10.12) then no arbitrage is possible. However, as many people believe that the price of a stock actually follows a geometric Brownian motion — that is, $X(t) = x_0 e^{Y(t)}$ where $Y(t)$ is Brownian motion with parameters μ and σ^2 — it has been suggested that it is natural to price the option according to the formula (10.12) with the parameter σ^2 taken equal to the estimated value (see the remark that follows) of the variance parameter under the assumption of a geometric Brownian motion model. When this is done, the formula (10.12) is known as the Black–Scholes option cost valuation. It is interesting that this valuation does not depend on the value of the drift parameter μ but only on the variance parameter σ^2.

If the option itself can be traded, then the formula of Equation (10.12) can be used to set its price in such a way so that no arbitrage is possible. If at time s the price of the stock is $X(s) = x_s$, then the price of a (t, K) option — that is, an option to purchase one unit of the stock at time t for a price K — should be set by replacing t by $t - s$ and x_0 by x_s in Equation (10.12).

Remark If we observe a Brownian motion process with variance parameter σ^2 over any time interval, then we could theoretically obtain an arbitrarily precise estimate of σ^2. For suppose we observe such a process $\{Y(s)\}$ for a time t. Then, for fixed h, Let $N = [t/h]$ and set

$$W_1 = Y(h) - Y(0),$$

$$W_2 = Y(2h) - Y(h),$$

$$\vdots$$

$$W_N = Y(Nh) - Y(Nh - h)$$

Then random variables $W_1, \ldots, W_N$ are independent and identically distributed normal random variables having variance $h\sigma^2$. We now use the fact (see Section 3.6.4) that $(N - 1)S^2/(\sigma^2 h)$ has a chi-squared distribution with $N - 1$ degrees of freedom, where S^2 is the sample variance defined by

$$S^2 = \sum_{i=1}^{N} (W_i - \bar{W})^2/(N - 1)$$

Since the expected value and variance of a chi-squared with k degrees of freedom are equal to k and $2k$, respectively, we see that

$$E[(N - 1)S^2/(\sigma^2 h)] = N - 1$$

and

$$\text{Var}[(N - 1)S^2/(\sigma^2 h)] = 2(N - 1)$$

From this, we see that

$$E[S^2/h] = \sigma^2$$

and

$$\text{Var}[S^2/h] = 2\sigma^4/(N - 1)$$

Hence, as we let h become smaller (and so $N = [t/h]$ becomes larger) the variance of the unbiased estimator of σ^2 becomes arbitrarily small. ✦

Equation (10.12) is not the only way in which options can be priced so that no arbitrage is possible. Let $\{X(t), 0 \leqslant t \leqslant T\}$ be any stochastic process satisfying, for $s < t$,

$$E[e^{-\alpha t}X(t) \mid X(u), 0 \leqslant u \leqslant s] = e^{-\alpha s}X(s) \tag{10.13}$$

[that is, Equation (10.9) is satisfied]. By setting c, the cost of an option to purchase one share of the stock at time t for price K, equal to

$$c = E[e^{-\alpha t}(X(t) - K)^+] \tag{10.14}$$

it follows that no arbitrage is possible.

Another type of stochastic process, aside from geometric Brownian motion, that satisfies Equation (10.13) is obtained as follows. Let $Y_1, Y_2, \ldots$ be a sequence of independent random variables having a common mean μ, and suppose that this process is independent of $\{N(t), t \geq 0\}$, which is a Poisson process with rate λ. Let

$$X(t) = x_0 \prod_{i=1}^{N(t)} Y_i$$

Using the identity

$$X(t) = x_0 \prod_{i=1}^{N(s)} Y_i \prod_{j=N(s)+1}^{N(t)} Y_j$$

and the independent increment assumption of the Poisson process, we see that, for $s < t$,

$$E[X(t) \mid X(u), \; 0 \leq u \leq s] = X(s) \, E\left[\prod_{j=N(s)+1}^{N(t)} Y_j \right]$$

Conditioning on the number of events between s and t yields

$$E\left[\prod_{j=N(s)+1}^{N(t)} Y_j \right] = \sum_{n=0}^{\infty} \mu^n e^{-\lambda(t-s)} [\lambda(t-s)]^n / n!$$

$$= e^{-\lambda(t-s)(1-\mu)}$$

Hence,

$$E[X(t) \mid X(u), \; 0 \leq u \leq s] = X(s) e^{-\lambda(t-s)(1-\mu)}$$

Thus, if we choose λ and μ to satisfy

$$\lambda(1 - \mu) = -\alpha$$

then Equation (10.13) is satisfied. Therefore, if for any value of λ we let the Y_i have any distributions with a common mean equal to $\mu = 1 + \alpha/\lambda$ and then price the options according to Equation (10.14), then no arbitrage is possible.

Remark If $\{X(t), \; t \geq 0\}$ satisfies Equation (10.13), then the process $\{e^{-\alpha t}X(t), \; t \geq 0\}$ is called a Martingale. Thus, any pricing of options for which the expected gain on the option is equal to 0 when $\{e^{-\alpha t}X(t)\}$ follows the probability law of some Martingale will result in no arbitrage possibilities.

That is, if we choose any Martingale process $\{Z(t)\}$ and let the cost of a (t, K) option be

$$c = E[e^{-\alpha t}(e^{\alpha t} Z(t) - K)^+]$$
$$= E[(Z(t) - Ke^{-\alpha t})^+]$$

then there is no sure win.

In addition, while we did not consider the type of wager where a stock that is purchased at time s is sold not at a fixed time t but rather at some random time that depends on the movement of the stock, it can be shown using results about Martingales that the expected return of such wagers is also equal to 0.

Remark A variation of the arbitrage theorem was first noted by de Finetti in 1937. A more general version of de Finetti's result, of which the arbitrage theorem is a special case, is given in Reference 3.

10.5. White Noise

Let $\{X(t), t \geq 0\}$ denote a standard Brownian motion process and let f be a function having a continuous derivative in the region $[a, b]$. The stochastic integral $\int_a^b f(t) \, dX(t)$ is defined as follows:

$$\int_a^b f(t) \, dX(t) \equiv \lim_{\substack{n \to \infty \\ \max(t_i - t_{i-1}) \to 0}} \sum_{i=1}^n f(t_{i-1})[X(t_i) - X(t_{i-1})] \qquad (10.15)$$

where $a = t_0 < t_1 < \cdots < t_n = b$ is a partition of the region $[a, b]$. Using the identity (the integration by parts formula applied to sums)

$$\sum_{i=1}^n f(t_{i-1})[X(t_i) - X(t_{i-1})] = f(b)X(b) - f(a)X(a) - \sum_{i=1}^n X(t_i)[f(t_i) - f(t_{i-1})]$$

we see that

$$\int_a^b f(t) \, dX(t) = f(b)X(b) - f(a)X(a) - \int_a^b X(t) \, df(t) \qquad (10.16)$$

Equation (10.16) is usually taken as the definition of $\int_a^b f(t) \, dX(t)$.

By using the right side of Equation (10.16) we obtain, upon assuming the interchangeability of expectation and limit, that

$$E\left[\int_a^b f(t) \, dX(t)\right] = 0$$

Also,

$$\mathrm{Var}\left(\sum_{i=1}^{n} f(t_{i-1})[X(t_i) - X(t_{i-1})] \right) = \sum_{i=1}^{n} f^2(t_{i-1}) \mathrm{Var}[X(t_i) - X(t_{i-1})]$$

$$= \sum_{i=1}^{n} f^2(t_{i-1})(t_i - t_{i-1})$$

where the top equality follows from the independent increments of Brownian motion. Hence, we obtain from Equation (10.15) upon taking limits of the preceding that

$$\mathrm{Var}\left[\int_a^b f(t)\, dX(t) \right] = \int_a^b f^2(t)\, dt$$

Remark   The above gives operational meaning to the family of quantities $\{dX(t), 0 \leqslant t < \infty\}$ by viewing it as an operator that carries functions f into the values $\int_a^b f(t)\, dX(t)$. This is called a white noise transformation, or more loosely $\{dX(t), 0 \leqslant t < \infty\}$ is called white noise since it can be imagined that a time varying function f travels through a white noise medium to yield the output (at time b) $\int_a^b f(t)\, dX(t)$.

Example 10.4   Consider a particle of unit mass that is suspended in a liquid and suppose that, due to the liquid, there is a viscous force that retards the velocity of the particle at a rate proportional to its present velocity. In addition, let us suppose that the velocity instantaneously changes according to a constant multiple of white noise. That is, if $V(t)$ denotes the particle's velocity at t, suppose that

$$V'(t) = -\beta V(t) + \alpha X'(t)$$

where $\{X(t),\ t \geqslant 0\}$ is standard Brownian motion. This can be written as follows:

$$e^{\beta t}[V'(t) + \beta V(t)] = \alpha e^{\beta t} X'(t)$$

or

$$\frac{d}{dt}\left[e^{\beta t} V(t) \right] = \alpha e^{\beta t} X'(t)$$

Hence, upon integration, we obtain

$$e^{\beta t} V(t) = V(0) + \alpha \int_0^t e^{\beta s} X'(s)\, ds$$

or

$$V(t) = V(0)e^{-\beta t} + \alpha \int_0^t e^{-\beta(t-s)} \, dX(s)$$

Hence, from Equation (10.16),

$$V(t) = V(0)e^{-\beta t} + \alpha \left[X(t) - \int_0^t X(s)\beta e^{-\beta(t-s)} \, ds \right] \quad \blacklozenge$$

10.6. Gaussian Processes

We start with the following definition.

Definition 10.2 A stochastic process $X(t)$, $t \geqslant 0$ is called a *Gaussian*, or a *normal*, process if $X(t_1), \ldots, X(t_n)$ has a multivariate normal distribution for all $t_1, \ldots, t_n$.

If $\{X(t),\ t \geqslant 0\}$ is a Brownian motion process, then because each of $X(t_1), X(t_2), \ldots, X(t_n)$ can be expressed as a linear combination of the independent normal random variables $X(t_1)$, $X(t_2) - X(t_1)$, $X(t_3) - X(t_2), \ldots,$ $X(t_n) - X(t_{n-1})$ it follows that Brownian motion is a Gaussian process.

Because a multivariate normal distribution is completely determined by the marginal mean values and the covariance values (see Section 2.6) it follows that standard Brownian motion could also be defined as a Gaussian process having $E[X(t)] = 0$ and, for $s \leqslant t$,

$$\begin{aligned}
\mathrm{Cov}(X(s),\ X(t)) &= \mathrm{Cov}(X(s),\ X(s) + X(t) - X(s)) \\
&= \mathrm{Cov}(X(s),\ X(s)) + \mathrm{Cov}(X(s),\ X(t) - X(s)) \\
&= \mathrm{Cov}(X(s),\ X(s)) \qquad \text{by independent increments} \\
&= s \qquad \text{since } \mathrm{Var}(X(s)) = s \qquad\qquad (10.17)
\end{aligned}$$

Let $\{X(t),\ t \geqslant 0\}$ be a standard Brownian motion process and consider the process values between 0 and 1 conditional on $X(1) = 0$. That is, consider the conditional stochastic process $\{X(t),\ 0 \leqslant t \leqslant 1 \,|\, X(1) = 0\}$. Since the conditional distribution of $X(t_1), \ldots, X(t_n)$ is multivariate normal it follows that this conditional process, known as the *Brownian bridge* (as it is tied down both at 0 and at 1), is a Gaussian process. Let us compute its covariance function. As, from Equation (10.4),

$$E[X(s) \,|\, X(1) = 0] = 0, \qquad \text{for } s < 1$$

we have that, for $s < t < 1$,

$$\text{Cov}[(X(s), X(t)) \mid X(1) = 0]$$

$$= E[X(s)X(t) \mid X(1) = 0]$$

$$= E[E[X(s)X(t) \mid X(t), X(1) = 0] \mid X(1) = 0]$$

$$= E[X(t)E[X(s) \mid X(t)] \mid X(1) = 0]$$

$$= E\left[X(t) \frac{s}{t} X(t) \mid X(1) = 0 \right] \quad \text{by (10.4)}$$

$$= \frac{s}{t} E[X^2(t) \mid X(1) = 0]$$

$$= \frac{s}{t} t(1 - t) \quad \text{by (10.4)}$$

$$= s(1 - t)$$

Thus, the Brownian bridge can be defined as a Gaussian process with mean value 0 and covariance function $s(1 - t)$, $s \leq t$. This leads to an alternative approach to obtaining such a process.

Proposition 10.1 If $\{X(t), t \geq 0\}$ is standard Brownian motion, then $\{Z(t), 0 \leq t \leq 1\}$ is a Brownian bridge process when $Z(t) = X(t) - tX(1)$.

Proof As it is immediate that $\{Z(t), t \geq 0\}$ is a Gaussian process, all we need verify is that $E[Z(t)] = 0$ and $\text{Cov}(Z(s), Z(t)) = s(1 - t)$, when $s \leq t$. The former is immediate and the latter follows from

$$\text{Cov}(Z(s), Z(t)) = \text{Cov}(X(s) - sX(1), X(t) - tX(1))$$

$$= \text{Cov}(X(s), X(t)) - t\,\text{Cov}(X(s), X(1))$$

$$\quad - s\,\text{Cov}(X(1), X(t)) + st\,\text{Cov}(X(1), X(1))$$

$$= s - st - st + st$$

$$= s(1 - t)$$

and the proof is complete. ✦

If $\{X(t), t \geq 0\}$ is Brownian motion, then the process $\{Z(t), t \geq 0\}$ defined by

$$Z(t) = \int_0^t X(s)\,ds \qquad (10.18)$$

is called *integrated Brownian motion*. As an illustration of how such a process may arise in practice, suppose we are interested in modeling the price of a commodity throughout time. Letting $Z(t)$ denote the price at t then, rather than assuming that $\{Z(t)\}$ is Brownian motion (or that $\log Z(t)$ is Brownian motion), we might want to assume that the rate of change of $Z(t)$ follows a Brownian motion. For instance, we might suppose that the rate of change of the commodity's price is the current inflation rate which is imagined to vary as Brownian motion. Hence,

$$\frac{d}{dt} Z(t) = X(t)$$

$$Z(t) = Z(0) + \int_0^t X(s)\, ds$$

It follows from the fact that Brownian motion is a Gaussian process that $\{Z(t),\ t \geqslant 0\}$ is also Gaussian. To prove this, first recall that $W_1, \ldots, W_n$ is said to have a multivariate normal distribution if they can be represented as

$$W_i = \sum_{j=1}^m a_{ij} U_j, \qquad i = 1, \ldots, n$$

where $U_j, j = 1, \ldots, m$ are independent normal random variables. From this it follows that any set of partial sums of $W_1, \ldots, W_n$ are also jointly normal. The fact that $Z(t_1), \ldots, Z(t_n)$ is multivariate normal can now be shown by writing the integral in Equation (10.18) as a limit of approximating sums.

As $\{Z(t),\ t \geqslant 0\}$ is Gaussian it follows that its distribution is characterized by its mean value and covariance function. We now compute these when $\{X(t),\ t \geqslant 0\}$ is standard Brownian motion.

$$E[Z(t)] = E\left[\int_0^t X(s)\, ds \right]$$

$$= \int_0^t E[X(s)]\, ds$$

$$= 0$$

For $s \leqslant t$,

$$\mathrm{Cov}[Z(s),\ Z(t)] = E[Z(s)Z(t)]$$

$$= E\left[\int_0^s \int_0^t X(y)\, dy \int_0^s X(u)\, du \right]$$

$$= E\left[\int_0^s \int_0^t X(y)X(u)\, dy\, du \right]$$

$$= \int_0^s \int_0^t E[X(y)X(u)] \, dy \, du$$

$$= \int_0^s \int_0^t \min(y, u) \, dy \, du \qquad \text{by (10.17)}$$

$$= \int_0^s \left(\int_0^u y \, dy + \int_u^t u \, dy \right) du = s^2 \left(\frac{t}{2} - \frac{s}{6} \right) \quad \blacklozenge$$

10.7. Stationary and Weakly Stationary Processes

A stochastic process $\{X(t), \ t \geq 0\}$ is said to be a *stationary process* if for all $n, s, t, \ldots, t_n$ the random vectors $X(t_1), \ldots, X(t_n)$ and $X(t_1 + s), \ldots, X(t_n + s)$ have the same joint distribution. In other words, a process is stationary if, in choosing any fixed point s as the origin, the ensuing process has the same probability law. Two examples of stationary processes are:

(i) An ergodic continuous-time Markov chain $\{X(t), \ t \geq 0\}$ when

$$P\{X(0) = j\} = P_j, \qquad j \geq 0$$

where $\{P_j, j \geq 0\}$ are the limiting probabilities.

(ii) $\{X(t), \ t \geq 0\}$ when $X(t) = N(t + L) - N(t), \ t \geq 0$, where $L > 0$ is a fixed constant and $\{N(t), \ t \geq 0\}$ is a Poisson process having rate λ.

The first one of the above processes is stationary for it is a Markov chain whose initial state is chosen according to the limiting probabilities, and it can thus be regarded as an ergodic Markov chain that one starts observing at time ∞. Hence the continuation of this process at time s after observation begins is just the continuation of the chain starting at time $\infty + s$, which clearly has the same probability for all s. That the second example — where $X(t)$ represents the number of events of a Poisson process that occur between t and $t + L$ — is stationary follows the stationary and independent increment assumption of the Poisson process which implies that the continuation of a Poisson process at any time s remains a Poisson process.

Example 10.5 (The Random Telegraph Signal Process): Let $\{N(t), \ t \geq 0\}$ denote a Poisson process, and let X_0 be independent of this process and be such that $P\{X_0 = 1\} = P\{X_0 = -1\} = \frac{1}{2}$. Defining $X(t) = X_0(-1)^{N(t)}$ then $\{X(t), \ t \geq 0\}$ is called *random telegraph signal* process. To see that it is stationary, note first that starting at any time t, no matter what the value of

$N(t)$, as X_0 is equally likely to be either plus or minus 1, it follows that $X(t)$ is equally likely to be either plus or minus 1. Hence, because the continuation of a Poisson process beyond any time remains a Poisson process, it follows that $\{X(t), \ t \geqslant 0\}$ is a stationary process.

Let us compute the mean and covariance function of the random telegraph signal

$$E[X(t)] = E[X_0(-1)^{N(t)}]$$

$$= E[X_0]E[(-1)^{N(t)}] \qquad \text{by independence}$$

$$= 0 \qquad \text{since } E[X_0] = 0,$$

$$\text{Cov}[X(t), \ X(t+s)] = E[X(t)X(t+s)]$$

$$= E[X_0^2(-1)^{N(t)+N(t+s)}]$$

$$= E[(-1)^{2N(t)}(-1)^{N(t+s)-N(t)}]$$

$$= E[(-1)^{N(t+s)-N(t)}]$$

$$= E[(-1)^{N(s)}]$$

$$= \sum_{i=0}^{\infty} (-1)^i e^{-\lambda s} \frac{(\lambda s)^i}{i!}$$

$$= e^{-2\lambda s} \qquad\qquad (10.19)$$

For an application of the random telegraph signal consider a particle moving at a constant unit velocity along a straight line and suppose that collisions involving this particle occur at a Poisson rate λ. Also suppose that each time the particle suffers a collision it reverses direction. Therefore, if X_0 represents the initial velocity of the particle, then its velocity at time t — call it $X(t)$ — is given by $X(t) = X_0(-1)^{N(t)}$, where $N(t)$ denotes the number of collisions involving the particle by time t. Hence, if X_0 is equally likely to be plus or minus 1, and is independent of $\{N(t), \ t \geqslant 0\}$, then $\{X(t), \ t \geqslant 0\}$ is a random telegraph signal process. If we now let

$$D(t) = \int_0^t X(s) \, ds$$

then $D(t)$ represents the displacement of the particle at time t from its position at time 0. The mean and variance of $D(t)$ are obtained as follows:

$$E[D(t)] = \int_0^t E[X(s)] \, ds = 0,$$

$$\text{Var}[D(t)] = E[D^2(t)]$$

$$= E\left[\int_0^t X(y) \, dy \int_0^t X(u) \, du\right]$$

$$= \int_0^t \int_0^t E[X(y)X(u)] \, dy \, du$$

$$= 2 \iint_{0<y<u<t} E[X(y)X(u)] \, dy \, du$$

$$= 2 \int_0^t \int_0^u e^{-2\lambda(u-y)} \, dy \, du \qquad \text{by (10.19)}$$

$$= \frac{1}{\lambda} \left(t - \frac{1}{2\lambda} + \frac{1}{2\lambda} e^{-2\lambda t} \right) \quad \blacklozenge$$

The condition for a process to be stationary is rather stringent and so we define the process $\{X(t), \ t \geqslant 0\}$ to be a *second-order stationary* or a *weakly stationary* process if $E[X(t)] = c$ and $\text{Cov}[X(t), X(t+s)]$ does not depend on t. That is, a process is second-order stationary if the first two moments of $X(t)$ are the same for all t and the covariance between $X(s)$ and $X(t)$ depends only on $|t-s|$. For a second-order stationary process, let

$$R(s) = \text{Cov}[X(t), \ X(t+s)]$$

As the finite dimensional distributions of a Gaussian process (being multivariate normal) are determined by their means and covariance, it follows that a second-order stationary Gaussian process is stationary.

Example 10.6 (The Ornstein–Uhlenbeck Process): Let $\{X(t), \ t \geqslant 0\}$ be a standard Brownian motion process, and define, for $\alpha > 0$,

$$V(t) = e^{-\alpha t/2} X(e^{\alpha t})$$

The process $\{V(t), \ t \geqslant 0\}$ is called the Ornstein–Uhlenbeck process. It has been proposed as a model for describing the velocity of a particle immersed in a liquid or gas, and as such is useful in statistical mechanics. Let us compute its mean and covariance function

$$E[V(t)] = 0,$$

$$\text{Cov}[V(t), \ V(t+s)] = e^{-\alpha t/2} e^{-\alpha(t+s)/2} \text{Cov}[X(e^{\alpha t}), \ X(e^{\alpha(t+s)})]$$

$$= e^{-\alpha t} e^{-\alpha s/2} e^{\alpha t} \qquad \text{by Equation (10.17)}$$

$$= e^{-\alpha s/2}$$

Hence, $\{V(t), \ t \geqslant 0\}$ is weakly stationary and as it is clearly a Gaussian process (since Brownian motion is Gaussian) we can conclude that it is stationary. It is interesting to note that (with $\alpha = 4\lambda$) it has the same mean and covariance function as the random telegraph signal process, thus illustrating that two quite different processes can have the same second-order properties. (Of course, if two Gaussian processes have the same mean and covariance functions then they are identically distributed.) ✦

As the following examples show, there are many types of second-order stationary processes that are not stationary.

Example 10.7 (An Autoregressive Process): Let $Z_0, Z_1, Z_2, \ldots$, be uncorrelated random variables with $E[Z_n] = 0$, $n \geq 0$ and

$$\text{Var}(Z_n) = \begin{cases} \sigma^2/(1 - \lambda^2) & n = 0 \\ \sigma^2, & n \geq 1 \end{cases}$$

where $\lambda^2 < 1$. Define

$$\begin{aligned} X_0 &= Z_0, \\ X_n &= \lambda X_{n-1} + Z_n, \qquad n \geq 1 \end{aligned} \tag{10.20}$$

The process $\{X_n, n \geq 0\}$ is called a *first-order autoregressive process*. It says that the state at time n (that is, X_n) is a constant multiple of the state at time $n - 1$ plus a random error term Z_n.

Iterating Equation (10.20) yields

$$\begin{aligned} X_n &= \lambda(\lambda X_{n-2} + Z_{n-1}) + Z_n \\ &= \lambda^2 X_{n-2} + \lambda Z_{n-1} + Z_n \\ &\ \ \vdots \\ &= \sum_{i=0}^{n} \lambda^{n-i} Z_i \end{aligned}$$

and so

$$\begin{aligned} \text{Cov}(X_n, X_{n+m}) &= \text{Cov}\left(\sum_{i=0}^{n} \lambda^{n-i} Z_i, \sum_{i=0}^{n+m} \lambda^{n+m-i} Z_i \right) \\ &= \sum_{i=0}^{n} \lambda^{n-i} \lambda^{n+m-i} \text{Cov}(Z_i, Z_i) \\ &= \sigma^2 \lambda^{2n+m} \left(\frac{1}{1 + \lambda^2} + \sum_{i=1}^{n} \lambda^{-2i} \right) \\ &= \frac{\sigma^2 \lambda^m}{1 - \lambda^2} \end{aligned}$$

where the preceding uses the fact that Z_i and Z_j are uncorrelated when $i \neq j$. As $E[X_n] = 0$, we see that $\{X_n, n \geq 0\}$ is weakly stationary (the definition for a discrete time process is the obvious analog of that given for continuous time processes). ✦

Example 10.8 If, in the random telegraph signal process, we drop the requirement that $P\{X_0 = 1\} = P\{X_0 = -1\} = \frac{1}{2}$ and only require that $E[X_0] = 0$, then the process $\{X(t),\ t \geq 0\}$ need no longer be stationary. (It will remain stationary if X_0 has a symmetric distribution in the sense that $-X_0$ has the same distribution as X_0.) However, the process will be weakly stationary since

$$E[X(t)] = E[X_0]E[(-1)^{N(t)}] = 0,$$

$$\mathrm{Cov}[X(t),\ X(t + s)] = E[X(t)X(t + s)]$$

$$= E[X_0^2]E[(-1)^{N(t) + N(t + s)}]$$

$$= E[X_0^2]e^{-2\lambda s} \qquad \text{from (10.19)} \quad ✦$$

Example 10.9 Let $W_0, W_1, W_2, \ldots$ be uncorrelated with $E[W_n] = \mu$ and $\mathrm{Var}(W_n) = \sigma^2$, $n \geq 0$, and for some positive integer k define

$$X_n = \frac{W_n + W_{n-1} + \cdots + W_{n-k}}{k + 1}, \qquad n \geq k$$

The process $\{X_n,\ n \geq k\}$, which at each time keeps track of the arithmetic average of the most recent $k + 1$ values of the Ws, is called a moving average process. Using the fact that the W_n, $n \geq 0$ are uncorrelated, we see that

$$\mathrm{Cov}(X_n,\ X_{n+m}) = \begin{cases} \dfrac{(k + 1 - m)\sigma^2}{(k + 1)^2}, & \text{if } 0 \leq m \leq k \\ 0, & \text{if } m > k \end{cases}$$

Hence, $\{X_n,\ n \geq k\}$ is a second-order stationary process. ✦

Let $\{X_n,\ n \geq 1\}$ be a second-order stationary process with $E[X_n] = \mu$. An important question is when, if ever, does $\bar{X}_n \equiv \sum_{i=1}^{n} X_i/n$ converge to μ? The following proposition, which we state without proof, shows that $E[(\bar{X}_n - \mu)^2] \to 0$ if and only if $\sum_{i=1}^{n} R(i)/n \to 0$. That is, the expected square of the difference between $\bar{X}_n$ and μ will converge to 0 if and only if the limiting average value of $R(i)$ converges to 0.

Proposition 10.2 Let $\{X_n,\ n \geq 1\}$ be a second-order stationary process having mean μ and covariance function $R(i) = \mathrm{Cov}(X_n, X_{n+i})$, and let $\bar{X}_n \equiv \sum_{i=1}^{n} X_i/n$. Then $\lim_{n\to\infty} E[(\bar{X}_n - \mu)^2] = 0$ and only if $\lim_{n\to\infty} \sum_{i=1}^{n} R(i)/n = 0$.

10.8. Harmonic Analysis of Weakly Stationary Processes

Suppose that the stochastic processes $\{X(t), \ -\infty < t < \infty\}$ and $\{Y(t), \ -\infty < t < \infty\}$ are related as follows:

$$Y(t) = \int_{-\infty}^{\infty} X(t - s)h(s) \, ds \qquad (10.21)$$

We can imagine that a signal, whose value at time t is $X(t)$, is passed through a physical system that distorts its value so that $Y(t)$, the received value at t, is given by Equation (10.21). The processes $\{X(t)\}$ and $\{Y(t)\}$ are called respectively the input and output processes. The function h is called the *impulse response* function. If $h(s) = 0$ whenever $s < 0$, then h is also called a weighting function since Equation (10.21) expresses the output at t as a weighted integral of all the inputs prior to t with $h(s)$ representing the weight given the input s time units ago.

The relationship expressed by Equation (10.21) is a special case of a time invariant linear filter. It is called a filter because we can imagine that the input process $\{X(t)\}$ is passed through a medium and then filtered to yield the output process $\{Y(t)\}$. It is a linear filter because if the input processes $\{X_i(t)\}$, $i = 1, 2$, result in the output processes $\{Y_i(t)\}$ —that is, if $Y_i(t) = \int_0^\infty X_i(t - s)h(s) \, ds$ —then the output process corresponding to the input process $\{aX_1(t) + bX_2(t)\}$ is just $\{aY_1(t) + bY_2(t)\}$. It is called time invariant since lagging the input process by a time τ —that is, considering the new input process $\bar{X}(t) = X(t + \tau)$ —results in a lag of τ in the output process since

$$\int_0^{\infty} \bar{X}(t - s)h(s) \, ds = \int_0^{\infty} X(t + \tau - s)h(s) \, ds = Y(t + \tau)$$

Let now suppose that the input process $\{X(t), \ -\infty < t < \infty\}$ is weakly stationary with $E[X(t)] = 0$ and covariance function

$$R_X(s) = \mathrm{Cov}[X(t), \ X(t + s)].$$

Let us compute the mean value and covariance function of the output process $\{Y(t)\}$.

Assuming that we can interchange the expectation and integration operations (a sufficient condition being that $\int |h(s)| \, ds < \infty$* and, for some $M < \infty$, $E|X(t)| < M$ for all t) we obtain

$$E[Y(t)] = \int E[X(t - s)]h(s) \, ds = 0$$

---

* The range of all integrals in this section is from $-\infty$ to $+\infty$.

Similarly,

$$\text{Cov}[Y(t_1), \ Y(t_2)] = \text{Cov}\left[\int X(t_1 - s_1)h(s_1)\,ds_1, \ \int X(t_2 - s_2)h(s_2)\,ds_2\right]$$

$$= \iint \text{Cov}[X(t_1 - s_1), \ X(t_2 - s_2)]h(s_1)h(s_2)\,ds_1\,ds_2$$

$$= \iint R_X(t_2 - s_2 - t_1 + s_1)h(s_1)h(s_2)\,ds_1\,ds_2 \qquad (10.22)$$

Hence, $\text{Cov}[Y(t_1), \ Y(t_2)]$ depends on t_1, t_2 only through $t_2 - t_1$; thus, showing that $\{Y(t)\}$ is also weakly stationary.

The preceding expression for $R_Y(t_2 - t_1) = \text{Cov}[Y(t_1), \ Y(t_2)]$ is, however, more compactly and usefully expressed in terms of Fourier transforms of R_X and R_Y. Let, for $i = \sqrt{-1}$,

$$\tilde{R}_X(w) = \int e^{-iws} R_X(s)\,ds$$

and

$$\tilde{R}_Y(w) = \int e^{-iws} R_Y(s)\,ds$$

denote the Fourier transforms respectively of R_X and R_Y. The function $\tilde{R}_X(w)$ is also called the *power spectral density* of the process $\{X(t)\}$. Also, let

$$\tilde{h}(w) = \int e^{-iws} h(s)\,ds$$

denote the Fourier transform of the function h. Then, from Equation (10.22),

$$\tilde{R}_Y(w) = \iiint e^{-iws} R_X(s - s_2 + s_1)h(s_1)h(s_2)\,ds_1\,ds_2\,ds$$

$$= \iiint e^{-iw(s - s_2 + s_1)} R_X(s - s_2 + s_1)\,ds\, e^{-iws_2}h(s_2)\,ds_2\, e^{iws_1} h(s_1)\,ds_1$$

$$= \tilde{R}_X(w)\tilde{h}(w)\tilde{h}(-w) \qquad (10.23)$$

Now, using the representation

$$e^{ix} = \cos x + i\sin x,$$

$$e^{-ix} = \cos(-x) + i\sin(-x) = \cos x - i\sin x$$

we obtain

$$\tilde{h}(w)\tilde{h}(-w) = \left[\int h(s)\cos(ws)\,ds - i\int h(s)\sin(ws)\,ds\right]$$
$$\times \left[\int h(s)\cos(ws)\,ds + i\int h(s)\sin(ws)\,ds\right]$$
$$= \left[\int h(s)\cos(ws)\,ds\right]^2 + \left[\int h(s)\sin(ws)\,ds\right]^2$$
$$= \left|\int h(s)e^{-iws}\,ds\right|^2 = |\tilde{h}(w)|^2$$

Hence, from Equation (10.23) we obtain

$$\tilde{R}_Y(w) = \tilde{R}_X(w)|\tilde{h}(w)|^2$$

In words, the Fourier transform of the covariance function of the output process is equal to the square of the amplitude of the Fourier transform of the impulse function multiplied by the Fourier transform of the covariance function of the input process.

Exercises

In the following exercises $\{B(t),\ t \geq 0\}$ is a standard Brownian motion process and T_a denotes the time it takes this process to hit a.

*1. What is the distribution of $B(s) + B(t)$, $s \leq t$?

2. Compute the conditional distribution of $B(s)$ given that $B(t_1) = A$, $B(t_2) = B$, where $0 < t_1 < s < t_2$.

*3. Compute $E[B(t_1)B(t_2)B(t_3)]$ for $t_1 < t_2 < t_3$.

4. Show that

$$P\{T_a < \infty\} = 1,$$
$$E[T_a] = \infty, \qquad a \neq 0$$

*5. What is $P\{T_1 < T_{-1} < T_2\}$?

6. Suppose you own one share of a stock whose price changes according to a standard Brownian motion process. Suppose that you purchased the stock at a price $b + c$, $c > 0$, and the present price is b. You have decided to sell the stock either when it reaches the price $b + c$ or when an additional time t goes by (whichever occurs first). What is the probability that you do not recover your purchase price?

7. Compute an expression for

$$P \left\{ \max_{t_1 \leqslant s \leqslant t_2} B(s) > x \right\}$$

8. Consider the random walk which in each Δt time unit either goes up or down the amount $\sqrt{\Delta t}$ with respective probabilities p and $1 - p$ where $p = \frac{1}{2}(1 + \mu \sqrt{\Delta t})$.

(a) Argue that as $\Delta t \to 0$ the resulting limiting process is a Brownian motion process with drift rate μ.
(b) Using part (a) and the results of the gambler's ruin problem (Section 4.5.1), compute the probability that a Brownian motion process with drift rate μ goes up A before going down B, $A > 0$, $B > 0$.

9. Let $\{X(t), \ t \geqslant 0\}$ be a Brownian motion with drift coefficient μ and variance parameter σ^2. What is the joint density function of $X(s)$ and $X(t)$, $s < t$?

*10. Let $\{X(t), \ t \geqslant 0\}$ be a Brownian motion with drift coefficient μ and variance parameter σ^2. What is the conditional distribution of $X(t)$ given that $X(s) = c$ when

(a) $s < t$?
(b) $t < s$?

11. A stock is presently selling at a price of $50 per share. After one time period, its selling price will (in present value dollars) be either $150 or $25. An option to purchase y units of the stock at time 1 can be purchased at cost cy.

(a) What should c be in order for there to be no sure win?
(b) If $c = 4$, explain how you could guarantee a sure win.
(c) If $c = 10$, explain how you could guarantee a sure win.

12. Verify the statement made in the remark following Example 10.2.

13. Use the arbitrage theorem to verify your answer to part (a) of Exercise 11.

14. The present price of a stock is 100. The price at time 1 will be either 50, 100, or 200. An option to purchase y shares of the stock at time 1 for the (present value) price ky costs cy.

(a) If $k = 120$, show than an arbitrage opportunity occurs if and only if $c > 80/3$.

(b) If $k = 80$, show that there is not an arbitrage opportunity if and only if $20 \leqslant c \leqslant 40$.

15. The current price of a stock is 100. Suppose that the logarithm of the price of the stock changes according to a Brownian motion with drift coefficient $\mu = 2$ and variance parameter $\sigma^2 = 1$. Give the Black–Scholes cost of an option to buy the stock at time 10 for a cost of

(a) 100 per unit.
(b) 120 per unit.
(c) 80 per unit.

Assume that the continuously compounded interest rate is 5 percent.

A stochastic process $\{Y(t), \; t \geqslant 0\}$ is said to be a *Martingale* process if, for $s < t$,

$$E[Y(t) \mid Y(u), \; 0 \leqslant u \leqslant s] = Y(s)$$

16. If $\{Y(t), \; t \geqslant 0\}$ is a Martingale, show that

$$E[Y(t)] = E[Y(0)]$$

17. Show that standard Brownian motion is a Martingale.

18. Show that $\{Y(t), \; t \geqslant 0\}$ is a Martingale when

$$Y(t) = B^2(t) - t$$

What is $E[Y(t)]$?

Hint: First compute $E[Y(t) \mid B(u), \; 0 \leqslant u \leqslant s]$.

*19. Show that $\{Y(t), \; t \geqslant 0\}$ is a Martingale when

$$Y(t) = \exp\{cB(t) - c^2 t/2\}$$

where c is an arbitrary constant. What is $E[Y(t)]$?

An important property of a Martingale is that if you continually observe the process and then stop at some time T, then, subject to some technical conditions (which will hold in the problems to be considered),

$$E[Y(T)] = E[Y(0)]$$

The time T usually depends on the values of the process and is known as a *stopping time* for the Martingale. This result, that the expected value of the stopped Martingale is equal to its fixed time expectation, is known as the *Martingale stopping theorem*.

*20. Let

$$T = \text{Min}\{t : B(t) = 2 - 4t\}$$

That is, T is the first time that standard Brownian motion hits the line $2 - 4t$. Use the Martingale stopping theorem to find $E[T]$.

21. Let $\{X(t),\ t \geqslant 0\}$ be Brownian motion with drift coefficient μ and variance parameter σ^2. That is,

$$X(t) = \sigma B(t) + \mu t$$

Let $\mu > 0$, and for a positive constant x let

$$T = \text{Min}\{t: X(t) = x\}$$

$$= \text{Min}\left\{t:\ B(t) = \frac{x - \mu t}{\sigma}\right\}$$

That is, T is the first time the process $\{X(t),\ t \geqslant 0\}$ hits x. Use the Martingale stopping theorem to show that

$$E[T] = x/\mu$$

22. Let $X(t) = \sigma B(t) + \mu t$, and for given positive constants A and B, let p denote the probability that $\{X(t),\ t \geqslant 0\}$ hits A before it hits $-B$.

(a) Define the stopping time T to be the first time the process hits either A or $-B$. Use this stopping time and the Martingale defined in Exercise 19 to show that

$$E[\exp\{c(X(T) - \mu T)/\sigma - c^2 T/2\}] = 1$$

(b) Let $c = -2\mu/\sigma$, and show that

$$E[\exp\{-2\mu X(T)/\sigma\}] = 1$$

(c) Use part (b) and the definition of T to find p.

Hint: What are the possible values of $\exp\{-2\mu X(T)/\sigma^2\}$?

23. Let $X(t) = \sigma B(t) + \mu t$, and define T to be the first time the process $\{X(t),\ t \geqslant 0\}$ hits either A or $-B$, where A and B are given positive numbers. Use the Martingale stopping theorem and part (c) of Exercise 22 to find $E[T]$.

*24. Let $\{X(t),\ t \geqslant 0\}$ be Brownian motion with drift coefficient μ and variance parameter σ^2. Suppose that $\mu > 0$. Let $x > 0$ and define the stopping time T (as in Exercise 21) by

$$T = \text{Min}\{t: X(t) = x\}$$

Use the Martingale defined in Exercise 18, along with the result of Exercise 21, to show that

$$\text{Var}(T) = x\sigma^2/\mu^3$$

25. Compute the mean and variance of

(a) $\int_0^1 t \, dB(t)$.
(b) $\int_0^1 t^2 \, dB(t)$.

26. Let $Y(t) = tB(1/t)$, $t > 0$ and $Y(0) = 0$.

(a) What is the distribution of $Y(t)$?
(b) Compare $\mathrm{Cov}(Y(s), \; Y(t))$.
(c) Argue that $\{Y(t), \; t \geq 0\}$ is a standard Brownian motion process.

*27. Let $Y(t) = B(a^2 t)/a$ for $a > 0$. Argue that $\{Y(t)\}$ is a standard Brownian motion process.

28. Let $\{Z(t), \; t \geq 0\}$ denote a Brownian bridge process. Show that if

$$Y(t) = (t+1)Z(t/(t+1))$$

then $\{Y(t), \; t \geq 0\}$ is a standard Brownian motion process.

29. Let $X(t) = N(t+1) - N(t)$ where $\{N(t), \; t \geq 0\}$ is a Poisson process with rate λ. Compute

$$\mathrm{Cov}[X(t), \; X(t+s)]$$

*30. Let $\{N(t), \; t \geq 0\}$ denote a Poisson process with rate λ and define $Y(t)$ to be the time from t until the next Poisson event.

(a) Argue that $\{Y(t), \; t \geq 0\}$ is a stationary process.
(b) Compute $\mathrm{Cov}[Y(t), \; Y(t+s)]$.

31. Let $\{X(t), \; -\infty < t < \infty\}$ be a weakly stationary process having covariance function $R_X(s) = \mathrm{Cov}[X(t), \; X(t+s)]$.

(a) Show that

$$\mathrm{Var}(X(t+s) - X(t)) = 2R_X(0) - 2R_X(t)$$

(b) If $Y(t) = X(t+1) - X(t)$ show that $\{Y(t), \; -\infty < t < \infty\}$ is also weakly stationary having a covariance function $R_Y(s) = \mathrm{Cov}[Y(t), \; Y(t+s)]$ that satisfies

$$R_Y(s) = 2R_X(s) - R_X(s-1) - R_X(s+1)$$

32. Let Y_1 and Y_2 be independent unit normal random varianbles and for some constant w set

$$X(t) = Y_1 \cos wt + Y_2 \sin wt, \qquad -\infty < t < \infty$$

(a) Show that $\{X(t)\}$ is a weakly stationary process.
(b) Argue that $\{X(t)\}$ is a stationary process.

33. Let $\{X(t), \ -\infty < t < \infty\}$ be weakly stationary with covariance function $R(s) = \text{Cov}(X(t), \ X(t + s))$ and let $\tilde{R}(w)$ denote the power spectral density of the process.

(i) Show that $\tilde{R}(w) = \tilde{R}(-w)$. It can be shown that

$$R(s) = \frac{1}{2\pi} \int_{-\infty}^{\infty} \tilde{R}(w) e^{iws} \, dw$$

(ii) Use the preceding to show that

$$\int_{-\infty}^{\infty} \tilde{R}(w) \, dw = 2\pi E[X^2(t)]$$

References

1. M. S. Bartlett, "An Introduction to Stochastic Processes," Cambridge University Press, London, 1954.
2. U. Grenander and M. Rosenblatt, "Statistical Analysis of Stationary Time Series," John Wiley, New York, 1957.
3. D. Heath and W. Sudderth, "On a Theorem of De Finetti, Oddsmaking, and Game Theory," *Ann. Math. Stat.* **43**, 2072–2077 (1972).
4. S. Karlin and H. Taylor, "A Second Course in Stochastic Processes," Academic Press, Orlando, FL, 1981.
5. L. H. Koopmans, "The Spectral Analysis of Time Series," Academic Press, Orlando, FL, 1974.
6. S. Ross, "Stochastic Processes," Second Edition, John Wiley, New York, 1996.

Simulation

11

♦♦♦

11.1. Introduction

Let $\mathbf{X} = (X_1, \ldots, X_n)$ denote a random vector having a given density function $f(x_1, \ldots, x_n)$ and suppose we are interested in computing

$$E[g(\mathbf{X})] = \int\int \cdots \int g(x_1, \ldots, x_n) f(x_1, \ldots, x_n)\, dx_1\, dx_2 \cdots dx_n$$

for some n-dimensional function g. For instance, g could represent the total delay in queue of the first $[n/2]$ customers when the X values represent the first $[n/2]$ interarrival and service times.* In many situations, it is not analytically possible either to compute the above multiple integral exactly or even to numerically approximate it within a given accuracy. One possibility that remains is to approximate $E[g(\mathbf{X})]$ by means of simulation.

To approximate $E[g(\mathbf{X})]$, start by generating a random vector $\mathbf{X}^{(1)} = (X_1^{(1)}, \ldots, X_n^{(1)})$ having the joint density $f(x_1, \ldots, x_n)$ and then compute $Y^{(1)} = g(\mathbf{X}^{(1)})$. Now generate a second random vector (independent of the first) $\mathbf{X}^{(2)}$ and compute $Y^{(2)} = g(\mathbf{X}^{(2)})$. Keep on doing this until r, a fixed number, of independent and identically distributed random variables $Y^{(i)} = g(\mathbf{X}^{(i)})$, $i = 1, \ldots, r$ have been generated. Now by the strong law of large numbers, we know that

$$\lim_{r \to \infty} \frac{Y^{(1)} + \cdots + Y^{(r)}}{r} = E[Y^{(i)}] = E[g(\mathbf{X})]$$

and so we can use the average of the generated Ys as an estimate of $E[g(\mathbf{X})]$. This approach to estimating $E[g(\mathbf{X})]$ is called the *Monte Carlo simulation* approach.

---

* We are using the notation $[a]$ to represent the largest integer less than or equal to a.

585

Clearly there remains the problem of how to generate, or *simulate*, random vectors having a specified joint distribution. The first step in doing this is to be able to generate random variables from a uniform distribution on (0, 1). One way to do this would be to take 10 identical slips of paper, numbered 0, 1, ..., 9, place them in a hat and then successively select n slips, with replacement, from the hat. The sequence of digits obtained (with a decimal point in front) can be regarded as the value of a uniform (0, 1) random variable rounded off to the nearest $(\frac{1}{10})^n$. For instance, if the sequence of digits selected is 3, 8, 7, 2, 1, then the value of the uniform (0, 1) random variable is 0.38721 (to the nearest 0.00001). Tables of the values of uniform (0, 1) random variables, known as random number tables, have been extensively published [for instance, see The RAND Corporation, *A Million Random Digits with 100,000 Normal Deviates* (New York: The Free Press, 1955)]. Table 11.1 is such a table.

However, the above is not the way in which digital computers simulate uniform (0, 1) random variables. In practice, they use pseudo random numbers instead of truly random ones. Most random number generators start with an initial value X_0, called the seed, and then recursively compute values by specifying positive integers a, c, and m, and then letting

$$X_{n+1} = (aX_n + c) \quad \text{modulo } m, \qquad n \geqslant 0$$

where the above means that $aX_n + c$ is divided by m and the remainder is taken as the value of X_{n+1}. Thus each X_n is either 0, 1, ..., $m - 1$ and the quantity X_n/m is taken as an approximation to a uniform (0, 1) random variable. It can be shown that subject to suitable choices for a, c, m, the above gives rise to a sequence of numbers that looks as if it were generated from independent uniform (0, 1) random variables.

As our starting point in the simulation of random variables from an arbitrary distribution, we shall suppose that we can simulate from the uniform (0, 1) distribution, and we shall use the term "random numbers" to mean independent random variables from this distribution. In Sections 11.2 and 11.3 we present both general and special techniques for simulating continuous random variables; and in Section 11.4 we do the same for discrete random variables. In Section 11.5 we discuss the simulation both of jointly distributed random variables and stochastic processes. Particular attention is given to the simulation of nonhomogeneous Poisson processes, and in fact three different approaches for this are discussed. Simulation of two-dimensional Poisson processes is discussed in Section 11.5.2. In Section 11.6 we discuss various methods for increasing the precision of the simulation estimates by reducing their variance; and in Section 11.7 we consider the problem of choosing the number of simulation runs needed to attain a desired level of precision. Before beginning this program, however, let us consider two applications of simulation to combinatorial problems.

Table 11.1 A Random Number Table

04839	96423	24878	82651	66566	14778	76797	14780	13300	87074
68086	26432	46901	20848	89768	81536	86645	12659	92259	57102
39064	66432	84673	40027	32832	61362	98947	96067	64760	64584
25669	26422	44407	44048	37937	63904	45766	66134	75470	66520
64117	94305	26766	25940	39972	22209	71500	64568	91402	42416
87917	77341	42206	35126	74087	99547	81817	42607	43808	76655
62797	56170	86324	88072	76222	36086	84637	93161	76038	65855
95876	55293	18988	27354	26575	08625	40801	59920	29841	80150
29888	88604	67917	48708	18912	82271	65424	69774	33611	54262
73577	12908	30883	18317	28290	35797	05998	41688	34952	37888
27958	30134	04024	86385	29880	99730	55536	84855	29080	09250
90999	49127	20044	59931	06115	20542	18059	02008	73708	83517
18845	49618	02304	51038	20655	58727	28168	15475	56942	53389
94824	78171	84610	82834	09922	25417	44137	48413	25555	21246
35605	81263	39667	47358	56873	56307	61607	49518	89356	20103
33362	64270	01638	92477	66969	98420	04880	45585	46565	04102
88720	82765	34476	17032	87589	40836	32427	70002	70663	88863
39475	46473	23219	53416	94970	25832	69975	94884	19661	72828
06990	67245	68350	82948	11398	42878	80287	88267	47363	46634
40980	07391	58745	25774	22987	80059	39911	96189	41151	14222
83974	29992	65381	38857	50490	83765	55657	14361	31720	57375
33339	31926	14883	24413	59744	92351	97473	89286	35931	04110
31662	25388	61642	34072	81249	35648	56891	69352	48373	45578
93526	70765	10592	04542	76463	54328	02349	17247	28865	14777
20492	38391	91132	21999	59516	81652	27195	48223	46751	22923
04153	53381	79401	21438	83035	92350	36693	31238	59649	91754
05520	91962	04739	13092	97662	24822	94730	06496	35090	04822
47498	87637	99016	71060	88824	71013	18735	20286	23153	72924
23167	49323	45021	33132	12544	41035	80780	45393	44812	12515
23792	14422	15059	45799	22716	19792	09983	74353	68668	30429
85900	98275	32388	52390	16815	69298	82732	38480	73817	32523
42559	78985	05300	22164	24369	54224	35083	19687	11062	91491
14349	82674	66523	44133	00697	35552	35970	19124	63318	29686
17403	53363	44167	64486	64758	75366	76554	31601	12614	33072
23632	27889	47914	02584	37680	20801	72152	39339	34806	08930

Example 11.1 (Generating a Random Permutation): Suppose we are interested in generating a permutation of the numbers $1, 2, \ldots, n$ that is such that all $n!$ possible orderings are equally likely. The following algorithm will accomplish this by first choosing one of the numbers $1, \ldots, n$ at random and then putting that number in position n; it then chooses at random one of the remaining $n - 1$ numbers and puts that number in position $n - 1$; it then chooses at random one of the remaining $n - 2$ numbers and puts it in position $n - 2$, and so on (where choosing a number at random means that each of the remaining numbers is equally likely to be chosen). However, so

that we do not have to consider exactly which of the numbers remain to be positioned, it is convenient and efficient to keep the numbers in an ordered list and then randomly choose the position of the number rather than the number itself. That is, starting with any initial ordering $p_1, p_2, \ldots, p_n$, we pick one of the positions $1, \ldots, n$ at random and then interchange the number in that position with the one in position n. Now we randomly choose one of the positions $1, \ldots, n-1$ and interchange the number in this position with the one in position $n-1$, and so on.

To implement the preceding, we need to be able to generate a random variable that is equally likely to take on any of the values $1, 2, \ldots, k$. To accomplish this, let U denote a random number — that is, U is uniformly distributed over $(0, 1)$ — and note that kU is uniform on $(0, k)$ and so

$$P\{i - 1 < kU < i\} = \frac{1}{k}, \qquad i = 1, \ldots, k$$

Hence, if we let $\text{Int}(kU)$ denote the largest integer less than or equal to kU, then the random variable $I = \text{Int}(kU) + 1$ will be such that

$$P\{I = i\} = P\{\text{Int}(kU) = i - 1\} = P\{i - 1 < kU < i\} = \frac{1}{k}$$

The preceding algorithm for generating a random permutation can now be written as follows:

Step 1: Let $p_1, p_2, \ldots, p_n$ be any permutation of $1, 2, \ldots, n$ (for instance, we can choose $p_j = j, j = 1, \ldots, n$).
Step 2: Set $k = n$.
Step 3: Generate a random number U and let $I = \text{Int}(kU) + 1$.
Step 4: Interchange the values of p_I and p_k.
Step 5: Let $k = k - 1$ and if $k > 1$ go to Step 3.
Step 6: $p_1, \ldots, p_n$ is the desired random permutation.

For instance, suppose $n = 4$ and the initial permutation is 1, 2, 3, 4. If the first value of I (which is equally likely to be either 1, 2, 3, 4) is $I = 3$, then the new permutation is 1, 2, 4, 3. If the next value of I is $I = 2$ then the new permutation is 1, 4, 2, 3. If the final value of I is $I = 2$, then the final permutation is 1, 4, 2, 3, and this is the value of the random permutation.

One very important property of the above algorithm is that it can also be used to generate a random subset, say of size r, of the integers $1, \ldots, n$. Namely, just follow the algorithm until the positions $n, n - 1, \ldots, n - r + 1$ are filled. The elements in these positions constitute the random subset. ✦

Example 11.2 (Estimating the Number of Distinct Entries in a Large List): Consider a list of n entries where n is very large, and suppose we are interested in estimating d, the number of distinct elements in the list. If we let m_i denote the number of times that the element in position i appears on the list, then we can express d by

$$d = \sum_{i=1}^{n} \frac{1}{m_i}$$

To estimate d, suppose that we generate a random value X equally likely to be either $1, 2, \ldots, n$ (that is, we take $X = [nU] + 1$) and then let $m(X)$ denote the number of times the element in position X appears on the list. Then

$$E\left[\frac{1}{m(X)}\right] = \sum_{i=1}^{n} \frac{1}{m_i} \frac{1}{n} = \frac{d}{n}$$

Hence, if we generate k such random variables $X_1, \ldots, X_k$ we can estimate d by

$$d \approx \frac{n \sum_{i=1}^{k} 1/m(X_i)}{k}$$

Suppose now that each item in the list has a value attached to it—$v(i)$ being the value of the ith element. The sum of the values of the distinct items—call it v—can be expressed as

$$v = \sum_{i=1}^{n} \frac{v(i)}{m(i)}$$

Now if $X = [nU] + 1$, where U is a random number, then

$$E\left[\frac{v(X)}{m(X)}\right] = \sum_{i=1}^{n} \frac{v(i)}{m(i)} \frac{1}{n} = \frac{v}{n}$$

Hence, we can estimate v by generating $X_1, \ldots, X_k$ and then estimating v by

$$v \approx \frac{n}{k} \sum_{i=1}^{k} \frac{v(X_i)}{m(X_i)}$$

For an important application of the above, let $A_i = \{a_{i,1}, \ldots, a_{i,n_i}\}$, $i = 1, \ldots, s$ denote events, and suppose we are interested in estimating $P(\bigcup_{i=1}^{s} A_i)$. Since

$$P\left(\bigcup_{i=1}^{s} A_i\right) = \sum_{a \in \cup A_i} P(a) = \sum_{i=1}^{s} \sum_{j=1}^{n_i} \frac{P(a_{i,j})}{m(a_{i,j})}$$

where $m(a_{i,j})$ is the number of events to which the point $a_{i,j}$ belongs, the above method can be used to estimate $P(\bigcup_1^s A_i)$.

Note that the above procedure for estimating v can be effected without prior knowledge of the set of values $\{v_1, \ldots, v_n\}$. That is, it suffices that we can determine the value of an element in a specific place and the number of times that element appears on the list. When the set of values is *a priori* known, there is a more efficient approach available as will be shown in Example 11.11. ✦

11.2. General Techniques for Simulating Continuous Random Variables

In this section we present three methods for simulating continuous random variables.

11.2.1. The Inverse Transformation Method

A general method for simulating a random variable having a continuous distribution — called the *inverse transformation method* — is based on the following proposition.

Proposition 11.1   Let U be a uniform $(0, 1)$ random variable. For any continuous distribution function F if we define the random variable X by

$$X = F^{-1}(U)$$

then the random variable X has distribution function F. $[F^{-1}(u)$ is defined to equal that value x for which $F(x) = u.]$

Proof

$$F_X(a) = P\{X \leqslant a\}$$

$$= P\{F^{-1}(U) \leqslant a\} \tag{11.1}$$

Now, since $F(x)$ is a monotone function, it follows that $F^{-1}(U) \leqslant a$ if and only if $U \leqslant F(a)$. Hence, from Equation (11.1), we see that

$$F_X(a) = P\{U \leqslant F(a)\}$$

$$= F(a) \quad ✦$$

Hence we can simulate a random variable X from the continuous distribution F, when F^{-1} is computable, by simulating a random number U and then setting $X = F^{-1}(U)$.

Example 11.3 (Simulating an Exponential Random Variable): If $F(x) = 1 - e^{-x}$, then $F^{-1}(u)$ is that value of x such that

$$1 - e^{-x} = u$$

or

$$x = -\log(1 - u)$$

Hence, if U is a uniform $(0, 1)$ variable, then

$$F^{-1}(U) = -\log(1 - U)$$

is exponentially distributed with mean 1. Since $1 - U$ is also uniformly distributed on $(0, 1)$ it follows that $-\log U$ is exponential with mean 1. Since cX is exponential with mean c when X is exponential with mean 1, it follows that $-c \log U$ is exponential with mean c. ✦

11.2.2. The Rejection Method

Suppose that we have a method for simulating a random variable having density function $g(x)$. We can use this as the basis for simulating from the continuous distribution having density $f(x)$ by simulating Y from g and then accepting this simulated value with a probability proportional to $f(Y)/g(Y)$.

Specifically let c be a constant such that

$$\frac{f(y)}{g(y)} \leqslant c \qquad \text{for all } y$$

We then have the following technique for simulating a random variable having density f.

Rejection Method

Step 1: Simulate Y having density g and simulate a random number U.
Step 2: If $U \leqslant f(Y)/cg(Y)$ set $X = Y$. Otherwise return to Step 1.

Proposition 11.2 The random variable X generated by the rejection method has density function f.

Proof Let X be the value obtained, and let N denote the number of necessary iterations. Then

$$P\{X \leqslant x\} = P\{Y_N \leqslant x\}$$

$$= P\{Y \leqslant x \,|\, U \leqslant f(Y)/cg(Y)\}$$

$$= \frac{P\{Y \leqslant x, U \leqslant f(Y)/cg(Y)\}}{K}$$

$$= \frac{\int P\{Y \leqslant x, U \leqslant f(Y)/cg(Y) \,|\, Y = y\}g(y)\,dy}{K}$$

$$= \frac{\int_{-\infty}^{x} (f(y)/cg(y))g(y)\,dy}{K}$$

$$= \frac{\int_{-\infty}^{x} f(y)\,dy}{Kc}$$

where $K = P\{U \leqslant f(Y)/cg(Y)\}$. Letting $x \to \infty$ shows that $K = 1/c$ and the proof is complete. ✦

Remarks (i) The preceding method was originally presented by Von Neumann in the special case where g was positive only in some finite interval (a, b), and Y was chosen to be uniform over (a, b). [That is, $Y = a + (b - a)U$.]

(ii) Note that the way in which we "accept the value Y with probability $f(Y)/cg(Y)$" is by generating a uniform $(0, 1)$ random variable U and then accepting Y if $U \leqslant f(Y)/cg(Y)$.

(iii) Since each iteration of the method will, independently, result in an accepted value with probability $P\{U \leqslant f(Y)/cg(Y)\} = 1/c$ it follows that the number of iterations is geometric with mean c.

(iv) Actually, it is not necessary to generate a new uniform random number when deciding whether or not to accept, since at a cost of some additional computation, a single random number, suitably modified at each iteration, can be used throughout. To see how, note that the actual value of U is not used — only whether or not $U < f(Y)/cg(Y)$. Hence, if Y is rejected — that is, if $U > f(Y)/cg(Y)$ — we can use the fact that, given Y,

$$\frac{U - f(Y)/cg(Y)}{1 - f(Y)/cg(Y)} = \frac{cUg(Y) - f(Y)}{cg(Y) - f(Y)}$$

is uniform on $(0, 1)$. Hence, this may be used as a uniform random number in the next iteration. As this saves the generation of a random number at the cost of the computation above, whether it is a net savings depends greatly upon the method being used to generate random numbers.

Example 11.4 Let us use the rejection method to generate a random variable having density function

$$f(x) = 20x(1 - x)^3, \qquad 0 < x < 1$$

Since this random variable (which is beta with parameters 2, 4) is concentrated in the interval $(0, 1)$, let us consider the rejection method with

$$g(x) = 1, \qquad 0 < x < 1$$

To determine the constant c such that $f(x)/g(x) \leqslant c$, we use calculus to determine the maximum value of

$$\frac{f(x)}{g(x)} = 20x(1 - x)^3$$

Differentiation of this quantity yields

$$\frac{d}{dx}\left[\frac{f(x)}{g(x)}\right] = 20[(1 - x)^3 - 3x(1 - x)^2]$$

Setting this equal to 0 shows that the maximal value is attained when $x = \frac{1}{4}$, and thus

$$\frac{f(x)}{g(x)} \leqslant 20\left(\frac{1}{4}\right)\left(\frac{3}{4}\right)^3 = \frac{135}{64} \equiv c$$

Hence,

$$\frac{f(x)}{cg(x)} = \frac{256}{27} x(1 - x)^3$$

and thus the rejection procedure is as follows:

Step 1: Generate random numbers U_1 and U_2.
Step 2: If $U_2 \leqslant \frac{256}{27} U_1(1 - U_1)^3$, stop and set $X = U_1$. Otherwise return to step 1.

The average number of times that step 1 will be performed is $c = \frac{135}{64}$. ✦

Example 11.5 (Simulating a Normal Random Variable): To simulate a unit normal random variable Z (that is, one with mean 0 and variance 1) note first that the absolute value of Z has density function

$$f(x) = \frac{2}{\sqrt{2\pi}} e^{-x^2/2}, \qquad 0 < x < \infty \qquad (11.2)$$

We will start by simulating from the above density by using the rejection method with

$$g(x) = e^{-x}, \qquad 0 < x < \infty$$

Now, note that

$$\frac{f(x)}{g(x)} = \sqrt{2e/\pi}\exp\{-(x-1)^2/2\} \leqslant \sqrt{2e/\pi}$$

Hence, using the rejection method we can simulate from Equation (11.2) as follows:

(a) Generate independent random variables Y and U, Y being exponential with rate 1 and U being uniform on $(0, 1)$.
(b) If $U \leqslant \exp\{-(Y-1)^2/2\}$, or equivalently, if

$$-\log U \geqslant (Y-1)^2/2$$

set $X = Y$. Otherwise return to step (a).

Once we have simulated a random variable X having density function (11.2) we can then generate a unit normal random variable Z by letting Z be equally likely to be either X or $-X$.

To improve upon the foregoing, note first that from Example 11.3 it follows that $-\log U$ will also be exponential with rate 1. Hence, steps (a) and (b) are equivalent to the following:

(a') Generate independent exponentials with rate 1, Y_1, and Y_2.
(b') Set $X = Y_1$ if $Y_2 \geqslant (Y_1 - 1)^2/2$. Otherwise return to (a').

Now suppose that we accept step (b'). It then follows by the lack of memory property of the exponential that the amount by which Y_2 exceeds $(Y_1 - 1)^2/2$ will also be exponential with rate 1.

Hence, summing up, we have the following algorithm which generates an exponential with rate 1 and an independent unit normal random variable.

Step 1: Generate Y_1, an exponential random variable with rate 1.
Step 2: Generate Y_2, an exponential with rate 1.
Step 3: If $Y_2 - (Y_1 - 1)^2/2 > 0$, set $Y = Y_2 - (Y_1 - 1)^2/2$ and go to step 4. Otherwise go to step 1.
Step 4: Generate a random number U and set

$$Z = \begin{cases} Y_1, & \text{if } U \leqslant \frac{1}{2} \\ -Y_1, & \text{if } U > \frac{1}{2} \end{cases}$$

The random variables Z and Y generated by the above are independent with Z being normal with mean 0 and variance 1 and Y being exponential with rate 1. (If we want the normal random variable to have mean μ and variance σ^2, just take $\mu + \sigma Z$). ✦

Remarks   (i) Since $c = \sqrt{2e/\pi} \approx 1.32$, the above requires a geometric distributed number of iterations of step 2 with mean 1.32.

(ii) The final random number of step 4 need not be separately simulated but rather can be obtained from the first digit of any random number used earlier. That is, suppose we generate a random number to simulate an exponential; then we can strip off the initial digit of this random number and just use the remaining digits (with the decimal point moved one step to the right) as the random number. If this initial digit is 0, 1, 2, 3, or 4 (or 0 if the computer is generating binary digits), then we take the sign of Z to be positive and take it to be negative otherwise.

(iii) If we are generating a sequence of unit normal random variables, then we can use the exponential obtained in step 4 as the initial exponential needed in step 1 for the next normal to be generated. Hence, on the average, we can simulate a unit normal by generating 1.64 exponentials and computing 1.32 squares.

11.2.3.   The Hazard Rate Method

Let F be a continuous distribution function with $\bar{F}(0) = 1$. Recall that $\lambda(t)$, the hazard rate function of F, is defined by

$$\lambda(t) = \frac{f(t)}{\bar{F}(t)}, \qquad t \geq 0$$

[where $f(t) = F'(t)$ is the density function]. Recall also that $\lambda(t)$ represents the instantaneous probability intensity that an item having life distribution F will fail at time t given it has survived to that time.

Suppose now that we are given a bounded function $\lambda(t)$, such that $\int_0^\infty \lambda(t) \, dt = \infty$, and we desire to simulate a random variable S having $\lambda(t)$ as its hazard rate function.

To do so let λ be such that

$$\lambda(t) \leq \lambda \qquad \text{for all } t \geq 0$$

To simulate from $\lambda(t)$, $t \geq 0$, we will

(a) simulate a Poisson process having rate λ. We will then only "accept" or "count" certain of these Poisson events. Specifically we will

(b) count an event that occurs at time t, independently of all else, with probability $\lambda(t)/\lambda$.

We now have the following proposition.

Proposition 11.3   The time of the first counted event—call it S—is a random variable whose distribution has hazard rate function $\lambda(t)$, $t \geq 0$.

Proof

$P\{t < S < t + dt \,|\, S > t\}$

$\quad = P\{\text{first counted event in } (t, t + dt) \,|\, \text{no counted events prior to } t\}$

$\quad = P\{\text{Poisson event in } (t, t + dt), \text{ it is counted} \,|\, \text{no counted events prior to } t\}$

$\quad = P\{\text{Poisson event in } (t, t + dt), \text{ it is counted}\}$

$\quad = [\lambda \, dt + o(dt)] \dfrac{\lambda(t)}{\lambda} = \lambda(t) \, dt + o(dt)$

which completes the proof. Note that the next to last equality follows from the independent increment property of Poisson processes.   ✦

Because the interarrival times of a Poisson process having rate λ are exponential with rate λ, it thus follows from Example 11.3 and the previous proposition that the following algorithm will generate a random variable having hazard rate function $\lambda(t)$, $t \geq 0$.

Hazard Rate Method for Generating S: $\lambda_s(t) = \lambda(t)$

Let λ be such that $\lambda(t) \leq \lambda$ for all $t \geq 0$. Generate pairs of random variables U_i, X_i, $i \geq 1$, with X_i being exponential with rate λ and U_i being uniform $(0, 1)$, stopping at

$$N = \min\left\{n\colon U_n \leq \lambda\left(\sum_{i=1}^{n} X_i\right)\middle/\lambda\right\}$$

Set

$$S = \sum_{i=1}^{N} X_i \quad \text{✦}$$

To compute $E[N]$ we need the result, known as Wald's equation, which states that if $X_1, X_2, \ldots$ are independent and identically distributed random variables that are observed in sequence up to some random time N then

$$E\left[\sum_{i=1}^{N} X_i\right] = E[N]E[X]$$

More precisely let $X_1, X_2, \ldots$ denote a sequence of independent random variables and consider the following definition.

Definition 11.1   An integer-valued random variable N is said to be a *stopping time* for the sequence $X_1, X_2 \ldots$ if the event $\{N = n\}$ is independent of $X_{n+1}, X_{n+2}, \ldots$ for all $n = 1, 2, \ldots$.

Intuitively, we observe the X_ns in sequential order and N denotes the number observed before stopping. If $N = n$, then we have stopped after observing $X_1, \ldots, X_n$ and before observing $X_{n+1}, X_{n+2}, \ldots$ for all $n = 1, 2, \ldots$.

Example 11.6   Let X_n, $n = 1, 2, \ldots$, be independent and such that

$$P\{X_n = 0\} = P\{X_n = 1\} = \tfrac{1}{2}, \qquad n = 1, 2, \ldots$$

If we let

$$N = \min\{n : X_1 + \cdots + X_n = 10\}$$

then N is a stopping time. We may regard N as being the stopping time of an experiment that successively flips a fair coin and then stops when the number of heads reaches 10.   ✦

Proposition 11.4   (Wald's Equation): If $X_1, X_2, \ldots$ are independent and identically distributed random variables having finite expectations, and if N is a stopping time for $X_1, X_2, \ldots$ such that $E[N] < \infty$, then

$$E\left[\sum_1^N X_n\right] = E[N]E[X]$$

Proof   Letting

$$I_n = \begin{cases} 1, & \text{if } N \geq n \\ 0, & \text{if } N < n \end{cases}$$

we have that

$$\sum_{n=1}^{N} X_n = \sum_{n=1}^{\infty} X_n I_n$$

Hence,

$$E\left[\sum_{n=1}^{N} X_n\right] = E\left[\sum_{n=1}^{\infty} X_n I_n\right] = \sum_{n=1}^{\infty} E[X_n I_n] \tag{11.3}$$

However, $I_n = 1$ if and only if we have not stopped after successively observing $X_1, \ldots, X_{n-1}$. Therefore, I_n is determined by $X_1, \ldots, X_{n-1}$ and is thus independent of X_n. From Equation (11.3) we thus obtain

$$E\left[\sum_{n=1}^{N} X_n\right] = \sum_{n=1}^{\infty} E[X_n]E[I_n]$$

$$= E[X] \sum_{n=1}^{\infty} E[I_n]$$

$$= E[X]E\left[\sum_{n=1}^{\infty} I_n\right]$$

$$= E[X]E[N] \quad \blacklozenge$$

Returning to the hazard rate method, we have that

$$S = \sum_{i=1}^{N} X_i$$

As $N = \min\{n: U_n \leqslant \lambda(\sum_1^n X_i)/\lambda\}$ it follows that the event that $N = n$ is independent of $X_{n+1}, X_{n+2}, \ldots$. Hence, by Wald's equation,

$$E[S] = E[N]E[X_i]$$

$$= \frac{E[N]}{\lambda}$$

or

$$E[N] = \lambda E[S]$$

where $E[S]$ is the mean of the desired random variable.

11.3. Special Techniques for Simulating Continuous Random Variables

Special techniques have been devised to simulate from most of the common continuous distributions. We now present certain of these.

11.3.1. The Normal Distribution

Let X and Y denote independent unit normal random variables and thus have the joint density function

$$f(x, y) = \frac{1}{2\pi} e^{-(x^2 + y^2)/2}, \qquad -\infty < x < \infty, -\infty < y < \infty$$

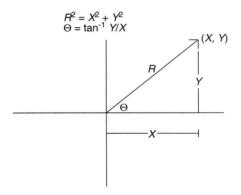

Figure 11.1.

Consider now the polar coordinates of the point (X, Y). As shown in Figure 11.1,

$$R^2 = X^2 + Y^2,$$

$$\Theta = \tan^{-1} Y/X$$

To obtain the joint density of R^2 and Θ, consider the transformation

$$d = x^2 + y^2, \qquad \theta = \tan^{-1} y/x$$

The Jacobian of this transformation is

$$J = \begin{vmatrix} \dfrac{\partial d}{\partial x} & \dfrac{\partial d}{\partial y} \\[2mm] \dfrac{\partial \theta}{\partial x} & \dfrac{\partial \theta}{\partial y} \end{vmatrix}$$

$$= \begin{vmatrix} 2x & 2y \\[2mm] \dfrac{1}{1+y^2/x^2}\left(\dfrac{-y}{x^2}\right) & \dfrac{1}{1+y^2/x^2}\left(\dfrac{1}{x}\right) \end{vmatrix} = 2 \begin{vmatrix} x & y \\[2mm] -\dfrac{y}{x^2+y^2} & \dfrac{x}{x^2+y^2} \end{vmatrix} = 2$$

Hence, from Section 2.5.3 the joint density of R^2 and Θ is given by

$$f_{R^2,\Theta}(d, \theta) = \frac{1}{2\pi} e^{-d/2} \frac{1}{2}$$

$$= \frac{1}{2} e^{-d/2} \frac{1}{2\pi}, \qquad 0 < d < \infty, 0 < \theta < 2\pi$$

Thus, we can conclude that R^2 and Θ are independent with R^2 having an exponential distribution with rate $\frac{1}{2}$ and Θ being uniform on $(0, 2\pi)$.

Let us now go in reverse from the polar to the rectangular coordinates. From the preceding if we start with W, an exponential random variable with rate $\frac{1}{2}$ (W plays the role of R^2) and with V, independent of W and uniformly distributed over $(0, 2\pi)$ (V plays the role of Θ) then $X = \sqrt{W} \cos V$, $Y = \sqrt{W} \sin V$ will be independent unit normals. Hence using the results of Example 11.3 we see that if U_1 and U_2 are independent uniform $(0, 1)$ random numbers, then

$$
\begin{aligned}
X &= (-2 \log U_1)^{1/2} \cos(2\pi U_2), \\
Y &= (-2 \log U_1)^{1/2} \sin(2\pi U_2)
\end{aligned}
\tag{11.4}
$$

are independent unit normal random variables.

Remark   The fact that $X^2 + Y^2$ has an exponential distribution with rate $\frac{1}{2}$ is quite interesting for, by the definition of the chi-square distribution, $X^2 + Y^2$ has a chi-squared distribution with 2 degrees of freedom. Hence, these two distributions are identical.

The preceding approach to generating unit normal random variables is called the Box–Muller approach. Its efficiency suffers somewhat from its need to compute the above sine and cosine values. There is, however, a way to get around this potentially time-consuming difficulty. To begin, note that if U is uniform on $(0, 1)$, then $2U$ is uniform on $(0, 2)$, and so $2U - 1$ is uniform on $(-1, 1)$. Thus, if we generate random numbers U_1 and U_2 and set

$$
\begin{aligned}
V_1 &= 2U_1 - 1, \\
V_2 &= 2U_2 - 1
\end{aligned}
$$

then (V_1, V_2) is uniformly distributed in the square of area 4 centered at $(0, 0)$ (see Figure 11.2).

Suppose now that we continually generate such pairs (V_1, V_2) until we obtain one that is contained in the circle of radius 1 centered at $(0, 0)$ — that is, until (V_1, V_2) is such that $V_1^2 + V_2^2 \leqslant 1$. It now follows that such a pair (V_1, V_2) is uniformly distributed in the circle. If we let $\bar{R}$, $\bar{\Theta}$ denote the polar coordinates of this pair, then it is easy to verify that $\bar{R}$ and $\bar{\Theta}$ are independent, with $\bar{R}^2$ being uniformly distributed on $(0, 1)$, and $\bar{\Theta}$ uniformly distributed on $(0, 2\pi)$.

Since

$$
\sin \bar{\Theta} = V_2 / \bar{R} = \frac{V_2}{\sqrt{V_1^2 + V_2^2}},
$$

$$
\cos \bar{\Theta} = V_1 / \bar{R} = \frac{V_1}{\sqrt{V_1^2 + V_2^2}}
$$

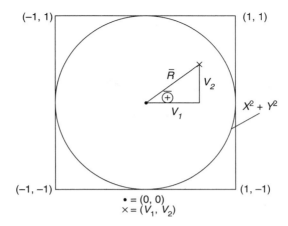

Figure 11.2.

it follows from Equation (11.4) that we can generate independent unit normals X and Y by generating another random number U and setting

$$X = (-2 \log U)^{1/2} V_1 / \bar{R},$$

$$Y = (-2 \log U)^{1/2} V_2 / \bar{R}$$

In fact, since (conditional on $V_1^2 + V_2^2 \leq 1$) $\bar{R}^2$ is uniform on $(0, 1)$ and is independent of $\bar{\Theta}$, we can use it instead of generating a new random number U; thus showing that

$$X = (-2 \log \bar{R}^2)^{1/2} V_1 / \bar{R} = \sqrt{\frac{-2 \log S}{S}} V_1,$$

$$Y = (-2 \log \bar{R}^2)^{1/2} V_2 / \bar{R} = \sqrt{\frac{-2 \log S}{S}} V_2$$

are independent unit normals, where

$$S = \bar{R}^2 = V_1^2 + V_2^2$$

Summing up, we thus have the following approach to generating a pair of independent unit normals:

Step 1: Generate random numbers U_1 and U_2.
Step 2: Set $V_1 = 2U_1 - 1$, $V_2 = 2U_2 - 1$, $S = V_1^2 + V_2^2$.
Step 3: If $S > 1$, return to step 1.

Step 4:   Return the independent unit normals

$$X = \sqrt{\frac{-2 \log S}{S}} \, V_1, \qquad Y = \sqrt{\frac{-2 \log S}{S}} \, V_2$$

The preceding is called the *polar method*. Since the probability that a random point in the square will fall within the circle is equal to $\pi/4$ (the area of the circle divided by the area of the square), it follows that, on average, the polar method will require $4/\pi = 1.273$ iterations of step 1. Hence, it will, on average, require 2.546 random numbers, 1 logarithm, 1 square root, 1 division, and 4.546 multiplications to generate 2 independent unit normals.

11.3.2. The Gamma Distribution

To simulate from a gamma distribution with parameters (n, λ), where n is an integer, we use the fact that the sum of n independent exponential random variables each having rate λ has this distribution. Hence, if $U_1, \ldots, U_n$ are independent uniform $(0, 1)$ random variables,

$$X = \frac{1}{\lambda} \sum_{i=1}^{n} \log U_i = -\frac{1}{\lambda} \log \left(\prod_{i=1}^{n} U_i \right)$$

has the desired distribution.

When n is large, there are other techniques available that do not require so many random numbers. One possibility is to use the rejection procedure with $g(x)$ being taken as the density of an exponential random variable with mean n/λ (as this is the mean of the gamma). It can be shown that for large n the average number of iterations needed by the rejection algorithm is $e[(n-1)/2\pi]^{1/2}$. In addition, if we wanted to generate a series of gammas, then, just as in Example 11.4, we can arrange things so that upon acceptance we obtain not only a gamma random variable but also, for free, an exponential random variable that can then be used in obtaining the next gamma (see Exercise 8).

11.3.3. The Chi-Squared Distribution

The chi-squared distribution with n degrees of freedom is the distribution of $\chi_n^2 = Z_1^2 + \cdots + Z_n^2$ where Z_i, $i = 1, \ldots, n$ are independent unit normals. Using the fact noted in the remark at the end of Section 3.1 we see that $Z_1^2 + Z_2^2$ has an exponential distribution with rate $\frac{1}{2}$. Hence, when n is even — say $n = 2k$ — χ_{2k}^2 has a gamma distribution with parameters $(k, \frac{1}{2})$. Hence, $-2 \log(\prod_{i=1}^{k} U_i)$ has a chi-squared distribution with $2k$ degrees of freedom. We can simulate a chi-squared random variable with $2k + 1$ degrees of freedom by first simulating a unit normal random variable Z and then adding Z^2 to the preceding. That is,

$$\chi^2_{2k+1} = Z^2 - 2\log\left(\prod_{i=1}^{k} U_i\right)$$

where $Z, U_1, \ldots, U_n$ are independent with Z being a unit normal and the others being uniform $(0, 1)$ random variables.

11.3.4. The Beta (n, m) Distribution

The random variable X is said to have a beta distribution with parameters n, m if its density is given by

$$f(x) = \frac{(n + m - 1)!}{(n - 1)!(m - 1)!} x^{n-1}(1 - x)^{m-1}, \qquad 0 < x < 1$$

One approach to simulating from the above distribution is to let $U_1, \ldots, U_{n+m-1}$ be independent uniform $(0, 1)$ random variables and consider the nth smallest value of this set — call it $U_{(n)}$. Now $U_{(n)}$ will equal x if, of the $n + m - 1$ variables,

(i) $n - 1$ are smaller than x,
(ii) one equals x,
(iii) $m - 1$ are greater than x.

Hence, if the $n + m - 1$ uniform random variables are partitioned into three subsets of sizes $n - 1$, 1, and $m - 1$ the probability (density) that each of the variables in the first set is less than x, the variable in the second set equals x, and all the variables in the third set are greater than x is given by

$$(P\{U < x\})^{n-1} f_u(x)(P\{U > x\})^{m-1} = x^{n-1}(1 - x)^{m-1}$$

Hence, as there are $(n + m - 1)!/(n - 1)!(m - 1)!$ possible partitions, it follows that $U_{(n)}$ is beta with parameters (n, m).

Thus, one way to simulate from the beta distribution is to find the nth smallest of a set of $n + m - 1$ random numbers. However, when n and m are large, this procedure is not particularly efficient.

For another approach consider a Poisson process with rate 1, and recall that given S_{n+m}, the time of the $(n + m)$th event, the set of the first $n + m - 1$ event times is distributed independently and uniformly on $(0, S_{n+m})$. Hence, given S_{n+m}, the nth smallest of the first $n + m - 1$ event times — that is S_n — is distributed as the nth smallest of a set of $n + m - 1$ uniform $(0, S_{n+m})$ random variables. But from the above we can thus conclude that S_n/S_{n+m} has a beta distribution with parameters (n, m).

Therefore, if $U_1, \ldots, U_{n+m}$ are random numbers,

$$\frac{-\log \Pi_{i=1}^{n} U_i}{-\log \Pi_{i=1}^{m+n} U_i} \quad \text{is beta with parameters } (n, m)$$

By writing the preceding as

$$\frac{-\log \Pi_{i=1}^{n} U_i}{-\log \Pi_1^{n} U_i - \log \Pi_{n+1}^{n+m} U_i}$$

we see that it has the same distribution as $X/(X + Y)$ where X and Y are independent gamma random variables with respective parameters $(n, 1)$ and $(m, 1)$. Hence, when n and m are large, we can efficiently simulate a beta by first simulating two gamma random variables.

11.3.5.   The Exponential Distribution — The Von Neumann Algorithm

As we have seen, an exponential random variable with rate 1 can be simulated by computing the negative of the logarithm of a random number. Most computer programs for computing a logarithm, however, involve a power series expansion, and so it might be useful to have at hand a second method that is computationally easier. We now present such a method due to Von Neumann.

To begin let $U_1, U_2, \ldots$ be independent uniform $(0, 1)$ random variables and define $N, N \geq 2$, by

$$N = \min\{n: U_1 \geq U_2 \geq \cdots \geq U_{n-1} < U_n\}$$

That is, N is the first random number that is greater than its predecessor. Let us now compute the joint distribution of N and U_1:

$$P\{N > n, U_1 \leq y\} = \int_0^1 P\{N > n, U_1 \leq y \mid U_1 = x\} \, dx$$

$$= \int_0^y P\{N > n \mid U_1 = x\} \, dx$$

Now, given that $U_1 = x$, N will be greater than n if $x \geq U_2 \cdots \geq U_n$ or, equivalently, if

(a) $$U_i \leq x, \qquad i = 2, \ldots, n$$

and

(b) $$U_2 \geq \cdots \geq U_n$$

Now, (a) has probability x^{n-1} of occurring and given (a), since all of the $(n-1)!$ possible rankings of $U_2, \ldots, U_n$ are equally likely, (b) has probability $1/(n-1)!$ of occurring. Hence,

$$P\{N > n \mid U_1 = x\} = \frac{x^{n-1}}{(n-1)!}$$

and so

$$P\{N > n, U_1 \leqslant y\} = \int_0^y \frac{x^{n-1}}{(n-1)!} \, dx = \frac{y^n}{n!}$$

which yields that

$$P\{N = n, U_1 \leqslant y\} = P\{N > n - 1, U_1 \leqslant y\} - P\{N > n, U_1 \leqslant y\}$$

$$= \frac{y^{n-1}}{(n-1)!} - \frac{y^n}{n!}$$

Upon summing over all the even integers, we see that

$$P\{N \text{ is even}, U_1 \leqslant y\} = y - \frac{y^2}{2!} + \frac{y^3}{3!} - \frac{y^4}{4!} - \cdots$$

$$= 1 - e^{-y} \tag{11.5}$$

We are now ready for the following algorithm for generating an exponential random variable with rate 1.

Step 1: Generate uniform random numbers $U_1, U_2, \ldots$ stopping at $N = \min\{n: U_1 \geqslant \cdots \geqslant U_{n-1} < U_n\}$.

Step 2: If N is even accept that run, and go to step 3. If N is odd reject the run, and return to step 1.

Step 3: Set X equal to the number of failed runs plus the first random number in the successful run.

To show that X is exponential with rate 1, first note that the probability of a successful run is, from Equation (11.5) with $y = 1$,

$$P\{N \text{ is even}\} = 1 - e^{-1}$$

Now, in order for X to exceed x, the first $[x]$ runs must all be unsuccessful and the next run must either be unsuccessful or be successful but have $U_1 > x - [x]$ (where $[x]$ is the largest integer not exceeding x). As

$$P\{N \text{ even}, U_1 > y\} = P\{N \text{ even}\} - P\{N \text{ even}, U_1 \leqslant y\}$$

$$= 1 - e^{-1} - (1 - e^{-y}) = e^{-y} - e^{-1}$$

we see that

$$P\{X > x\} = e^{-[x]}[e^{-1} + e^{-(x-[x])} - e^{-1}] = e^{-x}$$

which yields the result.

Let T denote the number of trials needed to generate a successful run. As each trial is a success with probability $1 - e^{-1}$ it follows that T is geometric with means $1/(1 - e^{-1})$. If we let N_i denote the number of uniform random variables used on the ith run, $i \geq 1$, then T (being the first run i for which N_i is even) is a stopping time for this sequence. Hence, by Wald's equation, the mean number of uniform random variables needed by this algorithm is given by

$$E\left[\sum_{i=1}^{T} N_i\right] = E[N]E[T]$$

Now,

$$E[N] = \sum_{n=0}^{\infty} P\{N > n\}$$

$$= 1 + \sum_{n=1}^{\infty} P\{U_1 \geq \cdots \geq U_n\}$$

$$= 1 + \sum_{n=1}^{\infty} 1/n! = e$$

and so

$$E\left[\sum_{i=1}^{T} N_i\right] = \frac{e}{1 - e^{-1}} \approx 4.3$$

Hence, this algorithm, which computationally speaking is quite easy to perform, requires on the average about 4.3 random numbers to execute.

11.4. Simulating from Discrete Distributions

All of the general methods for simulating from continuous distributions have analogs in the discrete case. For instance, if we want to simulate a random variable X having probability mass function

$$P\{X = x_j\} = P_j, \quad j = 1, \ldots, \quad \sum_j P_j = 1$$

We can use the following discrete time analog of the inverse transform technique.

> *To simulate X for which* $P\{X = x_j\} = P_j$
> let U be uniformly distributed over $(0, 1)$, and set

$$X = \begin{cases} x_1, & \text{if } U < P_1 \\ x_2, & \text{if } P_1 < U < P_1 + P_2 \\ \vdots & \\ x_j, & \text{if } \sum_1^{j-1} P_i < U < \sum_1^j P_i \\ \vdots & \end{cases}$$

As,

$$P\{X = x_j\} = P\left\{ \sum_1^{j-1} P_i < U < \sum_1^j P_i \right\} = P_j$$

we see that X has the desired distribution.

Example 11.7 (The Geometric Distribution): Suppose we want to simulate X such that

$$P\{X = i\} = p(1 - p)^{i-1}, \qquad i \geqslant 1$$

As

$$\sum_{i=1}^{j-1} P\{X = i\} = 1 - P\{X > j - 1\} = 1 - (1 - p)^{j-1}$$

we can simulate such a random variable by generating a random number U and then setting X equal to that value j for which

$$1 - (1 - p)^{j-1} < U < 1 - (1 - p)^j$$

or, equivalently, for which

$$(1 - p)^j < 1 - U < (1 - p)^{j-1}$$

As $1 - U$ has the same distribution as U, we can thus define X by

$$X = \min\{j : (1 - p)^j < U\} = \min\left\{ j : j > \frac{\log U}{\log(1 - p)} \right\}$$

$$= 1 + \left[\frac{\log U}{\log(1 - p)} \right] \quad \blacklozenge$$

As in the continuous case, special simulation techniques have been developed for the more common discrete distributions. We now present certain of these.

Example 11.8   (Simulating a Binomial Random Variable): A binomial (n, p) random variable can be most easily simulated by recalling that it can be expressed as the sum of n independent Bernoulli random variables. That is, if $U_1, \ldots, U_n$ are independent uniform $(0, 1)$ variables, then letting

$$X_i = \begin{cases} 1, & \text{if } U_i < p \\ 0, & \text{otherwise} \end{cases}$$

it follows that $X \equiv \Sigma_{i=1}^{n} X_i$ is a binomial random variable with parameters n and p.

One difficulty with the above procedure is that it requires the generation of n random numbers. To show how to reduce the number of random numbers needed, note first that the above procedure does not use the actual value of a random number U but only whether or not it exceeds p. Using this and the result that the conditional distribution of U given that $U < p$ is uniform on $(0, p)$ and the conditional distribution of U given that $U > p$ is uniform on $(p, 1)$, we now show how we can simulate a binomial (n, p) random variable using only a single random number:

Step 1:   Let $\alpha = 1/p$, $\beta = 1/(1 - p)$.
Step 2:   Set $k = 0$.
Step 3:   Generate a uniform random number U.
Step 4:   If $k = n$ stop. Otherwise reset k to equal $k + 1$.
Step 5:   If $U \leqslant p$ set $X_k = 1$ and reset U to equal αU. If $U > p$ set $X_k = 0$ and reset U to equal $\beta(U - p)$. Return to step 4.

This procedure generates $X_1, \ldots, X_n$ and $X = \Sigma_{i=1}^{n} X_i$ is the desired random variable. It works by noting whether $U_k \leqslant p$ or $U_k > p$; in the former case it takes U_{k+1} to equal U_k/p, and in the latter case it takes U_{k+1} to equal $(U_k - p)/(1 - p)$.*   ◆

Example 11.9   (Simulating a Poisson Random Variable): To simulate a Poisson random variable with mean λ, generate independent uniform $(0, 1)$ random variables $U_1, U_2, \ldots$ stopping at

$$N + 1 = \min \left\{ n: \prod_{i=1}^{n} U_i < e^{-\lambda} \right\}$$

---

* Because of computer round-off errors, a single random number should not be continuously used when n is large.

The random variable N has the desired distribution, which can be seen by noting that

$$N = \max \left\{ n: \sum_{i=1}^{n} -\log U_i < \lambda \right\}$$

But $-\log U_i$ is exponential with rate 1, and so if we interpret $-\log U_i$, $i \geq 1$, as the interarrival times of a Poisson process having rate 1, we see that $N = N(\lambda)$ would equal the number of events by time λ. Hence N is Poisson with mean λ.

When λ is large we can reduce the amount of computation in the above simulation of $N(\lambda)$, the number of events by time λ of a Poisson process having rate 1, by first choosing an integer m and simulating S_m, the time of the mth event of the Poisson process and then simulating $N(\lambda)$ according to the conditional distribution of $N(\lambda)$ given S_m. Now the conditional distribution of $N(\lambda)$ given S_m is as follows:

$$N(\lambda) \mid S_m = s \sim m + \text{Poisson}(\lambda - s), \qquad \text{if } s < \lambda$$

$$N(\lambda) \mid S_m = s \sim \text{Binomial}\left(m - 1, \frac{\lambda}{s} \right), \qquad \text{if } s > \lambda$$

where $\sim$ means "has the distribution of." This follows since if the mth event occurs at time s, where $s < \lambda$, then the number of events by time λ is m plus the number of events in (s, λ). On the other hand given that $S_m = s$ the set of times at which the first $m - 1$ events occur has the same distribution as a set of $m - 1$ uniform $(0, s)$ random variables (see Section 5.3.5). Hence, when $\lambda < s$, the number of these which occur by time λ is binomial with parameters $m - 1$ and λ/s. Hence, we can simulate $N(\lambda)$ by first simulating S_m and then simulate either $P(\lambda - S_m)$, a Poisson random variable with mean $\lambda - S_m$ when $S_m < \lambda$, or simulate $\text{Bin}(m - 1, \lambda/S_m)$, a binomial random variable with parameters $m - 1$, and λ/S_m, when $S_m > \lambda$; and then setting

$$N(\lambda) = \begin{cases} m + P(\lambda - S_m), & \text{if } S_m < \lambda \\ \text{Bin}(m - 1, \lambda/S_m), & \text{if } S_m > \lambda \end{cases}$$

In the preceding it has been found computationally effective to let m be approximately $\frac{7}{8}\lambda$. Of course, S_m is simulated by simulating from a gamma (m, λ) distribution via an approach that is computationally fast when m is large (see Section 11.3.3). ✦

There are also rejection and hazard rate methods for discrete distributions but we leave their development as exercises. However, there is a technique available for simulating finite discrete random variables — called the alias method—which, though requiring some setup time, is very fast to implement.

11.4.1. The Alias Method

In what follows, the quantities $\mathbf{P}$, $\mathbf{P}^{(k)}$, $\mathbf{Q}^{(k)}$, $k \leqslant n - 1$ will represent probability mass functions on the integers $1, 2, \ldots, n$—that is, they will be n-vectors of nonnegative numbers summing to 1. In addition, the vector $\mathbf{P}^{(k)}$ will have at most k nonzero components, and each of the $\mathbf{Q}^{(k)}$ will have at most two nonzero components. We show that any probability mass function $\mathbf{P}$ can be represented as an equally weighted mixture of $n - 1$ probability mass functions $\mathbf{Q}$ (each having at most two nonzero components). That is, we show that for suitably defined $\mathbf{Q}^{(1)}, \ldots, \mathbf{Q}^{(n-1)}$, $\mathbf{P}$ can be expressed as

$$\mathbf{P} = \frac{1}{n-1} \sum_{k=1}^{n-1} \mathbf{Q}^{(k)} \tag{11.6}$$

As a prelude to presenting the method for obtaining this representation, we will need the following simple lemma whose proof is left as an exercise.

Lemma 11.5   Let $\mathbf{P} = \{P_i, i = 1, \ldots, n\}$ denote a probability mass function, then

(a) there exists an i, $1 \leqslant i \leqslant n$, such that $P_i < 1/(n - 1)$, and
(b) for this i, there exists a j, $j \neq i$, such that $P_i + P_j \geqslant 1/(n - 1)$.

Before presenting the general technique for obtaining the representation of Equation (11.6), let us illustrate it by an example.

Example 11.10   Consider the three-point distribution $\mathbf{P}$ with $P_1 = \frac{7}{16}$, $P_2 = \frac{1}{2}$, $P_3 = \frac{1}{16}$. We start by choosing i and j such that they satisfy the conditions of Lemma 11.5. As $P_3 < \frac{1}{2}$ and $P_3 + P_2 > \frac{1}{2}$, we can work with $i = 3$ and $j = 2$. We will now define a 2-point mass function $\mathbf{Q}^{(1)}$ putting all of its weight on 3 and 2 and such that $\mathbf{P}$ will be expressible as an equally weighted mixture between $\mathbf{Q}^{(1)}$ and a second 2-point mass function $\mathbf{Q}^{(2)}$. Secondly, all of the mass of point 3 will be contained in $\mathbf{Q}^{(1)}$. As we will have

$$P_j = \tfrac{1}{2}(Q_j^{(1)} + Q_j^{(2)}), \qquad j = 1, 2, 3 \tag{11.7}$$

and, by the preceding, $Q_3^{(2)}$ is supposed to equal 0, we must therefore take

$$Q_3^{(1)} = 2P_3 = \tfrac{1}{8}, \qquad Q_2^{(1)} = 1 - Q_3^{(1)} = \tfrac{7}{8}, \qquad Q_1^{(1)} = 0$$

To satisfy Equation (11.7), we must then set

$$Q_3^{(2)} = 0, \qquad Q_2^{(2)} = 2P_2 - \tfrac{7}{8} = \tfrac{1}{8}, \qquad Q_1^{(2)} = 2P_1 = \tfrac{7}{8}$$

Hence, we have the desired representation in this case. Suppose now that the original distribution was the following 4-point mass function:

$$P_1 = \tfrac{7}{16}, \qquad P_2 = \tfrac{1}{4}, \qquad P_3 = \tfrac{1}{8}, \qquad P_4 = \tfrac{3}{16}$$

Now, $P_3 < \tfrac{1}{3}$ and $P_3 + P_1 > \tfrac{1}{3}$. Hence our initial 2-point mass function—$\mathbf{Q}^{(1)}$—will concentrate on points 3 and 1 (giving no weights to 2 and 4). As the final representation will give weight $\tfrac{1}{3}$ to $\mathbf{Q}^{(1)}$ and in addition the other $\mathbf{Q}^{(j)}$, $j = 2, 3$, will not give any mass to the value 3, we must have that

$$\tfrac{1}{3} Q_3^{(1)} = P_3 = \tfrac{1}{8}$$

Hence,

$$Q_3^{(1)} = \tfrac{3}{8}, \qquad Q_1^{(1)} = 1 - \tfrac{3}{8} = \tfrac{5}{8}$$

Also, we can write

$$\mathbf{P} = \tfrac{1}{3}\mathbf{Q}^{(1)} + \tfrac{2}{3}\mathbf{P}^{(3)}$$

where $\mathbf{P}^{(3)}$, to satisfy the above, must be the vector

$$\mathbf{P}_1^{(3)} = \tfrac{3}{2}(P_1 - \tfrac{1}{3}Q_1^{(1)}) = \tfrac{11}{32},$$

$$\mathbf{P}_2^{(3)} = \tfrac{3}{2}P_2 = \tfrac{3}{8},$$

$$\mathbf{P}_3^{(3)} = 0,$$

$$\mathbf{P}_4^{(3)} = \tfrac{3}{2}P_4 = \tfrac{9}{32}$$

Note that $\mathbf{P}^{(3)}$ gives no mass to the value 3. We can now express the mass function $\mathbf{P}^{(3)}$ as an equally weighted mixture of two point mass functions $\mathbf{Q}^{(2)}$ and $\mathbf{Q}^{(3)}$, and we will end up with

$$\mathbf{P} = \tfrac{1}{3}\mathbf{Q}^{(1)} + \tfrac{2}{3}(\tfrac{1}{2}\mathbf{Q}^{(2)} + \tfrac{1}{2}\mathbf{Q}^{(3)})$$

$$= \tfrac{1}{3}(\mathbf{Q}^{(1)} + \mathbf{Q}^{(2)} + \mathbf{Q}^{(3)})$$

(We leave it as an exercise for the reader to fill in the details.)   ✦

The preceding example outlines the following general procedure for writing the n-point mass function $\mathbf{P}$ in the form of Equation (11.6) where each of the $\mathbf{Q}^{(i)}$ are mass functions giving all their mass to at most 2 points. To start, we choose i and j satisfying the conditions of Lemma 11.5. We now define the mass function $\mathbf{Q}^{(1)}$ concentrating on the points i and j and which will contain all of the mass for point i by noting that, in the representation of Equation (11.6), $Q_i^{(k)} = 0$ for $k = 2, \ldots, n - 1$, implying that

$$Q_i^{(1)} = (n - 1)P_i, \qquad \text{and so } Q_j^{(1)} = 1 - (n - 1)P_i$$

Writing

$$\mathbf{P} = \frac{1}{n-1}\mathbf{Q}^{(1)} + \frac{n-2}{n-1}\mathbf{P}^{(n-1)} \tag{11.8}$$

where $\mathbf{P}^{(n-1)}$ represents the remaining mass, we see that

$P_i^{(n-1)} = 0,$

$$P_j^{(n-1)} = \frac{n-1}{n-2}\left(P_j - \frac{1}{n-1}Q_j^{(1)}\right) = \frac{n-1}{n-2}\left(P_i + P_j - \frac{1}{n-1}\right),$$

$$P_k^{(n-1)} = \frac{n-1}{n-2}P_k, \qquad k \neq i \text{ or } j$$

That the foregoing is indeed a probability mass function is easily checked—
for instance, the nonnegativity of $P_j^{(n-1)}$ follows from the fact that j was
chosen so that $P_i + P_j \geqslant 1/(n-1)$.

We may now repeat the foregoing procedure on the $(n-1)$-point probability mass function $\mathbf{P}^{(n-1)}$ to obtain

$$\mathbf{P}^{(n-1)} = \frac{1}{n-2}\mathbf{Q}^{(2)} + \frac{n-3}{n-2}\mathbf{P}^{(n-2)}$$

and thus from Equation (11.8) we have

$$\mathbf{P} = \frac{1}{n-1}\mathbf{Q}^{(1)} + \frac{1}{n-1}\mathbf{Q}^{(2)} + \frac{n-3}{n-1}\mathbf{P}^{(n-2)}$$

We now repeat the procedure on $\mathbf{P}^{(n-2)}$ and so on until we finally obtain

$$\mathbf{P} = \frac{1}{n-1}(\mathbf{Q}^{(1)} + \cdots + \mathbf{Q}^{(n-1)})$$

In this way we are able to represent $\mathbf{P}$ as an equally weighted mixture of
$n-1$ two-point mass functions. We can now easily simulate from $\mathbf{P}$ by first
generating a random integer N equally likely to be either $1, 2, \ldots, n-1$. If
the resulting value N is such that $\mathbf{Q}^{(N)}$ puts positive weight only on the
points i_N and j_N, then we can set X equal to i_N if a second random number
is less than $Q_{i_N}^{(N)}$ and equal to j_N otherwise. The random variable X will have
probability mass function $\mathbf{P}$. That is, we have the following procedure for
simulating from $\mathbf{P}$.

Step 1:   Generate U_1 and set $N = 1 + [(n-1)U_1]$.
Step 2:   Generate U_2 and set

$$X = \begin{cases} i_N, & \text{if } U_2 < Q_{i_N}^{(N)} \\ j_N, & \text{otherwise} \end{cases}$$

Remarks (i) The above is called the alias method because by a renumbering of the $\mathbf{Q}$s we can always arrange things so that for each k, $Q_k^{(k)} > 0$. (That is, we can arrange things so that the kth 2-point mass function gives positive weight to the value k.) Hence, the procedure calls for simulating N, equally likely to be $1, 2, \ldots, n-1$, and then if $N = k$ it either accepts k as the value of X, or it accepts for the value of X the "alias" of k (namely, the other value that $\mathbf{Q}^{(k)}$ gives positive weight).

(ii) Actually, it is not necessary to generate a new random number in step 2. Because $N - 1$ is the integer part of $(n-1)U_1$, it follows that the remainder $(n-1)U_1 - (N-1)$ is independent of U_1 and is uniformly distributed in $(0, 1)$. Hence, rather than generating a new random number U_2 in step 2, we can use $(n-1)U_1 - (N-1) = (n-1)U_1 - [(n-1)U_1]$.

Example 11.11 Let us return to the problem of Example 11.1 which considers a list of n, not necessarily distinct, items. Each item has a value — $v(i)$ being the value of the item in position i — and we are interested in estimating

$$v = \sum_{i=1}^{n} v(i)/m(i)$$

where $m(i)$ is the number of times the item in position i appears on the list. In words, v is the sum of the values of the (distinct) items on the list.

To estimate v, note that if X is a random variable such that

$$P\{X = i\} = v(i) \bigg/ \sum_{1}^{n} v(j), \qquad i = 1, \ldots, n$$

then

$$E[1/m(X)] = \frac{\Sigma_i v(i)/m(i)}{\Sigma_j v(j)} = v \bigg/ \sum_{j=1}^{n} v(j)$$

Hence, we can estimate v by using the alias (or any other) method to generate independent random variables $X_1, \ldots, X_k$ having the same distribution as X and then estimating v by

$$v \approx \sum_{j=1}^{n} v(j) \frac{\Sigma_{i=1}^{k} 1/m(X_i)}{k} \quad \blacklozenge$$

11.5. Stochastic Processes

One can easily simulate a stochastic process by simulating a sequence of random variables. For instance, to simulate the first t time units of a renewal process having interarrival distribution F we can simulate independent random variables $X_1, X_2, \ldots$ having distribution F, stopping at

$$N = \min\{n: X_1 + \cdots + X_n > t\}$$

The X_i, $i \geq 1$, represent the interarrival times of the renewal process and so the preceding simulation yields $N - 1$ events by time t—the events occurring at times $X_1, X_1 + X_2, \ldots, X_1 + \cdots + X_{N-1}$.

Actually there is another approach for simulating a Poisson process that is quite efficient. Suppose we want to simulate the first t time units of a Poisson process having rate λ. To do so, we can first simulate $N(t)$, the number of events by t, and then use the result that given the value of $N(t)$, the set of $N(t)$ event times is distributed as a set of n independent uniform $(0, t)$ random variables. Hence, we start by simulating $N(t)$, a Poisson random variable with mean λt (by one of the methods given in Example 11.9). Then, if $N(t) = n$, generate a new set of n random numbers—call them $U_1, \ldots, U_n$—and $\{tU_1, \ldots, tU_n\}$ will represent the set of $N(t)$ event times. If we could stop here this would be much more efficient than simulating the exponentially distributed interarrival times. However, we usually desire the event times in increasing order—for instance, for $s < t$,

$$N(s) = \text{number of } U_i: tU_i \leq s$$

and so to compute the function $N(s)$, $s \leq t$, it is best to first order the values U_i, $i = 1, \ldots, n$ before multiplying by t. However, in doing so one should not use an all-purpose sorting algorithm, such as quick sort (see Example 3.14), but rather one that takes into account that the elements to be sorted come from a uniform $(0, 1)$ population. Such a sorting algorithm, of n uniform $(0, 1)$ variables, is as follows: Rather than a single list to be sorted of length n we will consider n ordered, or linked, lists of random size. The value U will be put in list i if its value is between $(i - 1)/n$ and i/n—that is, U is put in list $[nU] + 1$. The individual lists are then ordered and the total linkage of all the lists is the desired ordering. As almost all of the n lists will be of relatively small size [for instance, if $n = 1000$ the mean number of lists of size greater than 4 is (using the Poisson approximation to the binomial) approximately equal to $1000(1 - \frac{65}{24}e^{-1}) \simeq 4$] the sorting of individual lists will be quite quick, and so the running time of such an algorithm will be proportional to n (rather than to $n \log n$ as in the best all-purpose sorting algorithms).

An extremely important counting process for modeling purposes is the nonhomogeneous Poisson process, which relaxes the Poisson process assumption of stationary increments. Thus it allows for the possibility that the arrival rate need not be constant but can vary with time. However, there are few analytical studies that assume a nonhomogeneous Poisson arrival process for the simple reason that such models are not usually mathematically tractable. (For example, there is no known expression for the average customer delay in the single-server exponential service distribution queueing model which assumes a nonhomogeneous arrival process.)* Clearly such models are strong candidates for simulation studies.

11.5.1.   Simulating a Nonhomogeneous Poisson Process

We now present three methods for simulating a nonhomogeneous Poisson process having intensity function $\lambda(t)$, $0 \leqslant t < \infty$.

Method 1.   Sampling a Poisson Process

To simulate the first T time units of a nonhomogeneous Poisson process with intensity function $\lambda(t)$, let λ be such that

$$\lambda(t) \leqslant \lambda \qquad \text{for all } t \leqslant T$$

Now as shown in Chapter 5, such a nonhomogeneous Poisson process can be generated by a random selection of the event times of a Poisson process having rate λ. That is, if an event of a Poisson process with rate λ that occurs at time t is counted (independently of what has transpired previously) with probability $\lambda(t)/\lambda$ then the process of counted events is a nonhomogeneous Poisson process with intensity function $\lambda(t)$, $0 \leqslant t \leqslant T$. Hence, by simulating a Poisson process and then randomly counting its events, we can generate the desired nonhomogeneous Poisson process. We thus have the following procedure:

Generate independent random variables X_1, U_1, X_2, $U_2, \ldots$ where the X_i are exponential with rate λ and the U_i are random numbers, stopping at

$$N = \min \left\{ n\colon \sum_{i=1}^{n} X_i > T \right\}$$

Now let, for $j = 1, \ldots, N - 1$,

$$I_j = \begin{cases} 1, & \text{if } U_j \leqslant \lambda \left(\sum_{i=1}^{j} X_i \Big/ \lambda \right) \\ 0, & \text{otherwise} \end{cases}$$

and set

$$J = \{ j\colon I_j = 1 \}$$

Thus, the counting process having events at the set of times $\{ \sum_{i=1}^{j} X_i \colon j \in J \}$ constitutes the desired process.

---

*One queueing model that assumes a nonhomogeneous Poisson arrival process and is mathematically tractable is the infinite server model.

The foregoing procedure, referred to as the thinning algorithm (because it "thins" the homogeneous Poisson points) will clearly be most efficient, in the sense of having the fewest number of rejected event times, when $\lambda(t)$ is near λ throughout the interval. Thus, an obvious improvement is to break up the interval into subintervals and then use the procedure over each subinterval. That is, determine appropriate values k, $0 < t_1 < t_2 < \cdots < t_k < T$, $\lambda_1, \ldots, \lambda_{k+1}$, such that

$$\lambda(s) \leqslant \lambda_i \qquad \text{when } t_{i-1} \leqslant s < t_i, \, i = 1, \ldots, k+1 \quad \text{(where } t_0 = 0, \, t_{k+1} = T)$$

$$(11.9)$$

Now simulate the nonhomogeneous Poisson process over the interval (t_{i-1}, t_i) by generating exponential random variables with rate λ_i and accepting the generated event occurring at time s, $s \in (t_{i-1}, t_i)$, with probability $\lambda(s)/\lambda_i$. Because of the memoryless property of the exponential and the fact that the rate of an exponential can be changed upon multiplication by a constant, it follows that there is no loss of efficiency in going from one subinterval to the next. In other words, if we are at $t \in [t_{i-1}, t_i)$ and generate X, an exponential with rate λ_i, which is such that $t + X > t_i$ then we can use $\lambda_i[X - (t_i - t)]/\lambda_{i+1}$ as the next exponential with rate λ_{i+1}. Thus, we have the following algorithm for generating the first t time units of a nonhomogeneous Poisson process with intensity function $\lambda(s)$ when the relations (11.9) are satisfied. In the algorithm, t will represent the present time and I the present interval (that is, $I = i$ when $t_{i-1} \leqslant t < t_i$).

Step 1: $t = 0, I = 1$.
Step 2: Generate an exponential random variable X having rate λ_I.
Step 3: If $t + X < t_I$, reset $t = t + X$, generate a random number U, and accept the event time t if $U \leqslant \lambda(t)/\lambda_I$. Return to step 2.
Step 4: Step reached if $t + X \geqslant t_I$. Stop if $I = k + 1$. Otherwise, reset $X = (X - t_I + t)\lambda_I/\lambda_{I+1}$. Also reset $t = t_I$ and $I = I + 1$, and go to step 3.

Suppose now that over some subinterval (t_{i-1}, t_i) it follows that $\underline{\lambda}_i > 0$ where

$$\underline{\lambda}_i \equiv \text{infimum}\{\lambda(s): t_{i-1} \leqslant s < t_i\}.$$

In such a situation, we should not use the thinning algorithm directly but rather should first simulate a Poisson process with rate $\underline{\lambda}_i$ over the desired interval and then simulate a nonhomogeneous Poisson process with the intensity function $\lambda(s) = \lambda(s) - \underline{\lambda}_i$ when $s \in (t_{i-1}, t_i)$. (The final exponential generated for the Poisson process, which carries one beyond the desired boundary, need not be wasted but can be suitably transformed so as to be reusable.) The superposition (or, merging) of the two processes yields the desired process over the interval. The reason for doing it this way is that it saves the need to generate uniform random variables for a Poisson distributed number, with mean $\lambda_i(t_i - t_{i-1})$ of the event times. For instance, consider the case where

$$\lambda(s) = 10 + s, \qquad 0 < s < 1$$

Using the thinning method with $\lambda = 11$ would generate an expected number of 11 events each of which would require a random number to determine whether or not to accept it. On the other hand, to generate a Poisson process with rate 10 and then merge it with a generated nonhomogeneous Poisson process with rate $\lambda(s) = s$, $0 < s < 1$, would yield an equally distributed number of event times but with the expected number needing to be checked to determine acceptance being equal to 1.

Another way to make the simulation of nonhomogeneous Poisson processes more efficient is to make use of superpositions. For instance, consider the process where

$$\lambda(t) = \begin{cases} \exp\{t^2\}, & 0 < t < 1.5 \\ \exp\{2.25\}, & 1.5 < t < 2.5 \\ \exp\{(4-t)^2\}, & 2.5 < t < 4 \end{cases}$$

A plot of this intensity function is given in Figure 11.3. One way of simulating this process up to time 4 is to first generate a Poisson process with rate 1 over this interval; then generate a Poisson process with rate $e - 1$

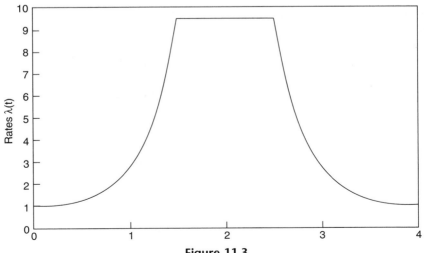

Figure 11.3.

over this interval and accept all events in $(1, 3)$ and only accept an event at time t which is not contained in $(1, 3)$ with probability $[\lambda(t) - 1]/(e - 1)$; then generate a Poisson process with rate $e^{2.25} - e$ over the interval $(1, 3)$, accepting all event times between 1.5 and 2.5 and any event time t outside this interval with probability $[\lambda(t) - e]/(e^{2.25} - e)$. The superposition of these processes is the desired nonhomogeneous Poisson process. In other words, what we have done is to break up $\lambda(t)$ into the following nonnegative parts:

$$\lambda(t) = \lambda_1(t) + \lambda_2(t) + \lambda_3(t), \qquad 0 < t < 4$$

where

$$\lambda_1(t) \equiv 1,$$

$$\lambda_2(t) = \begin{cases} \lambda(t) - 1, & 0 < t < 1 \\ e - 1, & 1 < t < 3 \\ \lambda(t) - 1, & 3 < t < 4 \end{cases}$$

$$\lambda_3(t) = \begin{cases} \lambda(t) - e, & 1 < t < 1.5 \\ e^{2.25} - e, & 1.5 < t < 2.5 \\ \lambda(t) - e, & 2.5 < t < 3 \\ 0, & 3 < t < 4 \end{cases}$$

and where the thinning algorithm (with a single interval in each case) was used to simulate the constituent nonhomogeneous processes.

Method 2. Conditional Distribution of the Arrival Times

Recall the result for a Poisson process having rate λ that given the number of events by time T the set of event times are independent and identically distributed uniform $(0, T)$ random variables. Now suppose that each of these events is independently counted with a probability that is equal to $\lambda(t)/\lambda$ when the event occurred at time t. Hence, given the number of counted events, it follows that the set of times of these counted events are independent with a common distribution given by $F(s)$, where

$$F(s) = P\{\text{time} \leqslant s \,|\, \text{counted}\}$$

$$= \frac{P\{\text{time} \leqslant s, \text{counted}\}}{P\{\text{counted}\}}$$

$$= \frac{\int_0^T P\{\text{time} \leqslant s, \text{counted} \,|\, \text{time} = x\} \, dx/T}{P\{\text{counted}\}}$$

$$= \frac{\int_0^s \lambda(x) \, dx}{\int_0^T \lambda(x) \, dx}$$

The preceding (somewhat heuristic) argument thus shows that given n events of a nonhomogeneous Poisson process by time T the n event times are independent with a common density function

$$f(s) = \frac{\lambda(s)}{m(T)}, \quad 0 < s < T, \quad m(T) = \int_0^T \lambda(s)\, ds \tag{11.10}$$

Since $N(T)$, the number of events by time T, is Poisson distributed with mean $m(T)$, we can simulate the nonhomogeneous Poisson process by first simulating $N(T)$ and then simulating $N(T)$ random variables from the density (11.10).

Example 11.12 If $\lambda(s) = cs$, then we can simulate the first T time units of the nonhomogeneous Poisson process by first simulating $N(T)$, a Poisson random variable having mean $m(T) = \int_0^T cs\, ds = CT^2/2$, and then simulating $N(T)$ random variables having distribution

$$F(s) = \frac{s^2}{T^2}, \quad 0 < s < T$$

Random variables having the preceding distribution either can be simulated by use of the inverse transform method (since $F^{-1}(U) = T\sqrt{U}$) or by noting that F is the distribution function of $\max(TU_1, TU_2)$ when U_1 and U_2 are independent random numbers. ◆

If the distribution function specified by Equation (11.10) is not easily invertible, we can always simulate from (11.10) by using the rejection method where we either accept or reject simulated values of uniform $(0, T)$ random variables. That is, let $h(s) = 1/T, 0 < s < T$. Then

$$\frac{f(s)}{h(s)} = \frac{T\lambda(s)}{m(T)} \leqslant \frac{\lambda T}{m(T)} \equiv C$$

where λ is a bound on $\lambda(s)$, $0 \leqslant s \leqslant T$. Hence, the rejection method is to generate random numbers U_1 and U_2 and then accept TU_1 if

$$U_2 \leqslant \frac{f(TU_1)}{C h(TU_1)}$$

or, equivalently, if

$$U_2 \leqslant \frac{\lambda(TU_1)}{\lambda}$$

Method 3. Simulating the Event Times

The third method we shall present for simulating a nonhomogeneous

Poisson process having intensity function $\lambda(t)$, $t \geqslant 0$ is probably the most basic approach — namely, to simulate the successive event times. So let $X_1, X_2, \ldots$ denote the event times of such a process. As these random variables are dependent we will use the conditional distribution approach to simulation. Hence, we need the conditional distribution of X_i given $X_1, \ldots, X_{i-1}$.

To start, note that if an event occurs at time x then, independent of what has occurred prior to x, the time until the next event has the distribution F_x given by

$$\bar{F}_x(t) = P\{0 \text{ events in } (x, x+t) \,|\, \text{event at } x\}$$

$$= P(0 \text{ events in } (x, x+t)\} \qquad \text{by independent increments}$$

$$= \exp\left\{-\int_0^t \lambda(x+y)\,dy\right\}$$

Differentiation yields that the density corresponding to F_x is

$$f_x(t) = \lambda(x+t)\exp\left\{-\int_0^t \lambda(x+y)\,dy\right\}$$

implying that the hazard rate function of F_x is

$$r_x(t) = \frac{f_x(t)}{\bar{F}_x(t)} = \lambda(x+t)$$

We can now simulate the event times $X_1, X_2, \ldots$ by simulating X_1 from F_0; then if the simulated value of X_1 is x_1, simulate X_2 by adding x_1 to a value generated from F_{x_1}, and if this sum is x_2 simulate X_3 by adding x_2 to a value generated from F_{x_2}, and so on. The method used to simulate from these distributions should depend, of course, on the form of these distributions. However, it is interesting to note that if we let λ be such that $\lambda(t) \leqslant \lambda$ and use the hazard rate method to simulate, then we end up with the approach of Method 1 (we leave the verification of this fact as an exercise). Sometimes, however, the distributions F_x can be easily inverted and so the inverse transform method can be applied.

Example 11.13   Suppose that $\lambda(x) = 1/(x+a)$, $x \geqslant 0$. Then

$$\int_0^t \lambda(x+y)\,dy = \log\left(\frac{x+a+t}{x+a}\right)$$

Hence,

$$F_x(t) = 1 - \frac{x+a}{x+a+t} = \frac{t}{x+a+t}$$

and so

$$F_x^{-1}(u) = (x + a)\frac{u}{1 - u}$$

We can, therefore, simulate the successive event times $X_1, X_2, \ldots$ by generating $U_1, U_2, \ldots$ and then setting

$$X_1 = \frac{aU_1}{1 - U_1},$$

$$X_2 = (X_1 + a)\frac{U_2}{1 - U_2} + X_1$$

and, in general,

$$X_j = (X_{j-1} + a)\frac{U_j}{1 - U_j} + X_{j-1}, \qquad j \geq 2 \quad \blacklozenge$$

11.5.2. Simulating a Two-Dimensional Poisson Process

A point process consisting of randomly occurring points in the plane is said to be a two-dimensional Poisson process having rate λ if

(a) the number of points in any given region of area A is Poisson distributed with mean λA; and
(b) the numbers of points in disjoint regions are independent.

For a given fixed point $\mathbf{O}$ in the plane, we now show how to simulate events occurring according to a two-dimensional Poisson process with rate λ in a circular region of radius r centered about $\mathbf{O}$. Let R_i, $i \geq 1$, denote the distance between $\mathbf{O}$ and its ith nearest Poisson point, and let $C(a)$ denote the circle of radius a centered at $\mathbf{O}$. Then

$$P\{\pi R_1^2 > b\} = P\left\{R_1 > \sqrt{\frac{b}{\pi}}\right\} = P\{\text{no points in } C(\sqrt{b/\pi})\} = e^{-\lambda b}$$

Also, with $C(a_2) - C(a_1)$ denoting the region between $C(a_2)$ and $C(a_1)$:

$$P\{\pi R_2^2 - \pi R_1^2 > b \mid R_1 = r\}$$

$$= P\{R_2 > \sqrt{(b + \pi r^2)/\pi} \mid R_1 = r\}$$

$$= P\{\text{no points in } C(\sqrt{(b + \pi r^2)/\pi}) - C(r) \mid R_1 = r\}$$

$$= P\{\text{no points in } C(\sqrt{(b + \pi r^2)/\pi}) - C(r)\} \qquad \text{by } (b)$$

$$= e^{-\lambda b}$$

In fact, the same argument can be repeated to obtain the following.

Proposition 11.6   With $R_0 = 0$,

$$\pi R_i^2 - \pi R_{i-1}^2, \qquad i \geqslant 1,$$

are independent exponentials with rate λ.

In other words, the amount of area that needs to be traversed to encompass a Poisson point is exponential with rate λ. Since, by symmetry, the respective angles of the Poisson points are independent and uniformly distributed over $(0, 2\pi)$, we thus have the following algorithm for simulating the Poisson process over a circular region of radius r about **O**:

Step 1:   Generate independent exponentials with rate 1, $X_1, X_2, \ldots,$ stopping at

$$N = \min \left\{ n: \frac{X_1 + \cdots + X_n}{\lambda \pi} > r^2 \right\}$$

Step 2:   If $N = 1$, stop. There are no points in $C(r)$. Otherwise, for $i = 1, \ldots, N - 1$, set

$$R_i = \sqrt{(X_1 + \cdots + X_i)/\lambda \pi}$$

Step 3:   Generate independent uniform $(0, 1)$ random variables $U_1, \ldots, U_{N-1}$.

Step 4:   Return the $N - 1$ Poisson points in $C(r)$ whose polar coordinates are

$$(R_i, 2\pi U_i), \qquad i = 1, \ldots, N - 1$$

The preceding algorithm requires, on average, $1 + \lambda \pi r^2$ exponentials and an equal number of uniform random numbers. Another approach to simulating points in $C(r)$ is to first simulate N, the number of such points, and then use the fact that, given N, the points are uniformly distributed in $C(r)$. This latter procedure requires the simulation of N, a Poisson random variable with mean $\lambda \pi r^2$; we must then simulate N uniform points on $C(r)$, by simulating R from the distribution $F_R(a) = a^2/r^2$ (see Exercise 25) and θ from uniform $(0, 2\pi)$ and must then sort these N uniform values in increasing order of R. The main advantage of the first procedure is that it eliminates the need to sort.

The preceding algorithm can be thought of as the fanning out of a circle centered at **O** with a radius that expands continuously from 0 to r. The successive radii at which Poisson points are encountered is simulated by noting that the additional area necessary to encompass a Poisson point is always, independent of the past, exponential with rate λ. This technique can be used to simulate the process over noncircular regions. For instance, consider a nonnegative function $g(x)$, and suppose we are interested in

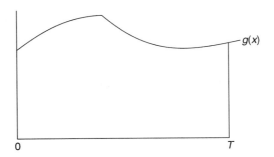

Figure 11.4.

simulating the Poisson process in the region between the x-axis and g with x going from 0 to T (see Figure 11.4). To do so we can start at the left-hand end and fan vertically to the right by considering the successive areas $\int_0^a g(x)\,dx$. Now if $X_1 < X_2 < \cdots$ denote the successive projections of the Poisson points on the x-axis, then analogous to Proposition 11.6, it will follow that (with $X_0 = 0$) $\lambda \int_{X_{i-1}}^{X_i} g(x)\,dx$, $i \geq 1$, will be independent exponentials with rate 1. Hence, we should simulate $\epsilon_1, \epsilon_2, \ldots$, independent exponentials with rate 1, stopping at

$$N = \min\left\{n : \epsilon_1 + \cdots + \epsilon_n > \lambda \int_0^T g(x)\,dx\right\}$$

and determine $X_1, \ldots, X_{N-1}$ by

$$\lambda \int_0^{X_1} g(x)\,dx = \epsilon_1,$$

$$\lambda \int_{X_1}^{X_2} g(x)\,dx = \epsilon_2,$$

$$\vdots$$

$$\lambda \int_{X_{N-2}}^{X_{N-1}} g(x)\,dx = \epsilon_{N-1}.$$

If we now simulate $U_1, \ldots, U_{N-1}$—independent uniform $(0,1)$ random numbers—then as the projection on the y-axis of the Poisson point whose x-coordinate is X_i, is uniform on $(0, g(X_i))$, it follows that the simulated Poisson points in the interval are $(X_i, U_i g(X_i))$, $i = 1, \ldots, N-1$.

Of course, the preceding technique is most useful when g is regular enough so that the foregoing equations can be solved for the X_i. For instance, if $g(x) = y$ (and so the region of interest is a rectangle), then

$$X_i = \frac{\epsilon_1 + \cdots + \epsilon_i}{\lambda y}, \qquad i = 1, \ldots, N - 1$$

and the Poisson points are

$$(X_i, yU_i), \qquad i = 1, \ldots, N - 1$$

11.6. Variance Reduction Techniques

Let $X_1, \ldots, X_n$ have a given joint distribution, and suppose we are interested in computing

$$\theta \equiv E[g(X_1, \ldots, X_n)]$$

where g is some specified function. It is often the case that it is not possible to analytically compute the above, and when such is the case we can attempt to use simulation to estimate θ. This is done as follows: Generate $X_1^{(1)}, \ldots, X_n^{(1)}$ having the same joint distribution as $X_1, \ldots, X_n$ and set

$$Y_1 = g(X_1^{(1)}, \ldots, X_n^{(1)})$$

Now, simulate a second set of random variables (independent of the first set) $X_1^{(2)}, \ldots, X_n^{(2)}$ having the distribution of $X_1, \ldots, X_n$ and set

$$Y_2 = g(X_1^{(2)}, \ldots, X_n^{(2)})$$

Continue this until you have generated k (some predetermined number) sets, and so have also computed $Y_1, Y_2, \ldots, Y_k$. Now, $Y_1, \ldots, Y_k$ are independent and identically distributed random variables each having the same distribution of $g(X_1, \ldots, X_n)$. Thus, if we let $\bar{Y}$ denote the average of these k random variables — that is,

$$\bar{Y} = \sum_{i=1}^{k} Y_i/k$$

then

$$E[\bar{Y}] = \theta,$$

$$E[(\bar{Y} - \theta)^2] = \text{Var}(\bar{Y})$$

Hence, we can use $\bar{Y}$ as an estimate of θ. As the expected square of the difference between $\bar{Y}$ and θ is equal to the variance of $\bar{Y}$, we would like this quantity to be as small as possible. [In the preceding situation, $\text{Var}(\bar{Y}) = \text{Var}(Y_i)/k$, which is usually not known in advance but must be estimated from the generated values $Y_1, \ldots, Y_n$.] We now present three general techniques for reducing the variance of our estimator.

11.6.1. Use of Antithetic Variables

In the preceding situation, suppose that we have generated Y_1 and Y_2, identically distributed random variables having mean θ. Now,

$$\text{Var}\left(\frac{Y_1 + Y_2}{2}\right) = \tfrac{1}{4}[\text{Var}(Y_1) + \text{Var}(Y_2) + 2\,\text{Cov}(Y_1, Y_2)]$$

$$= \frac{\text{Var}(Y_1)}{2} + \frac{\text{Cov}(Y_1, Y_2)}{2}$$

Hence, it would be advantageous (in the sense that the variance would be reduced) if Y_1 and Y_2 rather than being independent were negatively correlated. To see how we could arrange this, let us suppose that the random variables $X_1, \ldots, X_n$ are independent and, in addition, that each is simulated via the inverse transform technique. That is, X_i is simulated from $F_i^{-1}(U_i)$ where U_i is a random number and F_i is the distribution of X_i. Hence, Y_1 can be expressed as

$$Y_1 = g(F_1^{-1}(U_1), \ldots, F_n^{-1}(U_n))$$

Now, since $1 - U$ is also uniform over $(0, 1)$ whenever U is a random number (and is negatively correlated with U) it follows that Y_2 defined by

$$Y_2 = g(F_1^{-1}(1 - U_1), \ldots, F_n^{-1}(1 - U_n))$$

will have the same distribution as Y_1. Hence, if Y_1 and Y_2 were negatively correlated, then generating Y_2 by this means would lead to a smaller variance than if it were generated by a new set of random numbers. (In addition, there is a computational savings since rather than having to generate n additional random numbers, we need only subtract each of the previous n from 1.) The following theorem will be the key to showing that this technique—known as the use of antithetic variables—will lead to a reduction in variance whenever g is a monotone function.

Theorem 11.1 If $X_1, \ldots, X_n$ are independent, then, for any increasing functions f and g of n variables,

$$E[f(\mathbf{X})g(\mathbf{X})] \geq E[f(\mathbf{X})]E[g(\mathbf{X})] \tag{11.11}$$

where $\mathbf{X} = (X_1, \ldots, X_n)$.

Proof   The proof is by induction on n. To prove it when $n = 1$, let f and g be increasing functions of a single variable. Then, for any x and y,

$$(f(x) - f(y))(g(x) - g(y)) \geq 0$$

since if $x \geq y$ $(x \leq y)$ then both factors are nonnegative (nonpositive). Hence, for any random variables X and Y,

$$(f(X) - f(Y))(g(X) - g(Y)) \geq 0$$

implying that

$$E[(f(X) - f(Y))(g(X) - g(Y))] \geq 0$$

or, equivalently,

$$E[f(X)g(X)] + E[f(Y)g(Y)] \geq E[f(X)g(Y)] + E[f(Y)g(X)]$$

If we suppose that X and Y are independent and identically distributed then, as in this case,

$$E[f(X)g(X)] = E[f(Y)g(Y)],$$

$$E[f(X)g(Y)] = E[f(Y)g(X)] = E[f(X)]E[g(X)]$$

we obtain the result when $n = 1$.

So assume that (11.11) holds for $n - 1$ variables, and now suppose that $X_1, \ldots, X_n$ are independent and f and g are increasing functions. Then

$E[f(\mathbf{X})g(\mathbf{X}) \mid X_n = x_n]$

$$= E[f(X_1, \ldots, X_{n-1}, x_n)g(X_1, \ldots, X_{n-1}, x_n) \mid X_n = x]$$

$$= E[f(X_1, \ldots, X_{n-1}, x_n)g(X_1, \ldots, X_{n-1}, x_n)] \qquad \text{by independence}$$

$$\geq E[f(X_1, \ldots, X_{n-1}, x_n)]E[g(X_1, \ldots, X_{n-1}, x_n)]$$

$$\text{by the induction hypothesis}$$

$$= E[f(\mathbf{X}) \mid X_n = x_n]E[g(\mathbf{X}) \mid X_n = x_n]$$

Hence,

$$E[f(\mathbf{X})g(\mathbf{X}) \mid X_n] \geq E[f(\mathbf{X}) \mid X_n]E[g(\mathbf{X}) \mid X_n]$$

and, upon taking expectations of both sides,

$$E[f(\mathbf{X})g(\mathbf{X})] \geq E[E[f(\mathbf{X}) \mid X_n]E[g(\mathbf{X}) \mid X_n]]$$

$$\geq E[f(\mathbf{X})]E[g(\mathbf{X})]$$

The last inequality follows because $E[f(\mathbf{X}) \mid X_n]$ and $E[g(\mathbf{X}) \mid X_n]$ are both increasing functions of X_n, and so, by the result for $n = 1$,

$$E[E[f(\mathbf{X}) \mid X_n]E[g(\mathbf{X}) \mid X_n]] \geq E[E[f(\mathbf{X}) \mid X_n]]E[E[g(\mathbf{X}) \mid X_n]]$$
$$= E[f(\mathbf{X})]E[g(\mathbf{X})] \quad \blacklozenge$$

Corollary 11.7 If $U_1, \ldots, U_n$ are independent, and k is either an increasing or decreasing function, then

$$\text{Cov}(k(U_1, \ldots, U_n), k(1 - U_1, \ldots, 1 - U_n)) \leq 0$$

Proof Suppose k is increasing. As $-k(1 - U_1, \ldots, 1 - U_n)$ is increasing in $U_1, \ldots, U_n$, then, from Theorem 11.1,

$$\text{Cov}(k(U_1, \ldots, U_n), -k(1 - U_1, \ldots, 1 - U_n)) \geq 0$$

When k is decreasing just replace k by its negative. $\blacklozenge$

Since $F_i^{-1}(U_i)$ is increasing in U_i (as F_i, being a distribution function, is increasing) it follows that $g(F_1^{-1}(U_1), \ldots, F_n^{-1}(U_n))$ is a monotone function of $U_1, \ldots, U_n$ whenever g is monotone. Hence, if g is monotone the antithetic variable approach of twice using each set of random numbers $U_1, \ldots, U_n$ by first computing $g(F_1^{-1}(U_1), \ldots, F_n^{-1}(U_n))$ and then $g(F_1^{-1}(1 - U_1), \ldots, F_n^{-1}(1 - U_n))$ will reduce the variance of the estimate of $E[g(X_1, \ldots, X_n)]$. That is, rather than generating k sets of n random numbers, we should generate $k/2$ sets and use each set twice.

Example 11.14 (Simulating the Reliability Function): Consider a system of n components in which component i, independently of other components, works with probability p_i, $i = 1, \ldots, n$. Letting

$$X_i = \begin{cases} 1, & \text{if component } i \text{ works} \\ 0, & \text{otherwise} \end{cases}$$

suppose there is a monotone structure function ϕ such that

$$\phi(X_1, \ldots, X_n) = \begin{cases} 1, & \text{if the system works under } X_1, \ldots, X_n \\ 0, & \text{otherwise} \end{cases}$$

We are interested in using simulation to estimate

$$r(p_1, \ldots, p_n) \equiv E[\phi(X_1, \ldots, X_n)] = P\{\phi(X_1, \ldots, X_n) = 1\}$$

Now, we can simulate the X_i by generating uniform random numbers

$U_1, \ldots, U_n$ and then setting

$$X_i = \begin{cases} 1, & \text{if } U_i < p_i \\ 0, & \text{otherwise} \end{cases}$$

Hence, we see that

$$\phi(X_1, \ldots, X_n) = k(U_1, \ldots, U_n)$$

where k is a decreasing function of $U_1, \ldots, U_n$. Hence,

$$\text{Cov}(k(\mathbf{U}), k(\mathbf{1} - \mathbf{U})) \leq 0$$

and so the antithetic variable approach of using $U_1, \ldots, U_n$ to generate both $k(U_1, \ldots, U_n)$ and $k(1 - U_1, \ldots, 1 - U_n)$ results in a smaller variance than if an independent set of random numbers was used to generate the second k. ✦

Example 11.15 (Simulation a Queueing System): Consider a given queueing system, and let D_i denote the delay in queue of the ith arriving customer, and suppose we are interested in simulating the system so as to estimate

$$\theta = E[D_1 + \cdots + D_n]$$

Let $X_1, \ldots, X_n$ denote the first n interarrival times and $S_1, \ldots, S_n$ the first n service times of this system, and suppose these random variables are all independent. Now in most systems $D_1 + \cdots + D_n$ will be a function of $X_1, \ldots, X_n, S_1, \ldots, S_n$ — say,

$$D_1 + \cdots + D_n = g(X_1, \ldots, X_n, S_1, \ldots, S_n)$$

Also g will usually be increasing in S_i and decreasing in X_i, $i = 1, \ldots, n$. If we use the inverse transform method to simulate $X_i, S_i, i = 1, \ldots, n$ — say, $X_i = F_i^{-1}(1 - U_i)$, $S_i = G_i^{-1}(\bar{U}_i)$ where $U_1, \ldots, U_n, \bar{U}_1, \ldots, \bar{U}_n$ are independent uniform random numbers — then we may write

$$D_1 + \cdots + D_n = k(U_1, \ldots, U_n, \bar{U}_1, \ldots, \bar{U}_n)$$

where k is increasing in its variates. Hence, the antithetic variable approach will reduce the variance of the estimator of θ. [Thus, we would generate U_i, $\bar{U}_i, i = 1, \ldots, n$ and set $X_i = F_i^{-1}(1 - U_i)$ and $Y_i = G_i^{-1}(\bar{U}_i)$ for the first run, and $X_i = F_i^{-1}(U_i)$ and $Y_i = G_i^{-1}(1 - \bar{U}_i)$ for the second.] As all the U_i and $\bar{U}_i$ are independent, however, this is equivalent to setting $X_i = F_i^{-1}(U_i)$, $Y_i = G_i^{-1}(\bar{U}_i)$ in the first run and using $1 - U_i$ for U_i and $1 - \bar{U}_i$ for $\bar{U}_i$ in the second. ✦

11.6.2. Variance Reduction by Conditioning

Let us start by recalling (see Exercise 43 of Chapter 3) the conditional variance formula

$$\text{Var}(Y) = E[\text{Var}(Y|Z)] + \text{Var}(E[Y|Z]) \tag{11.12}$$

Now suppose we are interested in estimating $E[g(X_1, \ldots, X_n)]$ by simulating $\mathbf{X} = (X_1, \ldots, X_n)$ and then computing $Y = g(X_1, \ldots, X_n)$. Now, if for some random variable Z we can compute $E[Y|Z]$ then, as $\text{Var}(Y|Z) \geqslant 0$, it follows from the conditional variance formula that

$$\text{Var}(E[Y|Z]) \leqslant \text{Var}(Y)$$

implying, since $E[E[Y|Z]] = E[Y]$, that $E[Y|Z]$ is a better estimator of $E[Y]$ than is Y.

In many situations, there are a variety of Z_i that can be conditioned on to obtain an improved estimator. Each of these estimators $E[Y|Z_i]$ will have mean $E[Y]$ and smaller variance than does the raw estimator Y. We now show that for any choice of weights λ_i, $\lambda_i \geqslant 0$, $\Sigma_i \lambda_i = 1$, $\Sigma_i \lambda_i E[Y|Z_i]$ is also an improvement over Y.

Proposition 11.8   For any $\lambda_i \geqslant 0$, $\Sigma_{i=1}^{\infty} \lambda_i = 1$,

(a) $E\left[\sum_i \lambda_i E[Y|Z_i] \right] = E[Y]$

(b) $\text{Var}\left(\sum_i \lambda_i E[Y|Z_i] \right) \leqslant \text{Var}(Y)$

Proof   The proof of (a) is immediate. To prove (b), let N denote an integer valued random variable independent of all the other random variables under consideration and such that

$$P\{N = i\} = \lambda_i, \qquad i \geqslant 1$$

Applying the conditional variance formula twice yields

$$\text{Var}(Y) \geqslant \text{Var}(E[Y|N, Z_N])$$
$$\geqslant \text{Var}(E[E[Y|N, Z_N]|Z_1, \ldots])$$
$$= \text{Var} \sum_i \lambda_i E[Y|Z_i] \quad \blacklozenge$$

Example 11.16   Consider a queueing system having Poisson arrivals and suppose that any customer arrriving when there are already N others in the system is lost. Suppose that we are interested in using simulation to estimate the expected number of lost customers by time t. The raw simulation approach would be to simulate the system up to time t and determine L, the number of lost customers for that run. A better estimate, however, can be obtained by conditioning on the total time in $[0, t]$ that the system is at capacity. Indeed, if we let T denote the time in $[0, t]$ that there are N in the system, then

$$E[L \mid T] = \lambda T$$

where λ is the Poisson arrival rate. Hence, a better estimate for $E[L]$ than the average value of L over all simulation runs can be obtained by multiplying the average value of T per simulation run by λ. If the arrival process were a nonhomogeneous Poisson process, then we could improve over the raw estimator L by keeping track of those time periods for which the system is at capacity. If we let $I_1, \ldots, I_C$ denote the time intervals in $[0, t]$ in which there are N in the system, then

$$E[L \mid I_1, \ldots, I_C] = \sum_{i=1}^{C} \int_{I_i} \lambda(s) \, ds$$

where $\lambda(s)$ is the intensity function of the nonhomogeneous Poisson arrival process. The use of the right side of the preceding would thus lead to a better estimate of $E[L]$ than the raw estimator L. ✦

Example 11.17   Suppose that we wanted to estimate the expected sum of the times in the system of the first n customers in a queueing system. That is, if W_i is the time that the ith customer spends in the system, then we are interested in estimating

$$\theta = E\left[\sum_{i=1}^{n} W_i \right]$$

Let Y_i denote the "state of the system" at the moment at which the ith customer arrives. It can be shown[*] that for a wide class of models the estimator $\sum_{i=1}^{n} E[W_i \mid Y_i]$ has (the same mean and) a smaller variance than the estimator $\sum_{i=1}^{n} W_i$. (It should be noted that whereas it is immediate that $E[W_i \mid Y_i]$ has smaller variance than W_i, because of the covariance terms involved it is not immediately apparent that $\sum_{i=1}^{n} E[W_i \mid Y_i]$ has smaller variance than $\sum_{i=1}^{n} W_i$.) For instance, in the model $G/M/1$

$$E[W_i \mid Y_i] = (N_i + 1)/\mu$$

where N_i is the number in the system encountered by the ith arrival and $1/\mu$

[*] S. M. Ross, "Simulating Average Delay—Variance Reduction by Conditioning," *Probability in the Engineering and Informational Sciences* 2(3), (1988), pp. 309–312.

is the mean service time; the result implies that $\Sigma_{i=1}^n (N_i + 1)/\mu$ is a better estimate of the expected total time in the system of the first n customers than is the raw estimator $\Sigma_{i=1}^n W_i$. ✦

Example 11.18 (Estimating the Renewal Function by Simulation): Consider a queueing model in which customers arrive daily in accordance with a renewal process having interarrival distribution F. However, suppose that at some fixed time T, for instance 5 P.M., no additional arrivals are permitted and those customers that are still in the system are serviced. At the start of the next, and each succeeding day, customers again begin to arrive in accordance with the renewal process. Suppose we are interested in determining the average time that a customer spends in the system. Upon using the theory of renewal reward processes (with a cycle starting every T time units), it can be shown that

average time that a customer spends in the system

$$= \frac{E[\text{sum of the times in the system of arrivals in } (0, T)]}{m(T)}$$

where $m(T)$ is the expected number of renewals in $(0, T)$.

If we were to use simulation to estimate the preceding quantity, a run would consist of simulating a single day, and as part of a simulation run, we would observe the quantity $N(T)$, the number of arrivals by time T. Since $E[N(T)] = m(T)$, the natural simulation estimator of $m(T)$ would be the average (over all simulated days) value of $N(T)$ obtained. However, $\text{Var}(N(T))$ is, for large T, proportional to T (its asymptotic form being $T\sigma^2/\mu^3$, where σ^2 is the variance and μ the mean of the interarrival distribution F), and so, for large T, the variance of our estimator would be large. A considerable improvement can be obtained by using the analytic formula (see Section 7.3)

$$m(T) = \frac{T}{\mu} - 1 + \frac{E[Y(T)]}{\mu} \tag{11.13}$$

where $Y(T)$ denotes the time from T until the next renewal—that is, it is the excess life at T. Since the variance of $Y(T)$ does not grow with T (indeed, it converges to a finite value provided the moments of F are finite), it follows that for T large, we would do much better by using the simulation to estimate $E[Y(T)]$ and then use Equation (11.13) to estimate $m(T)$.

However, by employing conditioning, we can improve further on our estimate of $m(T)$. To do so, let $A(T)$ denote the age of the renewal process at time T—that is, it is the time at T since the last renewal. Then, rather than using the value of $Y(T)$, we can reduce the variance by considering

Figure 11.5. $A(T) = x$.

$E[Y(T) \mid A(T)]$. Now knowing that the age at T is equal to x is equivalent to knowing that there was a renewal at time $T - x$ and the next interarrival time X is greater than x. Since the excess at T will equal $X - x$ (see Figure 11.5), it follows that

$$E[Y(T) \mid A(T) = x] = E[X - x \mid X > x]$$
$$= \int_0^\infty \frac{P\{X - x > t\}}{P\{X > x\}}\, dt$$
$$= \int_0^\infty \frac{[1 - F(t + x)]}{1 - F(x)}\, dt$$

which can be numerically evaluated if necessary.

As an illustration of the preceding note that if the renewal process is a Poisson process with rate λ, then the raw simulation estimator $N(T)$ will have variance λT; since $Y(T)$ will be exponential with rate λ, the estimator based on (11.13) will have variance $\lambda^2 \, \mathrm{Var}(Y(T)) = 1$. On the other hand, since $Y(T)$ will be independent of $A(T)$ (and $E[Y(T) \mid A(T)] = 1/\lambda$), it follows that the variance of the improved estimator $E[Y(T) \mid A(T)]$ is 0. That is, conditioning on the age at time T yields, in this case, the exact answer. ◆

Example 11.19 (Reliability): Suppose as in Example 11.14 that X_j, $j = 1, \ldots, n$ are independent with $P\{X_j = 1\} = P_j = 1 - P\{X_j = 0\}$, and suppose we are interested in estimating $E[\phi(X_1, \ldots, X_n)]$, where ϕ is a monotone binary function. If we simulate $X_1, \ldots, X_n$, an improvement over the raw estimator, $\phi(X_1, \ldots, X_n)$ is to take its conditional expectation given all the X_j except one. That is, for fixed i, $E[\phi(\mathbf{X}) \mid \epsilon_i(\mathbf{X})]$ is an improved estimator, where $\mathbf{X} = (X_1, \ldots, X_n)$ and $\epsilon_i(\mathbf{X}) = (X_1, \ldots, X_{i-1}, X_{i+1}, \ldots, X_n)$. $E[\phi(\mathbf{X}) \mid \epsilon_i(X)]$ will have three possible values — either it will equal 1 [if $\epsilon_i(\mathbf{X})$ is such that the system will function even if $X_i = 0$], or 0 [if $\epsilon_i(\mathbf{X})$ is such that the system will be failed even if $X_i = 1$], or P_i[if $\epsilon_i(\mathbf{X})$ is such that the system will function if $X_i = 1$ and will be failed otherwise]. Also, by Proposition 11.8 any estimator of the form

$$\sum_i \lambda_i E[\phi(\mathbf{X}) \mid \epsilon_i(\mathbf{X})], \qquad \sum_i \lambda_i = 1, \qquad \lambda_i \geq 0$$

is an improvement over $\phi(\mathbf{X})$. ◆

11.6.3. Control Variates

Again suppose we want to use simulation to estimate $E[g(\mathbf{X})]$ where $\mathbf{X} = (X_1, \ldots, X_n)$. But now suppose that for some function f the expected value of $f(\mathbf{X})$ is known — say, $E[f(\mathbf{X})] = \mu$. Then for any constant a we can also use

$$W = g(\mathbf{X}) + a(f(\mathbf{X}) - \mu)$$

as an estimator of $E[g(\mathbf{X})]$. Now,

$$\text{Var}(W) = \text{Var}(g(\mathbf{X})) + a^2 \,\text{Var}(f(\mathbf{X})) + 2a \,\text{Cov}(g(\mathbf{X}), f(\mathbf{X}))$$

Simple calculus shows that the preceding is minimized when

$$a = \frac{-\text{Cov}(f(\mathbf{X}), g(\mathbf{X}))}{\text{Var}(f(\mathbf{X}))}$$

and, for this value of a,

$$\text{Var}(W) = \text{Var}(g(\mathbf{X})) - \frac{[\text{Cov}(f(\mathbf{X}), g(\mathbf{X}))]^2}{\text{Var}(f(\mathbf{X}))}$$

Because $\text{Var}(f(\mathbf{X}))$ and $\text{Cov}(f(\mathbf{X}), g(\mathbf{X}))$ are usually unknown, the simulated data should be used to estimate these quantities.

Example 11.20 (A Queueing System): Let D_{n+1} denote the delay in queue of the $n + 1$ customer in a queueing system in which the interarrival times are independent and identically distributed (i.i.d.) with distribution F having mean μ_F and are independent of the service times which are i.i.d. with distribution G having mean μ_G. If X_i is the interarrival time between arrival i and $i + 1$, and if S_i is the service time of customer i, $i \geqslant 1$, we may write

$$D_{n+1} = g(X_1, \ldots, X_n, S_1, \ldots, S_n)$$

To take into account the possibility that the simulated variables X_i, S_i may by chance be quite different from what might be expected we can let

$$f(X_1, \ldots, X_n, S_1, \ldots, S_n) = \sum_{i=1}^{n} (S_i - X_i)$$

As $E[f(\mathbf{X}, \mathbf{S})] = n(\mu_G - \mu_F)$ we could use

$$g(\mathbf{X}, \mathbf{S}) + a[f(\mathbf{X}, \mathbf{S}) - n(\mu_G - \mu_F)]$$

as an estimator of $E[D_{n+1}]$. Since D_{n+1} and f are both increasing functions of S_i, $-X_i$, $i = 1, \ldots, n$ it follows from Theorem 11.1 that $f(\mathbf{X}, \mathbf{S})$ and D_{n+1} are positively correlated, and so the simulated estimate of a should turn out to be negative.

If we wanted to estimate the expected sum of the delays in queue of the first $N(T)$ arrivals (see Example 11.18 for the motivation), then we could use $\sum_{i=1}^{N(T)} S_i$ as our control variable. Indeed as the arrival process is usually assumed independent of the service times, it follows that

$$E\left[\sum_{i=1}^{N(T)} S_i\right] = E[S]E[N(T)]$$

where $E[N(T)]$ can either be computed by the method suggested in Section 7.8 or it can be estimated from the simulation as in Example 11.18. This control variable could also be used if the arrival process were a non-homogeneous Poisson with rate $\lambda(t)$; in this case,

$$E[N(T)] = \int_0^T \lambda(t)\, dt \quad \blacklozenge$$

11.6.4. Importance Sampling

Let $\mathbf{X} = (X_1, \ldots, X_n)$ denote a vector of random variables having a joint density function (or joint mass function in the discrete case) $f(\mathbf{x}) = f(x_1, \ldots, x_n)$, and suppose that we are interested in estimating

$$\theta = E[h(\mathbf{X})] = \int h(\mathbf{x})\, f(\mathbf{x})\, d\mathbf{x}$$

where the preceding is an n-dimensional integral. (If the X_i are discrete, then interpret the integral as an n-fold summation.)

Suppose that a direct simulation of the random vector $\mathbf{X}$, so as to compute values of $h(\mathbf{X})$, is inefficient, possibly because (a) it is difficult to simulate a random vector having density function $f(\mathbf{x})$, or (b) the variance of $h(\mathbf{X})$ is large, or (c) a combination of (a) and (b).

Another way in which we can use simulation to estimate θ is to note that if $g(\mathbf{x})$ is another probability density such that $f(\mathbf{x}) = 0$ whenever $g(\mathbf{x}) = 0$, then we can express θ as

$$\theta = \int \frac{h(\mathbf{x})\, f(\mathbf{x})}{g(\mathbf{x})} g(\mathbf{x})\, d\mathbf{x}$$

$$= E_g\left[\frac{h(\mathbf{X})\, f(\mathbf{X})}{g(\mathbf{X})}\right] \tag{11.14}$$

where we have written E_g to emphasize that the random vector $\mathbf{X}$ has joint density $g(\mathbf{x})$.

It follows from Equation (11.14) that θ can be estimated by successively generating values of a random vector $\mathbf{X}$ having density function $g(\mathbf{x})$ and then using as the estimator the average of the values of $h(\mathbf{X}) f(\mathbf{X})/g(\mathbf{X})$. If a density function $g(\mathbf{x})$ can be chosen so that the random variable $h(\mathbf{X}) f(\mathbf{X})/g(\mathbf{X})$ has a small variance then this approach — referred to as *importance sampling* — can result in an efficient estimator of θ.

Let us now try to obtain a feel for why importance sampling can be useful. To begin, note that $f(\mathbf{X})$ and $g(\mathbf{X})$ represent the respective likelihoods of obtaining the vector $\mathbf{X}$ when $\mathbf{X}$ is a random vector with respective densities f and g. Hence, if $\mathbf{X}$ is distributed according to g, then it will usually be the case that $f(\mathbf{X})$ will be small in relation to $g(\mathbf{X})$ and thus when $\mathbf{X}$ is simulated according to g the likelihood ratio $f(\mathbf{X})/g(\mathbf{X})$ will usually be small in comparison to 1. However, it is easy to check that its mean is 1:

$$E_g\left[\frac{f(\mathbf{X})}{g(\mathbf{X})}\right] = \int \frac{f(\mathbf{x})}{g(\mathbf{x})} g(\mathbf{x})\, dx = \int f(\mathbf{x})\, d\mathbf{x} = 1$$

Thus we see that even though $f(\mathbf{X})/g(\mathbf{X})$ is usually smaller than 1, its mean is equal to 1; thus implying that it is occasionally large and so will tend to have a large variance. So how can $h(\mathbf{X}) f(\mathbf{X})/g(\mathbf{X})$ have a small variance? The answer is that we can sometimes arrange to choose a density g such that those values of $\mathbf{x}$ for which $f(\mathbf{x})/g(\mathbf{x})$ is large are precisely the values for which $h(\mathbf{x})$ is exceedingly small, and thus the ratio $h(\mathbf{X}) f(\mathbf{X})/g(\mathbf{X})$ is always small. Since this will require that $h(\mathbf{x})$ sometimes be small, importance sampling seems to work best when estimating a small probability; for in this case the function $h(\mathbf{x})$ is equal to 1 when $\mathbf{x}$ lies in some set and is equal to 0 otherwise.

We will now consider how to select an appropriate density g. We will find that the so-called tilted densities are useful. Let $M(t) = E_f[e^{tX}] = \int e^{tx} f(x)\, dx$ be the moment generating function corresponding to a one-dimensional density f.

Definition 11.2   A density function

$$f_t(x) = \frac{e^{tx} f(x)}{M(t)}$$

is called a *tilted* density of f, $-\infty < t < \infty$.

A random variable with density f_t tends to be larger than one with density f when $t > 0$ and tends to be smaller when $t < 0$.

In certain cases the tilted distributions f_t have the same parametric form as does f.

Example 11.21 If f is the exponential density with rate λ then

$$f_t(x) = Ce^{tx}\lambda e^{-\lambda x} = \lambda Ce^{-(\lambda - t)x}$$

where $C = 1/M(t)$ does not depend on x. Therefore, for $t \leqslant \lambda$, f_t is an exponential density with rate $\lambda - t$.

If f is a Bernoulli probability mass function with parameter p, then

$$f(x) = p^x(1 - p)^{1-x}, \qquad x = 0, 1$$

Hence, $M(t) = E_f[e^{tX}] = pe^t + 1 - p$ and so

$$f_t(x) = \frac{1}{M(t)}(pe^t)^x(1 - p)^{1-x}$$

$$= \left(\frac{pe^t}{pe^t + 1 - p}\right)^x \left(\frac{1 - p}{pe^t + 1 - p}\right)^{1-x}$$

That is, f_t is the probability mass function of a Bernoulli random variable with parameter

$$p_t = \frac{pe^t}{pe^t + 1 - p}$$

We leave it as an exercise to show that if f is a normal density with parameters μ and σ^2 then f_t is a normal density mean $\mu + \sigma^2 t$ and variance σ^2. ✦

In certain situations the quantity of interest is the sum of the independent random variables $X_1, \ldots, X_n$. In this case the joint density f is the product of one-dimensional densities. That is,

$$f(x_1, \ldots, x_n) = f_1(x_1) \cdots f_n(x_n)$$

where f_i is the density function of X_i. In this situation it is often useful to generate the X_i according to their tilted densities, with a common choice of t employed.

Example 11.22 Let $X_1, \ldots, X_n$ be independent random variables having respective probability density (or mass) functions f_i, for $i = 1, \ldots, n$. Suppose we are interested in approximating the probability that their sum is at least as large as a, where a is much larger than the mean of the sum. That is, we are interested in

$$\theta = P\{S \geqslant a\}$$

where $S = \sum_{i=1}^n X_i$, and where $a > \sum_{i=1}^n E[X_i]$. Letting $I\{S \geqslant a\}$ equal 1 if

$S \geqslant a$ and letting it be 0 otherwise, we have that

$$\theta = E_{\mathbf{f}}[I\{S \geqslant a\}]$$

where $\mathbf{f} = (f_1, \ldots, f_n)$. Suppose now that we simulate X_i according to the tilted mass function $f_{i,t}$, $i = 1, \ldots, n$, with the value of t, $t > 0$ left to be determined. The importance sampling estimator of θ would then be

$$\hat{\theta} = I\{S \geqslant a\} \prod \frac{f_i(X_i)}{f_{i,t}(X_i)}$$

Now,

$$\frac{f_i(X_i)}{f_{i,t}(X_i)} = M_i(t)e^{-tX_i}$$

and so

$$\hat{\theta} = I\{S \geqslant a\}M(t)e^{-tS}$$

where $M(t) = \prod M_i(t)$ is the moment generating function of S. Since $t > 0$ and $I\{S \geqslant a\}$ is equal to 0 when $S < a$, it follows that

$$I\{S \geqslant a\}e^{-tS} \leqslant e^{-ta}$$

and so

$$\hat{\theta} \leqslant M(t)e^{-ta}$$

To make the bound on the estimator as small as possible we thus choose t, $t > 0$, to minimize $M(t)e^{-ta}$. In doing so, we will obtain an estimator whose value on each iteration is between 0 and $\min_t M(t)e^{-ta}$. It can be shown that the minimizing t, call it t^*, is such that

$$E_{t^*}[S] = E_{t^*}\left[\sum_{i=1}^{n} X_i\right] = a$$

where, in the preceding, we mean that the expected value is to be taken under the assumption that the distribution of X_i is f_{i,t^*} for $i = 1, \ldots, n$.

For instance, suppose that $X_1, \ldots, X_n$ are independent Bernoulli random variables having respective parameters p_i, for $i = 1, \ldots, n$. Then, if we generate the X_i according to their tilted mass functions $p_{i,t}$, $i = 1, \ldots, n$ then the importance sampling estimator of $\theta = P\{S \geqslant a\}$ is

$$\hat{\theta} = I\{S \geqslant a\}e^{-tS} \prod_{i=1}^{n} (p_i e^t + 1 - p_i)$$

Since $p_{i,t}$ is the mass function of a Bernoulli random variable with parameter $p_i e^t/(p_i e^t + 1 - p_i)$ it follows that

$$E_t\left[\sum_{i=1}^{n} X_i\right] = \sum_{i=1}^{n} \frac{p_i e^t}{p_i e^t + 1 - p_i}$$

The value of t that makes the preceding equal to a can be numerically approximated and then utilized in the simulation.

As an illustration, suppose that $n = 20$, $p_i = 0.4$, and $a = 16$. Then

$$E_t[S] = 20 \frac{0.4e^t}{0.4e^t + 0.6}$$

Setting this equal to 16 yields, after a little algebra,

$$e^{t^*} = 6$$

Thus, if we generate the Bernoullis using the parameter

$$\frac{0.4e^{t^*}}{0.4e^{t^*} + 0.6} = 0.8$$

then because

$$M(t^*) = (0.4e^{t^*} + 0.6)^{20} \qquad \text{and} \qquad e^{-t^*S} = (1/6)^S$$

we see that the importance sampling estimator is

$$\hat{\theta} = I\{S \geqslant 16\}(1/6)^S 3^{20}$$

It follows from the preceding that

$$\hat{\theta} \leqslant (1/6)^{16} 3^{20} = 81/2^{16} = 0.001236$$

That is, on each iteration the value of the estimator is between 0 and 0.001236. Since, in this case, θ is the probability that a binomial random variable with parameters 20, 0.4 is at least 16, it can be explicitly computed with the result $\theta = 0.000317$. Hence, the raw simulation estimator I, which on each iteration takes the value 0 if the sum of the Bernoullis with parameter 0.4 is less than 16 and takes the value 1 otherwise, will have variance

$$\text{Var}(I) = \theta(1 - \theta) = 3.169 \times 10^{-4}$$

On the other hand, it follows from the fact that $0 \leqslant \hat{\theta} \leqslant 0.001236$ that (see Exercise 33)

$$\text{Var}(\hat{\theta}) \leqslant 2.9131 \times 10^{-7} \quad \blacklozenge$$

Example 11.23 Consider a single-server queue in which the times between successive customer arrivals have density function f and the service times have density g. Let D_n denote the amount of time that the nth arrival spends waiting in queue and suppose we are interested in estimating $\alpha = P\{D_n \geqslant a\}$ when a is much larger than $E[D_n]$. Rather than generating the successive interarrival and service times according to f and g respectively, they should be generated according to the densities f_{-t} and g_t, where t is a positive number to be determined. Note that using these distributions as opposed to f and g will result in smaller interarrival times (since $-t < 0$) and larger service times. Hence, there will be a greater chance that $D_n > a$ than if we had simulated using the densities f and g. The importance sampling estimator of α would then be

$$\hat{\alpha} = I\{D_n > a\}e^{t(S_n - Y_n)}[M_f(-t)M_g(t)]^n$$

where S_n is the sum of the first n interarrival times, Y_n is the sum of the first n service times, and M_f and M_g are the moment generating functions of the densities f and g respectively. The value of t used should be determined by experimenting with a variety of different choices. ✦

11.7. Determining the Number of Runs

Suppose that we are going to use simulation to generate r independent and identically distributed random variables $Y^{(1)}, \ldots, Y^{(r)}$ having mean μ and variance σ^2. We are then going to use

$$\bar{Y}_r = \frac{Y^{(1)} + \cdots + Y^{(r)}}{r}$$

as an estimate of μ. The precision of this estimate can be measured by its variance

$$\mathrm{Var}(\bar{Y}_r) = E[(\bar{Y}_r - \mu)^2]$$
$$= \sigma^2/r$$

Hence we would want to choose r, the number of necessary runs, large enough so that σ^2/r is acceptably small. However, the difficulty is that σ^2 is not known in advance. To get around this, one should initially simulate k runs (where $k \geqslant 30$) and then use the simulated values $Y^{(1)}, \ldots, Y^{(k)}$ to estimate σ^2 by the sample variance

$$\sum_{i=1}^{k} (Y^{(i)} - \bar{Y}_k)^2/(k - 1)$$

Based on this estimate of σ^2 the value of r that attains the desired level of precision can now be determined and an additional $r - k$ runs can be generated.

Exercises

*1. Suppose it is relatively easy to simulate from the distributions F_i, $i = 1, \ldots, n$. If n is small, how can we simulate from

$$F(x) = \sum_{i=1}^{n} P_i F_i(x), \qquad P_i \geqslant 0, \qquad \sum_i P_i = 1?$$

Give a method for simulating from

$$F(x) = \begin{cases} \dfrac{1 - e^{-2x} + 2x}{3}, & 0 < x < 1 \\[2mm] \dfrac{3 - e^{-2x}}{3}, & 1 < x < \infty \end{cases}$$

2. Give a method for simulating a negative binomial random variable.

*3. Give a method for simulating a hypergeometric random variable.

4. Suppose we want to simulate a point located at random in a circle of radius r centered at the origin. That is, we want to simulate X, Y having joint density

$$f(x, y) = \frac{1}{\pi r^2}, \qquad x^2 + y^2 \leqslant r^2$$

(a) Let $R = \sqrt{X^2 + Y^2}$, $\theta = \tan^{-1} Y/X$ denote the polar coordinates. Compute the joint density of R, θ and use this to give a simulation method. Another method for simulating X, Y is as follows:

Step 1: Generate independent random numbers U_1, U_2 and set $Z_1 = 2rU_1 - r$, $Z_2 = 2rU_2 - r$. Then Z_1, Z_2 is uniform in the square whose sides are of length $2r$ and which enclose the circle of radius r (see Figure 11.6).

Step 2: If (Z_1, Z_2) lies in the circle of radius r—that is, if $Z_1^2 + Z_2^2 \leqslant r^2$—set $(X, Y) = (Z_1, Z_2)$. Otherwise return to step 1.

(b) Prove that this method works, and compute the distribution of the number of random numbers it requires.

5. Suppose it is relatively easy to simulate from F_i for each $i = 1, \ldots, n$. How can we simulate from
(a) $F(x) = \prod_{i=1}^{n} F_i(x)$?

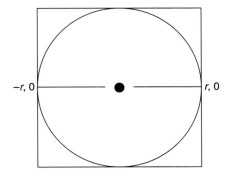

Figure 11.6.

(b) $F(x) = 1 - \Pi_{i=1}^{n}(1 - F_i(x))$?
(c) Give two methods for simulating from the distribution $F(x) = x^n$, $0 < x < 1$.

*6. In Example 11.4 we simulated the absolute value of a unit normal by using the Von Neumann rejection procedure on exponential random variables with rate 1. This raises the question of whether we could obtain a more efficient algorithm by using a different exponential density—that is, we could use the density $g(x) = \lambda e^{-\lambda x}$. Show that the mean number of iterations needed in the rejection scheme is minimized when $\lambda = 1$.

7. Give an algorithm for simulating a random variable having density function

$$f(x) = 30(x^2 - 2x^3 + x^4), \qquad 0 < x < 1$$

8. Consider the technique of simulating a gamma (n, λ) random variable by using the rejection method with g being an exponential density with rate λ/n.

(a) Show that the average number of iterations of the algorithm needed to generate a gamma is $n^n e^{1-n}/(n - 1)!$.
(b) Use Stirling's approximation to show that for large n the answer to part (a) is approximately equal to $e[(n - 1)/(2\pi)]^{1/2}$.
(c) Show that the procedure is equivalent to the following:

Step 1: Generate Y_1 and Y_2, independent exponentials with rate 1.
Step 2: If $Y_1 < (n - 1)[Y_2 - \log(Y_2) - 1]$, return to step 1.
Step 3: Set $X = nY_2/\lambda$.

(d) Explain how to obtain an independent exponential along with a gamma from the preceding algorithm.

9. Set up the alias method for simulating from a binomial random variable with parameters $n = 6$, $p = 0.4$.

10. Explain how we can number the $Q^{(k)}$ in the alias method so that k is one of the two points that $Q^{(k)}$ gives weight.

Hint: Rather than name the initial Q, $Q^{(1)}$ what else could we call it?

11. Complete the details of Example 11.10.

12. Let $X_1, \ldots, X_k$ be independent with

$$P\{X_i = j\} = \frac{1}{n}, \quad j = 1, \ldots, n, \ i = 1, \ldots, k$$

If D is the number of distinct values among $X_1, \ldots, X_k$ show that

$$E[D] = n\left[1 - \left(\frac{n-1}{n}\right)^k\right]$$

$$\approx k - \frac{k^2}{2n} \quad \text{when } \frac{k^2}{n} \text{ is small}$$

13. *The Discrete Rejection Method:* Suppose we want to simulate X having probability mass function $P\{X = i\} = P_i$, $i = 1, \ldots, n$ and suppose we can easily simulate from the probability mass function Q_i, $\Sigma_i Q_i = 1$, $Q_i \geqslant 0$. Let C be such that $P_i \leqslant CQ_i$, $i = 1, \ldots, n$. Show that the following algorithm generates the desired random variable:

Step 1: Generate Y having mass function Q and U an independent random number.

Step 2: If $U \leqslant P_Y/CQ_Y$, set $X = Y$. Otherwise return to step 1.

14. *The Discrete Hazard Rate Method:* Let X denote a nonnegative integer valued random variable. The function $\lambda(n) = P\{X = n \mid X \geqslant n\}$, $n \geqslant 0$, is called the discrete hazard rate function.

(a) Show that $P\{X = n\} = \lambda(n) \prod_{i=0}^{n-1} (1 - \lambda(i))$.

(b) Show that we can simulate X by generating random numbers $U_1, U_2, \ldots$ stopping at

$$X = \min\{n : U_n \leqslant \lambda(n)\}$$

(c) Apply this method to simulating a geometric random variable. Explain, intuitively, why it works.

(d) Suppose that $\lambda(n) \leqslant p < 1$ for all n. Consider the following algorithm for simulating X and explain why it works: Simulate X_i, U_i, $i \geqslant 1$ where X_i is geometric with mean $1/p$ and U_i is a random number. Set

$S_k = X_1 + \cdots + X_k$ and let

$$X = \min\{S_k : U_k \leqslant \lambda(S_k)/p\}$$

15. Suppose you have just simulated a normal random variable X with mean μ and variance σ^2. Give an easy way to generate a second normal variable with the same mean and variance that is negatively correlated with X.

*16. Suppose n balls having weights $w_1, w_2, \ldots, w_n$ are in an urn. These balls are sequentially removed in the following manner: At each selection, a given ball in the urn is chosen with a probability equal to its weight divided by the sum of the weights of the other balls that are still in the urn. Let $I_1, I_2, \ldots, I_n$ denote the order in which the balls are removed — thus $I_1, \ldots, I_n$ is a random permutation with weights.

(a) Give a method for simulating $I_1, \ldots, I_n$.
(b) Let X_i be independent exponentials with rates w_i, $i = 1, \ldots, n$. Explain how X_i can be utilized to simulate $I_1, \ldots, I_n$.

17. *Order Statistics:* Let $X_1, \ldots, X_n$ be i.i.d. from a continuous distribution F, and let $X_{(i)}$ denote the ith smallest of $X_1, \ldots, X_n$, $i = 1, \ldots, n$. Suppose we want to simulate $X_{(1)} < X_{(2)} < \cdots < X_{(n)}$. One approach is to simulate n values from F, and then order these values. However, this ordering, or *sorting*, can be time consuming when n is large.

(a) Suppose that $\lambda(t)$, the hazard rate function of F, is bounded. Show how the hazard rate method can be applied to generate the n variables in such a manner that no sorting is necessary.

Suppose now that F^{-1} is easily computed.

(b) Argue that $X_{(1)}, \ldots, X_{(n)}$ can be generated by simulating $U_{(1)} < U_{(2)} < \cdots < U_{(n)}$ — the ordered values of n independent random numbers — and then setting $X_{(i)} = F^{-1}(U_{(i)})$. Explain why this means that $X_{(i)}$ can be generated from $F^{-1}(\beta_i)$ where β_i is beta with parameters i, $n + i + 1$.
(c) Argue that $U_{(1)}, \ldots, U_{(n)}$ can be generated, without any need for sorting, by simulating i.i.d. exponentials $Y_1, \ldots, Y_{n+1}$ and then setting

$$U_{(i)} = \frac{Y_1 + \cdots + Y_i}{Y_1 + \cdots + Y_{n+1}}, \qquad i = 1, \ldots, n$$

Hint: Given the time of the $(n + 1)$st event of a Poisson process, what can be said about the set of times of the first n events?

(d) Show that if $U_{(n)} = y$ then $U_{(1)}, \ldots, U_{(n-1)}$ has the same joint distribution as the order statistics of a set of $n - 1$ uniform $(0, y)$ random variables.
(e) Use part (d) to show that $U_{(1)}, \ldots, U_{(n)}$ can be generated as follows:

Step 1:   Generate random numbers $U_1, \ldots, U_n$.
Step 2:   Set

$$U_{(n)} = U_1^{1/n}, \qquad\qquad U_{(n-1)} = U_{(n)}(U_2)^{1/(n-1)},$$

$$U_{(j-1)} = U_{(j)}(U_{n-j+2})^{1/(j-1)}, \qquad j = 2, \ldots, n-1$$

18.   Let $X_1, \ldots, X_n$ be independent exponential random variables each having rate 1. Set

$$W_1 = X_1/n,$$

$$W_i = W_{i-1} + \frac{X_i}{n-i+1}, \qquad i = 2, \ldots, n$$

Explain why $W_1, \ldots, W_n$ has the same joint distribution as the order statistics of a sample of n exponentials each having rate 1.

19.   Suppose we want to simulate a large number n of independent exponentials with rate 1—call them $X_1, X_2, \ldots, X_n$. If we were to employ the inverse transform technique we would require one logarithmic computation for each exponential generated. One way to avoid this is to first simulate S_n, a gamma random variable with parameters $(n, 1)$ (say, by the method of Section 11.3.3). Now interpret S_n as the time of the nth event of a Poisson process with rate 1 and use the result that given S_n the set of the first $n-1$ event times is distributed as the set of $n-1$ independent uniform $(0, S_n)$ random variables. Based on this, explain why the following algorithm simulates n independent exponentials:

Step 1:   Generate S_n, a gamma random variable with parameters $(n, 1)$.
Step 2:   Generate $n-1$ random numbers $U_1, U_2, \ldots, U_{n-1}$.
Step 3:   Order the U_i, $i = 1, \ldots, n-1$ to obtain $U_{(1)} < U_{(2)} < \cdots < U_{(n-1)}$.
Step 4:   Let $U_{(0)} = 0$, $U_{(n)} = 1$, and set $X_i = S_n(U_{(i)} - U_{(i-1)})$, $i = 1, \ldots, n$.

When the ordering (step 3) is performed according to the algorithm described in Section 11.5, the above is an efficient method for simulating n exponentials when all n are simultaneously required. If memory space is limited, however, and the exponentials can be employed sequentially, discarding each exponential from memory once it has been used, then the above may not be appropriate.

20.   Consider the following procedure for randomly choosing a subset of size k from the numbers $1, 2, \ldots, n$: Fix p and generate the first n time units of a renewal process whose interarrival distribution is geometric with mean $1/p$—that is, $P\{\text{interarrival time} = k\} = p(1-p)^{k-1}, k = 1, 2, \ldots$. Suppose events occur at times $i_1 < i_2 < \cdots < i_m \leqslant n$. If $m = k$, stop; $i_1, \ldots, i_m$ is the desired set. If $m > k$, then randomly choose (by some method) a subset of size k from $i_1, \ldots, i_m$ and then stop. If $m < k$, take $i_1, \ldots, i_m$ as part of the subset of size k and then select (by some method) a random subset of size $k - m$ from the set $\{1, 2, \ldots, n\} - (i_1, \ldots, i_m)$. Explain why this algorithm works. As $E[N(n)] = np$ a reasonable choice of p is to take $p \approx k/n$. (This approach is due to Dieter.)

21. Consider the following algorithm for generating a random permutation of the elements $1, 2, \ldots, n$. In this algorithm, $P(i)$ can be interpreted as the element in position i

Step 1:   Set $k = 1$.
Step 2:   Set $P(1) = 1$.
Step 3:   If $k = n$, stop. Otherwise, let $k = k + 1$.
Step 4:   Generate a random number U, and let

$$P(k) = P([kU] + 1),$$

$$P([kU] + 1) = k.$$

Go to step 3

(a) Explain in words what the algorithm is doing.
(b) Show that at iteration k — that is, when the value of $P(k)$ is initially set — that $P(1)$, $P(2)$, $\ldots$, $P(k)$ is a random permutation of $1, 2, \ldots, k$.

Hint: Use induction and argue that

$$P_k\{i_1, i_2, \ldots, i_{j-1}, k, i_j, \ldots, i_{k-2}, i\}$$

$$= P_{k-1}\{i_1, i_2, \ldots, i_{j-1}, i, i_j, \ldots, i_{k-2}\} \frac{1}{k}$$

$$= \frac{1}{k!} \qquad \text{by the induction hypothesis}$$

The preceding algorithm can be used even if n is not initially known.

22. Verify that if we use the hazard rate approach to simulate the event times of a nonhomogeneous Poisson process whose intensity function $\lambda(t)$ is such that $\lambda(t) \leqslant \lambda$, then we end up with the approach given in method 1 of Section 11.5.

*23. For a nonhomogeneous Poisson process with intensity function $\lambda(t)$, $t \geqslant 0$, where $\int_0^\infty \lambda(t)\, dt = \infty$, let $X_1, X_2, \ldots$ denote the sequence of times at which events occur.

(a) Show that $\int_0^{X_1} \lambda(t)\, dt$ is exponential with rate 1.
(b) Show that $\int_{X_{i-1}}^{X_i} \lambda(t)\, dt$, $i \geqslant 1$, are independent exponentials with rate 1, where $X_0 = 0$.

In words, independent of the past, the additional amount of hazard that must be experienced until an event occurs is exponential with rate 1.

24. Give an efficient method for simulating a nonhomogeneous Poisson process with intensity function

$$\lambda(t) = b + \frac{1}{t + a}, \qquad t \geqslant 0$$

25. Let (X, Y) be uniformly distributed in a circle of radius r about the origin. That is, their joint density is given by

$$f(x, y) = \frac{1}{\pi r^2}, \qquad 0 \leqslant x^2 + y^2 \leqslant r^2$$

Let $R = \sqrt{X^2 + Y^2}$ and $\theta = \arctan Y/X$ denote their polar coordinates. Show that R and θ are independent with θ being uniform on $(0, 2\pi)$ and $P\{R < a\} = a^2/r^2, 0 < a < r$.

26. Let R denote a region in the two-dimensional plane. Show that for a two-dimensional Poisson process, given that there are n points located in R, the points are independently and uniformly distributed in R—that is, their density is $f(x, y) = c, (x, y) \in R$ where c is the inverse of the area of R.

27. Let $X_1, \ldots, X_n$ be independent random variables with $E[X_i] = \theta$, $\text{Var}(X_i) = \sigma_i^2$ $i = 1, \ldots, n$, and consider estimates of θ of the form $\Sigma_{i=1}^n \lambda_i X_i$ where $\Sigma_{i=1}^n \lambda_i = 1$. Show that $\text{Var}(\Sigma_{i=1}^n \lambda_i X_i)$ is minimized when $\lambda_i = (1/\sigma_i^2)/(\Sigma_{j=1}^n 1/\sigma_j^2), i = 1, \ldots, n$.

Possible Hint: If you cannot do this for general n, try it first when $n = 2$.

The following two problems are concerned with the estimation of $\int_0^1 g(x) \, dx = E[g(U)]$ where U is uniform $(0, 1)$.

28. *The Hit–Miss Method:* Suppose g is bounded in $[0, 1]$—for instance, suppose $0 \leqslant g(x) \leqslant b$ for $x \in [0, 1]$. Let U_1, U_2 be independent random numbers and set $X = U_1, Y = bU_2$—so the point (X, Y) is uniformly distributed in a rectangle of length 1 and height b. Now set

$$I = \begin{cases} 1, & \text{if } Y < g(X) \\ 0, & \text{otherwise} \end{cases}$$

That is accept (X, Y) if it falls in the shaded area of Figure 11.7.

(a) Show that $E[bI] = \int_0^1 g(x) \, dx$.
(b) Show that $\text{Var}(bI) \geqslant \text{Var}(g(U))$, and so hit–miss has larger variance than simply computing g of a random number.

29. *Stratified Sampling:* Let $U_1, \ldots, U_n$ be independent random numbers

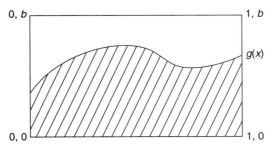

Figure 11.7.

and set $\bar{U}_i = (U_i + i - 1)/n$, $i = 1, \ldots, n$. Hence, $\bar{U}_i$, $i \geqslant 1$, is uniform on $((i - 1)/n, i/n)$. $\sum_{i=1}^{n} g(\bar{U}_i)/n$ is called the stratified sampling estimator of $\int_0^1 g(x)\, dx$.

(a) Show that $E[\sum_{i=1}^{n} g(\bar{U}_i)/n] = \int_0^1 g(x)\, dx$.
(b) Show that $\text{Var}[\sum_{i=1}^{n} g(\bar{U}_i)/n] \leqslant \text{Var}[\sum_{i=1}^{n} g(U_i)/n]$.

Hint: Let U be uniform $(0, 1)$ and define N by $N = i$ if $(i - 1)/n < U < i/n$, $i = 1, \ldots, n$. Now use the conditional variance formula to obtain

$$\text{Var}(g(U)) = E[\text{Var}(g(U) \mid N)] + \text{Var}(E[g(U) \mid N])$$

$$\geqslant E[\text{Var}(g(U) \mid N)]$$

$$= \sum_{i=1}^{n} \frac{\text{Var}(g(U) \mid N = i)}{n} = \sum_{i=1}^{n} \frac{\text{Var}[g(\bar{U}_i)]}{n}$$

30. If f is the density function of a normal random variable with mean μ and variance σ^2, show that the tilted density f_t is the density of a normal random variable with mean $\mu + \sigma^2 t$ and variance σ^2.

31. Consider a queueing system in which each service time, independent of the past, has mean μ. Let W_n and D_n denote respectively the amounts of time customer n spends in the system and in queue respectively. Hence, $D_n = W_n - S_n$ where S_n is the service time of customer n. Therefore,

$$E[D_n] = E[W_n] - \mu$$

If we use simulation to estimate $E[D_n]$, should we

(a) use the simulated data to determine D_n which is then used as an estimate of $E[D_n]$; or

(b) use the simulated data to determine W_n and then use this quantity minus μ as an estimate of $E[D_n]$.

Repeat if we want to estimate $E[W_n]$.

*32. Show that if X and Y have the same distribution then

$$\text{Var}((X + Y)/2) \leqslant \text{Var}(X)$$

Hence, conclude that the use of antithetic variables can never increase variance (though it need not be as efficient as generating an independent set of random numbers).

33. If $0 \leqslant X \leqslant a$, show that

(a) $E[X^2] \leqslant aE[X]$,
(b) $\text{Var}(X) \leqslant E[X](a - E[X])$,
(c) $\text{Var}(X) \leqslant a^2/4$.

References

1. J. Banks and J. Carson, "Discrete Event System Simulation," Prentice Hall, Englewood Cliffs, New Jersey, 1984.
2. G. Fishman, "Principles of Discrete Event Simulation," John Wiley, New York, 1978.
3. D. Knuth, "Semi Numerical Algorithms," Vol. 2 of *The Art of Computer Programming*, Second Edition, Addison-Wesley, Reading, Massachusetts, 1981.
4. A. Law and W. Kelton, "Simulation Modelling and Analysis," Second Edition, McGraw-Hill, New York, 1992.
5. S. Ross, "Simulation," Academic Press, San Diego, 1997.
6. R. Rubenstein, "Simulation and the Monte Carlo Method," John Wiley, New York, 1981.

Appendix
Solutions to Starred Exercises
++

Chapter 1

2. $S = \{(r, g), (r, b), (g, r), (g, b), (b, r), (b, g)\}$ where, for instance, (r, g) means that the first marble drawn was red and the second one green. The probability of each one of these outcomes is $\frac{1}{6}$.

5. $\frac{3}{4}$. If he wins, he only wins \$1, while if he loses, he loses \$3.

9. $F = E \cup FE^c$, implying since E and FE^c are disjoint that $P(F) = P(E) + P(FE^c)$.

17.
$$P\{end\} = 1 - P\{continue\}$$
$$= 1 - [\text{Prob}(H, H, H) + \text{Prob}(T, T, T)]$$

Fair coin:
$$P\{end\} = 1 - \left[\frac{1}{2} \cdot \frac{1}{2} \cdot \frac{1}{2} + \frac{1}{2} \cdot \frac{1}{2} \cdot \frac{1}{2}\right]$$
$$= \frac{3}{4}$$

Biased coin:
$$P\{end\} = 1 - \left[\frac{1}{4} \cdot \frac{1}{4} \cdot \frac{1}{4} + \frac{3}{4} \cdot \frac{3}{4} \cdot \frac{3}{4}\right]$$
$$= \frac{9}{16}$$

19. $E = $ event at least 1 six

$$P(E) = \frac{\text{number of ways to get } E}{\text{number of samples pts}} = \frac{11}{36}$$

$D = $ event two faces are different

$$P(D) = 1 - P(\text{two faces the same}) = 1 - \frac{6}{36} = \frac{5}{6}$$

$$P(E \mid D) = \frac{P(ED)}{P(D)} = \frac{10/36}{5/6} = \frac{1}{3}$$

25.   (a) $P\{\text{pair}\} = P\{\text{second card is same denomination as first}\}$

$$= \tfrac{3}{51}$$

  (b) $P\{\text{pair} \mid \text{different suits}\} = \dfrac{P\{\text{pair, different suits}\}}{P\{\text{different suits}\}}$

$$= \frac{P\{\text{pair}\}}{P\{\text{different suits}\}}$$

$$= \frac{3/51}{39/51} = \frac{1}{13}$$

27. $P(E_1) = 1$

$$P(E_2 \mid E_1) = \tfrac{39}{51}$$

since 12 cards are in the ace of spades pile and 39 are not.

$$P(E_3 \mid E_1 E_2) = \tfrac{26}{50}$$

since 24 cards are in the piles of the two aces and 26 are in the other two piles.

$$P(E_4 \mid E_1 E_2 E_3) = \tfrac{13}{49}$$

So

$$P\{\text{each pile has an ace}\} = (\tfrac{39}{51})(\tfrac{26}{50})(\tfrac{13}{49})$$

30.   (a) $P\{\text{George} \mid \text{exactly 1 hit}\} = \dfrac{P\{\text{George, not Bill}\}}{P\{\text{exactly 1}\}}$

$$= \frac{P\{G, \text{ not } B\}}{P\{G, \text{ not } B\} + P\{B, \text{ not } G\}}$$

$$= \frac{(0.4)(0.3)}{(0.4)(0.3) + (0.7)(0.6)}$$

$$= \frac{2}{9}$$

(b)
$$P\{G|\text{hit}\} = \frac{P\{G, \text{hit}\}}{P\{\text{hit}\}}$$

$$= \frac{P\{G\}}{P\{\text{hit}\}} = \frac{0.4}{1 - (0.3)(0.6)} = \frac{20}{41}$$

32. Let E_i = event person i selects own hat.

$P(\text{no one selects own hat})$

$$= 1 - P(E_1 \cup E_2 \cup \cdots \cup E_n)$$

$$= 1 - \left[\sum_{i_i} P(E_{i_1}) - \sum_{i_1 < i_2} P(E_{i_1} E_{i_2}) + \cdots + (-1)^{n+1} P(E_1 E_2 \cdots E_n)\right]$$

$$= 1 - \sum_{i_1} P(E_{i_1}) - \sum_{i_1 < i_2} P(E_{i_1} E_{i_2}) - \sum_{i_1 < i_2 < i_3} P(E_{i_1} E_{i_2} E_{i_3}) + \cdots$$

$$+ (-1)^n P(E_1 E_2 \cdots E_n)$$

Let $k \in \{1, 2, \ldots, n\}$. $P(E_{i_1} E_{i_2} E_{i_k})$ = number of ways k specific men can select own hats $\div$ total number of ways hats can be arranged $= (n - k)!/n!$. Number of terms in summation $\sum_{i_1 < i_2 < \cdots < i_k}$ = number of ways to choose k variables out of n variables $= \binom{n}{k} = n!/k!(n - k)!$. Thus,

$$\sum_{i_1 < \cdots < i_k} P(E_{i_1} E_{i_2} \cdots E_{i_k}) = \sum_{i_1 < \cdots < i_k} \frac{(n - k)!}{n!}$$

$$= \binom{n}{k} \frac{(n - k)!}{n!} = \frac{1}{k!}$$

$$\therefore P(\text{no one selects own hat}) = 1 - \frac{1}{1!} + \frac{1}{2!} - \frac{1}{3!} + \cdots + (-1)^n \frac{1}{n!}$$

$$= \frac{1}{2!} - \frac{1}{3!} + \cdots + (-1)^n \frac{1}{n!}$$

40. (a) F = event fair coin flipped; U = event two-headed coin flipped.

$$P(F|H) = \frac{P(H|F)P(F)}{P(H|F)P(F) + P(H|U)P(U)}$$

$$= \frac{\frac{1}{2} \cdot \frac{1}{2}}{\frac{1}{2} \cdot \frac{1}{2} + 1 \cdot \frac{1}{2}} = \frac{\frac{1}{4}}{\frac{3}{4}} = \frac{1}{3}$$

(b)
$$P(F|HH) = \frac{P(HH|F)P(F)}{P(HH|F)P(F) + P(HH|U)P(U)}$$

$$= \frac{\frac{1}{4} \cdot \frac{1}{2}}{\frac{1}{4} \cdot \frac{1}{2} + 1 \cdot \frac{1}{2}} = \frac{\frac{1}{8}}{\frac{5}{8}} = \frac{1}{5}$$

(c) $$P(F \mid HHT) = \frac{P(HHT \mid F)P(F)}{P(HHT \mid F)P(F) + P(HHT \mid U)P(U)}$$

$$= \frac{P(HHT \mid F)P(F)}{P(HHT \mid F)P(F) + 0} = 1$$

since the fair coin is the only one that can show tails.

45. Let B_i = event ith ball is black; R_i = event ith ball is red.

$$P(B_1 \mid R_2) = \frac{P(R_2 \mid B_1)P(B_1)}{P(R_2 \mid B_1)P(B_1) + P(R_2 \mid R_1)P(R_1)}$$

$$= \frac{\dfrac{r}{b+r+c} \cdot \dfrac{b}{b+r}}{\dfrac{r}{b+r+c} \cdot \dfrac{b}{b+r} + \dfrac{r+c}{b+r+c} \cdot \dfrac{r}{b+r}}$$

$$= \frac{rb}{rb + (r+c)r}$$

$$= \frac{b}{b+r+c}$$

Chapter 2

4. (i) 1, 2, 3, 4, 5, 6.
   (ii) 1, 2, 3, 4, 5, 6.
   (iii) 2, 3, ..., 11, 12.
   (iv) $-5, -4, \ldots, 4, 5$.

11. $\binom{4}{2}(\frac{1}{2})^2(\frac{1}{2})^2 = \frac{3}{8}$.

16. $1 - (0.95)^{52} - 52(0.95)^{51}(0.05)$.

23. In order for X to equal n, the first $n-1$ flips must have $r-1$ heads, and then the nth flip must land heads. By independence the desired probability is thus

$$\binom{n-1}{r-1} p^{r-1}(1-p)^{n-r} \times p$$

27. $\qquad P\{\text{same number of heads}\} = \sum_i P\{A = i, B = i\}$

$$= \sum_i \binom{k}{i}\left(\frac{1}{2}\right)^k \binom{n-k}{i}\left(\frac{1}{2}\right)^{n-k}$$

$$= \sum_i \binom{k}{i}\binom{n-k}{i}\left(\frac{1}{2}\right)^n$$

$$= \sum_i \binom{k}{k-i}\binom{n-k}{i}\left(\frac{1}{2}\right)^n$$

$$= \binom{n}{k}\left(\frac{1}{2}\right)^n$$

Another argument is as follows:

$\qquad P\{\# \text{ heads of } A = \# \text{ heads of } B\}$

$\qquad\qquad = P\{\# \text{ tails of } A = \# \text{ heads of } B\} \qquad \text{since coin is fair}$

$\qquad\qquad = P\{k - \# \text{ heads of } A = \# \text{ heads of } B\}$

$\qquad\qquad = P\{k = \text{total } \# \text{ heads}\}$

38. $\qquad\qquad c = 2, \qquad P\{X > 2\} = \displaystyle\int_2^\infty 2e^{-2x}\,dx = e^{-4}$

46. Let X_i be 1 if trial i is a success and 0 otherwise.

(a) The largest value is 0.6. If $X_1 = X_2 = X_3$, then

$$1.8 = E[X] = 3E[X_1] = 3P\{X_1 = 1\}$$

and so $P\{X = 3\} = P\{X_1 = 1\} = 0.6$. That this is the largest value is seen by Markov's inequality which yields that

$$P\{X \geqslant 3\} \leqslant E[X]/3 = 0.6$$

(b) The smallest value is 0. To construct a probability scenario for which $P\{X = 3\} = 0$, let U be a uniform random variable on $(0, 1)$, and define

$$X_1 = \begin{cases} 1 & \text{if } U \leqslant 0.6 \\ 0 & \text{otherwise} \end{cases}$$

$$X_2 = \begin{cases} 1 & \text{if } U \geqslant 0.4 \\ 0 & \text{otherwise} \end{cases}$$

$$X_3 = \begin{cases} 1 & \text{if either } U \leqslant 0.3 \text{ or } U \geqslant 0.7 \\ 0 & \text{otherwise} \end{cases}$$

it is easy to see that

$$P\{X_1 = X_2 = X_3 = 1\} = 0$$

48. $E[X^2] - (E[X])^2 = \text{Var}(X) = E[(X - E[X])^2] \geq 0$. Equality when $\text{Var}(X) = 0$, that is, when X is constant.

56. If X is binomial with parameters n and p, then

$$E[X^2] = \sum_{i=1}^{n} i^2 \binom{n}{i} p^i (1-p)^{n-i}$$

Writing $i^2 = i(i-1) + i$, we have

$$E[X^2] = \sum_{i=0}^{n} i(i-1) \binom{n}{i} p^i (1-p)^{n-i} + E[X]$$

$$= \sum_{i=2}^{n} \frac{n!}{(n-i)!(i-2)!} p^i (1-p)^{n-i} + E[X]$$

$$= n(n-1)p^2 \sum_{i=2}^{n} \frac{(n-2)!}{(n-i)!(i-2)!} p^{i-2} (1-p)^{n-i} + E[X]$$

$$= n(n-1)p^2 \sum_{j=0}^{n-2} \binom{n-2}{j} p^j (1-p)^{n-2-j} + E[X] \qquad \text{(by } j = i-2\text{)}$$

$$= n(n-1)p^2 [p + (1-p)]^{n-2} + E[X]$$

$$= n(n-1)p^2 + E[X]$$

Because $E[X] = np$, we arrive at

$$\text{Var}(X) = E[X^2] - (E[X])^2$$

$$= n(n-1)p^2 + np - n^2 p^2$$

$$= np(1-p)$$

64. See Section 5.23 of Chapter 5. Another way is to use moment generating functions. The moment generating function of the sum of n independent exponentials with rate λ is equal to the product of their moment generating functions. That is, it is $[\lambda/(\lambda - t)]^n$. But this is precisely the moment generating function of a gamma with parameters n and λ.

70. Let X_i be Poisson with mean 1. Then

$$P\left\{\sum_{1}^{n} X_i \leq n\right\} = e^{-n} \sum_{k=0}^{n} \frac{n^k}{k!}$$

But for n large $\sum_1^n X_i - n$ has approximately a normal distribution with mean 0, and so the result follows.

72.   For the matching problem, letting $X = X_1 + \cdots + X_N$, where

$$X_i = \begin{cases} 1 & \text{if } i\text{th man selects his own hat} \\ 0 & \text{otherwise} \end{cases}$$

we obtain

$$\text{Var}(X) = \sum_{i=1}^N \text{Var}(X_i) + 2 \sum\sum_{i<j} \text{Cov}(X_i, X_j)$$

Since $P\{X_i = 1\} = 1/N$, we see

$$\text{Var}(X_i) = \frac{1}{N}\left(1 - \frac{1}{N}\right) = \frac{N-1}{N^2}$$

Also

$$\text{Cov}(X_i, X_j) = E[X_i X_j] - E[X_i]E[X_j]$$

Now,

$$X_i X_j = \begin{cases} 1 & \text{if the } i\text{th and } j\text{th men both select their own hats} \\ 0 & \text{otherwise} \end{cases}$$

and thus

$$\begin{aligned} E[X_i X_j] &= P\{X_i = 1, X_j = 1\} \\ &= P\{X_i = 1\}P\{X_j = 1 \mid X_i = 1\} \\ &= \frac{1}{N}\frac{1}{N-1} \end{aligned}$$

Hence,

$$\text{Cov}(X_i, X_j) = \frac{1}{N(N-1)} - \left(\frac{1}{N}\right)^2 = \frac{1}{N^2(N-1)}$$

and

$$\begin{aligned} \text{Var}(X) &= \frac{N-1}{N} + 2\binom{N}{2}\frac{1}{N^2(N-1)} \\ &= \frac{N-1}{N} + \frac{1}{N} \\ &= 1 \end{aligned}$$

Chapter 3

2.   Intuitively it would seem that the first head would be equally likely to occur on any of trials $1, \ldots, n-1$. That is, it is intuitive that

$$P\{X_1 = i \mid X_1 + X_2 = n\} = \frac{1}{n-1}, \qquad i = 1, \ldots, n-1$$

Formally,

$$P\{X_1 = i \mid X_1 + X_2 = n\} = \frac{P\{X_1 = i, \ X_1 + X_2 = n\}}{P\{X_1 + X_2 = n\}}$$

$$= \frac{P\{X_1 = i, \ X_2 = n - i\}}{P\{X_1 + X_2 = n\}}$$

$$= \frac{p(1-p)^{i-1} p(1-p)^{n-i-1}}{\binom{n-1}{1} p(1-p)^{n-2} p}$$

$$= \frac{1}{n-1}$$

In the above, the next to last equality uses the independence of X_1 and X_2 to evaluate the numerator and the fact that $X_1 + X_2$ has a negative binomial distribution to evaluate the denominator.

6.

$$p_{X \mid Y}(1 \mid 3) = \frac{P\{X = 1, \ Y = 3\}}{P\{Y = 3\}}$$

$$= \frac{P\{1 \text{ white, 3 black, 2 red}\}}{P\{3 \text{ black}\}}$$

$$= \frac{\dfrac{6!}{1!3!2!} \left(\dfrac{3}{14}\right)^1 \left(\dfrac{5}{14}\right)^3 \left(\dfrac{6}{14}\right)^2}{\dfrac{6!}{3!3!} \left(\dfrac{5}{14}\right)^3 \left(\dfrac{9}{14}\right)^3}$$

$$= \tfrac{4}{9}$$

$$p_{X \mid Y}(0 \mid 3) = \tfrac{8}{27}$$

$$p_{X \mid Y}(2 \mid 3) = \tfrac{2}{9}$$

$$p_{X \mid Y}(3 \mid 3) = \tfrac{1}{27}$$

$$E[X \mid Y = 1] = \tfrac{5}{3}$$

13. The conditional density of X given that $X > 1$ is

$$f_{X|X>1}(X) = \frac{f(x)}{P\{X>1\}} = \frac{\lambda \exp^{-\lambda x}}{e^{-\lambda}} \qquad \text{when } x > 1$$

$$E[X|X>1] = e^{\lambda} \int_1^{\infty} x\lambda\, e^{-\lambda x}\, dx = 1 + 1/\lambda$$

by integration by parts. This latter result also follows immediately by the lack of memory property of the exponential.

19.
$$\int E[X|Y=y]f_Y(y)\,dy = \iint x f_{X|Y}(x\,|\,y)\,dx\, f_Y(y)\,dy$$

$$= \iint x\,\frac{f(x,\,y)}{f_Y(y)}\,dx\, f_Y(y)\,dy$$

$$= \int x \int f(x,\,y)\,dy\,dx$$

$$= \int x f_X(x)\,dx$$

$$= E[X]$$

23. Let X denote the first time a head appears. Let us obtain an equation for $E[N|X]$ by conditioning on the next two flips after X. This gives

$$E[N|X] = E[N|X, h, h]p^2 + E[N|X, h, t]pq + E[N|X, t, h]pq$$
$$+ E[N|X, t, t]q^2$$

where $q = 1 - p$. Now

$$E[N|X, h, h] = X + 1, \qquad E[N|X, h, t] = X + 1$$
$$E[N|X, t, h] = X + 2, \qquad E[N|X, t, t] = X + 2 + E[N]$$

Substituting back gives

$$E[N|X] = (X+1)(p^2 + pq) + (X+2)pq + (X + 2 + E[N])q^2$$

Taking expectations, and using the fact that X is geometric with mean $1/p$, we obtain

$$E[N] = 1 + p + q + 2pq + q^2/p + 2q^2 + q^2 E[N]$$

Solving for $E[N]$ yields

$$E[N] = \frac{2 + 2q + q^2/p}{1 - q^2}$$

44.
$$\text{Var}\left(\sum_1^N X_i \,|\, N\right) = N\,\text{Var}(X)$$

$$E\left(\sum_1^N X_i \,|\, N\right) = NE(X)$$

Hence,

$$E\left[\text{Var}\left(\sum X_i \,|\, N\right)\right] = E[N]\text{Var}(X)$$

$$\text{Var}\left[E\left(\sum_1^N X_i \,|\, N\right)\right] = \text{Var}(N)E^2[X]$$

and the result follows from Exercise 43.

47.
$$E[X^2 Y^2 \,|\, X] = X^2 E[Y^2 \,|\, X]$$
$$\geq X^2 (E[Y \,|\, X])^2 = X^2$$

The inequality follows since for any random variable U, $E[U^2] \geq (E[U])^2$ and this remains true when conditioning on some other random variable X. Taking expectations of the above shows that

$$E[(XY)^2] \geq E[X^2]$$

As

$$E[XY] = E[E[XY \,|\, X]] = E[XE[Y \,|\, X]] = E[X]$$

the results follows.

53.
$$P\{X = n\} = \int_0^\infty P\{X = n \,|\, \lambda\} e^{-\lambda} \, d\lambda$$

$$= \int_0^\infty \frac{e^{-\lambda}\lambda^n}{n!} e^{-\lambda} \, d\lambda$$

$$= \int_0^\infty e^{-2\lambda} \lambda^n \frac{d\lambda}{n!}$$

$$= \int_0^\infty e^{-t} t^n \frac{dt}{n!} \left(\frac{1}{2}\right)^{n+1}$$

The results follows since $\int_0^\infty e^{-t} t^n \, dt = \Gamma(n+1) = n!$

57. (a) Intuitive that $f(p)$ is increasing in p, since the larger p is the greater is the advantage of going first.
(b) 1
(c) $\frac{1}{2}$ since the advantage of going first becomes nil.

(d) Condition on the outcome of the first flip:

$$f(p) = P\{I \text{ wins} \mid h\}p + P\{I \text{ wins} \mid t\}(1 - p)$$
$$= p + [1 - f(p)](1 - p)$$

Therefore,

$$f(p) = \frac{1}{2 - p}$$

70. Condition on the value of the sum prior to going over 100. In all cases the most likely value is 101. (For instance, if this sum is 98 then the final sum is equally likely to be either 101, 102, 103, or 104. If the sum prior to going over is 95, then the final sum is 101 with certainty.)

Chapter 4

1.
$$P_{01} = 1, \; P_{10} = \tfrac{1}{9}, \; P_{21} = \tfrac{4}{9}, \; P_{32} = 1$$
$$P_{11} = \tfrac{4}{9}, \; P_{22} = \tfrac{4}{9}$$
$$P_{12} = \tfrac{4}{9}, \; P_{23} = \tfrac{1}{9}$$

4.   Let the state space be $S = \{0, 1, 2, \bar{0}, \bar{1}, \bar{2}\}$, where state $i(\bar{i})$ signifies that the present value is i, and the present day is even (odd).

12.   If P_{ij} were (strictly) positive, then P_{ji}^n would be 0 for all n (otherwise, i and j would communicate). But then the process, starting in i, has a positive probability of at least P_{ij} of never returning to i. This contradicts the recurrence of i. Hence $P_{ij} = 0$.

17.   (a)

$$P = \begin{vmatrix} 0 & p & 0 & 0 & 1-p \\ 1-p & 0 & p & 0 & 0 \\ 0 & 1-p & 0 & p & 0 \\ 0 & 0 & 1-p & 0 & p \\ p & 0 & 0 & 1-p & 0 \end{vmatrix}$$

(b)   As the column sums all equal 1, we can use the results of Exercise 15 to conclude that $\pi_i = \tfrac{1}{5}$ for all $i = 0, 1, 2, 3, 4$.

23.   The limiting probabilities are obtained from

$$\pi_0 = \tfrac{1}{9}\pi_1,$$

$$\pi_1 = \pi_0 + \tfrac{4}{9}\pi_1 + \tfrac{4}{9}\pi_2,$$

$$\pi_2 = \tfrac{4}{9}\pi_1 + \tfrac{4}{9}\pi_2 + \pi_3,$$

$$\pi_0 + \pi_1 + \pi_2 + \pi_3 = 1$$

and the solution is $\pi_0 = \pi_3 = \tfrac{1}{20}$. $\pi_1 = \pi_2 = \tfrac{9}{20}$.

28.   With the state being the number of on switches this is a three-state Markov chain. The equations for the long-run proportions are

$$\pi_0 = \tfrac{1}{16}\pi_0 + \tfrac{1}{4}\pi_1 + \tfrac{9}{16}\pi_2,$$

$$\pi_1 = \tfrac{3}{8}\pi_0 + \tfrac{1}{2}\pi_1 + \tfrac{3}{8}\pi_2,$$

$$\pi_0 + \pi_1 + \pi_2 = 1$$

This gives the solution

$$\pi_0 = \tfrac{2}{7}, \qquad \pi_1 = \tfrac{3}{7}, \qquad \pi_2 = \tfrac{2}{7}$$

32.   (a) The number of transitions into state i by time n, the number of transitions originating from state i by time n, and the number of time periods the chain is in state i by time n, all differ by at most 1. Thus, their long-run proportions must be equal.
  (b) $\pi_i P_{ij}$ is the long-run proportion of transitions that go from state i to state j.
  (c) $\Sigma_i \pi_i P_{ij}$ is the long-run proportion of transitions that are into state j.
  (d) Since π_j is also the long-run proportion of transitions that are into state j, it follows that $\pi_j = \Sigma_i \pi_i P_{ij}$.

38. $\{Y_n, n \geqslant 1\}$ is a Markov chain with states (i, j).

$$P_{(i,j),(k,l)} = \begin{cases} 0, & \text{if } j \neq k \\ P_{jl}, & \text{if } j = k \end{cases}$$

where P_{jl} is the transition probability for $\{X_n\}$.

$$\lim_{n \to \infty} P\{Y_n = (i, j)\} = \lim_{n} P\{X_n = i, X_{n+1} = j\}$$

$$= \lim_{n} [P\{X_n = i\}P_{ij}]$$

$$= \pi_i P_{ij}$$

41. (b) $64 + E[\text{time until } HH] = 64 + 4 + E[\text{time until } H]$

$$= 70$$

51. (a) Since $\pi_i = \frac{1}{5}$ is equal to the inverse of the expected number of transitions to return to state i, it follows that the expected number of steps to return to the original position is 5.

(b) Condition on the first transition. Suppose it is to the right. In this case the probability is just the probability that a gambler who always bets 1 and wins each bet with probability p will, when starting with 1, reach 4 before going broke. By the gambler's ruin problem this probability is equal to

$$\frac{1 - q/p}{1 - (q/p)^4}$$

Similarly, if the first move is to the left then the problem is again the same gambler's ruin problem but with p and q reversed. The desired probability is thus

$$\frac{p - q}{1 - (q/p)^4} + \frac{q - p}{1 - (p/q)^4}$$

57. (a) $$\sum_i \pi_i Q_{ij} = \sum_i \pi_j P_{ji} = \pi_j \sum_i P_{ji} = \pi_j$$

(b) Whether perusing the sequence of states in the forward direction of time or in the reverse direction, the proportion of time the state is i will be the same.

Chapter 5

7. $$P\{X_1 < X_2 \mid \min(X_1, X_2) = t\}$$

$$= \frac{P\{X_1 < X_2, \ \min(X_1, X_2) = t\}}{P\{\min(X_1, X_2) = t\}}$$

$$= \frac{P\{X_1 = t, \ X_2 > t\}}{P\{X_1 = t, \ X_2 > t\} + P\{X_2 = t, \ X_1 > t\}}$$

$$= \frac{f_1(t)[1 - F_2(t)]}{f_1(t)[1 - F_2(t)] + f_2(t)[1 - F_1(t)]}$$

Dividing through by $[1 - F_1(t)][1 - F_2(t)]$ yields the result. (Of course, f_i and F_i are the density and distribution function of X_i, $i = 1, 2$.) To make the preceeding derivation rigorous, we should replace "$= t$" by $\in(t, t + \varepsilon)$ throughout and then let $\varepsilon \to 0$.

10.   (a) $P\{\min(X, Y) > t\} = P\{X > t, Y > t\}$

$$= P\{X > t\}P\{Y > t\} \qquad \text{by independence}$$

$$= e^{-\lambda_1 t}e^{-\lambda_2 t} = e^{-(\lambda_1 + \lambda_2)t}$$

Hence, Z is exponential with rate $\lambda_1 + \lambda_2$. Another way of seeing this is to use the lack of memory property. If we are given that Z (equal to the minimum of the two exponentials) is larger than s then we are given that each of the exponentials is larger than s. But by the lack of memory property the amounts of time by which these exponentials exceed s have the same distributions as the original exponentials. Therefore, the minimum is memoryless and so must be exponential. To see that its rate is $\lambda_1 + \lambda_2$, note that if both exponentials are greater than s (and so Z is greater than s) then the probability that at least one will die in the next ds time units is $\lambda_1 \, ds + \lambda_2 \, ds = (\lambda_1 + \lambda_2) \, ds$, which shows that the failure rate of Z is $\lambda_1 + \lambda_2$.

(b) $P\{Z \in (t, t + \varepsilon) \mid Z = X\} = P\{t < X < t + \varepsilon \mid X < Y\}$

$$= \frac{P\{t < X < t + \varepsilon, \ X < Y\}}{P\{X < Y\}}$$

$$\simeq \frac{P\{t < x < t + \varepsilon, \ t < y\}}{P\{x < y\}}$$

$$= \frac{P\{t < X < t < \varepsilon\}P\{t < Y\}}{P\{X < Y\}}$$

$$\simeq \frac{\varepsilon \lambda_1 e^{-\lambda_1 t}e^{-\lambda_2 t}}{\lambda_1/(\lambda_1 + \lambda_2)}$$

$$= \varepsilon(\lambda_1 + \lambda_2)e^{-(\lambda_1 + \lambda_2)t}$$

Hence, given $Z = X$, Z is (still) exponential with rate $\lambda_1 + \lambda_2$.

(c) Given that $Z = X$, $Y - Z$ represents the additional life from X onward from Y. Hence, by the lack of memory of the exponential, it must be exponential with mean $1/\lambda_2$.

18.   (a) $1/(2\mu)$.

(b) $1/(4\mu^2)$, since the variance of an exponential is its mean squared.

(c) and (d). By the lack of memory property of the exponential it follows that A, the amount by which $X_{(2)}$ exceeds $X_{(1)}$, is exponentially distributed with rate μ and is independent of $X_{(1)}$. Therefore,

$$E[X_{(2)}] = E[X_{(1)} + A] = \frac{1}{2\mu} + \frac{1}{\mu}$$

$$\text{Var}(X_{(2)}) = \text{Var}(X_{(1)} + A) = \frac{1}{4\mu^2} + \frac{1}{\mu^2} = \frac{5}{4\mu^2}$$

23. (a) $\frac{1}{2}$.
(b) $(\frac{1}{2})^{n-1}$. Whenever battery 1 is in use and a failure occurs the probability is $\frac{1}{2}$ that it is not battery 1 that has failed.
(c) $(\frac{1}{2})^{n-i+1}$, $i > 1$.
(d) T is the sum of $n - 1$ independent exponentials with rate 2μ (since each time a failure occurs the time until the next failure is exponential with rate 2μ).
(e) Gamma with parameters $n - 1$ and 2μ.

34. $P\{N(t + s) = 0\} = P\{N(t) = 0, N(t + s) - N(t) = 0\}$

$$= P\{N(t) = 0\}P\{N(t + s) - N(t) = 0\}$$

$$= P\{N(t) = 0\}P\{N(s) = 0\}$$

Therefore, $P\{N(t) = 0\} = e^{-\lambda t}$. Hence,

$$P\{N(h) \geq 2\} = 1 - P\{N(h) = 0\} - P\{N(h) = 1\}$$

$$= 1 - e^{-\lambda h} - \lambda h + o(h) \qquad \text{by (iii)}$$

$$= 1 - [1 - \lambda h + o(h)] - \lambda h + o(h)$$

$$= o(h)$$

38. The easiest way is to use Definition 3.1. It is easy to see that $\{N(t), t \geq 0\}$ will also possess stationary and independent increments. Since the sum of two independent Poisson random variables is also Poisson, it follows that $N(t)$ is a Poisson random variable with mean $(\lambda_1 + \lambda_2)t$.

54. (a) e^{-2}.
(b) 2 P.M.
(c) $1 - 5e^{-4}$.

57. (a) $\frac{1}{9}$.
(b) $\frac{5}{9}$.

61. (a) Since, given $N(t)$, each arrival is uniformly distributed on $(0, t)$ it follows that

$$E[X \mid N(t)] = N(t) \int_0^t (t - s) \frac{ds}{t} = N(t) \frac{t}{2}$$

(b) Let $U_1, U_2, \ldots$ be independent uniform $(0, t)$ random variables. Then

$$\text{Var}(X \mid N(t) = n) = \text{Var}\left[\sum_{i=1}^{n} (t - U_i) \right]$$

$$= n \, \text{Var}(U_i) = n \frac{t^2}{12}$$

(c) By parts (a) and (b) and the condition variance formula,

$$\text{Var}(X) = \text{Var}\left(\frac{N(t) \, t}{2} \right) + E\left[\frac{N(t) \, t^2}{12} \right]$$

$$= \frac{\lambda t t^2}{4} + \frac{\lambda t t^2}{12} = \frac{\lambda t^3}{3}$$

71. Consider a Poisson process with rate λ in which an event at time t is counted with probability $\lambda(t)/\lambda$ independently of the past. Clearly such a process will have independent increments. In addition,

$$P\{2 \text{ or more counted events in } (t, t + h)\}$$

$$\leqslant P\{2 \text{ or more events in } (t, t + h)\}$$

$$= o(h)$$

and

$$P\{1 \text{ counted event in } (t, t + h)\}$$

$$= P\{1 \text{ counted} \mid 1 \text{ event}\} P(1 \text{ event})$$

$$+ P\{1 \text{ counted} \mid \geqslant 2 \text{ events}\} P\{\geqslant 2\}$$

$$= \int_t^{t+h} \frac{\lambda(s)}{\lambda} \frac{ds}{h} (\lambda h + o(h)) + o(h)$$

$$= \frac{\lambda(t)}{\lambda} \lambda h + o(h)$$

$$= \lambda(t) \, h + o(h)$$

76. There is a record whose value is between t and $t + dt$ if the first X larger than t lies between t and $t + dt$. From this we see that, independent of all record values less than t, there will be one between t and $t + dt$ with probability $\lambda(t)\,dt$ where $\lambda(t)$ is the failure rate function given by

$$\lambda(t) = \frac{f(t)}{1 - F(t)}$$

Since the counting process of record values has, by the above, independent increments we can conclude (since there cannot be multiple record values because the X_i are continuous) that it is a nonhomogeneous Poisson process with intensity function $\lambda(t)$. When f is the exponential density, $\lambda(t) = \lambda$ and so the counting process of record values becomes an ordinary Poisson process with rate λ.

83.   To begin, note that

$$P\left\{X_1 > \sum_2^n X_i\right\} = P\{X_1 > X_2\}P\{X_1 - X_2 > X_3 \,|\, X_1 > X_2\}$$

$$\times P\{X_1 - X_2 - X_3 > X_4 \,|\, X_1 > X_2 + X_3\} \cdots$$

$$\times P\{X_1 - X_2 \cdots - X_{n-1} > X_n \,|\, X_1 > X_2 + \cdots + X_{n-1}\}$$

$$= (\tfrac{1}{2})^{n-1} \qquad \text{by lack of memory}$$

Hence,

$$P\left\{M > \sum_{i=1}^n X_i - M\right\} = \sum_{i=1}^n P\left\{X_i > \sum_{j \neq i} X_j\right\} = \frac{n}{2^{n-1}}$$

Chapter 6

2.   Let $N_A(t)$ be the member of organisms in state A and let $N_B(t)$ be the number of organisms in state B. Then $\{N_A(t),\ N_B(t)\}$ is a continuous-Markov chain with

$$v_{\{n,m\}} = \alpha n + \beta m$$

$$P_{\{n,m\},\{n-1,m+1\}} = \frac{\alpha n}{\alpha n + \beta m}$$

$$P_{\{n,m\},\{n+2,m-1\}} = \frac{\beta m}{\alpha n + \beta m}$$

4.   Let $N(t)$ denote the number of customers in the station at time t. Then $\{N(t)\}$ is a birth and death process with

$$\lambda_n = \lambda \alpha_n, \qquad \mu_n = \mu$$

7.   (a) Yes!
   (b) For $\mathbf{n} = (n_1, \ldots, n_i, n_{i+1}, \ldots, n_{k-1})$ let

$$S_i(\mathbf{n}) = (n_1, \ldots, n_i - 1, n_{i+1} + 1, \ldots, n_{k-1}) \qquad i = 1, \ldots, k-2$$

$$S_{k-1}(\mathbf{n}) = (n_1, \ldots, n_i, n_{i+1}, \ldots, n_{k-1} - 1),$$

$$S_0(\mathbf{n}) = (n_1 + 1, \ldots, n_i, n_{i+1}, \ldots, n_{k-1}).$$

Then

$$q_{\mathbf{n}, S_i(\mathbf{n})} = n_i \mu \qquad i = 1, \ldots, k-1$$

$$q_{\mathbf{n}, S_0(n)} = \lambda$$

11.   (b) Follows from the hint about using the lack of memory property and the fact that ε_i, the minimum of $j - (i-1)$ independent exponentials with rate λ, is exponential with rate $(j - i + 1)\lambda$.
   (c) From parts (a) and (b)

$$P\{T_1 + \cdots + T_j \leqslant t\} = P\left\{\max_{1 \leqslant i \leqslant j} X_i \leqslant t\right\} = (1 - e^{-\lambda t})^j$$

   (d) With all probabilities conditional on $X(0) = 1$

$$\begin{aligned} P_{1j}(t) &= P\{X(t) = j\} \\ &= P\{X(t) \geqslant j\} - P\{X(t) \geqslant j + 1\} \\ &= P\{T_1 + \cdots + T_j \leqslant t\} - P\{T_1 + \cdots + T_{j+1} \leqslant t\} \end{aligned}$$

   (e) The sum of i independent geometrics, each having parameter $p = e^{-\lambda t}$, is a negative binomial with parameters i, p. The result follows since starting with an initial population of i is equivalent to having i independent Yule processes, each starting with a single individual.

16.   Let the state be

   2: an acceptable molecule is attached
   0: no molecule attached
   1: an unacceptable molecule is attached.

Then this is a birth and death process with balance equations

$$\mu_1 P_1 = \lambda(1 - \alpha)P_0$$

$$\mu_2 P_2 = \lambda \alpha P_0$$

Since $\Sigma_0^2 P_i = 1$, we get

$$P_2 = \left[1 + \frac{\mu_2}{\lambda \alpha} + \frac{1 - \alpha}{\alpha}\frac{\mu_2}{\mu_1}\right]^{-1} = \frac{\lambda \alpha \mu_1}{\lambda \alpha \mu_1 + \mu_1 \mu_2 + \lambda(1 - \alpha)\mu_2}$$

where P_2 is the percentage of time the site is occupied by an acceptable molecule. The percentage of time the site is occupied by an unacceptable molecule is

$$P_1 = \frac{1 - \alpha}{\alpha} \frac{\mu_2}{\mu_1} P_1 = \frac{\lambda(1 - \alpha)\mu_2}{\lambda\alpha\mu_1 + \mu_1 u_2 + \lambda(1 - \alpha)\mu_2}$$

19. There are 4 states. Let state 0 mean that no machines are down, state 1 that machine 1 is down and 2 is up, state 2 that machine 1 is up and 2 is down, and 3 that both machines are down. The balance equations are as follows:

$$(\lambda_1 + \lambda_2)P_0 = \mu_1 P_1 + \mu_2 P_2$$
$$(\mu_1 + \lambda_2)P_1 = \lambda_1 P_0$$
$$(\lambda_1 + \mu_2)P_2 = \lambda_2 P_0 + \mu_1 P_3$$
$$\mu_1 P_3 = \lambda_2 P_1 + \lambda_1 P_2$$
$$P_0 + P_1 + P_2 + P_3 = 1$$

The equations are easily solved and the proportion of time machine 2 is down is $P_2 + P_3$.

24. We will let the state be the number of taxis waiting. Then, we get a birth and death process with $\lambda_n = 1$, $\mu_n = 2$. This is an $M/M/1$. Therefore:

(a) Average number of taxis waiting $= \dfrac{1}{\mu - \lambda} = \dfrac{1}{2 - 1} = 1.$

(b) The proportion of arriving customers that gets taxis is the proportion of arriving customers that find at least one taxi waiting. The rate of arrival of such customers is $2(1 - P_0)$. The proportion of such arrivals is therefore

$$\frac{2(1 - P_0)}{2} = 1 - P_0 = 1 - \left(1 - \frac{\lambda}{\mu}\right) = \frac{\lambda}{\mu} = \frac{1}{2}$$

28. Let P_{ij}^x, v_i^x denote the parameters of the $X(t)$ and P_{ij}^y, v_i^y of the $Y(t)$ process; and let the limiting probabilities be P_i^x, P_i^y, respectively. By independence we have that for the Markov chain $\{X(t), Y(t)\}$ its parameters are

$$v_{(i,l)} = v_i^x + v_l^y$$

$$P_{(i,l),(j,l)} = \frac{v_i^x}{v_i^x + v_l^y} P_{ij}^x$$

$$P_{(i,l),(i,k)} = \frac{v_l^y}{v_i^x + v_l^y} P_{lk}^y$$

and

$$\lim_{t \to \infty} P\{(X(t),\ Y(t)) = (i, j)\} = P_i^x P_j^y$$

Hence, we need to show that

$$P_i^x P_l^y v_i^x P_{ij}^x = P_j^x P_l^y v_j^x P_{ji}^x.$$

[That is, rate from (i, l) to (j, l) equals the rate from (j, l) to (i, l).] But this follows from the fact that the rate from i to j in $X(t)$ equals the rate from j to i; that is,

$$P_i^x v_i^x P_{ij}^x = P_j^x v_j^x P_{ji}^x$$

The analysis is similar in looking at pairs (i, l) and (i, k).

33.   Suppose first that the waiting room is of infinite size. Let $X_i(t)$ denote the number of customers at server i, $i = 1, 2$. Then since each of the $M/M/1$ processes $\{X_i(t)\}$ is time reversible, it follows from Exercise 28 that the vector process $\{(X_1(t),\ X_2(t)),\ t \geq 0\}$ is a time reversible Markov chain. Now the process of interest is just the truncation of this vector process to the set of states A where

$$A = \{(0,\ m) : m \leq 4\} \cup \{(n,\ 0) : n \leq 4\} \cup \{(n,\ m) : nm > 0,\ n + m \leq 5\}$$

Hence, the probability that there are n with server 1 and m with server 2 is

$$P_{n,m} = k \left(\frac{\lambda_1}{\mu_1}\right)^n \left(1 - \frac{\lambda_1}{\mu_1}\right) \left(\frac{\lambda_2}{\mu_2}\right)^m \left(1 - \frac{\lambda_2}{\mu_2}\right)$$

$$= C \left(\frac{\lambda_1}{\mu_1}\right)^n \left(\frac{\lambda_2}{\mu_2}\right)^m, \qquad (n,\ m) \in A$$

The constant C is determined from

$$\sum P_{n,m} = 1$$

where the sum is over all $(n,\ m)$ in A.

42.   (a) The matrix $\mathbf{P}^*$ can be written as

$$\mathbf{P}^* = \mathbf{I} + \mathbf{R}/v$$

and so P_{ij}^{*n} can be obtained by taking the i, j element of $(\mathbf{I} + \mathbf{R}/v)^n$, which gives the result when $v = n/t$.
(b) Uniformization shows that $P_{ij}(t) = E[P_{ij}^{*N}]$, where N is independent of the Markov chain with transition probabilities P_{ij}^* and is Poisson distributed with mean vt. Since a Poisson random variable

with mean vt has standard deviation $(vt)^{1/2}$, it follows that for large values of vt it should be near vt. (For instance, a Poisson random variable with mean 10^6 has standard deviation 10^3 and thus will, with high probability, be within 3000 of 10^6.) Hence, since for fixed i and j, P_{ij}^{*m} should not vary much for values of m about vt where vt is large, it follows that, for large vt,

$$E[P_{ij}^{*N}] \approx P_{ij}^{*n} \qquad \text{where } n = vt$$

Chapter 7

3. By the one-to-one correspondence of $m(t)$ and F, it follows that $\{N(t), t \geq 0\}$ is a Poisson process with rate $\frac{1}{2}$. Hence,

$$P\{N(5) = 0\} = e^{-5/2}$$

6. (a) Consider a Poisson process having rate λ and say that an event of the renewal process occurs whenever one of the events numbered r, $2r$, $3r,\ldots$ of the Poisson process occurs. Then

$$P\{N(t) \geq n\} = P\{nr \text{ or more Poisson events by } t\}$$

$$= \sum_{i=nr}^{\infty} e^{-\lambda t}(\lambda t)^i/i!$$

(b)
$$E[N(t) = \sum_{n=1}^{\infty} P\{N(t) \geq n\} = \sum_{n=1}^{\infty} \sum_{i=nr}^{\infty} e^{-\lambda t}(\lambda t)^i/i!$$

$$= \sum_{i=r}^{\infty} \sum_{n=1}^{[i/r]} e^{-\lambda t}(\lambda t)^i/i! = \sum_{i=r}^{\infty} [i/r]e^{-\lambda t}(\lambda t)^i/i!$$

8. (a) The number of replaced machines by time t constitutes a renewal process. The time between replacements equals T, if lifetime of new machine is $\geq T$; x, if lifetime of new machine is x, $x < T$. Hence,

$$E[\text{time between replacements}] = \int_0^T xf(x)\,dx + T[1 - F(T)]$$

and the result follows by Proposition 3.1.
(b) The number of machines that have failed in use by time t constitutes a renewal process. The mean time between in-use failures, $E[F]$, can be calculated by conditioning on the lifetime of the initial machine as $E[F] = E[E[F \mid \text{lifetime of initial machine}]]$. Now

$$E[F \mid \text{lifetime of machine is } x] = \begin{cases} x, & \text{if } x \leq T \\ T + E[F], & \text{if } x > T \end{cases}$$

Hence,

$$E[F] = \int_0^T xf(x)\,dx + (T + E[F])[1 - F(T)]$$

or

$$E[F] = \frac{\int_0^T xf(x)\,dx + T[1 - F(T)]}{F(T)}$$

and the result follows from Proposition 3.1.

17.   We can imagine that a renewal corresponds to a machine failure, and each time a new machine is put in use its life distribution will be exponential with rate μ_1 with probability p, and exponential with rate μ_2 otherwise. Hence, if our state is the index of the exponential life distribution of the machine presently in use, then this is a two-state continuous-time Markov chain with intensity rates

$$q_{1,2} = \mu_1(1 - p) \qquad q_{2,1} = \mu_2 p$$

Hence,

$$P_{11}(t) = \frac{\mu_1(1 - p)}{\mu_1(1 - p) + \mu_2 p} \exp\{-[\mu_1(1 - p) + \mu_2 p]t\}$$

$$+ \frac{\mu_2 p}{\mu_1(1 - p) + \mu_2 p}$$

with similar expressions for the other transition probabilities [$P_{12}(t) = 1 - P_{11}(t)$, and $P_{22}(t)$ is the same with $\mu_2 p$ and $\mu_1(1 - p)$ switching places]. Conditioning on the initial machine now gives

$$E[Y(t)] = pE[Y(t) \mid X(0) = 1] + (1 - p)E[Y(t) \mid X(0) = 2]$$

$$= p\left[\frac{P_{11}(t)}{\mu_1} + \frac{P_{12}(t)}{\mu_2}\right] + (1 - p)\left[\frac{P_{21}(t)}{\mu_1} + \frac{P_{22}(t)}{\mu_2}\right]$$

Finally, we can obtain $m(t)$ from

$$\mu[m(t) + 1] = t + E[Y(t)]$$

where

$$\mu = p/\mu_1 + (1 - p)/\mu_2$$

is the mean interarrival time.

21. Cost of a cycle $= C_1 + C_2 I - R(T)(1 - I)$

$$I = \begin{cases} 1, & \text{if } X < T \\ 0, & \text{if } X \geq T \end{cases} \quad \text{where } X = \text{life of car}$$

Hence,

$$E[\text{cost of a cycle}] = C_1 + C_2 H(T) - R(T)[1 - H(T)]$$

Also,

$$E[\text{time of cycle}] = \int E[\text{time} \mid X = x] h(x) \, dx$$

$$= \int_0^T x h(x) \, dx + T[1 - H(T)]$$

Thus the average cost per unit time is given by

$$\frac{C_1 + C_2 H(T) - R(T)[1 - H(T)]}{\int_0^T x h(x) \, dx + T[1 - H(T)]}$$

28.
$$\frac{A(t)}{t} = \frac{t - S_{N(t)}}{t}$$

$$= 1 - \frac{S_{N(t)}}{t}$$

$$= 1 - \frac{S_{N(t)}}{N(t)} \frac{N(t)}{t}$$

The result follows since $S_{N(t)}/N(t) \to \mu$ (by the strong law of large numbers) and $N(t)/t \to 1/\mu$.

31. (a) We can view this as an $M/G/\infty$ system where a satellite launching corresponds to an arrival and F is the service distribution. Hence,

$$P\{X(t) = k\} = e^{-\lambda(t)} [\lambda(t)]^k / k!$$

where $\lambda(t) = \lambda \int_0^t (1 - F(s)) \, ds$.
(b) By viewing the system as an alternating renewal process that is on when there is at least one satellite orbiting, we obtain

$$\lim P\{X(t) = 0\} = \frac{1/\lambda}{1/\lambda + E[T]}$$

where T, the on time in a cycle, is the quantity of interest. From part (a)

$$\lim P\{X(t) = 0\} = e^{-\lambda \mu}$$

where $\mu = \int_0^\infty (1 - F(s)) \, ds$ is the mean time that a satellite orbits.

Hence,

$$e^{-\lambda\mu} = \frac{1/\lambda}{1/\lambda + E[T]}$$

so

$$E[T] = \frac{1 - e^{-\lambda\mu}}{\lambda e^{-\lambda\mu}}$$

38. (a)

$$F_e(x) = \frac{1}{\mu}\int_0^x e^{-y/\mu}\,dy = 1 - e^{-x/\mu}$$

(b)

$$F_e(x) = \frac{1}{c}\int_0^x dy = \frac{x}{c} \qquad 0 \leqslant x \leqslant c$$

(c) You will receive a ticket if, starting when you park, an official appears within 1 hour. From Example 7.23 the time until the official appears has the distribution F_e which, by part (a), is the uniform distribution on $(0, 2)$. Thus, the probability is equal to $\frac{1}{2}$.

45. Think of each interarrival time as consisting of n independent phases—each of which is exponentially distributed with rate λ—and consider the semi-Markov process whose state at any time is the phase of the present interarrival time. Hence, this semi-Markov process goes from state 1 to 2 to 3...to n to 1, and so on. Also the time spent in each state has the same distribution. Thus, clearly the limiting probability of this semi-Markov chain is $P_i = 1/n$, $i = 1, \ldots, n$. To compute $\lim P\{Y(t) < x\}$, we condition on the phase at time t and note that if it is $n - i + 1$, which will be the case with probability $1/n$, then the time until a renewal occurs will be sum of i exponential phases, which will thus have a gamma distribution with parameters i and λ.

Chapter 8

2. This problem can be modeled by an $M/M/1$ queue in which $\lambda = 6$, $\mu = 8$. The average cost rate will be

$10 per hour per machine $\times$ average number of broken machines

The average number of broken machines is just L, which can be computed from Equation (3.2):

$$L = \frac{\lambda}{\mu - \lambda}$$

$$= \frac{6}{2} = 3$$

Hence, the average cost rate $= \$30/\text{hour}$.

7. To compute W for the $M/M/2$, set up balance equations as follows:

$$\lambda P_0 = \mu P_1 \qquad \text{(each server has rate } \mu\text{)}$$

$$(\lambda + \mu)P_1 = \lambda P_0 + 2\mu P_2$$

$$(\lambda + 2\mu)P_n = \lambda P_{n-1} + 2\mu P_{n+1} \qquad n \geqslant 2$$

These have solutions $P_n = \rho^n/2^{n-1}P_0$ where $\rho = \lambda/\mu$. The boundary condition $\Sigma_{n=0}^{\infty} P_n = 1$ implies

$$P_0 = \frac{1 - \rho/2}{1 + \rho/2} = \frac{(2 - \rho)}{(2 + \rho)}$$

Now we have P_n, so we can compute L, and hence W from $L = \lambda W$:

$$L = \sum_{n=0}^{\infty} nP_n = \rho P_0 \sum_{n=0}^{\infty} n\left(\frac{\rho}{2}\right)^{n-1}$$

$$= 2P_0 \sum_{n=0}^{\infty} n\left(\frac{\rho}{2}\right)^{n}$$

$$= 2\frac{(2 - \rho)}{(2 + \rho)} \frac{(\rho/2)}{(1 - \rho/2)^2} \qquad \text{[See derivation of Eq. (3.2).]}$$

$$= \frac{4\rho}{(2 + \rho)(2 - \rho)}$$

$$= \frac{4\mu\lambda}{(2\mu + \lambda)(2\mu - \lambda)}$$

From $L = \lambda W$ we have

$$W = W(M/M/2) = \frac{4\mu}{(2\mu + \lambda)(2\mu - \lambda)}$$

The $M/M/1$ queue with service rate 2μ has

$$W(M/M/1) = \frac{1}{2\mu - \lambda}$$

from Equation (3.3). We assume that in the $M/M/1$ queue, $2\mu > \lambda$ so that the queue is stable. But then $4\mu > 2\mu + \lambda$, or $4\mu/(2\mu + \lambda) > 1$, which implies $W(M/M/2) > W(M/M/1)$. The intuitive explanation is that if one finds the queue empty in the $M/M/2$ case, it would do no good to have two servers. One would be better off with one faster server. Now let $W_Q^1 = W_Q(M/M/1)$ and $W_Q^2 = W_Q(M/M/2)$. Then,

$$W_Q^1 = W(M/M/1) - 1/2\mu$$

$$W_Q^2 = W(M/M/2) - 1/\mu$$

So,

$$W_Q^1 = \frac{\lambda}{2\mu(2\mu - \lambda)} \qquad (3.3)$$

and

$$W_Q^2 = \frac{\lambda^2}{\mu(2\mu - \lambda)(2\mu + \lambda)}$$

Then,

$$W_Q^1 > W_Q^2 \Leftrightarrow \frac{1}{2} > \frac{\lambda}{(2\mu + \lambda)}$$

$$\lambda < 2\mu$$

Since we assume $\lambda < 2\mu$ for stability in the $M/M/1$ case, $W_Q^2 < W_Q^1$ whenever this comparison is possible, that is, whenever $\lambda < 2\mu$.

13. (a)
$$\lambda P_0 = \mu P_1$$
$$(\lambda + \mu)P_1 = \lambda P_0 + 2\mu P_2$$
$$(\lambda + 2\mu)P_n = \lambda P_{n-1} + 2\mu P_{n+1} \qquad n \geq 2$$

These are the same balance equations as for the $M/M/2$ queue and have solution

$$P_0 = \left(\frac{2\mu - \lambda}{2\mu + \lambda}\right), \qquad P_n = \frac{\lambda^n}{2^{n-1}\mu^n} P_0$$

(b) The system goes from 0 to 1 at rate

$$\lambda P_0 = \frac{\lambda(2\mu - \lambda)}{(2\mu + \lambda)}$$

The system goes from 2 to 1 at rate

$$2\mu P_2 = \frac{\lambda^2}{\mu} \frac{(2\mu - \lambda)}{(2\mu + \lambda)}$$

(c) Introduce a new state cl to indicate that the stock clerk is checking by himself. The balance equation for P_{cl} is

$$(\lambda + \mu)P_{cl} = \mu P_2$$

Hence

$$P_{cl} = \frac{\mu}{\lambda + \mu} P_2 = \frac{\lambda^2}{2\mu(\lambda + \mu)} \frac{(2\mu - \lambda)}{(2\mu + \lambda)}$$

Finally, the proportion of time the stock clerk is checking is

$$P_{cl} + \sum_{n=2}^{\infty} P_n = P_{cl} + \frac{2\lambda^2}{\mu(2\mu - \lambda)}$$

21. (a) $\lambda_1 P_{10}$
(b) $\lambda_2(P_0 + P_{10})$
(c) $\lambda_1 P_{10}/[\lambda_1 P_{10} + \lambda_2(P_0 + P_{10})]$
(d) This is equal to the fraction of server 2's customers that are type 1 multiplied by the proportion of time server 2 is busy. (This is true since the amount of time server 2 spends with a customer does not depend on which type of customer it is.) By (c) the answer is thus

$$\frac{(P_{01} + P_{11})\lambda_1 P_{10}}{\lambda_1 P_{10} + \lambda_2(P_0 + P_{10})}$$

24. The states are now $n, n \geqslant 0$, and n', $n \geqslant 1$ where the state is n when there are n in the system and no breakdown, and it is n' when there are n in the system an a breakdown is in progress. The balance equations are

$$\lambda P_0 = \mu P_1$$

$$(\lambda + \mu + \alpha)P_n = \lambda P_{n-1} + \mu P_{n+1} + \beta P_{n'} \qquad n \geqslant 1$$

$$(\beta + \lambda)P_{1'} = \alpha P_1$$

$$(\beta + \lambda)P_{n'} = \alpha P_n + \lambda P_{(n-1)'} \qquad n \geqslant 2$$

$$\sum_{n=0}^{\infty} P_n + \sum_{n=1}^{\infty} P_{n'} = 1$$

In terms of the solution to the above,

$$L = \sum_{n=1}^{\infty} n(P_n + P_{n'})$$

and so

$$W = \frac{L}{\lambda_a} = \frac{L}{\lambda}$$

28. If a customer leaves the system busy, the time until the next departure is the time of a service. If a customer leaves the system empty, the time until the next departure is the time until an arrival *plus* the time of a service.

Using moment generating functions we get

$$E\{e^{sD}\} = \frac{\lambda}{\mu} E\{e^{sD} | \text{system left busy}\}$$

$$+ \left(1 - \frac{\lambda}{\mu}\right) E\{e^{sD} | \text{system left empty}\}$$

$$= \left(\frac{\lambda}{\mu}\right)\left(\frac{\mu}{\mu - s}\right) + \left(1 - \frac{\lambda}{\mu}\right) E\{e^{s(X+Y)}\}$$

where X has the distribution of interarrival times, Y has the distribution of service times, and X and Y are independent. Then

$$E[e^{s(X+Y)}] = E[e^{sX}e^{sY}]$$

$$= E[e^{sX}]E[e^{sY}] \qquad \text{by independence}$$

$$= \left(\frac{\lambda}{\lambda - s}\right)\left(\frac{\mu}{\mu - s}\right)$$

So,

$$E\{e^{sD}\} = \left(\frac{\lambda}{\mu}\right)\left(\frac{\mu}{\mu - s}\right) + \left(1 - \frac{\lambda}{\mu}\right)\left(\frac{\lambda}{\lambda - s}\right)\left(\frac{\mu}{\mu - s}\right)$$

$$= \frac{\lambda}{(\lambda - s)}$$

By the uniqueness of generating functions, it follows that D has an exponential distribution with parameter λ.

36. The distributions of the queue size and busy period are the same for all three disciplines; that of the waiting time is different. However, the means are identical. This can be seen by using $W = L/\lambda$, since L is the same for all. The smallest variance in the waiting time occurs under first-come, first-served and the largest under last-come, first-served.

39. (a) $a_0 = P_0$ due to Poisson arrivals. Assuming that each customer pays 1 per unit time while in service the cost identity of Equation (2.1) states that

$$\text{average number in service} = \lambda E[S]$$

or

$$1 - P_0 = \lambda E[S]$$

(b) Since a_0 is the proportion of arrivals that have service distribution G_1 and $1 - a_0$ the proportion having service distribution G_2, the result follows.

(c) We have

$$P_0 = \frac{E[I]}{E[I] + E[B]}$$

and $E[I] = 1/\lambda$ and thus,

$$E[B] = \frac{1 - P_0}{\lambda P_0}$$

$$= \frac{E[S]}{1 - \lambda E[S]}$$

Now from parts (a) and (b) we have

$$E[S] = (1 - \lambda E[S])E[S_1] + \lambda E[S]E[S_2]$$

or

$$E[S] = \frac{E[S_1]}{1 + \lambda E[S_1] + \lambda E[S_2]}$$

Substituting into $E[B] = E[S]/(1 - \lambda E[S])$ now yields the result.
(d) $a_0 = 1/E[C]$, implying that

$$E[C] = \frac{E[S_1] + 1/\lambda - E[S_2]}{1/\lambda - E[S_2]}$$

45.   By regarding any breakdowns that occur during a service as being part of that service, we see that this is an $M/G/1$ model. We need to calculate the first two moments of a service time. Now the time of a service is the time T until something happens (either a service completion or a breakdown) plus any additional time A. Thus,

$$E[S] = E[T + A]$$
$$= E[T] + E[A]$$

To compute $E[A]$, we condition upon whether the happening is a service or a breakdown. This gives

$$E[A] = E[A \mid \text{service}]\frac{\mu}{\mu + \alpha} + E[A \mid \text{breakdown}]\frac{\alpha}{\mu + \alpha}$$

$$= E[A \mid \text{breakdown}]\frac{\alpha}{\mu + \alpha}$$

$$= \left(\frac{1}{\beta} + E[S]\right)\frac{\alpha}{\mu + \alpha}$$

Since $E[T] = 1/(\alpha + \mu)$ we obtain that

$$E[S] = \frac{1}{\alpha + \mu} + \left(\frac{1}{\beta} + E[S]\right)\frac{\alpha}{\mu + \alpha}$$

or

$$E[S] = \frac{1}{\mu} + \frac{\alpha}{\mu\beta}$$

We also need $E[S^2]$, which is obtained as follows.

$$E[S^2] = E[(T + A)^2]$$
$$= E[T^2] + 2E[AT] + E[A^2]$$
$$= E[T^2] + 2E[A]E[T] + E[A^2]$$

The independence of A and T follows because the time of the first happening is independent of whether the happening was a service or a breakdown. Now,

$$E[A^2] = E[A^2 \mid \text{breakdown}] \frac{\alpha}{\mu + \alpha}$$

$$= \frac{\alpha}{\mu + \alpha} E[(\text{downtime} + S^*)^2]$$

$$= \frac{\alpha}{\mu + \alpha} \{E[\text{down}^2] + 2E[\text{down}]E[S] + E[S^2]\}$$

$$= \frac{\alpha}{\mu + \alpha} \left\{ \frac{2}{\beta^2} + \frac{2}{\beta}\left[\frac{1}{\mu} + \frac{\alpha}{\mu\beta}\right] + E[S^2] \right\}$$

Hence,

$$E[S^2] = \frac{2}{(\mu + \beta)^2} + 2\left[\frac{\alpha}{\beta(\mu + \alpha)} + \frac{\alpha}{\mu + \alpha}\left(\frac{1}{\mu} + \frac{\alpha}{\mu\beta}\right)\right]$$

$$+ \frac{\alpha}{\mu + \alpha}\left\{ \frac{2}{\beta^2} + \frac{2}{\beta}\left[\frac{1}{\mu} + \frac{\alpha}{\mu\beta}\right] + E[S^2] \right\}$$

Now solve for $E[S^2]$. The desired answer is

$$W_Q = \frac{\lambda E[S^2]}{2(1 - \lambda E[S])}$$

In the above, S^* is the additional service needed after the breakdown is over and S^* has the same distribution as S. The above also uses the fact that the expected square of an exponential is twice the square of its mean.

Another way of calculating the moments of S is to use the representation

$$S = \sum_{i=1}^{N} (T_i + B_i) + T_{N+1}$$

where N is the number of breakdowns while a customer is in service, T_i is the time starting when service commences for the ith time until a happening occurs, and B_i is the length of the ith breakdown. We now use the fact that, given N, all of the random variables in the representation are independent exponentials with the T_i having rate $\mu + \alpha$ and the B_i having rate β. This yields

$$E[S \mid N] = \frac{N+1}{\mu + \alpha} + \frac{N}{\beta}$$

$$\text{Var}(S \mid N) = \frac{N+1}{(\mu + \alpha)^2} + \frac{N}{\beta^2}$$

Therefore, since $1 + N$ is geometric with mean $(\mu + \alpha)/\mu$ [and variance $\alpha(\alpha + \mu)/\mu^2$] we obtain

$$E[S] = \frac{1}{\mu} + \frac{\alpha}{\mu\beta}$$

and, using the conditional variance formula,

$$\text{Var}(S) = \left[\frac{1}{\mu + \alpha} + \frac{1}{\beta}\right]^2 \frac{\alpha(\alpha + \mu)}{\mu^2} + \frac{1}{\mu(\mu + \alpha)} + \frac{\alpha}{\mu\beta^2}$$

52. S_n is the service time of the nth customer; T_n is the time between the arrival of the nth and $(n + 1)$st customer.

Chapter 9

4. (a) $\phi(x) = x_1 \max(x_2, x_3, x_4)x_5$.
   (b) $\phi(x) = x_1 \max(x_2 x_4, x_3 x_5)x_6$.
   (c) $\phi(x) = \max(x_1, x_2 x_3)x_4$.

6. A minimal cut set has to contain at least one component of each minimal path set. There are six minimal cut sets: $\{1, 5\}, \{1, 6\}, \{2, 5\}, \{2, 3, 6\}, \{3, 4, 6\}, \{4, 5\}$.

12. The minimal path sets are $\{1, 4\}, \{1, 5\}, \{2, 4\}, \{2, 5\}, \{3, 4\}, \{3, 5\}$. With $q_i = 1 - p_i$, the reliability function is

$$r(\mathbf{p}) = P\{\text{either of } 1, 2, \text{ or } 3 \text{ works}\}P\{\text{either of } 4 \text{ or } 5 \text{ works}\}$$

$$= (1 - q_1 q_2 q_3)(1 - q_4 q_5)$$

17. $E[N^2] = E[N^2 \mid N > 0]P\{N > 0\}$

$\qquad\qquad \geqslant (E[N \mid N > 0])^2 P\{N > 0\}$, since $E[X^2] \geqslant (E[X])^2$

Thus,

$$E[N^2]P\{N > 0\} \geqslant (E[N \mid N > 0]P[N > 0])^2$$

$$= (E[N])^2$$

Let N denote the number of minimal path sets having all of its components functioning. Then $r(p) = P\{N > 0\}$. Similarly, if we define N as the number of minimal cut sets having all of its components failed, then $1 - r(p) = P\{N > 0\}$. In both cases we can compute expressions for $E[N]$ and $E[N^2]$ by writing N as the sum of indicator (i.e., Bernoulli) random variables. Then we can use the inequality to derive bounds on $r(p)$.

22.   (a)
$$\bar{F}_t(a) = P\{X > t + a \,|\, X > t\}$$
$$= \frac{P[X > t + a]}{P\{X > t\}} = \frac{\bar{F}(t + a)}{\bar{F}(t)}$$

(b) Suppose $\lambda(t)$ is increasing. Recall that
$$\bar{F}(t) = e^{-\int_0^t \lambda(s)\, ds}$$

Hence,

$$\frac{\bar{F}(t + a)}{\bar{F}(t)} = \exp\left\{ -\int_t^{t+a} \lambda(s)\, ds \right\}$$

which decreases in t since $\lambda(t)$ is increasing. To go the other way, suppose $\bar{F}(t + a)/\bar{F}(t)$ decreases in t. Now when a is small

$$\frac{\bar{F}(t + a)}{\bar{F}(t)} \approx e^{-a\lambda(t)}$$

Hence, $e^{-a\lambda(t)}$ must decrease in t and thus $\lambda(t)$ increases.

25.   For $x \geqslant \xi$,

$$1 - p = \bar{F}(\xi) = \bar{F}(x(\xi/x)) \geqslant [\bar{F}(x)]^{\xi/x}$$

since IFRA. Hence, $\bar{F}(x) \leqslant (1 - p)^{x/\xi} = e^{-\theta x}$.
   For $x \leqslant \xi$,

$$\bar{F}(x) = \bar{F}(\xi(x/\xi)) \geqslant [\bar{F}(\xi)]^{x/\xi}$$

since IFRA. Hence, $\bar{F}(x) \geqslant (1 - p)^{x/\xi} = e^{-\theta x}$.

30. $r(p) = p_1 p_2 p_3 + p_1 p_2 p_4 + p_1 p_3 p_4 + p_2 p_3 p_4 - 3p_1 p_2 p_3 p_4$

$$r(1 - F(t)) = \begin{cases} 2(1 - t)^2(1 - t/2) + 2(1 - t)(1 - t/2)^2 \\ -3(1 - t)^2\ (1 - t/2)^2, \quad 0 \leqslant t \leqslant 1 \\ 0, \hspace{4.2cm} 1 \leqslant t \leqslant 2 \end{cases}$$

$$E[\text{lifetime}] = \int_0^1 [2(1-t)^2(1-t/2) + 2(1-t)(1-t/2)^2 - 3(1-t)^2(1-t/2)^2]\ dt$$

$$= \frac{31}{60}$$

Chapter 10

1. $B(s) + B(t) = 2B(s) + B(t) - B(s)$. Now $2B(s)$ is normal with mean 0 and variance $4s$ and $B(t) - B(s)$ is normal with mean 0 and variance $t - s$.

Because $B(s)$ and $B(t) - B(s)$ are independent, it follows that $B(s) + B(t)$ is normal with mean 0 and variance $4s + t - s = 3s + t$.

3.

$$E[B(t_1)B(t_2)B(t_3)] = E[E[B(t_1)B(t_2)B(t_3) \mid B(t_1), B(t_2)]]$$
$$= E[B(t_1)B(t_2)E[B(t_3) \mid B(t_1), B(t_2)]]$$
$$= E[B(t_1)B(t_2)B(t_2)]$$
$$= E[E[B(t_1)B^2(t_2) \mid B(t_1)]]$$
$$= E[B(t_1)E[B^2(t_2) \mid B(t_1)]] \qquad (*)$$
$$= E[B(t_1)\{(t_2 - t_1) + B^2(t_1)\}]$$
$$= E[B^3(t_1)] + (t_2 - t_1)E[B(t_1)]$$
$$= 0$$

where the equality $(*)$ follows since given $B(t_1)$, $B(t_2)$ is normal with mean $B(t_1)$ and variance $t_2 - t_1$. Also, $E[B^3(t)] = 0$ since $B(t)$ is normal with mean 0.

5.

$$P\{T_1 < T_{-1} < T_2\} = P\{\text{hit 1 before } -1 \text{ before } 2\}$$
$$= P\{\text{hit 1 before } -1\}$$
$$\times P\{\text{hit } -1 \text{ before 2} \mid \text{hit 1 before } -1\}$$
$$= \tfrac{1}{2}P\{\text{down 2 before up 1}\}$$
$$= \tfrac{1}{2}\tfrac{1}{3} = \tfrac{1}{6}$$

The next to last equality follows by looking at the Brownian motion when it first hits 1.

10. (a) Writing $X(t) = X(s) + X(t) - X(s)$ and using independent increments, we see that given $X(s) = c$, $X()$ is distributed as $c + X(t) - X(s)$. By stationary increments this has the same distribution as $c + X(t - s)$, and is thus normal with mean $c + \mu(t - s)$ and variance $(t - s)\sigma^2$.

(b) Use the representation $X(t) = \sigma B(t) + \mu t$, where $\{B(t)\}$ is standard Brownian motion. Using Equation (1.4), but reversing s and t, we see that the conditional distribution of $B(t)$ given that $B(s) = (c - \mu s)/\sigma$ is normal with mean $t(c - \mu s)/(\sigma s)$ and variance $t(s - t)/s$. Thus, the conditional distribution of $X(t)$ given that $X(s) = c$, $s > t$, is normal with mean

$$\sigma\left[\frac{t(c - \mu s)}{\sigma s}\right] + \mu t = \frac{(c - \mu s)t}{s} + \mu t$$

and variance

$$\frac{\sigma^2 t(s - t)}{s}$$

19. Since knowing the value of $Y(t)$ is equivalent to knowing $B(t)$, we have

$$E[Y(t) \mid Y(u), \ 0 \leqslant u \leqslant s] = e^{-c^2 t/2} E[e^{cB(t)} \mid B(u), \ 0 \leqslant u \leqslant s]$$

$$= e^{-c^2 t/2} E[e^{cB(t)} \mid B(s)]$$

Now, given $B(s)$, the conditional distribution of $B(t)$ is normal with mean $B(s)$ and variance $t - s$. Using the formula for the moment generating function of a normal random variable we see that

$$e^{-c^2 t/2} E[e^{cB(t)} \mid B(s)] = e^{-c^2 t/2} e^{cB(s) + (t - s)c^2/2}$$

$$= e^{-c^2 s/2} e^{cB(s)}$$

$$= Y(s)$$

Thus $\{Y(t)\}$ is a Martingale.

$$E[Y(t)] = E[Y(0)] = 1$$

20. By the Martingale stopping theorem

$$E[B(T)] = E[B(0)] = 0$$

However, $B(T) = 2 - 4T$ and so $2 - 4E[T] = 0$, or $E[T] = \frac{1}{2}$.

24. It follows from the Martingale stopping theorem and the result of Exercise 18 that

$$E[B^2(T) - T] = 0$$

where T is the stopping time given in this problem and

$$B(t) = \frac{X(t) - \mu t}{\sigma}$$

Therefore,

$$E\left[\frac{(X(T) - \mu T)^2}{\sigma^2} - T\right] = 0$$

However, $X(T) = x$ and so the above gives that

$$E[(x - \mu T)^2] = \sigma^2 E[T]$$

But, from Exercise 21, $E[T] = x/\mu$ and so the above is equivalent to

$$\mathrm{Var}(\mu T) = \sigma^2 \frac{x}{\mu} \qquad \text{or} \qquad \mathrm{Var}(T) = \sigma^2 \frac{x}{\mu^3}$$

27. $E[X(a^2t)/a] = (1/a)E[X(a^2t)] = 0.$ For $s < t$,

$$\text{Cov}(Y(s),\ Y(t)) = \frac{1}{a^2}\text{Cov}(X(a^2s),\ X(a^2t))$$

$$= \frac{1}{a^2}a^2s = s$$

Because $\{Y(t)\}$ is clearly Gaussian, the result follows.

30. (a) Starting at any time t the continuation of the Poisson process remains a Poisson process with rate λ.

(b) $$E[Y(t)Y(t+s)] = \int_0^\infty E[Y(t)Y(t+s)\mid Y(t) = y]\lambda e^{-\lambda y}\,dy$$

$$= \int_0^s yE[Y(t+s)\mid Y(t) = y]\lambda e^{-\lambda y}\,dy$$

$$+ \int_s^\infty y(y-s)\lambda e^{-\lambda y}\,dy$$

$$= \int_0^s y\frac{1}{\lambda}\lambda e^{-\lambda y}\,dy + \int_s^\infty y(y-s)\lambda e^{-\lambda y}\,dy$$

where the above used that

$$E[Y(t)Y(t+s)\mid Y(t) = y] = \begin{cases} yE(Y(t+s)) = \dfrac{y}{\lambda}, & \text{if } y < s \\[2mm] y(y-s), & \text{if } y > s \end{cases}$$

Hence,

$$\text{Cov}(Y(t),\ Y(t+s)) = \int_0^s ye^{-\lambda y}\,dy + \int_s^\infty y(y-s)\lambda e^{-\lambda y}\,dy - \frac{1}{\lambda^2}$$

Chapter 11

1. (a) Let U be a random number. If $\Sigma_{j=1}^{i-1}P_j < U \leqslant \Sigma_{j=1}^{i}P_j$ then simulate from F_i. (In the above $\Sigma_{j=1}^{i-1}P_j \equiv 0$ when $i = 1$.)

(b) Note that

$$F(x) = \tfrac{1}{3}F_1(x) + \tfrac{2}{3}F_2(x)$$

where

$$F_1(x) = 1 - e^{-2x}, \qquad 0 < x < \infty$$

$$F_2(x) = \begin{cases} x, & 0 < x < 1 \\ 1, & 1 < x \end{cases}$$

Hence, using part (a), let U_1, U_2, U_3 be random numbers and set

$$X = \begin{cases} \dfrac{-\log U_2}{2}, & \text{if } U_1 < \tfrac{1}{3} \\ U_3, & \text{if } U_1 > \tfrac{1}{3} \end{cases}$$

The above uses the fact that $-\log U_2/2$ is exponential with rate 2.

3. If a random sample of size n is chosen from a set of $N + M$ items of which N are acceptable, then X, the number of acceptable items in the sample, is such that

$$P\{X = k\} = \binom{N}{k}\binom{M}{n-k} \bigg/ \binom{N+M}{k}$$

To simulate X, note that if

$$I_j = \begin{cases} 1, & \text{if the } j\text{th selection is acceptable} \\ 0, & \text{otherwise} \end{cases}$$

then

$$P\{I_j = 1 \mid I_1, \ldots, I_{j-1}\} = \frac{N - \Sigma_1^{j-1} I_i}{N + M - (j-1)}$$

Hence, we can simulate $I_1, \ldots, I_n$ by generating random numbers $U_1, \ldots, U_n$ and then setting

$$I_j = \begin{cases} 1, & \text{if } U_j < \dfrac{N - \Sigma_1^{j-1} I_i}{N + M - (j-1)} \\ 0, & \text{otherwise} \end{cases}$$

and $X = \Sigma_{j=1}^n I_j$ has the desired distribution.

Another way is to let

$$X_j = \begin{cases} 1, & \text{the } j\text{th acceptable item is in the sample} \\ 0, & \text{otherwise} \end{cases}$$

and then simulate $X_1, \ldots, X_N$ by generating random numbers $U_1, \ldots, U_N$ and then setting

$$X_j = \begin{cases} 1, & \text{if } U_j < \dfrac{n - \Sigma_{i=1}^{j-1} X_i}{N + M - (j-1)} \\ 0, & \text{otherwise} \end{cases}$$

and $X = \Sigma_{j=1}^N X_j$ then has the desired distribution.

The former method is preferable when $n \leqslant N$ and the latter when $N \leqslant n$.

6. Let

$$c(\lambda) = \max_x \left\{ \frac{f(x)}{\lambda e^{-\lambda x}} \right\} = \frac{2}{\lambda\sqrt{2\pi}} \max_x \left[\exp \left\{ \frac{-x^2}{2} + \lambda x \right\} \right]$$

$$= \frac{2}{\lambda\sqrt{2\pi}} \exp \left\{ \frac{\lambda^2}{2} \right\}$$

Hence,

$$\frac{d}{d\lambda} c(\lambda) = \sqrt{2/\pi} \exp \left\{ \frac{\lambda^2}{2} \right\} \left[1 - \frac{1}{\lambda^2} \right]$$

Hence $(d/d\lambda)c(\lambda) = 0$ when $\lambda = 1$ and it is easy to check that this yields the minimal value of $c(\lambda)$.

16. (a) They can be simulated in the same sequential fashion in which they are defined. That is, first generate the value of a random variable I_1 such that

$$P\{I_1 = i\} = \frac{w_i}{\sum_{j=1}^{n} w_j}, \qquad i = 1, \ldots, n$$

Then, if $I_1 = k$, generate the value of I_2 where

$$P\{I_2 = i\} = \frac{w_i}{\sum_{j \neq k} w_j}, \qquad i \neq k$$

and so on. However, the approach given in part (b) is more efficient.
(b) Let I_j denote the index of the jth smallest X_i.

23. Let $m(t) = \int_0^t \lambda(s) \, ds$, and let $m^{-1}(t)$ be the inverse function. That is, $m(m^{-1}(t)) = t$.

(a) $$P\{m(X_1) > x\} = P\{X_1 > m^{-1}(x)\}$$

$$= P\{N(m^{-1}(x)) = 0\}$$

$$= e^{-m(m^{-1}(x))}$$

$$= e^{-x}$$

(b) $$P\{m(X_i) - m(X_{i-1}) > x \mid m(X_1), \ldots, m(X_{i-1}) - m(X_{i-2})\}$$

$$= P\{m(X_i) - m(X_{i-1}) > x \mid X_1, \ldots, X_{i-1}\}$$

$$= P\{m(X_i) - m(X_{i-1}) > x \mid X_{i-1}\}$$

$$= P\{m(X_i) - m(X_{i-1}) > x \mid m(X_{i-1})\}$$

Now,

$$P\{m(X_i) - m(X_{i-1}) > x \mid X_{i-1} = y\}$$

$$= P\left\{\int_y^{X_i} \lambda(t)\,dt > x \mid X_{i-1} = y\right\}$$

$$= P\{X_i > c \mid X_{i-1} = y\} \qquad \text{where } \int_y^c \lambda(t)\,dt = x$$

$$= P\{N(c) - N(y) = 0 \mid X_{i-1} = y\}$$

$$= P\{N(c) - N(y) = 0\}$$

$$= \exp\left\{-\int_y^c \lambda(t)\,dt\right\}$$

$$= e^{-x}$$

32. $$\text{Var}[(X + Y)/2] = \tfrac{1}{4}[\text{Var}(X) + \text{Var}(Y) + 2\,\text{Cov}(X, Y)]$$

$$= \frac{\text{Var}(X) + \text{Cov}(X, Y)}{2}$$

Now it is always true that

$$\frac{\text{Cov}(V, W)}{\sqrt{\text{Var}(V)\,\text{Var}(W)}} \leqslant 1$$

and so when X and Y have the same distribution $\text{Cov}(X, Y) \leqslant \text{Var}(X)$.

Index